Lecture Notes in Networks and Systems 1398

The series “Lecture Notes in Networks and Systems” publishes the latest developments in Networks and Systems—quickly, informally and with high quality. Original research reported in proceedings and post-proceedings represents the core of LNNS.

Volumes published in LNNS embrace all aspects and subfields of, as well as new challenges in, Networks and Systems.

The series contains proceedings and edited volumes in systems and networks, spanning the areas of Cyber-Physical Systems, Autonomous Systems, Sensor Networks, Control Systems, Energy Systems, Automotive Systems, Biological Systems, Vehicular Networking and Connected Vehicles, Aerospace Systems, Automation, Manufacturing, Smart Grids, Nonlinear Systems, Power Systems, Robotics, Social Systems, Economic Systems and other. Of particular value to both the contributors and the readership are the short publication timeframe and the world-wide distribution and exposure which enable both a wide and rapid dissemination of research output.

The series covers the theory, applications, and perspectives on the state of the art and future developments relevant to systems and networks, decision making, control, complex processes and related areas, as embedded in the fields of interdisciplinary and applied sciences, engineering, computer science, physics, economics, social, and life sciences, as well as the paradigms and methodologies behind them.

Indexed by SCOPUS, EI Compendex, INSPEC, WTI Frankfurt eG, zbMATH, SCImago.

All books published in the series are submitted for consideration in Web of Science.

For proposals from Asia please contact Aninda Bose (aninda.bose@springer.com).

Souvik Pal · Álvaro Rocha
Editors

Proceedings of 4th International Conference on Mathematical Modeling and Computational Science

ICMMCS 2025 Volume 1

Springer

Editors
Souvik Pal
Department of Computer Science and Engineering
Sister Nivedita University
Kolkata, West Bengal, India

Álvaro Rocha
ISEG
Universidade de Lisboa
Lisbon, Portugal

ISSN 2367-3370 ISSN 2367-3389 (electronic)
Lecture Notes in Networks and Systems
ISBN 978-3-031-90997-9 ISBN 978-3-031-90998-6 (eBook)
https://doi.org/10.1007/978-3-031-90998-6

This Springer imprint is published by the registered company Springer Nature Switzerland AG
The registered company address is: Gewerbestrasse 11, 6330 Cham, Switzerland

Preface and acknowledgement

The main goal of this proceedings book is to bring together top academic scientists, researchers, and research scholars so they can share their experiences and research results on all aspects of intelligent ecosystems, data sciences, and mathematics. ICMMCS 2025 is a conference that aims to bring together academic scientists, professors, research scholars, and students who work in different areas of engineering and technology. On ICMMCS 2025, the researchers had the chance to meet some of the best researchers in the world, learn about some new ideas and developments in research around the world, and get a feel for new Science-Technology trends. The conference is having authors, research scholars, and people who attended the chance to work together with universities and institutions across the country and around the world to promote research and develop technologies on a global scale. The goal of this conference is to make it easier for basic research to be used in institutional and industrial research and for applied research to be used in real life.

4th International Conference on Mathematical Modeling and Computational Science [ICMMCS 2025] organized by Society for Intelligent Systems in association with National Taipei University of Business, Taiwan; Berlin School of Business and Innovation, Germany, National University of Science and Technology, Iraq; and Université Ibn Tofail, Morocco, in Hybrid mode (Physical mode and Google Meet Platform) on 10 and 11 January, 2025, in Thailand. The conference brought together researchers from all regions around the world working on a variety of fields and provided a stimulating forum for them to exchange ideas and report on their researches. The proceeding of ICMMCS 2025 Vol. 1 consists of 42 best selected papers which were submitted to the conferences, and peer-reviewed by conference committee members and international reviewers. The presenters have shown their slides either virtually or in person. Experts in the field of education have gathered from all over the world, including India, Malaysia, Vietnam, Iraq, Germany, Saudi Arabia, and Egypt to discuss how to better prepare the next generation of leaders through education. Knowledge domains from many countries' research cultures were brought together at this meeting. Academic conferences rely heavily on its authors and presenters for their credibility. In light of the current global pandemic, we appreciate the authors' decision to present their works at this conference.

We are very grateful to Almighty for always being there for us, through good times and bad, and for giving us ways to help ourselves. From the Call for Papers to the finalization of the chapters, everyone on the team worked together well, which is a good sign of a strong team. The editors and organizers of the conference are very grateful to all the members of Springer, especially Prof. Thomas Ditzinger, for his helpful suggestions and for giving us the chance to finish the conference proceedings. We also appreciate the support of Sylvia Schneider. We are also thankful to Mr. Naveen Kumar Madhan Mohan, who works for Springer as a project coordinator, for his help. We are grateful that reviewers from all over the world and from different parts of the world gave their support and stuck to their goal of getting good chapters submitted.

Last but not least, we want to wish all of the participants luck with their presentations and social networking. This conference cannot go well without your strong support. We hope that the people who went to the conference enjoyed both the technical program and the speakers and delegates who were there virtually. We hope you have a productive and fun time at ICMMCS 2025 in Thailand.

Souvik Pal
Álvaro Rocha

About the Book

This conference proceedings book serves as an inclusive repository of current research findings. The primary objective of this volume is to consolidate various computing and communication technologies onto a single platform. ICMMCS 2025 aims to provide the scientific community with an opportunity to exchange insights on the latest advancements in Engineering Mathematics and Computational Science, fostering collaboration. This event brings together leading academic scientists, researchers, industry professionals, and scholars to share their expertise, experiences, and research outcomes in computer science, information technology, and mathematics. The conference explores cutting-edge topics, including mathematical modeling and computational science. The scope of mathematical modeling covers areas such as mathematical logic, numerical analysis, neural networks, fuzzy set theory, coding theory, higher algebra, number theory, graph theory, computation in complex networks, calculus, differential equations, soft computing, knowledge engineering, and machine learning. Computational science topics include artificial intelligence, big data, high-performance computing, cybersecurity, and the Internet of things (IoT). Additionally, this volume provides researchers, practitioners, and educators a platform to discuss the latest trends, challenges, and innovations in computer science and engineering mathematics. Designed for experts, researchers, graduate students, and professionals in these fields, this book is divided into distinct chapters, making it accessible and adaptable to diverse reader needs.

Organizing Committee and Key Members

Conference Committee Members

Conference Honorary Chair

Álvaro Rocha	University of Lisbon, Lisbon, Portugal

Conference General Chair

Jou-Ming Chang	National Taipei University of Business, Taiwan

Program Convener

Noor Zaman Jhanjhi	Taylor's University, Malaysia
Fathi Amsaad	College of Engineering & Computer Science, Wright State University, USA
Hanaa Hachimi	Secretary General of Sultan Moulay Slimane University USMS of Beni Mellal. Morocco

Conference Organizing Chairs

D. Balaganesh	Berlin School of Business and Innovation, Germany
Souvik Pal	Sister Nivedita University, India

Conference Convenor

D. Akila	Saveetha College of Liberal Arts and Sciences, SIMATS Deemed to be University, India
K. Kavitha	Mother Teresa Women's University, India

Program Chairs

Ahmed J. Obaid	Faculty of Computer Science and Mathematics, University of Kufa, Iraq
R. Esther Felicia	Shri Krishnaswamy College for Women, India

International Advisory Board

Ahmed A. Elnger	Beni-Suef University, Egypt
Alok Satsangi, Director	NSHM Knowledge Campus, Durgapur, India
Abdelilah Jraifi	Cadi Ayad university of Marrakech, Morocco
Bikramjit Sarkar	JIS College of Engineering, India
Chakib El Mokhi	Cadi Ayyad university of Marrakech, Morocco
Dac-Nhuong Le	Haiphong University, Vietnam.
Debashis De	Maulana Abul Kalam Azad University of Technology, India
Gia Nhu Nguyen	Duy Tan University, Vietnam
J. M. Chang	National Taipei University of Business, Taiwan
John Petearson Anzola	Los Libertadores University, Colombia
Faiez Gargouri	University of Sfex, Tunisia
Kusum Yadav	University of Hail, Kingdom of Saudi Arabia
M. J. Diván	National University of La Pampa, Argentina
Mohamed Amine Boutiche	University of Sciences and Technology Houari Boumediene, Algeria
Mohammed Kaicer	Ibn Tofail University of Kenitra, Morocco
Noor Zaman	Taylor's University, Malaysia
Mostafa Ezziyyani	Abdelmalek Essaâdi University, Morrocco
Srinath Doss	Botho University, Botswana
S. K. Hoskere	Shanghai United International School, China
Sanjeevikumar Padmanaban	Aalborg University, Denmark
Vasaki Ponnusamy	Universiti Tunku Abdul Rahman, Malaysia
Vicente García Díaz	University of Oviedo, Spain

Technical Program Committee Members

Anand Paul	Kyungpook National University, South Korea
Abdullah	Chandigarh University, India
Amanpreet Kaur Sandhu	Chandigarh University, India
Ashish Mishra	Gyan Ganga Institute of Technology and Science, India
Abhishek Dhar	Swami Vivekananda University, India
Bijoy Kumar Mandal	NSHM Knowledge Campus, Durgapur, India
B. Jagadhesan	Dhanraj Baid Jain College, Chennai, India
C. K. Lin	Fuzhou University, China
Dac-Nhuong Le	Haiphong University, Haiphong, Vietnam
Debabrata Roy	NSHM Knowledge Campus, Durgapur, India
D. Napoleon	Bharathiar University, India
Dibyendu Mukherjee	NSHM Knowledge Campus, Durgapur, India

E. Ramaraj	Alagappa University, India
K. Kavitha	Chellammal College for Women, India
K. J. Pai	Ming Chi University, Taiwan
K. Hema shankari	Women's Christian College, India
Kamal Wadhwa	Govt PG College Pipariya, India
K. Kavitha	Mother Teresa Women's University, India
Kusum Yadav	University of Hail, Kingdom of Saudi Arabia
Jarnail Singh	Chandigarh University, India
J. kamalakumari	Agurchand Manmull Jain College, Chennai, India
L. J. Hung	National Taipei University of Business, Taiwan
N. Pradeep	Bapuji Institute of Engineering and Technology, India
Mohammed Kaicer	Ibn Tofail University, Morocco
Manisha Malhotra	Chandigarh University, India
Pradeep Laxkar	Mandsaur University, India
P R S Choudhry	Govt Model Science College, India
P. Kandan	Annamalai University, India
R. Amutha	Tiruvalluvar University College Arts & Science, India
Ranbir Singh Batth	Lovely Professional University, India
R. Kalaiarasi	Tamil Nadu Open University, India
R. Velmurugan	Presidency College, India
Sayed Ameeddnuddin Irfan	Universiti Teknologi Petronas, Malaysia
Srinath Doss	Botho University, Botswana
S. Vimal	Jeppiaar Engineering College, Chennai, India
Suresh Rasappan	University of Technology and Applied Sciences–Ibri, Sultanate of Oman
S. Mathivilasini	Ethiraj College for Women, India
S. M. Tang	National Defense University, Taiwan
Saurabh Adhikari	Swami Vivekananda University, India
Sonjoy Pan	Swami Vivekananda University, India
T. Nathiya	New Prince Shri Bhvani Arts and Science College, Chennai, India
T.Nagarathinam	MASS College of Arts and Science, Thanjavur, India
Vicente García Díaz	University of Oviedo, Spain
V. R. Elangovan	Agurchand Manmull Jain College, Chennai, India
V. Vijayalakshmi	SRM Institute of Science and Technology, Kattankulathur, India
Vasaki Ponnusamy	Universiti Tunku Abdul Rahman, Malaysia

General Chair

Jou-Ming Chang, National Taipei University of Business, Taiwan

Jou-Ming Chang is currently working as a professor at Institute of Information and Decision Sciences, National Taipei University of Business, Taipei, Taiwan. He has been awarded PhD in 2001 from National Central University. He has more than 30 Years of teaching and research experience. He has more than 200 SCI Publications in his credit. His research interests are algorithm development, combinatorial optimization, discrete mathematics, graph theory, parallel and distributed computing.

Honorary General Chair

Álvaro Rocha, University of Lisbon, Lisbon, Portugal

Álvaro Rocha is Professor of Information Systems at the University of Lisbon-ISEG, researcher at the the ISEG Centre for Advanced Research in Management (ADVANCE), and a collaborator researcher at both Laboratory of Artificial Intelligence and Computer Science (LIACC) and Center for Research in Health Technologies and Information Systems (CINTESIS). His main research interests are maturity models, information systems quality, online service quality, requirements engineering, intelligent information systems, e-Government, e-Health, and information technology in education. He is also Vice-Chair of the IEEE Portugal Section Systems, Man, and Cybernetics Society Chapter, and Founder and Editor-in-Chief of two Scopus and SCIMago journals: JISEM-Journal of Information Systems Engineering & Management; and RISTI-Revista Ibérica de Sistemas e Tecnologias de Informação / Iberian Journal of Information Systems and Technologies.

Keynote Speakers

Sheng-Lung Peng, National Taipei University of Business, Taiwan

Sheng-Lung Peng is Professor at the Department of Creative Technologies and Product Design and Dean of the College of Innovative Design and Management at National Taipei University of Business in Taiwan. He received a PhD degree from the Computer Science Department of National Tsing Hua University in Taiwan. He is an honorary professor at Beijing Information Science and Technology University and a visiting professor at Ningxia Institute of Science and Technology in China. He is also an adjunct professor at National Dong Hwa University in Taiwan and Mandsaur University and Kazi Nazrul University in India. Dr. Peng has edited several special issues in journals, such as Frontiers in Public Health, Journal of Internet Technology, IEEE Internet of Things Magazine, Computers and Electrical Engineering, Sensors, and so on. His research interests are algorithm design in the fields of artificial intelligence, bioinformatics, combinatorics, data mining, and networking.

Noor Zaman Jhanjhi, Taylor's University, Malaysia

Noor Zaman Jhanjhi (N.Z Jhanjhi) is currently working as a professor in Computer Science (Cybersecurity and AI), program director for the Postgraduate Research Degree Programmes in computer science, and director of the Center for Smart Society (CSS5) at the School of Computer Science at Taylor's University, Malaysia. He has been nominated as the world's top 2% research scientist globally; he is among the top five computer science researchers in Malaysia and was nominated as an Outstanding Faculty Member by the MDEC Malaysia for the year 2022. He has highly indexed publications in WoS/ISI/SCI/Scopus. He received the Outstanding Associate Editor award for IEEE ACCESS. Active reviewer for a series of top-tier journals has been awarded globally as a top 1% reviewer by Publons (Web of Science). He has vast experience in academic qualifications, including ABET, NCAAA, and NCEAC, for ten years. His research areas include cybersecurity, IoT security, wireless security, data science, software engineering, and UAVs.

Sanpawat Kantabutra, Chiang Mai University, Thailand

Sanpawat Kantabutra is currently Associate Professor in the Theory of Computation Group in the Faculty of Engineering in Chiang Mai University, Thailand. He is a two-time Thailand Research Fund Career Award researcher and is the head of the Government-Funded Center of Excellence in Quantum Technology in Chiang Mai University. In 2001, he earned his PhD in theoretical computer science from Tufts University in Boston in the USA and, since then, has collaborated internationally on research and has regularly published his research findings in some of the world's top journals in theoretical computer science. His current research interests are in approximation and randomized algorithms and quantum computation and information. He has regularly been invited to give research talks on these topics both in Thailand and abroad. All of his research has been funded by units of the Thai government and universities.

Session Chair

- Dr. Nopadon Juneam, Kasetsart University, Bangkok, Thailand.
- Dr. Ton Quang Cuong, VNU University of Education (VNU-ED), Vietnam National University, Hanoi, Vietnam.
- Dr. Srinath Doss, Faculty of Engineering and Technology, Botho University, Botswana.
- Dr. Mohammed Ayad Alkhafaji, College of Technical Engineering, National University of Science and Technology, Dhi Qar, Iraq .
- Dr. Mohit Kumar, Dr B R Ambedkar NIT Jalandhar, India.
- Dr. Shakila Devi G T, Department of Computational Mathematics, SITMAS Engineering, India.
- Dr. S. Jayalaksshmi, Department of Information Technology, Vels Institute of Science Technology and Advanced Studies, India.
- Dr.V.R. Elangovan, Department of Computer Applications, College of Science and Humanities, SRM Institute of Science and Technology (SRMIST), India.

Contents

About the Editor(s)

Dr. Souvik Pal is Professor in the Department of Computer Science and Engineering at Sister Nivedita University (Techno India Group), Kolkata, India. Prior to that, he was associated with Global Institute of Management and Technology; Brainware University, Kolkata; JIS College of Engineering, Nadia; Elitte College of Engineering, Kolkata; and Nalanda Institute of Technology, Bhubaneswar, India. Dr. Pal received his MTech and PhD degrees in the field of Computer Science and Engineering from KIIT University, Bhubaneswar, India. He has more than a decade of academic experience. He is an author or co-editor of more than 20 books from reputed publishers, including Elsevier, Springer, CRC Press, and Wiley, and he holds three patents. He is serving as a series editor for "Advances in Learning Analytics for Intelligent Cloud-IoT Systems", published by Scrivener-Wiley Publishing (Scopus-indexed); "Internet of Things: Data-Centric Intelligent Computing, Informatics, and Communication", published CRC Press, Taylor & Francis Group, USA; "Conference Proceedings Series on Intelligent Systems, Data Engineering, and Optimization", published CRC Press, Taylor & Francis Group, USA; Dr. Pal has published a number of research papers in Scopus / SCI/SCIE Journals and conferences. He is the organizing chair of RICE 2019, Vietnam; RICE 2020 Vietnam; ICICIT 2019, Tunisia. He has been invited as a keynote speaker at ICICCT 2019, Turkey, and ICTIDS 2019, 2021, Malaysia. He has also served as Proceedings Editor of ICICCT 2019, 2020; ICMMCS 2020, 2021, 2023; ICWSNUCA 2021, India. His professional activities include roles as Associate Editor, Guest Editor, and Editorial Board Member for more than 100+ international journals and conferences of high repute and impact. His research area includes cloud computing, big data, Internet of things, wireless sensor network, and data analytics. He is a member of many professional organizations, including MIEEE; MCSI; MCSTA/ACM, USA; MIAENG, Hong Kong; MIRED, USA; MACEEE, New Delhi; MIACSIT, Singapore; and MAASCIT, USA.

Álvaro Rocha holds the title of Honorary Professor, and holds a D.Sc. in Information Science, Ph.D. in Information Systems and Technologies, M.Sc. in Information Management, and BCs in Computer Science.

He is a Professor of Information Systems at the University of Lisbon-ISEG, researcher at the ADVANCE (the ISEG Centre for Advanced Research in Management), and a collaborator researcher at both LIACC (Laboratory of Artificial Intelligence and Computer Science) and CINTESIS (Center for Research in Health Technologies and Information Systems). His main research interests are maturity models, information systems quality, online service quality, requirements engineering, intelligent information systems, e-Government, e-Health, and information technology in education. He is also Vice-Chair of the IEEE Portugal Section Systems, Man, and Cybernetics Society Chapter, and Founder and Editor-in-Chief of two Scopus and SCIMago journals: JISEM-Journal of Information Systems Engineering & Management; and RISTI-Revista

Ibérica de Sistemas e Tecnologias de Informação / Iberian Journal of Information Systems and Technologies. Additionally, he is Scientific Manager of the Information Systems Engineering & Management book series (https://www.springer.com/series/17396) at Springer-Verlag, the world's leading publisher of publications in Science, Technology and Health.

Moreover, he has served as Vice-Chair of Experts for the European Commission's Horizon 2020 Program, and as Expert at the COST-Intergovernmental Framework for European Cooperation in Science and Technology, at the European Commission's Horizon Europe Program, at the Government of Italy's Ministry of Universities and Research, at the Government of Latvia's Ministry of Finance, at the Government of Mexico's National Council of Science and Technology, at the Government of Polish's National Science Centre, and at the Government of Cyprus's Research and Innovation Foundation.

Enhancing Virtual Reality Interaction: GazeSense™ - An Innovative Algorithm Transforming Gaze Movement Accuracy and User Engagement

M. Abinaya(✉) and G. Vadivu

Department of Data Science and Business Systems, School of Computing, SRM Institute of Science and Technology, Kattankulathur, Chennai 603203, India
{am0150,vadivug}@srmist.edu.in

Abstract. The ability to accurately detect and understand user gaze movements is crucial in the ever-changing world of virtual reality (VR) for enabling intuitive and immersive experiences. In this paper, we present GazeSenseTM, an innovative algorithm that will change the way people think about identifying gaze movements in virtual reality settings. By effortlessly combining adaptive pattern recognition and machine learning, GazeSense™ addresses existing challenges, promising heightened accuracy and real-time responsiveness. The algorithm's adaptability to different virtual reality (VR) setups is made possible by its compatibility with all major software and hardware platforms, which is why it has been widely used. Presenting GazeSenseTM, we outline its fundamental principles, unique features, and a possible implementation strategy. We provide information on the revolutionary effect of GazeSenseTM plans for the virtual reality environment by talking about the practicality, possible advantages, and anticipated results. The ability to accurately interpret gaze behavior in dynamic scenarios is one of its notable contributions, along with its adaptability to diverse user interactions. The importance of GazeSenseTM as a catalyst for improving virtual reality interaction technology, increasing user satisfaction, and establishing a new benchmark for precise gaze movement recognition is highlighted in this paper. As we explore the cutting edge of immersive technology, GazeSenseTM is more than just an algorithm; it's a game-changer that's opening up new possibilities for human-computer interaction in the fascinating world of virtual reality.

Keywords: Artificial Intelligence · Machine Learning · Neural Networks · Virtual Reality

1 Introduction

Thoroughly identifying and analyzing users' gaze movements is essential to un- derstand and enhance the immersive experience in virtual reality (VR) [1]. A user's gaze movement in a virtual environment is a direct reflection of their visual attention and can be used to identify areas of interest, engagement, and possible points of interaction [2].

S. Pal and Á. Rocha (Eds.): ICMMCS 2025, LNNS 1398, pp. 1–10, 2025.
https://doi.org/10.1007/978-3-031-90998-6_1

Through gaze movement tracking and interpretation, virtual reality systems can gather important information about user preferences, behavior, and cognitive processes [3]. This information is essential for customizing VR content, improving user interfaces, and building more responsive and intuitive virtual worlds [4]. It is crucial because it may bridge the gap between user intention and system response, which will raise the VR experience's immersion and engagement factor [5]. The paper's objective is to present a brand-new algorithm called GazeSenseTM that was created especially for identifying gaze movements in virtual reality settings [6]. The principal aim is to tackle the current constraints in gaze identification techniques and introduce novel approach that warrants improved precision, effectiveness, and flexibility. To transform how VR systems, perceive and react to user gaze, GazeSenseTM aims to open the door to more natural and customized interactions within the virtual environment [7].

The launch of GazeSenseTM has the potential to significantly improve user experiences in a variety of virtual reality applications. The algorithm's wide-ranging applications, from gaming and education to healthcare and beyond, stem from its ability to precisely identify and interpret gaze movements [6]. GazeSenseTM can enable more realism, adaptive scenarios, and dynamic control mechanisms in games. By figuring out where users concentrate their attention, it can help create personalized learning experiences in educational settings [8]. GazeSenseTM can also be used in healthcare applications to track cognitive responses and administer therapeutic interventions [9]. GazeSenseTM has the potential to significantly change how people interact with and navigate virtual environments in the areas of accessibility, user interface design, and content optimization.

Objective of the study

1. Examine and talk about possible sources of information about gaze behavior in virtual reality (VR) environments, such as VR devices that can track eye movements.
2. Examine how eye-tracking technology is developing and think about working with VR industry partners to access or replicate gaze data for analysis.
3. Assess the suitability of artificial intelligence methods for examining gaze patterns, such as deep neural networks and machine learning algorithms.
4. Evaluate current AI models critically and determine whether they are appropriate for processing real or simulated gaze data to extract valuable insights.

2 Literature Survey

Today's VR gaze movement identification techniques have come a long way, using a variety of methods like computer vision algorithms and infrared eye-tracking [2, 10]. The common technique used in these approaches is to map eye movement data to virtual environment points of interest. Despite significant advancement in current methods, a thorough comprehension of user gaze behavior particularly in intricate and dynamic virtual reality scenarios remains a formidable obstacle [11, 12]. While gaze movement identification has advanced, there are still several drawbacks. Extant techniques might find it difficult to deliver precise and instantaneous tracking, especially when faced with situations involving swift head movements or a variety of user interactions [13]. Additionally, there may be limitations to how well-suited the current algorithms are to various VR software and hardware setups. Moreover, the complex gaze behavior in immersive environments necessitates more sophisticated algorithms that can more accurately interpret user intent. The shortcomings that have been found highlight the urgent need for a fresh and creative method for identifying gaze movements in virtual reality [14]. To effectively tackle these challenges, a novel algorithm like GazeSenseTM is essential. The novelty resides in its capacity to address the drawbacks of existing techniques, providing improved precision, flexibility to various VR situations, and real-time tracking features [15]. This work establishes the groundwork for the introduction of a ground-breaking solution that transcends the limitations of current methods, offering a revolutionary leap in VR interaction technology, by demonstrating the need for a new viewpoint on gaze movement identification [16].

3 GazeSenseTM Algorithm

3.1 Novel Idea of Algorithm

GazeSenseTM is an advanced algorithm that will change the way gaze movements are recognized in Virtual Reality (VR) environments. The main idea behind the algorithm is to use advanced computer vision and machine learning techniques together in a few different ways. VR headsets collect high-resolution-gaze data that GazeSenseTM uses to focus on accurate eye-tracking data. Then, a bunch of complex algorithms are run on this data to find intricate patterns in how people look at things. GazeSenseTM is different from other methods because it can adapt instantly to different VR situations, tracking head movements very accurately in real time. Figure 1 depicts the Pseudo code for the GazeSense Algorithm.

3.2 Methodology

Adaptive gaze behavior analysis is what GazeSenseTM is based on. It uses computer vision and machine learning to do this. The first step of the algorithm is to take raw eye-tracking data from VR headsets and pull out important features that show how people look. GazeSenseTM has machine learning models that learn and change based on how each user acts, which lets the algorithm interpret eye movements in real-time. This flexible method makes sure that GazeSenseTM can pick up on subtle patterns in real-time, giving a more accurate and quicker picture of what the user is trying to do in the VR world. GazeSenseTM is great at quickly adapting to new situations and user interactions, while older methods may have trouble tracking in real-time and not being flexible enough. A unique feature of this product is that it works with all VR hardware and software setups, making it even more unique. GazeSenseTM is different from rigid gaze identification methods because it provides a flexible and personalized answer that fixes the problems with current methods.

3.2.1 Strategies for Improving Robustness

GazeSenseTM's effectiveness in dynamic VR scenarios, especially with complex user interactions or fast head motions, requires a complete approach. For precision and responsiveness, advanced motion prediction systems anticipate and react to quick head movements. Adaptive machine learning can dynamically adjust tracking parameters to match user activity, improving performance quickly. Real-time input requires gaze tracking algorithm fine-tuning, although predictive tracking minimizes latency and smooth tracking during rapid head movements. Depth sensing and feature extraction will help GazeSenseTM adapt to VR environments with changing lighting and congestion. To assess effectiveness, identify faults, and optimize and improve user experience in dynamic virtual reality, extensive user research is needed. GazeSenseTM tracks gaze precisely and reliably using these approaches and algorithm iterations, improving user experience and virtual reality immersion.

3.2.2 Hardware Variability

Multiple solutions can be used to overcome hardware limitations and provide consistent gaze tracking across VR systems. Starting with extensive testing on several VR hardware devices to ensure compatibility and optimize GazeSense™'s performance for various specs. For accurate gaze tracking across devices, standardized calibration should be implemented. Dynamic parameter adjustment methods can optimize tracking settings depending on recognized hardware specifications for each device configuration. Cross-platform integration is essential for interoperability with major VR device brands and models, enabling widespread adoption. Monitoring hardware variability trends and updating software regularly are necessary to handle new issues and improve compatibility.

3.2.3 Real-Time Gaze Sensing for VR Interaction

To improve real-time processing and reduce latency and responsiveness, computing efficiency is optimized to give VR consumers a seamless and immersive experience. This optimization uses advanced algorithms and approaches to prioritize key computations, simplify data processing, and maximize resource utilization using parallel computing architectures. GazeSense™ minimizes latency for instantaneous, synchronized gaze monitoring, enhancing user immersion in virtual reality.

3.2.4 Evaluation

To confirm GazeSenseTM's versatility across a range of VR applications, comprehensive user research is necessary. To guarantee the algorithm's reliable operation in practical applications, these studies try to evaluate its efficacy in a range of user interactions and environmental settings. By gathering input from users interacting with various virtual reality scenarios, developers may pinpoint the advantages, disadvantages, and opportunities for development, resulting in iterative improvements and optimization of GazeSenseTM.

3.3 Key Features

```
#GazeSense™ Pseudocode

# Initialize parameters and models
initialize GazeSenseModel
initialize CalibrationModel
initialize FeatureExtractionModel

# Main loop for real-time gaze processing
while user_interacts_with_VR:
  # Capture raw eye-tracking data from a VR headset
  raw_gaze_data = capture_raw_gaze_data()

  # Preprocess raw gaze data
  preprocessed_gaze_data = preprocess_gaze_data(raw_gaze_data)

  # Calibrate the algorithm for user-specific gaze behavior
  calibrated_gaze_data = calibrate_gaze_data(preprocessed_gaze_data, CalibrationModel)

  # Extract features from calibrated gaze data
  gaze_features = extract_features(calibrated_gaze_data, FeatureExtractionModel)

  # Update GazeSense™ machine learning model
  update_model(GazeSenseModel, gaze_features)

  # Obtain real-time predictions for user gaze behavior
  real_time_prediction = predict_gaze_behavior(GazeSenseModel, gaze_features)

  # Provide real-time feedback to VR system based on predictions
  provide_real_time_feedback(real_time_prediction)

# End of loop
```

Fig. 1. Pseudocode for the Gaze Sense Algorithm

1. Processing Eye-Tracking Data GazeSenseTM starts by processing raw eye-tracking data. It uses complicated algorithms to pull out important features like fixation points, saccades, and dwell time.
2. Adaptive Pattern Recognition The algorithm has adaptive pattern recognition features that let it understand gaze behavior in real time and change based on how the user interacts with the VR environment.
3. Machine Learning Integration GazeSenseTM uses machine learning models to keep getting a better sense of what each user likes and how their gaze changes over time, making sure that each user has a unique and natural experience.
4. Compatible with All VR Hardware and Software One of the best things about GazeSenseTM is that it works well with all VR hardware and software setups. This makes sure that the algorithm can be used on a lot of differentVR platforms.
5. Real-Time Feedback GazeSenseTM gives VR systems instant feedback, so they can respond quickly and accurately to the user's gaze. This cuts down on latency and improves the overall user experience.

GazeSenseTM is a big step forward in figuring out what someone is looking at. It's a flexible, adaptable, real-time solution that goes beyond what other algorithms can do. Its complex parts and functions come together to make a dynamic and user-centered algorithm that looks like it will change the way people interact with virtual reality through the gaze.

4 Possible Implications

4.1 An Analysis of Feasibility

It's possible to use GazeSenseTM in virtual reality (VR) worlds using recent progress in eye-tracking technology and machine learning, it is thought that GazeSenseTM could be used in virtual reality (VR) environments. GazeSenseTM can easily be added to the data collection process since many modern VR headsets have high-resolution eye-tracking sensors. The adaptability of the algorithm makes it even more possible because it can handle different VR scenarios and user interactions. Efficient algorithm design and optimization techniques help solve problems with computing power and real-time processing. Identifying Possible Problems and Suggestions for Solutions Implementing GazeSenseTM may be hard because hardware specs can vary and environmental factors can affect how well it tracks your gaze. This is why the algorithm includes a strong calibration mechanism that lets users customize GazeSenseTM to their unique gaze behaviors and makes up for differences in hardware. Real-time processing problems can also be lessened by using parallel computing and optimizing algorithms. Continuous improvements and updates based on user feedback help get around possible problems and make sure the algorithm works well in a variety of VR settings.

4.2 The Ability to Work with Existing Systems

Adding new VR hardware and software to existing ones: The goal with GazeSenseTM is to make it easy to use with existing VR hardware and software. It was designed so that the algorithm could work with any hardware, including different eye-tracking sensors and VR headset setups. Standards-based interfaces and protocols make sure that all devices can work together. This makes it easy to add GazeSenseTM to existing VR systems without making many changes. Developers and people who make VR content will find it easy to add GazeSenseTM to their projects thanks to the simple integration process. Concerning compatibility and possible partnerships, GazeSenseTM takes an open approach, which makes it easier for VR hardware makers, software developers, and content creators to work together. GazeSenseTM supports teamwork to improve its compatibility with a wide range of VR systems by following industry standards and giving well-documented APIs. Research institutions can work together, which lets the algorithm change as eye-tracking technologies get better and new VR trends come up. This framework for working together makes GazeSenseTM a flexible tool that can help the VR field grow and become more innovative. Figure 2 depicts the Step by step-by-step procedure using new algorithm for accurate gaze prediction in VR.

5 Result and Benefit

5.1 Anticipated Results of GazeSenseTM Implementation

GazeSenseTM is expected to produce several significant results when fully deployed, radically changing the field of gaze movement identification in virtual reality (VR). First, a significant improvement in the algorithm's ability to interpret gaze behavior is anticipated. GazeSenseTM seeks to deliver an accurate and instantaneous comprehension of user intent, even in dynamic virtual reality situations, by means of adaptive pattern recognition and ongoing machine learning enhancement. Moreover, it is anticipated that the adaptability of the algorithm will facilitate smooth integration across various VR hardware configurations, guaranteeing broad applicability and adoption ease.

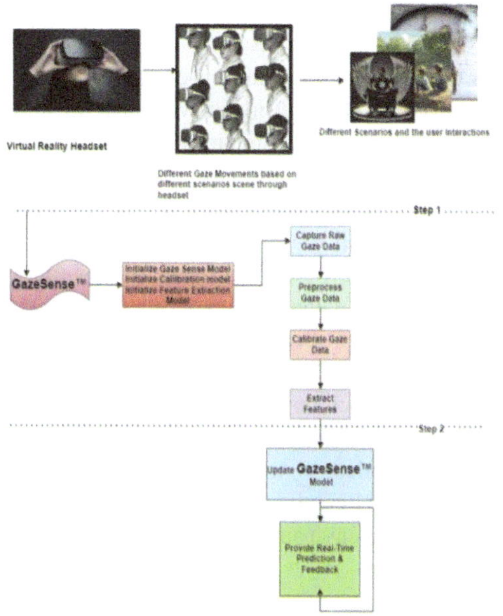

Fig. 2. Step by Step Procedure using new algorithm for accurate gaze prediction in virtual reality

5.2 Possible Gains in User Experience, Efficiency, and Accuracy

Accuracy Improvement: GazeSenseTM has the potential to improve gaze movement identification accuracy greatly. The integration of machine learning with adaptive pattern recognition mechanisms results in a more sophisticated comprehension of user gaze behavior. This increased precision in tracking small eye movements results in a more accurate portrayal of the user's intentions in the virtual reality environment. Efficiency Gains: GazeSenseTM seeks to optimize computational efficiency by reducing latency and streamlining the identification of gaze movements through real-time processing and dynamic adaptation. By removing delays in system responsiveness, this efficiency gain helps users and VR applications interact more smoothly and responsively. Enhanced User Experience: Improving the VR environment's user experience is GazeSenseTM's main objective. With real-time, personalized feedback derived from precise gaze behavior interpretation, users can engage and navigate virtual environments with greater ease. Because of this personalization, users will have a more engaging and pleasurable VR experience, which makes GazeSenseTM an essential tool for increasing user satisfaction. Universal Applicability: GazeSenseTM can be easily incorporated into a variety of VR hardware and software configurations thanks to the algorithm's universal compatibility. Because of its universal applicability, GazeSenseTM benefits are now more widely available to a wider audience while also making adoption easier for developers and VR content creators.

6 Conclusion

This research presents GazeSenseTM, a cutting-edge algorithm that has the potential to revolutionize gaze movement recognition in virtual reality (VR). The algorithm's machine learning integration, universal compatibility, and adaptive pattern recognition are major contributions that address shortcomings in existing techniques and promise increased accuracy and adaptability. GazeSenseTM is a novel approach that improves user experiences in various virtual reality scenarios and raises the accuracy of real-time gaze behavior interpretation. With the release of this innovative algorithm, VR interaction technology is about to undergo a revolutionary change. It establishes a new benchmark for gaze movement identification and highlights its potential for widespread adoption across a range of VR hardware and software setups. The field is called to action to advance further and unlock new dimensions of interaction within virtual environments. It encourages exploration of refinements, validation through user studies, and co-operative efforts with industry stakeholders. In addition to being a technological advancement, GazeSenseTM catalyzes the continued development of gaze-driven interactions in immersive virtual environments.

7 Future Work

The following research ought to concentrate on increasing GazeSenseTM's adaptation to more VR contexts and calibration methods for user-specific gaze behavior interpretation. Simplifying real-time processing and reducing computing resource needs will help VR device combinations integrate seamlessly. GazeSenseTM's functionality may be improved by testing it with many users and diverse environments and verifying the GazeSenseTM and assess its usefulness via extensive user surveys. Working with industry stakeholders and researching its implications in healthcare, education, and human-computer interaction will generate important knowledge. The algorithm's changing function in gaze movement detection in virtual reality, ethical issues, privacy concerns, and inclusion should be addressed in future study. This iterative strategy ensures that GazeSenseTM will evolve and contribute to the fast-expanding area of virtual reality interactions using technical advancements and user-centric ideas. Working with industry stakeholders is crucial for enhancing GazeSense™ and responding to new VR trends and technologies. GazeSense™ developers may enhance capabilities and compatibility by collaborating with VR gear makers, software developers, and content producers for feedback and insights. Responsible development and deployment of GazeSense™ in virtual reality requires addressing ethical concerns such as privacy and inclusiveness. This needs informed user consent, robust privacy measures, and inclusive design and execution. Developers could improve user and stakeholder trust and VR ethics by proactively investigating these issues.

Acknowledgment. The Full-time Ph.D. - IT/IT Enabled Services (IT/ITES)is supported by the Ministry of Electronics and Information Technology, Government of India initiated "Visvesvaraya Ph.D. Scheme for Electronics and IT" Funding Statement: The authors received no specific funding for this study and also thanks to Little Scholars Matriculation Higher Secondary School, Thanjavur District, Tamilnadu, India.

References

1. Ghosh, S., Dhall, A., Hayat, M., Knibbe, J., Ji, Q.: Automatic gaze analysis: a survey of deep learning-based approaches. IEEE Trans. Pattern Anal. Mach. Intell. **46**(1), 61–84 (2023)
2. Zhou, Q., Suraworachet, W., Cukurova, M.: Detecting non-verbal speech and gaze behaviours with multimodal data and computer vision to interpret effective collaborative learning interactions. Educ. Inf. Technol. **29**(1), 1071–1098 (2024)
3. Fehlmann, B., et al.: Virtual reality gaze exposure treatment reduces state anxiety during public speaking in individuals with public speaking anxiety: a randomized controlled trial. J. Affect. Disorders Rep. **14**, 100627 (2023)
4. Qi, J., Tang, H., Zhu, Z.: Exploring an affective and responsive virtual environment to improve remote learning. Virtual Worlds **2**(1), 53–74 (2023)
5. Mauri, M., Rancati, G., Riva, G., Gaggioli, A.: Comparing the effects of immersive and non-immersive real estate experience on behavioral intentions. Comput. Hum. Behav. **150**, 107996 (2024)
6. Vanoncini, M., Hoehl, S., Elsner, B., Wallot, S., Boll-Avetisyan, N., Kayhan, E.: Mother-infant social gaze dynamics relate to infant brain activity and word segmentation. Dev. Cogn. Neurosci. **65**, 101331 (2024)
7. Cebeci, B., Askin, M.B., Capin, T.K., Celikcan, U.: Gaze-directed and saliency-guided approaches of stereo camera control in interactive virtual reality. Comput. Graph. **118**, 23–32 (2024)
8. Miguel-Alonso, I., Checa, D., Guillen-Sanz, H., Bustillo, A.: Evaluation of the novelty effect in immersive Virtual Reality learning experiences. Virtual Reality **28**(1), 27 (2024)
9. Bloch, C., et al.: Interacting with autistic virtual characters: intrapersonal synchrony of non-verbal behavior affects participants' perception. Eur. Arch. Psychiatry Clin. Neurosci. 1–15 (2024)
10. Stone, S.A., et al.: Generating accurate 3D gaze vectors using synchronized eye tracking and motion capture. Behav. Res. Methods **56**(1), 18–31 (2024)
11. Zhang, H., Yin, L., Zhang, H.: A real-time camera-based gaze-tracking system involving dual interactive modes and its application in gaming. Multimedia Syst. **30**(1), 1–14 (2024)
12. Pichierri, G., Murer, K., de Bruin, E.D.: A cognitive-motor intervention using a dance video game to enhance foot placement accuracy and gait under dual task conditions in older adults: a randomized controlled trial. BMC Geriatr. **12**, 1–14 (2012)
13. Harris, D., Donaldson, R., Bray, M., Arthur, T., Wilson, M., Vine, S.: Attention computing for enhanced visuomotor skill performance: testing the effectiveness of gaze-adaptive cues in virtual reality golf putting. Multimedia Tools Appl. 1–19 (2024)
14. Yi-yan, W., et al.: Design recommendations of target size and tracking speed under circular and square trajectories for smooth pursuit with Euclidean algorithm in an eye-control system. Displays **81**, 102608 (2024)

15. Iglesias, I., Sanchez, A., Silva, F.J.: Robotic path compensation training method for optimizing face milling operations based on non-contact CMM techniques. Robot. Comput.-Integr. Manufact. **85**, 102623 (2024)
16. Barañano, A.R., Barrett, B.T., Buckley, J.G.: Measuring gaze angle changes to maintain fixation upon a small target during motion: 3D motion tracking versus wearable eye-tracker. Measurement **225**, 113971 (2024)

Data-Driven Pedagogy: Empowering Educational Excellence in Saudi Higher Institutions Through the Integration of AI and ML

Abeer Aljohani(✉)

Department of Computer Science, Applied College, Taibah University, Medina 42353, Saudi Arabia
aahjohani@taibahu.edu.sa

Abstract. We are experiencing a tremendous transition in the modern educational prospect. The advent of information technologies in individual lives coupled with AI and machine learning being advanced tools of data management have attracted different types of enterprises to the sector of education. The current research glazed through Higher Educational Institutions (HEIs) which are specifically recognized as focal locations for implementing AI and ML technologies to deliver this transformational shift. Saudi Arabia's story is a case study of a nation undergoing a high level of technical transition. The Kingdom is moving through a digital transformation beyond economic fields and is positioned at the forefront. It has promised to "change the way of life" and use technology as a catalyst for it. The research examines the AI and ML evolution in the learning space of Saudi Arabia, bringing the focus towards possible gains as well as certain troubles associated with data-driven pedagogy application. This research is a part of the broader conversation about the role of machine learning in educational quality by gathering evidence from different educational contexts and using the specific situation of Saudi Arabia as a reference. The results are meant to inform educational administrators, the government, and other stakeholders about the revolutionary effect of data-driven pedagogy and the ultimate aim of promoting a more flexible and responsive higher education environment in the time of technological development.

Keywords: Data-Driven Pedagogy · Machine Learning · Educational Technology · Artificial Intelligence · Higher Educational Institutions

1 Introduction

In today's educational scene, the combination of technological advances as well as creative teaching approaches is changing conventional models of learning and teaching. With the development of digital materials, online platforms, and data-driven approaches, teachers now have unparalleled potential to adapt their teaching methods to individual learners' specific needs and preferences. At the forefront of this pedagogical revolution

S. Pal and Á. Rocha (Eds.): ICMMCS 2025, LNNS 1398, pp. 11–21, 2025.
https://doi.org/10.1007/978-3-031-90998-6_2

are artificial intelligence (AI) as well as machine learning (ML), advanced data management and analysis systems with enormous promise for improving teaching and learning activities [1–4]. AI-driven educational programmes is transforming traditional teaching methods and redefining the industry landscape. AI solutions for education provide personalised and flexible learning experiences suited to specific student needs by leveraging advanced algorithms to analyse large datasets [5–7].

Based on P&S Intelligence's Global AI in Education Market Report, the worldwide AI in education market is expanding rapidly, with an anticipated value of USD 5,440.5 million in 2024 and a projected compound annual growth rate of 47.2% between 2024 and 2030. This significant development is attributable to a number of major factors, including greater investment in AI technology, expanding demand for personalised education using AI, and the growing need to automate administrative activities within educational organisations. The COVID-19 epidemic has triggered a major change in the educational landscape, driving schools throughout the world to use various technologies including virtual reality (VR), chatbots, learning management systems (LMS) as well as automated coursework assignments to improve their educational procedures. Additionally, various service companies are developing novel AI solutions for academic usage, typically in conjunction with educational institutions. Particularly, cooperation among tech businesses and schools or institutions are resulting in new training systems to foster AI expertise, according to the Shanghai Daily in December 2021. Global adherence to AI techniques in education is being driven by the increasing demand to automate administrative tasks like exam assessment, student registration as well as policy maintenance. This is because educators want to streamline administrative procedures so they can focus more of their time on instructing and interacting with students. Figure 1 depicts the dynamic growth trajectory of the global AI in education market, including its estimated size and expected increase from 2024 to 2030 [8].

Fig. 1. Global AI in Education Market Size (Source: P&S Intelligence)

Mordor Intelligence Industry Reports recently provided the information that the Saudi Arabia Big Data and AI Market is going up really fast, and it is forecasted to generate revenue of USD 0.38 billion in 2024 and reach revenue of USD. 2.19 billion by 2029, with an awesome compound annual growth rate of 34.24% between 2024 and 2029. The main factors driving this growth are the increased use of data-based AI solutions and the improvements of machine learning technology across the different sectors of the region. Saudi Arabia is creating an AI-nurturing environment by swiftly spreading its high-speed internet and 5G networks which will let data access, security, and communication to be more effective and secure. The Communications and Information Technology Commission (CITC)'s recent implementation of fifth-generation (5G) technology is an important milestone forward in creating a technologically sophisticated infrastructure suitable to AI innovation and development. Furthermore, the region's focus on e-commerce, governance measures, and intellectual property protection is increasing demand for web-based as well as mobile services, creating attractive prospects for developers and service providers.

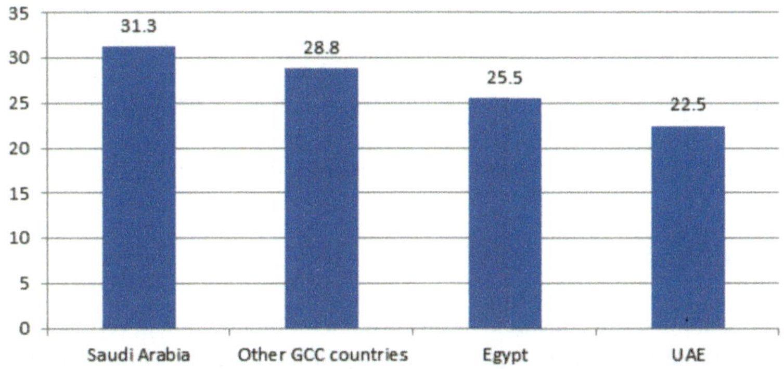

Fig. 2. Contribution of AI to GDP Growth in the MENA Region (2018–2030) by Country and Percentage (Source: General Authority for Statistics (Saudi Arabia)

AI's increasing significance in solving a wide range of difficulties, including oil price instability, urbanisation, and water shortages, highlights its strategic significance in Saudi Arabia's socioeconomic development plan. Despite increased knowledge among college students, technology enthusiasts, and academics about the benefits of AI, there is still a need for more public awareness. Moreover, the COVID-19 epidemic has increased demand for broadband and data services, necessitating the development of information and communication technologies (ICTs) to enable uninterrupted communication as well as enhance the quality of life for Kingdom citizens. Figure 2 depicts the estimated contribution of AI to GDP growth in the Middle East and North Africa (MENA) area from 2018 to 2030, with the anticipated effect divided by country and percentage [9].

1.1 Objective

This paper explores the developing topic of data-driven pedagogy, specifically how the integration of AI and ML technologies might enhance educational excellence in Saudi

Arabia's higher education system. Saudi Arabia, as one of the Middle East's leading proponents of technological advancement, is on the verge of a digital transformation that will go far beyond economics, offering to alter every aspect of societal involvement and progress [10, 11]. Within this backdrop, incorporating AI and ML into higher education institutions (HEIs) is a critical step towards modernising pedagogical approaches and creating a more dynamic and flexible learning environment. This study is significant because it investigates the consequences of AI and machine learning for educational procedures within Saudi Arabia's distinct socio-cultural setting. While various studies have examined the impact of these technologies on education in worldwide contexts, there is a scarcity of research that focuses exclusively on the Saudi higher education scene. By filling this gap, our research hopes to shed light on the possible benefits and obstacles of implementing data-driven pedagogy in the Kingdom, giving useful insights for school administrators, legislators, and other stakeholders.

By conducting a thorough analysis of current research and empirical data, the objective is to clarify the ways in which artificial intelligence (AI) and machine learning (ML) might be utilised to optimise learning environments, improve directions, and raise student achievement at Saudi higher education institutions. This study aims to add to the current conversation on utilising machine learning for superior teaching by combining research from various educational contexts with Saudi Arabia's unique socio-cultural background. The ultimate goal is to provide information for strategic decision-making and policy formulation that supports the creation of a more responsive, flexible and resilient higher education environment in the era of rapid technology advancement.

2 Related Works

The use of technology in education has received a lot of attention in recent years, especially in Saudi Arabia, where extensive efforts to improve education and implement digital transformation are being carried out. With an emphasis on the Saudi context, this section offers an overview of multiple important studies that offer insightful information about the nexus of technology, education, and pedagogy. In the Asir region of Saudi Arabia, Melweth et al. [11] carried out a cross-sectional survey investigation to gauge the burnout levels, technological confidence, as well as opinions on AI integration among secondary school teachers. Their research showed that teachers had varied degrees of technical competency in addition to significant levels of emotional weariness. Teachers were cautiously optimistic about AI's ability to improve student learning and lighten workloads, but they also highlighted exhaustion and the importance of professional development. In his critical analysis of the possible negative effects of using AI in educational settings, Al-Zahrani [12] emphasised the significance of resolving concerns about human connection, data privacy, transparency, as well as ethical considerations. Using a multi-phase investigation that included survey investigation and content analysis, the investigation created a theoretical model that highlighted several areas of concern and promoted all-encompassing tactics to reduce potential harmful effects.

Gollapalli et al. [13] introduced the Sustainable University Field Training (SUNFIT) framework, an educational approach that uses machine learning to improve computing students' practical knowledge development during field training programmes. The framework uses machine learning algorithms to monitor students' performance against worldwide accreditation criteria, with the goal of ensuring quality education delivery and improving learning outcomes in the IT industry. Bhutoria [14] investigated the possibilities for AI-driven personalisation in education, emphasising the role of Big Data, Machine Learning, as well as AI in meeting individual student requirements and preferences. The study conducted a systematic evaluation of literature from China, India, as well as the United States to determine AI's success in meeting certain learning needs while addressing data privacy and accessibility problems.

Elshamly and Gameel [15] looked into the complex interplay of AI, Big Data Analytics (BDA), and other socio-cultural aspects in education, emphasising the importance of user-centered design as well as cultural diversity in technological integration. Their survey-based research found that AI-induced anxiety and prior technological engagement were key factors in affecting people's views towards technology use in education. Radif and Hameed [16] investigated the influence of AI technologies on e-learning systems, emphasising its ability to personalise learning experiences, increase student engagement, as well as improve material delivery techniques. Although admitting the obstacles and ethical concerns involved with AI adoption, their research emphasised AI's revolutionary potential in improving educational methods.

Islam and Khan [17] investigated the relationship between the digital learning environment, academic accomplishment, psychological empowerment, as well as sustainability commitment at Saudi Arabian universities. Their statistical survey-based research demonstrated the favourable influence of digital learning environments on academic attainment while also emphasising the significance of promoting environmental consciousness and diversity in education. Fadlelmula and Qadhi [18] undertook a systematic review of AI research in higher education throughout the Gulf Cooperation Council (GCC) countries, including Saudi Arabia. Their findings demonstrated that AI applications can improve educational outcomes as well as institutional structures, while also revealing constraints linked to technical skills, infrastructure, and reluctance to conventional educational practices.

Table 1. Meta-Analysis of Studies on the Integration of Artificial Intelligence (AI) and Technology in Education

Study	Focus	Methodology	Key Findings
Melweth et al. [11]	Teacher burnout, technology integration, AI	Cross-sectional survey study	- High emotional exhaustion among teachers - Varying technology proficiency levels - Cautious optimism about AI's potential benefits - Need for professional development

(continued)

Table 1. (*continued*)

Study	Focus	Methodology	Key Findings
Al-Zahrani [12]	Potential drawbacks of AI implementation	Content analysis, survey research	- Concerns about human connection, data privacy, transparency, and ethical issues - Development of a theoretical model highlighting areas of concern - Advocacy for comprehensive strategies to mitigate negative consequences
Gollapalli et al. [13]	SUNFIT framework for computing students	Proposal of SUNFIT framework, machine learning models	- Proposal of SUNFIT framework to enhance practical skills acquisition - Use of machine learning to monitor student performance against accreditation standards
Bhutoria [14]	AI-driven personalization in education	Systematic review of literature	- AI's success in catering to specific learning requirements - Challenges related to data privacy and accessibility
Elshamly & Gameel [15]	Complex interplay between AI, BDA, and socio-cultural factors	Structured questionnaire survey	- AI-induced anxiety and previous technological experience influence technology adoption - Importance of user-centered design and cultural diversity
Radif & Hameed [16]	Impact of AI on e-learning platforms	Analysis of Likert scale questionnaire data	- AI's potential to customize learning experiences and enhance student engagement - Challenges and ethical considerations associated with AI implementation
Islam & Khan [17]	Interconnections among digital learning environment, academic achievement, psychological empowerment, sustainability motivation	Quantitative surveys	- Positive impact of digital learning environments on academic achievement - Importance of fostering environmental responsibility and diversity

(*continued*)

Table 1. *(continued)*

Study	Focus	Methodology	Key Findings
Fadlelmula & Qadhi [18]	AI research in higher education across GCC countries	Systematic review of AI research	- Benefits of AI applications in improving educational outcomes and institutional systems - Barriers related to technology skills, infrastructure, and resistance to traditional educational approaches

As per the meta-analysis shown in Table 1 these studies offer helpful perspectives into the benefits and problems of incorporating AI and technology into education, notably in Saudi Arabia. They contribute to the ongoing discussion about using technology to improve educational excellence as well as promote inclusive, learner-centered pedagogies by tackling challenges such as teacher burnout, technological proficiency, ethical considerations, practical knowledge acquisition, personalised learning, cultural diversity, as well as institutional obstacles.

3 AI and ML in Saudi Higher Education Pedagogy

The integration of AI and machine learning (ML) technology has important implications for data-driven teaching in Saudi higher education institutions. This section looks at how AI and machine learning may advance teaching and learning processes, as well as tackle the unique difficulties and opportunities that exist in the Saudi educational setting.

3.1 Personalized Learning Experiences

AI and ML systems can analyse large volumes of educational data, such as student demographics, learning preferences, and performance indicators. Using this data, instructors can tailor educational experiences to individual students' needs and interests. In Saudi higher education institutions, where classrooms are frequently characterised by varied student populations with different styles of learning and histories, personalised learning approaches enabled by AI and ML may encourage increased engagement, preservation, as well as academic achievement.

3.2 Adaptive Instructional Design

AI-powered adaptive educational platforms may dynamically modify instructional content and pace based on individual student growth and accomplishments. By analysing interactions among learners and learning results in real time, these systems may identify areas of advantages and disadvantages and provide tailored teaching strategies to help students learn. In Saudi higher education institutions, where huge class sizes and limited resources can make individualised instruction difficult, adaptive educational technologies empowered by AI and ML provide scalable approaches to improve teaching effectiveness as well as student learning outcomes.

3.3 Data-Driven Decision Making

AI and ML algorithms can analyse educational data to create useful knowledge and support evidence-based decision making at the institutional, departmental, as well as classroom levels. Understanding trends, patterns, as well as correlations in student performance data allows educators as well as administrators to make more well-versed decisions about syllabus design, teaching tactics and resource allocation. In the setting of Saudi higher education institutions, where reforming educational initiatives are taking place to improve quality and relevance, data-driven decision making facilitated by AI and ML may help fulfil national educational objectives and aspirations.

3.4 Predictive Analytics for Student Success

AI and ML can use past statistics and patterns to forecast student outcomes like as academic achievement, preservation, along with graduation rates. Educators can enhance student success rates through recognising at-risk pupils early on and offering focused interventions. In Saudi HEIs, where successful completion and retention percentages are crucial criteria for measuring educational quality and transparency, AI and ML-powered predictive analytics can help improve student retention as well as commitment towards their graduation.

3.5 Enhanced Accessibility and Inclusivity

AI-powered technology can improve educational accessibility and inclusivity by providing adaptable tools and resources for students with a wide range of learning requirements and abilities. For example, AI-powered linguistic translation tools may support non-native Arabic presenters gain access to educational content, as well as AI-driven assistive technologies may help students with impairments access and engage with course materials. In Saudi higher education institutions, where diversity and inclusivity are increasingly recognised as critical principles of educational fairness and social justice, AI-enabled accessibility solutions may encourage greater participation and success for all students.

3.6 Challenges and Considerations

While incorporating AI and ML into data-driven teaching has obvious advantages, it also introduces obstacles and considerations which must be handled. These may include concerns about data privacy and security, algorithmic prejudice as well as fairness, technological literacy and skill development, ethical usage of AI technologies, as well as the need for continual professional development and training for teachers. In the Saudi setting, where social and cultural customs may impact opinions regarding technology adoption as well as innovation, it is essential to make sure that AI and machine learning activities are linked with societal standards and educational goals. The consequences of AI and ML for data-driven teaching in Saudi higher education organisations are far-reaching and transformational. Educators and administrators can revolutionise teaching and learning practices, encourage student achievement, and improve the standard and usefulness of education in Saudi Arabia by leveraging AI and ML technology. However, realising the entire potential of AI and ML for educational purposes requires careful planning, targeted expenditure, as well as continuing collaboration across stakeholders to address issues and capitalise on possibilities for growth and enhancement.

The adoption of AI and ML technologies in Saudi higher education institutions is consistent with broader government policies aimed at supporting innovation and technical advancement. Saudi Arabia is on track to create an AI-friendly environment with the rapid rollout of high-speed broadband and 5G networks. The recent deployment of fifth-generation (5G) technology is a big step forward in this respect, demonstrating the country's commitment to developing a technologically capable infrastructure to enable AI research and growth. Through automating typical administrative tasks, AI technologies allow educators to devote more time and resources to teaching and engaging students. Furthermore, predictive analytics technologies may detect at-risk students early and give tailored interventions to promote their academic progress, resulting in higher retention and percentages of graduates. However, significant implementation of AI and ML in Saudi higher education institutions is not without hurdles. Considerations around data privacy, security, and algorithmic prejudice must be resolved to guarantee that these technologies are used ethically and responsibly. In addition, initiatives to improve digital literacy and development of abilities among teachers and pupils are critical to obtaining all the advantages of AI incorporation in education.

4 Conclusion

The integration of AI and ML technologies in Saudi higher education institutions has enormous potential to transform teaching and learning methods, increase student participation, and assist broader national goals for innovation as well as technological growth. Personalised learning experiences, flexible instructional design, as well as conversational AI tools enable instructors to respond to individual students' different requirements and preferences. Furthermore, the construction of high-speed internet and 5G networks in Saudi Arabia demonstrates the country's commitment to fostering an AI-friendly ecosystem for research, development, and implementation. AI-driven education can address important difficulties for Saudi higher education institutions by automating administrative processes and applying predictive analytics. Yet, significant implementation of AI and machine learning technologies in education must be supported by measures to address ethical problems, improve digital literacy, and assure responsible use of these tools. Saudi higher education institutions can unleash new prospects for development, creativity, and excellence in the digital era by making significant investments in AI-driven education as well as forming collaborative relationships with academia, industry, and the government. As Saudi Arabia progresses towards establishing a knowledge-based economy, AI and ML will play a more significant part in determining the future of education and promoting advancement and wealth for the country and its citizens. Future studies could investigate the ethical implications as well as data privacy concerns linked with AI integration in the educational perspective of Saudi Arabia.

References

1. Martin, S.M.: Artificial Intelligence, Mixed Reality, and the Redefinition of the Classroom. Rowman & Littlefield (2019)
2. Sadiku, M.N., Musa, S.M., Chukwu, U.C.: Artificial intelligence in education. iUniverse (2022)
3. Almusaed, A., Almssad, A., Yitmen, I., Homod, R.Z.: Enhancing student engagement: harnessing "AIED"'s power in hybrid education—a review analysis. Educ. Sci. **13**(7), 632 (2023)
4. Roll, I., McNamara, D., Sosnovsky, S., Luckin, R., Dimitrova, V.: Artificial Intelligence in Education. Springer, Heidelberg (2021). https://doi.org/10.1007/978-3-642-21869-9
5. Peng, S.L., Pal, S., Amsaad, F.: Proceedings of 3rd International Conference on Mathematical Modeling and Computational Science: ICMMCS 2023. Springer, Heidelberg (2023). https://doi.org/10.1007/978-981-99-3611-3
6. Peng, S.-L., Lin, C.-K., Pal, S. (eds.): Proceedings of 2nd International Conference on Mathematical Modeling and Computational Science. Advances in Intelligent Systems and Computing. Springer, Singapore (2022).https://doi.org/10.1007/978-981-19-0182-9
7. Peng, S.-L., Hao, R.-X., Pal, S. (eds.): Proceedings of First International Conference on Mathematical Modeling and Computational Science. Advances in Intelligent Systems and Computing. Springer Singapore (2021). https://doi.org/10.1007/978-981-33-4389-4
8. AI in education market size, latest trends and Growth Outlook. P&S Intelligence. (2024). https://www.psmarketresearch.com/market-analysis/ai-in-education-market
9. Saudi Arabia big data and Artificial Intelligence Market Size & Share Analysis - Industry Research Report - Growth trends. Saudi Arabia Big Data and Artificial Intelligence Market Size & Share Analysis - Industry Research Report - Growth Trends (2024). https://www.mordorintelligence.com/industry-reports/saudi-arabia-big-data-and-artificial-intelligence-market
10. Fandy, M.: Information technology, trust, and social change in the Arab world. Middle East J. 378–394 (2000)
11. Melweth, H.M.A., Alkahtani, A.S., Al Mdawi, A.M.M.: The impact of artificial intelligence on improving the quality of education and reducing future anxiety among a sample of teachers in Saudi Arabia. Kurdish Stud. **12**(2), 5741–5758 (2024)
12. Al-Zahrani, A.M.: Unveiling the Shadows: Beyond the Hype of AI in Education. Heliyon (2024)
13. Gollapalli, M., et al.: SUNFIT: a machine learning-based sustainable university field training framework for higher education. Sustainability **15**(10), 8057 (2023)
14. Bhutoria, A.: Personalized education and artificial intelligence in the United States, China, and India: A systematic review using a human-in-the-loop model. Comput. Educ.: Artif. Intell. **3**, 100068 (2022)
15. Elshamly, A., Gameel, Z.A.A.: AI and BDA impact on stakeholders' responses to education technology adoption. Migr. Lett. **20**(8), 1041–1067 (2023)
16. Radif, M., Hameed, O.M.: AI-driven innovations in e-learning: transforming educational paradigms for enhanced learning outcomes. Arts Educ. **38** (2024)

17. Islam, Q., Khan, S.M.F.A.: Sustainability-infused learning environments: investigating the role of digital technology and motivation for sustainability in achieving quality education. Int. J. Learn. Teach. Educ. Res. **23**(1), 519–548 (2024)
18. Fadlelmula, F.K., Qadhi, S.M.: A systematic review of research on artificial intelligence in higher education: Practice, gaps, and future directions in the GCC. J. Univ. Teach. Learn. Pract. **21**(06) (2024)

Current Security Issues and Vulnerabilities Associated with Mobile Application

Basheer Riskhan[1(✉)], Abdullahi Adewole Zakariyah[1], Md. Farman Ali[1], Md Amin Ullah Sheikh[1], and Rajan Verma[2,3]

[1] School of Computing and Informatics, Albukhary International University, Alor Setar, Kedah, Malaysia
b.riskhan@aiu.edu.my, {abdullahi.zakariyah,farman.ali, amin.sheikh}@student.aiu.edu.my

[2] University Centre of Research and Development, Chandigarh University, Mohali 140413, India

[3] Division of Research and Development, Lovely Professional University, Phagwara 144411, India

Abstract. Mobile applications have become a crucial part of modern life, facilitating everything from social interactions to financial transactions. However, this ubiquity also presents a significant security challenge. The landscape of mobile app security is fraught with vulnerabilities and threats that can compromise user data, privacy, and the overall integrity of applications. This paper delves into the security issues and vulnerabilities affecting mobile applications. It provides an in-depth analysis of various attack methods used by cybercriminals to extract user data and explores how users can protect their information from being compromised. The discussion will cover common tactics employed by attackers, the potential consequences of losing mobile application data to these threats, and the critical measures necessary for safeguarding user data against such attacks.

Keywords: Mobile Application · Security Challenge · social networking

1 Introduction

Statistics show that people spent an astonishing 3.8 trillion hours on mobile apps in 2022. Despite this extensive use, only 2.6% of Android app users and 4.3% of iPhone app users continue using an app 30 days after downloading it. Mobile apps are projected to generate $932 billion in revenue in 2023. However, this widespread usage also introduces significant security concerns. The key security issues associated with mobile apps include data confidentiality, data privacy, and non-compliance with security standards and regulations. This paper will delve into each of these security issues in detail in the literature review. Each issue will be analyzed independently, and strategies for prevention and resolution will be discussed [1].

Mobile applications have become integral to modern life, offering unprecedented convenience and accessibility. However, this widespread adoption has also opened the

S. Pal and Á. Rocha (Eds.): ICMMCS 2025, LNNS 1398, pp. 22–33, 2025.
https://doi.org/10.1007/978-3-031-90998-6_3

door to numerous security issues and vulnerabilities. As mobile devices increasingly store sensitive personal and financial information, they have become attractive targets for cybercriminals [2]. The rapid pace of app development, often prioritizing functionality over security, exacerbates the risk.

Furthermore, the diverse and fragmented ecosystem of mobile operating systems and devices introduces additional complexity in ensuring robust security measures. This introduction explores the current landscape of security threats and vulnerabilities in mobile applications. It highlights the need for comprehensive strategies to protect users and their data in an ever-evolving digital environment [3].

By addressing these challenges, it is possible to work towards creating a safer and more secure mobile app ecosystem for all users. This study aims to contribute to the ongoing efforts to enhance mobile app security and provide actionable insights for developers and security professionals.

1.1 Background

Recent statistics note that about 90% of the global internet population uses a mobile device to go online. For hackers, this means more people to victimize, making endpoint security for mobile devices increasingly vital. In this paper, the focus is on addressing the security issues associated with mobile applications. Before proceeding further, the relevant terminologies will be described. Mobile applications are software programs designed to operate on smartphones, tablets, or other mobile devices. These applications span various categories, including entertainment, social networking, productivity, and gaming, among others. The widespread use of mobile applications has transformed how we interact with technology, providing convenient access to services and information on the go [4].

Security issues or challenges regarding mobile applications can be defined as different approaches in which mobile application data can be compromised or breached. Data integrity is important, and data must be protected from unauthorized users. Due to the data loss from mobile applications, a decision was made to delve into the security issues associated with mobile applications [5]. Mobile application security refers to the technologies and security procedures that protect mobile applications against cyberattacks and data theft. An all-in-one mobile app security framework automates mobile application security testing on platforms like iOS, Android, and others [6].

Mobile device usage has steadily increased in recent years. Recent statistics note that about 90% of the global internet population uses a mobile device to go online. For hackers, this means more people to victimize, making endpoint security for mobile devices increasingly vital. Data security is very important for personal data. To secure data from being compromised by unauthorized users, there should be a study on the vulnerabilities of the data and how it can be compromised. This applies to mobile applications; for mobile application data to be secured, there needs to be a study of the vulnerability and the security issues that may affect it. A deliberate investigation into the current security landscape surrounding mobile applications is essential to mitigate risks effectively, address existing vulnerabilities, and safeguard users' digital experiences [7].

1.2 Objectives

- To identify prevalent security issues affecting mobile applications.
- To analyze the vulnerabilities commonly associated with mobile applications.
- To assess the impact of these vulnerabilities on users' data privacy and device security.

1.3 Scope

This study will identify the common issues affecting mobile applications (objective 1). Examining the vulnerabilities and security issues associated with mobile apps aims to identify and analyze the various threats, risks, and challenges posed by them (Objective 2). It will also focus on the impact and the consequences that lack of security would cause on users' data privacy and device security (Objective 3).

2 Literature Review

According to Tao et al. (2020), the rapid growth of mobile applications in sectors like banking, healthcare, and e-commerce has heightened security concerns, necessitating robust measures such as authentication, authorization, and data confidentiality [8]. These applications extend corporate networks, increasing exposure to security threats. Security testing is challenging, making mobile apps not inherently more secure than desktop or web applications. Khurana and Baraković (2016) both emphasized the need for robust security measures in wireless networks, with Khurana specifically discussing the merits and demerits of existing methods [9, 13, 15, 32].

Basavala et al. (2013) focused on the rapid growth of mobile applications in sectors like banking, healthcare, and e-commerce, which has heightened security concerns, necessitating robust measures such as authentication, authorization, and data confidentiality. These applications extend corporate networks, increasing exposure to security threats. Security testing is challenging, making mobile apps not inherently more secure than desktop or web applications [3, 21, 22, 34].

According to Mishra and Kokate(2014), the specific vulnerabilities and security issues in ad-hoc networks, with Mishra focusing on attacks against routing protocols and Kokate discussing the potential use of mobile ad hoc networks in real-life applications. These studies collectively underscore the importance of addressing security concerns in mobile applications. The increasing use of Android-based smartphones has made them a prime target for hackers, leading to a rise in vulnerabilities and attacks. These attacks include privilege escalation and privacy breaches, which can compromise the security of these devices [23, 24, 33].

The rapid growth of Android mobile applications has further exacerbated these vulnerabilities, with attackers exploiting them to access sensitive data. To address these issues, it is crucial to understand the security architecture of Android devices and implement effective countermeasures [22, 25]. Himanshu et al. (2016) proposed that the behavior of users' browsing and inability to see all the links on mobile phones had been one of the biggest exposures [6, 13, 15, 26]. The lack of ability to see full links in mobile phones due to the small screen has paved an easy path for phishing links to be effective. They have given examples of how social networking sites like email or text messages on devices with heavy reliance on URL links make it impossible for the average user to notice whether links are safe or not, so that way scammers get their hands on user's data [14, 27, 28]. This situation is happening now, and within ourselves, so it's more accurate and doesn't need an experiment to prove it [27, 34].

Dr. Robert P. Griffin Jr (2018) proposed that applications for Android and iOS have been found to save sensitive information like usernames and passwords and plaintext log files that may be read by attackers [11, 17, 29, 30]. The instances were recorded in the Common Vulnerabilities and Exposures (CVE) database (CVE-2012-2630 and CVE-2014-0647), revealing the Twitter credentials and Starbucks usernames, passwords, and email account information. Although this threat had been mitigated, the fact it had been a threat before shouldn't be forgotten.

Johannah Jebaraj (2016) stated that some of the applications request more permissions than their requirement for their functioning [18, 27, 31]. It is called Privilege Escalation, and such permissions are done by embedded advertising libraries within the apps. They can be misused by the agencies to collect personal information without the user's consent. These threats are counter-attacks by a mobile's security mechanism, antivirus software for malicious detection, and application firewalls and analysis tools. It is mentioned that there are other security tools and models like Kirin, SASQUATCH, Dexteroid, and a few others. All these tools can add more security to mobile applications and on mobile phones themselves [32, 33].

The study highlights the significant security challenges faced by mobile applications, including data breaches, unauthorized access, and malware attacks. These issues are exacerbated by inadequate encryption, insecure data storage, and a lack of security standards. Key vulnerabilities identified include weak authentication mechanisms, excessive permissions, and insecure communication channels. These vulnerabilities make mobile applications vulnerable to attacks like phishing, privilege escalation, and data interception. The study emphasizes the need for better user education and stringent security measures. Effective mitigation strategies include robust authentication protocols, regular security assessments, and adherence to security standards. Advanced security tools and frameworks can enhance security measures. The study contributes to developing improved security frameworks for mobile applications, ensuring better protection of user data, and maintaining trust in an increasingly connected world [20, 32, 33]. Recent studies have explored security vulnerabilities and challenges in mobile applications and software development [35, 36], with a focus on cybersecurity threats and vulnerability [37, 38].

3 Methodology

The methodology employed in this study is based on qualitative research. Data collection occurred through a structured questionnaire, incorporating open-ended questions to capture detailed and nuanced responses. This approach explores participants' experiences and perceptions in-depth, providing comprehensive insights into the subject matter. By focusing on qualitative data, the study seeks to uncover underlying patterns and themes that quantitative methods might overlook. The structured questionnaire is designed to elicit rich, detailed responses, ensuring a thorough understanding of the participants' viewpoints and experiences. This method enables a deeper exploration of the complexities surrounding mobile application security issues. The Research process steps are shown in Fig. 1.

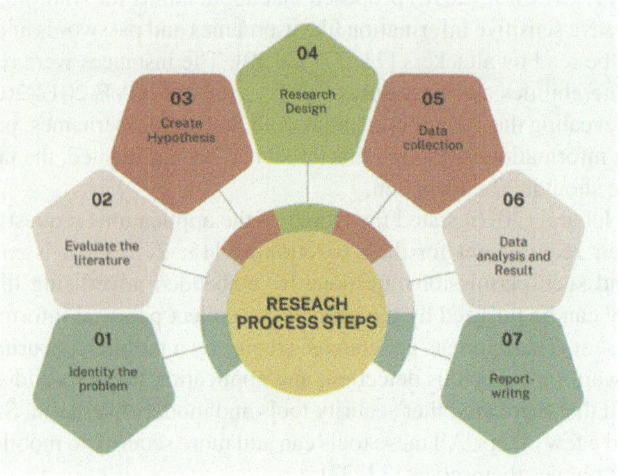

Fig. 1. Research Process Steps

3.1 Data Analysis

For this research paper on the security and vulnerabilities of mobile applications, we gathered comprehensive data through an extensive literature review and a structured questionnaire survey. Initially, we thoroughly analyzed existing research papers related to our topic. We drew insights from many scholarly articles, technical reports, and other documents, which we sourced from Google Scholar, Academia, and other renowned article publishing websites. These research papers provided an essential understanding of our topic. Furthermore, we conducted a questionnaire survey to obtain primary responses, which laid the groundwork for our theories. The survey included questions about personal experiences with mobile application security and malicious attacks encountered while using them. These two methods paved the way for data collection and provided more insights into our research.

3.2 Questionnaire Design

A survey was conducted to seek people's opinions on the security vulnerabilities associated with mobile applications. The questionnaire consisted of both open-ended and closed-ended questions. Many criteria were considered when selecting the respondents, including age, faculty, level of education, and knowledge about security vulnerabilities in mobile applications. Each respondent assessed our questionnaire, and we recorded and tracked their responses. After selecting the respondents, we shared a link to the questionnaire for them to fill out, and their feedback was generated in the form of a pie chart for better visualization. By designing the questionnaire with these components, we aimed to gather comprehensive data that would provide valuable insights into the security issues and vulnerabilities associated with mobile applications.

3.3 Justification of Choices of Methodology

We chose these methodologies because the issue we are addressing is a continually occurring event in this digital age. This research focuses on identifying the vulnerabilities associated with mobile applications and obtaining data from people who have experienced various security issues on mobile apps, which is crucial for this research. We learned about the different ways hackers attacked people to compromise their data and the multiple approaches those hackers used.

3.4 User Perceptions of Mobile Application Safety

The survey responses show that security concerns are a significant issue among mobile application users. Out of 26 respondents, a substantial majority (76.9%) expressed concern about the security of the mobile applications they use. Conversely, only 19.2% of the respondents indicated that they were not worried about app security, as presented in Fig. 2. This data underscores the critical need for developers to prioritize security features and reassure users about the safety of their personal information and data while using mobile applications.

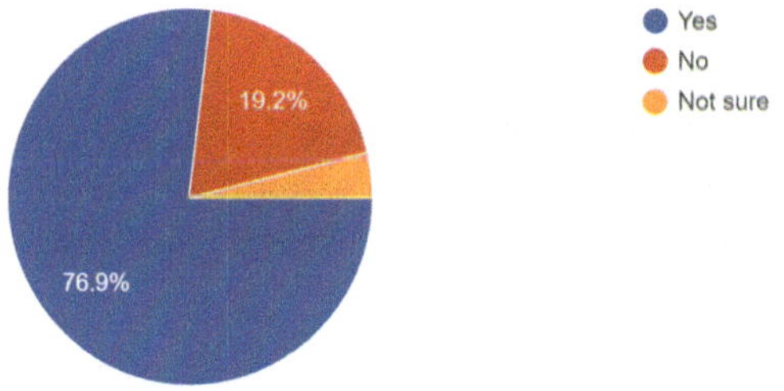

Fig. 2. Significant Issue Among Mobile Application Users

3.5 Privacy Issues in Mobile Applications

Based on the survey data collected from 27 respondents, it was found that a significant % of users, 59.3%, have experienced a security breach or privacy issue while using a mobile application. In contrast, 40.7% of the respondents reported not having encountered problems such as those presented in Fig. 3. This data highlights the prevalent concerns regarding mobile application security among users, indicating that over half of the surveyed population has faced security or privacy challenges, underscoring the importance of enhancing mobile app security measures.

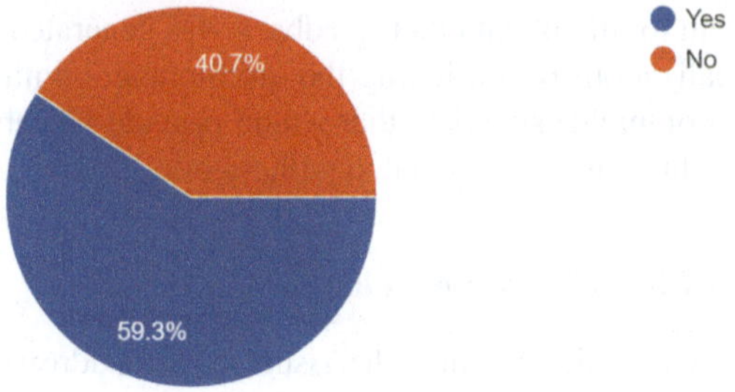

Fig. 3. User Experienced a Security Breach or Privacy Issues

3.6 Most Frequently Used Types of Mobile Applications

Figure 4 presents the social media applications that are the most frequently used among respondents, with a significant 92.6% indicating regular use. Following social media, communication apps are also widely used, with 44.4% of respondents using them frequently. Finance apps, such as banking, are utilized by 25.9% of the participants, while entertainment apps are used by 14.8%. Health and fitness apps account for 11.1% of frequent usage, and productivity apps are used by 7.4% of respondents. Other applications are used less frequently, with only 3.7% indicating their regular use. This data highlights the dominance of social media and communication apps in users' daily routines while showing varied usage patterns across other app categories.

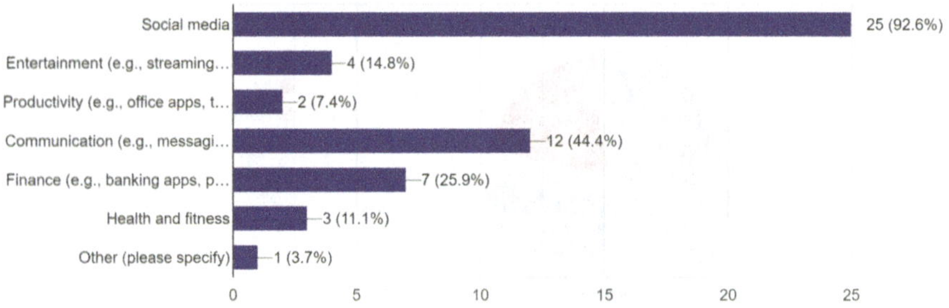

Fig. 4. Frequently Used Social Media Applications

3.7 Frequency of Mobile Application Usage

According to the survey data on mobile app usage frequency, a vast % of respondents, 88.9%, use mobile applications daily. In contrast, 11.1% of the respondents reported using mobile applications several times a week. This data illustrates the pervasive role of mobile apps in users' daily lives, with most individuals engaging with various applications daily, as shown in Fig. 5.

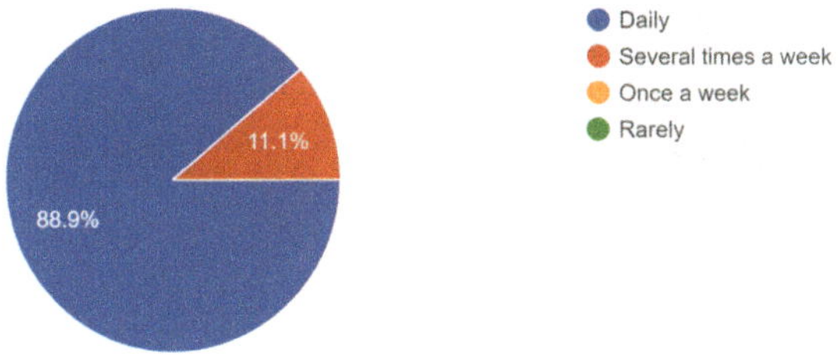

Fig. 5. Daily Usage of Mobile Applications

3.8 Awareness of Security Vulnerabilities in Mobile Applications

This part indicates varying levels of awareness regarding security vulnerabilities associated with mobile applications. Among the 27 respondents, 18.5% reported being familiar with these security issues. The majority, 66.7%, described themselves as somewhat familiar, indicating a general but not comprehensive understanding of mobile app security vulnerabilities. Meanwhile, 14.8% of the respondents admitted to being unfamiliar with such issues in Fig. 6. This data suggests that while most users are aware of mobile app security risks, there remains a need for increased education and information to enhance user understanding and vigilance.

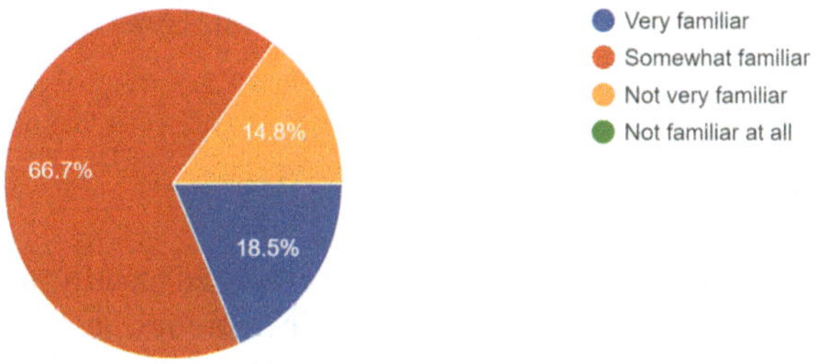

Fig. 6. Security Vulnerabilities Awareness

3.9 Pre-download Habits: Reviews and Security Reports

The data reveals varying behaviors among users regarding reading app reviews or security reports before downloading a mobile application. Among the 27 respondents, 55.6% indicated that they sometimes read reviews or security reports, while 25.9% stated that they always do so. Conversely, 14.8% of respondents reported that they rarely read these materials. This data indicates that while most users occasionally consider app reviews or security information, a smaller yet significant group consistently reviews this information, and another portion infrequently engages in this practice, highlighting the varied levels of diligence and caution exercised by mobile app users (Fig. 7).

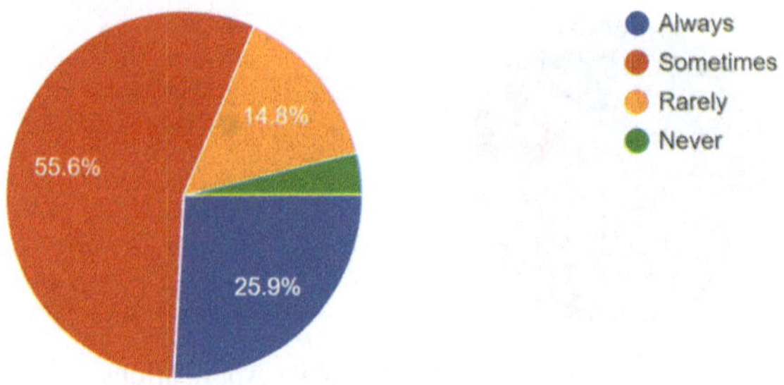

Fig. 7. Reading Reviews or Security Reports before Downloading App

4 Results

The survey results revealed several key findings regarding mobile application security. A significant portion of respondents (76.9%) expressed concern about the mobile applications' security, with 59.3% having experienced a security breach or privacy issue. The most commonly used types of mobile applications were social media (92.6%) and communication apps (44.4%), highlighting their pervasive presence in users' daily lives. Additionally, 88.9% of respondents reported using mobile applications daily. Despite high usage, there was a notable lack of familiarity with security vulnerabilities; only 18.5% of respondents were familiar with the issues. This gap in awareness was further underscored by the fact that 55.6% of users only sometimes read app reviews or security reports before downloading an application.

5 Discussion

The data collected underscores the critical need for enhanced security measures in mobile applications. Users identified various security vulnerabilities, including unauthorized access, insecure data storage, and phishing attacks. The impact of these vulnerabilities on user trust is significant, with 48.1% of respondents indicating that security issues increase their distrust in using mobile applications. Furthermore, the findings highlight a discrepancy between users' high usage rates and low awareness of security practices. This suggests that while mobile applications are integral to everyday activities, users remain vulnerable due to insufficient security knowledge and practices. Consequently, there is a pressing need for mobile app developers to implement robust security measures, such as strong encryption, regular security updates, and user education on security best practices, to mitigate these risks and enhance user trust and safety.

6 Conclusion

In conclusion, this research shows different ways data can be compromised by attackers from user mobile apps. It also illustrates different vulnerabilities that can occur to one's mobile application, guiding how to detect different scam approaches and how data can be protected against attackers. The types of security attacks discussed include phishing, malware, and other common types of security. This research elaborates on different security issues that generally occur in mobile applications and how to avoid or prevent these security issues. This paper explores the complex world of mobile application security, focusing on mobile applications as essential smartphones. The study of security issues involved in mobile applications is important because it provides vital insights into improving security mechanisms and protocols, which are necessary for protecting user data and privacy in the context of mobile devices; it also exposes the vulnerabilities involved. The study guides the vulnerabilities and how they should be addressed to avoid compromised or lost data. After studying the security issues involved in mobile applications, we are expected to be able to identify the vulnerabilities involved, develop security standards and best practices, and enhance security measures.

7 Future Work

There are limitations to this study, as it states theoretical insights into the usage of mobile apps, and it is crucial to acknowledge these limitations when interpreting the study's results. It is important to note that study findings might get slimed due to the absence of empirical experiments. Challenges like data security, data management, privacy concerns, and others in IoT realms further obstruct study findings. The dynamic changes and developments in technology can make a hindrance over time to its relevance and accuracy.

References

1. Abdullah, H., Zeebaree, S.R.M.: Android Mobile Applications Vulnerabilities and Prevention Methods: A Review. semanticscholar.org (2021). https://www.semanticscholar.org/paper/Android-Mobile-Applications-Vulnerabilities-and-A-Abdullah-Zeebaree/66f217c39116abdf9ab2f89455d228cd5579f110
2. Ahamad, S.: Smartphone security and protection practices, (PDF) Smartphone Security and Protection Practices | Shahanawaj Ahamad - Academia.edu (2022). http://www.academia.edu/70442415/Smartphone_Security_and_Protection_Practices?rhid=28519263243&swp=rr-rw-wc-30094607
3. Ahvanooey, M.T.: A survey on smartphone security: Software vulnerabilities, malware, and attacks. Int. J. Adv. Comput. Sci. Appl. (2020). https://www.academia.edu/43779422/A_Survey_on_Smartphones_Security_Software_Vulnerabilities_Malware_and_Attacks?rhid=28519395220&swp=rr-rw-wc-70442415
4. Goode, A.: Managing mobile security: how are we doing?(2010). https://doi.org/10.1016/S1353-4858(10)70025-8
5. Cifuentes, Y., Beltrán, L., Ramírez, L.: Analysis of security vulnerabilities for mobile health applications (2015). https://www.semanticscholar.org/paper/Analysis-of-Security-Vulnerabilities-for-Mobile-Cifuentes-Beltr%C3%A1n/6a27e716123bb59d24f7a418b2c6fdd3917ed361
6. Jhanjhi, N.Z., Shah, I.A. (eds.): Cybersecurity Measures for Logistics Industry Framework. IGI Global (2024)
7. Zinkus, M., Jois, T.M., Green, M.: Data security on mobile devices: current state of the art, open problems, and proposed solutions (2021). arXiv preprint arXiv:2105.12613

8. Mohan, D., et al.: Does having a mobile phone matter? Linking phone access among women to health in India: an exploratory analysis of the National family health survey. PLoS ONE **15**(7), e0236078 (2020)
9. Hur, J.B., Shamsi, J.A.: A survey on security issues, vulnerabilities, and attacks in Android-based smartphones. In: 2017 International Conference on Information and Communication Technologies (ICICT) (2017)
10. Graffiti: App Security (2020). https://www.grafiati.com/en/literature-selections/app-security/
11. Jebaraj, J.: An exploratory study on security issues about smartphone applications, Academia.edu (2016). https://www.academia.edu/30094607/An_Exploratory_Study_on_Security_Issues_Pertaining_to_Smartphone_Applications?rhid=28519166230&swp=rr-rw-wc-39804973
12. Laverty, J.P., Wood, D.F., Kohun, F.G., Turchek, J.: Comparative analysis of mobile application development and security models (2011). https://doi.org/10.48009/1_iis_2011_301-312
13. Khurana, P., Sharma, A., Singh, P.K.: A systematic analysis on mobile application software vulnerabilities: issues and challenges. Ind. J. Sci. Technol. **9**(32) (2016)
14. Mejia-Miranda, J., Melchor-Velasquez, R.E., Munoz-Mata, M.A.: Vulnerability detection in smartphones: a systematic literature review. In: 2017 12th Iberian Conference on Information Systems and Technologies (CISTI) (2017)
15. Riskhan, B., Safuan, H.A.J., Hussain, K., Elnour, A.A.H., Abdelmaboud, A.: An adaptive distributed denial of service attack prevention technique in a distributed environment. Sensors **23**, 6574 (2023). https://doi.org/10.3390/s23146574
16. Zaman, G., Mahdin, H., Hussain, K., Abawajy, J., Mostafa, S.A.: An ontological framework for information extraction from diverse scientific sources. IEEE Access (2021). https://doi.org/10.1109/ACCESS.2021.3063181
17. Basheer, R., Aziz, M.A.A., Arifa, H.: Convolutional neural network based Malaysian sign-language recognition web application (2024)
18. McGraw, G.: Mobile application security. McGraw-Hill (2010). https://www.amazon.com/Mobile-Application-Security-Himanshu-Dwivedi/dp/0071611016
19. Mishra, R., Sharma, S., Agrawal, R.: Vulnerabilities and security for ad-hoc networks. In: 2010 International Conference on Networking and Information Technology (2010)
20. Li, Q., Clark, G.: Mobile security: a look ahead. IEEE Secur. Priv. **11**(1), 78–81 (2013)
21. Drake, J.J., Lanier, Z., Mulliner, C., Fora, P.O., Ridley, S.A., Wicherski, G.: Android Hacker's Handbook. John Wiley & Sons (2014)
22. Al-Dhaqm, A., Abd Razak, S., Ikuesan, R.A., Kebande, V.R., Siddique, K.: A review of mobile forensic investigation process models. IEEE Access **8**, 173359–173375 (2020)
23. Mazuera-Rozo, A., Bautista-Mora, J., Linares-Vásquez, M., Rueda, S., Bavota, G.: The Android OS stack and its vulnerabilities: an empirical study. Empir. Softw. Eng. **24**, 2056–2101 (2019)
24. Weichbroth, P., Łysik, Ł: Mobile security: threats and best practices. Mob. Inf. Syst. **2020**(1), 8828078 (2020)
25. Weichbroth, P. and Łysik, Ł.: Mobile Security: Threats and Best Practices (2020). https://doi.org/10.1155/2020/8828078
26. Khurana1, P., Sharma, A., Singh, P.K.: Juit (2016). http://www.ir.juit.ac.in:8080/jspui/bitstream/123456789/8356/1/A%20Systematic%20Analysis%20on%20Mobile%20Application%20Software%20Vulnerabilities%20Issues%20and%20Challenges.pdf
27. Li, Q., Clark, G.: Mobile Security: A Look Ahead (2013). https://doi.org/10.1109/MSP.2013.15
28. Shoro, S.: Mobile Computing and its security, *(PDF) Mobile Computing and its Security* | saima shoro - Academia.edu (2023). http://www.academia.edu/94157460/Mobile_Computing_and_its_Security?rhid=28519263243&swp=rr-rw-wc-30094607

29. Ahvanooey, M.T., Li, Q., Rabbani, M., Rajput, A.R.: A survey on smartphones security: software vulnerabilities, malware, and attacks. arXiv preprint arXiv:2001.09406 (2020)
30. Tao, C., Guo, H., Huang, Z.: Identifying security issues for mobile applications based on user review summarization. Inf. Softw. Technol. **122**, 1–13 (2020)
31. Holla, S., Katti, M.M.: Android based mobile application development and its security (2012). https://citeseerx.ist.psu.edu/document?repid=rep1&type=pdf&doi=7ccdfdd3138014400a854954325d54dbb7447ead
32. Riskhan, B., Raufi, A.M., Usmáni, M.H.: Physical security to cybersecurity (challenges and implications in the modern digital landscape). J. Electr. Syst. **20**(4s), 692–702 (2024)
33. Hussain, K., Rahmatyar, A. R., Riskhan, B., Sheikh, M.A.U., Sindiramutty, S. R.: Threats and Vulnerabilities of Wireless Networks in the Internet of Things (IoT). In: 2024 IEEE 1st Karachi Section Humanitarian Technology Conference (KHI-HTC), pp. 1–8. IEEE (2024)
34. Almassri, S., Hussain, K., Abd, H.B., Riskhan, B.: Study of mobile e-communities effectiveness for society and social networking development (2023)
35. Jayakumar, P., Brohi, S.N., Jhanjhi, N.Z.: Artificial intelligence and military applications: innovations, cybersecurity challenges & open research areas (2021)
36. Humayun, M., Niazi, M., Jhanjhi, N.Z., Alshayeb, M., Mahmood, S.: Cyber security threats and vulnerabilities: a systematic mapping study. Arab. J. Sci. Eng. **45**, 3171–3189 (2020)
37. Shah, I.A., Rajper, S., ZamanJhanjhi, N.: Using ML and data-mining techniques in automatic vulnerability software discovery. Int. J. Adv. Trends Comput. Sci. Eng. **10**(3) (2021)
38. Humayun, M., Jhanjhi, N., Almufareh, M.F., Khalil, M.I.: Security threat and vulnerability assessment and measurement in secure software development. Comput. Mater. Contin **71**, 5039–5059 (2022)

An Intelligent and Secure Fog Computing-Based System for Predicting and Preventing Yellow Fever Outbreaks

Prabal Verma(✉) and Shamsul Haq

Department of Information Technology, National Institute of Technology Srinagar, Srinagar, Jammu and Kashmir, India

{prabal.verma,shamsul_2022phaite002}@nitsri.ac.in

Abstract. Yellow fever is a viral, hemorrhagic, vector-borne illness that often has a shorter life span. Large epidemics are caused by highly populated areas with high mosquito densities and low immunisation rates. The disease's toxic phase is lethal. Disease identification using traditional approaches was a laborious process. Improved internet connectivity, cloud computing, mobile technology, and fog computing advancements have improved the quality of distant healthcare services. Fog computing is being used to construct healthcare systems, and this is showing promise as an effective result. Fog computing offers low delay, minimum reaction time, high mobility, improved service quality, site/position awareness, and notification services at the network edge. This research proposes fog-based healthcare systems and mosquito sensors to detect and manage the yellow fever pandemic. The fuzzy k-NN is used to identify infected and uninfected users and quickly generates emergency reminders for users on fog servers. Furthermore, on cloud servers, information granulation and GPS (geographic positioning/location system) are used to represent each yellow fever user on Google Maps so that the isolation of infected users can be enforced. This will prevent further exposure to citizens, thereby improving development by government agencies and researchers in a better way.

Keywords: Fuzzy-KNN · Cloud Computing · Edge Computing · Fog Computing

1 Introduction

The effective management of healthcare is a critical and highly concerning issue for the growth and development of any nation. One of the most paramount challenges for any government management, Healthcare organization/agencies or researchers is safeguarding its citizens from infection outbursts that can outstretch from one citizen/nation to another through direct or indirect communication. Yellow fever is one of the most significant public health concerns, particularly in Africa and South America, where it is endemic and responsible for high mortality rates; effective management of such infectious diseases is crucial for any nation's development. Information technology (IT)-based

S. Pal and Á. Rocha (Eds.): ICMMCS 2025, LNNS 1398, pp. 34–44, 2025.
https://doi.org/10.1007/978-3-031-90998-6_4

healthcare services are increasingly being used to address these challenges. Technologies such as cloud-computing and fog-computing provide impressive, efficient, and cost-effective solutions for supervising large amounts of health data. This paper explores the potential of fog computing to enhance healthcare services, particularly for detecting and managing yellow fever outbreaks. The proposed system aims to leverage the strengths of fog computing to provide real-time data processing, low latency, and high-quality service (QoS) for health monitoring applications.

1.1 Motivation and Contribution

The motivation behind this research is the pressing need for an intelligent IT system to contain infectious illness epidemics, such as yellow fever outbreaks. These outbreaks have the capacity to seriously injure people and dramatically raise the death rate. The main contribution of this paper, in contrast to current approaches, is that the suggested system integrates multi-level security measures, uses GPS-based risk assessment for targeted outbreak prediction and prevention, uses fog computing for real-time data processing, and uses advanced Fuzzy K-NN for nuanced data classification. An extensive assessment of the literature, a thoughtfully suggested architecture, and an experimental evaluation are carried out on Amazon EC2. The ultimate goal is to develop a scalable, cost-effective solution that can eventually be employed to treat other infectious illnesses.

1.2 Paper Layout

Here is an outline of the remaining sections of the paper: A review of fog computing in healthcare and yellow fever is presented in Sect. 2. The architecture for a fog computing-based yellow fever monitoring and management system is presented in Sect. 3. The experimental setup and outcomes of the suggested framework using Amazon EC2 are described in detail in Sect. 4. Finally, this paper ends by summarizing its contributions and offering suggestions for further research or examination.

2 Literature Review

The various cloud-based health monitoring systems that employ distinct data mining and enhanced processing techniques for different healthcare environments are presented in this section. In order to contribute to social well-being, these systems seek to monitor and control diseases that are either communicable or non-communicable. Table 1 shows a related summary with the legends "MC (major contributions), AD (applications domain), SEN (sensors), CCC (cloud computing), FC (fog computing), PM (prediction model), RTP (real-time perspective), OD (output display), AG (alert generation), SM (security mechanism), and CID (comparison with infectious diseases)." The research on health monitoring systems, including managing and preventing infectious diseases, is summarized in this table. Each study's contributions are analyzed regarding specific methodologies and technologies used, focusing on cloud and fog computing integration, real-time data processing, and predictive modeling to improve healthcare outcomes.

Table 1. Related Survey

References	MC	AD	SEN	CID	CC	FC	PM	RT	AG	OD	SM
[1]	WBAN architecture enabled by the cloud in ubiquitous healthcare systems	Advocated for the widespread use of ubiquitous healthcare apps	Yes	No	Yes	No	No	No	Yes	Statistical	No
[2]	An automated forecasting system for epileptic episodes utilising body sensors and cloud computing	Seizures alert approach for epileptic patients	Yes	No	Yes	No	Yes	Yes	Yes	Statistical	No
[3]	Wireless body area network deployment in the healthcare sector	Proposed a model to obtain physiological data of patients from sensor nodes	Yes	No	Yes	No	No	Yes	No	Statistical	No
[4]	Enabled framework for health monitoring	ECG measurement methods are analyzed for patients healthcare data to avoid preventable deaths	Yes	No	Yes	No	No	Yes	No	Statistical	No
[5, 6]	Real-time health analytics for historical data streams	Monitoring health data security	No	Yes	No	No	No	No	No	Statistical	No
.[6, 7]	An effective approach for predicting and refraining MERS-CoV infection outbreaks	Use of a Geographical location system to represent MERS-CoV patients	No	Yes	Yes	No	No	Yes	No	Statistical	Yes
[8, 9]	Chikungunya virus patterns changing: resurgence of a zoonotic arbovirus	No	No	No	No	No	No	No	No	Statistical	No

3 Proposed Model

The proposed architecture aims to leverage fog computing for real-time monitoring, detection, and management of yellow fever outbreaks. This section details the system's design, emphasizing its three-layer structure: front-end (mobile devices), middle (fog computing), and back-end (cloud computing). The different layers of the proposed architecture are shown in Fig. 1. Each layer has specific roles in the data collection, processing, and analysis workflow. The detailed subsections are described as follows:

3.1 Data Collection

By submitting personal information via a mobile application and obtaining a registration number (RNO), the user registers with the system. Yellow fever symptoms are collected periodically. Data on prone areas is obtained from the fog server. Patient health readings are calculated over time. Personal details include PID, name, age, sex, DOB, address, email, and phone number. Symptoms include fever, seizures, abdominal pain, loss of appetite, headaches, muscle aches, joint aches, and decreased urination. Details regarding patient and symptom attributes are shown in Table 2. Additionally, with the help of characteristics from Table 3, mosquito risk zones can be located, enabling real-time mosquito sensor protection. In locations lacking sensors, users can enter data using the camera on their phone. Mobile apps also record and store GPS coordinates and pertinent data in a cloud database.

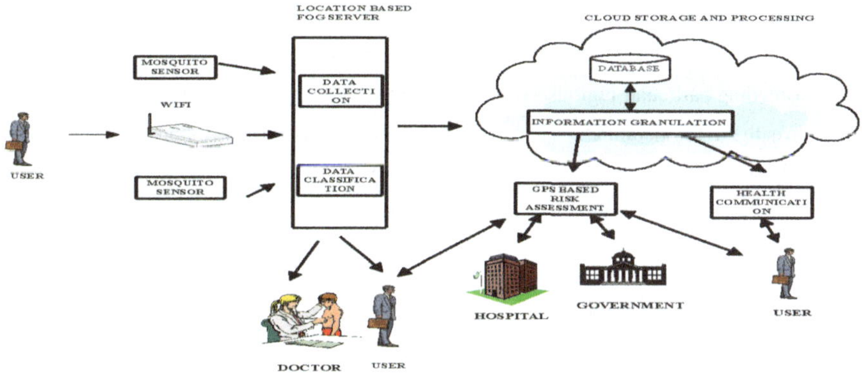

Fig. 1. Proposed Model

3.2 Fog Component Layer

The approach uniquely integrates fog computing to improve system security and service quality for real-time applications by integrating a cloud access security broker (CASB) with health fog. Mobile apps provide data to fog servers for on-site processing, and the cloud analyses the wide range of data that fog servers collect for further analysis.

Table 2. Personal characteristics and symptoms of yellow fever patients

S.no	Attributes	Description	Symptoms attributes	Response
1	PID	Person/patient identification number	Fever	Yes/no
2	Name	Name of person/patient	Joint aches	Yes /no
3	Age	Age of person/patient	Abdominal pain	Yes/no
4	Sex	Male/Female (M/F)	Loss of appetite	Yes/no
5	DOB	Date of birth of a person/patient	Headaches	Yes/no
6	Address	Complete home address of person/patient	Muscle aches	Yes/no
7	Email id	Personal email ID	Seizures	Yes/no
8	Phone number	Contact number of person/patient	Decreased urination	Yes/no

Table 3. Attributes of risk-prone areas' locations

S.no	Attributes	Location Description
1	Location of a mosquito-filled site	GPS locations/coordinates of the mosquito-filled area
2	Mosquitoes density	The number of Insects (Mosquitoes) measured by the detector/sensor
3	Breeding particular regions	GPS area/location of the breed area
4	Weather / Temperature	Temperature near presently standing water
5	Moisture	Humidity
6	Carbon dioxide (CO_2)	The carbon dioxide value

3.3 Fuzzy K-Nearest Neighbor

The standard KNN algorithm is integrated with fuzzy set theory, enabling partial data point membership in several classes, especially where symptoms are difficult to classify, like in yellow fever. Compared to traditional KNN, F-KNN uses fuzzy membership to classify health data more effectively. It assigns data instances a degree of belonging to the Infected Class for those with health issues needing immediate assistance and the Normal Class for individuals with normal health that do not require intervention. This allows for faster and more nuanced classification. For the fuzzy-KNN implementation, the following equations are used:

$$u_i(x) = \left(\sum\nolimits_{j=1}^{k} \left(u_{ij}\left(1/||x_i - x_j||^{2/(p-1)}\right)\right)\right) \div \left(\sum\nolimits_{j=1}^{k} \left(1/||x_i - x_j||^{2/(p-1)}\right)\right) \tag{1}$$

This equation calculates the fuzzy membership ($u\_i(x)$) for a data point (x), considering its distance to the (k) nearest neighbours (x_j) and a fuzzification parameter (p).

$$C(x) = max(u_i(x)) \tag{2}$$

The second equation for determining the class is the following one: The data point, (x)'s class (C(x)) is defined as the significant fuzzy membership value throughout all classes. In order to quickly recognize medical illnesses, the F-KNN algorithm is able to precisely identify data points by taking into account the fuzziness that exists in real-world data. It takes into consideration variables including the number of mosquitoes, the locations of breeding sites, temperature, humidity, and carbon dioxide levels, making it appropriate for complex datasets and medical situations.

3.4 Cloud Storage

The multifaceted strategy supported by cloud computing in this suggested part is an important step in public health informatics, with the goal of reducing the effect of infectious diseases by means of better data management and communication strategies. There, representations of health data from numerous fog nodes are combined using cloud storage. Then, in order to derive new application rules for the fog nodes and extract useful insights, this data is rigorously analyzed in conjunction with data from other sources. In order to handle sensitive data, the model uses a secret sharing system and information granulation to protect environmental and personal data from unwanted access. The GPS-based risk assessment part of the model, shown in Fig. 2, uses geographic location data to identify and control yellow fever-prone locations, facilitating the identification and removal of mosquito-dense areas. Furthermore, Algorithm 1 facilitates the dynamic creation and updating of Google Maps to visualize the spread of infection and high-risk zones, enhancing the model's responsiveness to changing data. The health communication component of the cloud securely stores patient information and enables the distribution of health education and alert messages, which is vital for the prevention and control of yellow fever outbreaks.

Algorithm 1. Creation and update of Google map

```
Read the patient's location address and RNO
  If RNO already exists, then
      Add updated location information to the record
  Else
      Create a new record using the user's RNO
  End if
  Read and Inspect the user's workplace and home addresses
  Read and Examine the locations of breeding grounds and high-density areas in your area for mosquitoes
Update the Google map with these locations indicated/pinpointed
```

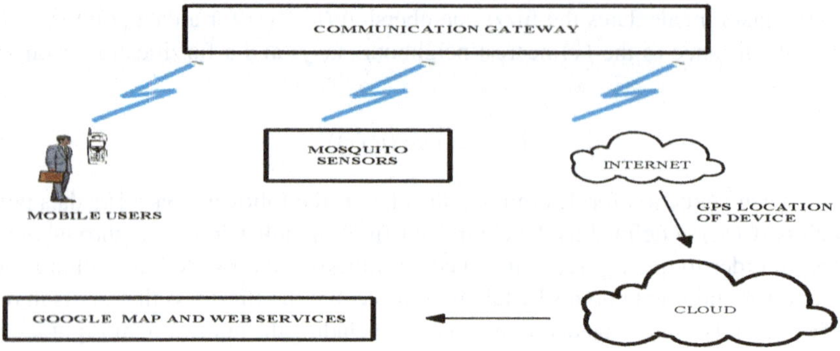

Fig. 2. GPS-based location tracking model

4 Experimental Design

In this section, the experimental setup, results and discussion are mentioned in the following subsections:

4.1 Experimental Setup

The creation of a synthetic dataset for a study on yellow fever, a vector-borne disease, is described. Due to the sensitive nature of medical data, which involves privacy, security, and confidentiality concerns, the researchers opted for a simulated environment rather than using real patient data. Whereas, we sourced an initial dataset of 427 patient records from 2017 related to yellow fever from the Humanitarian Data Exchange website at http://data.humdata.org/dataset?tags=yellowfever. This dataset was then used as a foundation to create a larger, simulated dataset through a process called bootstrapping. This was done to ensure a sufficiently large and varied dataset to test the proposed health diagnosis system effectively. Moreover, the probabilities of the presence of each yellow fever symptom in any newly generated case are outlined and shown in Table 4. The probability is necessary to build synthetic cases in simulations that adequately mirror the range of symptoms experienced by yellow fever patients. Weka 3.7 and R Studio were used to implements fuzzy-KNN and other state of the art classifiers for testing and training, which aids in assessing the performance and effectiveness of the proposed diagnosis system.

The performance of the system was measured using accuracy, sensitivity, specificity, and response time, calculated using the Eq. (3-6) as shown below:

$$Accuracy = \frac{TP + TN}{TP + TN + FP + FN} \tag{3}$$

$$Sensitivity = \frac{TP}{TP + FN} \tag{4}$$

$$Specificity = \frac{TN}{TN + FP} \tag{5}$$

Table 4. Probability for yellow fever

Primary symptoms		Secondary symptoms		Advanced symptoms	
High fever	0.25	Flushing	0.08	Decreased urination	0.01
Joint aches	0.12	Loss of appetite	0.14	Abdominal pains	0.01
Headaches	0.13	Vomiting	0.09	Seizures	0.02
No symptoms	0.50	Nausea	0.09	Delirium	0.03
NA	NA	No symptoms	0.60	Jaundice	0.12
NA	NA	NA	NA	No symptoms	0.80

$$Average\ response\ time = \frac{1}{n}\sum\nolimits_{i=1}^{n} T_i \tag{6}$$

Here, true positives, false positives, true negatives, and false negatives are indicated by the terms TP, TN, FP, and FN, respectively, and n denote the number of folds, which T_i is the response time taken by the classification algorithm for the *ith* fold.

4.2 Results

Based on a stratified 10-fold cross-validation, the results showed that the fuzzy-KNN algorithm performed the best among the classifiers, with the highest accuracy, sensitivity, and specificity. However, it also has a slighter response time. The classification time comparison between the fog and cloud layers, as shown in Fig. 7**,** indicated that the fog layer is more efficient in delivering critical health information. The performance of each classifier during different folds is depicted in Figs. 3, 4, 5 and 6.

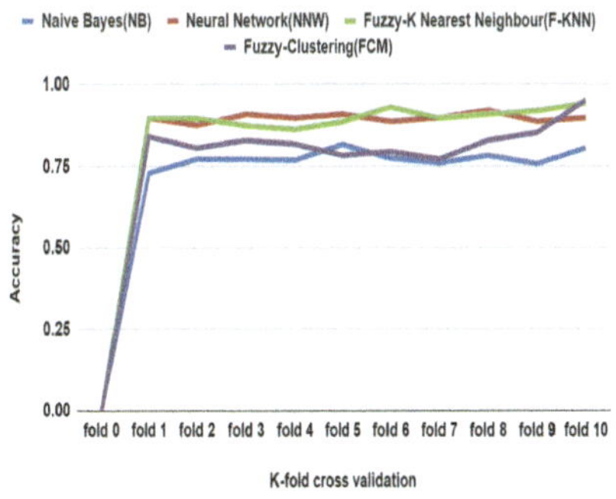

Fig. 3. Specificity vs K-Folds

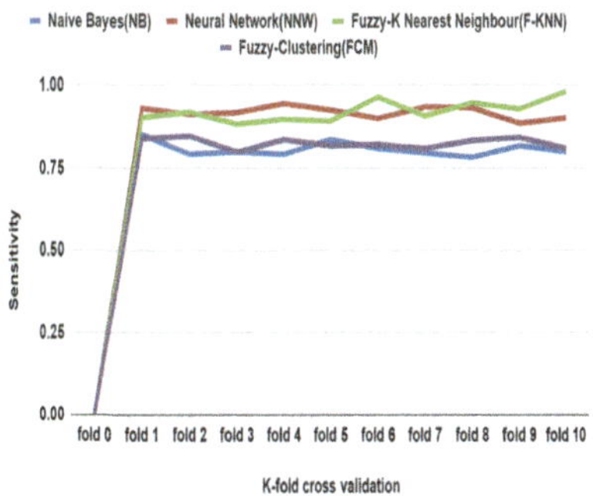

Fig. 4. Sensitivity vs K-Folds

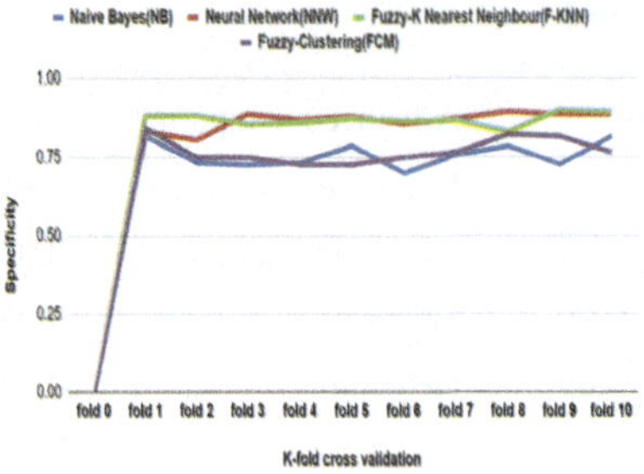

Fig. 5. Accuracy vs K-Folds

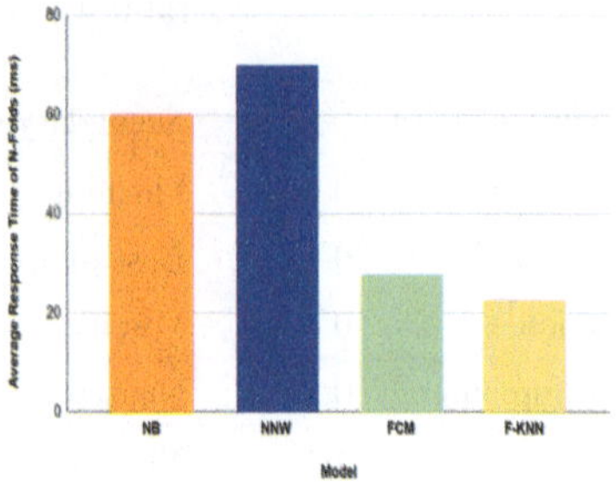

Fig. 6. Comparison (Model vs Average Response Time)

4.3 Discussion

The F-KNN method outperforms other models in terms of accuracy, sensitivity, and specificity while displaying a slightly slower response time when it comes to the suggested model's ability to diagnose yellow fever patients. The fog layer's delivery of information underscores its suitability for critical health situations. The classification time at both the fog and cloud layers is represented by Eq. **7** for delay time, and a comparative analysis is shown in Fig. 7. The statistical analysis of alert generation, which assesses the accuracy of the alert system, is summarized in Table 5, indicating a low percentage of false positive alerts and high overall accuracy in alert generation.

$$Delaytime = T_{doctor} T_{(THI(TGS(I), \Delta t) > \alpha)} \tag{7}$$

where T_{doctor} is the timing (period) at which information/data is supplied to the concerned doctor/medic and $T_{(THI\ (TGS\ (I),\ \Delta t)\ >\ \alpha)}$ is the time needed for calculating the sensitive relation value.

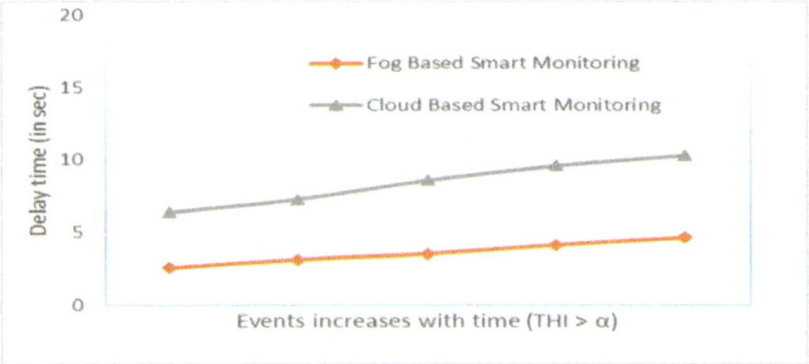

Fig. 7. Comparative Evaluation of Power Consumption Rate between Cloud Computing and Fog Computing

Table 5. Statistical Results

S.no	Parameters	Value in %
1	Alert for false positives	4.145
2	Root Relative Squared Error	4.438
3	Sensitivity	85.437
4	Specificity	94.217
5	Relative Absolute Error	7.834
6	Precision	90.233
7	Coverage	95.132
8	Root Mean Square Error	3.533
9	Mean Absolute Error	4.114

5 Conclusion

With advancements in medical IoT, fog computing, and cloud computing, infections can be controlled effectively. This paper focuses on predicting and preventing yellow fever using an intelligent and secure system. The proposed fog-based system accurately diagnoses yellow fever and sends alerts to users' mobile devices. Key features include information granulation, alert-based processes, and the use of GPS. The system ensures data security, timely alerts, and location representation. Real-time patient data is processed to issue timely warnings to healthcare providers and agencies. The utility of the system is validated through k-cross-validation, demonstrating its effective retrieval of time-sensitive information. The future scope of the proposed architecture includes the potential integration of Internet of Things (IoT) based smart appliances or devices to monitor the health status of each user. Additionally, rigorous security mechanisms can be implemented to enhance the communication security of the proposed architecture. Furthermore, adaptive data mining can be utilized to protect uninfected users from exposure to H1N1-like infections.

References

1. Anirudh Reddy, R., Venkatram, N.: Intelligent energy efficient routing in wireless body area network with mobile sink nodes using horse electric fish optimization. Peer-to-Peer Netw. Appl. **17**, 1–25 (2024)
2. Ein Shoka, A.A., Dessouky, M.M., El-Sayed, A., Hemdan, E.E.D.: EEG seizure detection: concepts, techniques, challenges, and future trends. Multimedia Tools Appl. **82**(27), 42021–42051 (2023)
3. Vinothkumar, S., Amutharaj, J.: Hybrid cryptosystem based healthcare data sharing with access control policy in cloud environment. Int. J. Inf. Technol. **16**(5), 3349–3357 (2024)
4. Fong, E.M., Chung, W.Y.: Mobile cloud-computing-based healthcare service by noncontact ECG monitoring, 2 December 2013. https://doi.org/10.3390/s131216451
5. Hao, Y., Foster, R.: Wireless body sensor networks for health-monitoring applications. Physiol. Meas. **29**(11), R27 (2008)

6. Hossain, M.S., Muhammad, G.: Cloud-assisted industrial internet of things (IIoT)–enabled framework for health monitoring. Comput. Netw. **101**, 192–202 (2016)
7. Berry, A., Milosevic, Z.: Real-time analytics for legacy data streams in health: monitoring health data quality. Date Added to IEEE Xplore, 14 November 2013. https://doi.org/10.1109/EDOC.2013.19
8. Sandhu, R., Sood, S.K., Kaur, G.: An intelligent system for predicting and preventing MERS-CoV infection outbreak. J. Supercomput. **72**, 3033–3056 (2016)
9. Chandrakanta, Prajapati, B.G., Sah, A.N.: Chikungunya fever: epidemiology, clinical manifestation, and management. In: Rising Contagious Diseases: Basics, Management, and Treatments, pp. 181–198 (2024)

The Effect of Topological Properties on the Rotational Movement of DNA and RNA by Using Genetic Algorithms

Dalael Saad Abdul-Zahra[1](✉), Fatimah Hussein Al-Bakaa[2], Shatha Haider Thyab[3], Zahraa N. Salman[4], and Mohammed Ayad Alkhafaji[5]

[1] Babil Technical Institute, Al-Furat Al-Awsat Technical University, Babil, Iraq
dalael20175@gmail.com
[2] Najaf Engineering Technical College, Al-Furat Al-Awsat Technical University, Najaf, Iraq
fatema.albakaa@atu.edu.iq
[3] Department of Mathematic, Al_Mustansiriya University, Baghdad, Iraq
shatha.haider1994@uomustansiriyah.edu.iq
[4] Department of Medical Physics, Hilla University College, Babylon, Iraq
zahraaalhusainy@hilla-unc.edu.iq
[5] National University of Science and Technology, DhiQar, Iraq
mohammed.alkhafaji@nust.edu.iq

Abstract. In this research, some new topological properties about the regular rotational movement of RNA and DNA were studied, and these properties were studied in one of the cells inside the liver programmatically using by using genetic algorithms in *X*–change Lab program, based on some mathematical equations for the level of concentration change in RNA and DNA. And their circular motion resulting from regular rotation using fixed equilibrium values and some important physical properties found in the topological structural structure of those cells.

Keywords: genetic algorithm · RNA · DNA · X–change Lab program

1 Introduction

Although topology is included into the framework of the majority of practical mathematics disciplines, including calculus, differential equations, and probability theory, topology is regarded as one of the most significant areas in research mathematics [1]. Topological ideas do, however, directly affect our daily lives. For instance, our present research will examine digital topology in great detail [2]. When studying geometric structures, topological qualities are always employed to ensure that the shape will not change even if it is subjected to stress, rupture, conduction, or bi-continuous changes. This is known as topological homomorphism. The length of the (RNA) (DNA) is in a tensile topological conformation without rupture or conduction, which is an illustration of the rotating movement that takes place in the (RNA) and (DNA) inside the cells in the liver. In each step of the gene expression cascade, protein nucleic acid interactions

S. Pal and Á. Rocha (Eds.): ICMMCS 2025, LNNS 1398, pp. 45–56, 2025.
https://doi.org/10.1007/978-3-031-90998-6_5

are essential [3]. Different regions on the surfaces of nucleic acid (NA) binding proteins that interact with their partners' DNA or RNA exhibit a variety of chemical, physical, and geometrical characteristics [4]. In this paper, we present a unique method to identify and forecast protein NA binding surfaces. We successfully differentiate between proteins that bind single-stranded RNA (ssRNA) and double-stranded DNA (dsDNA) with 83% accuracy using the approach on experimentally solved three-dimensional protein structures [5–14]. We demonstrate that the technique may be used to predict protein function from scratch while being insensitive to conformational changes brought on by binding. Surprisingly, We conclude by presenting a novel approach to protein surface characterization that can precisely distinguish dsDNA from a ssRNA binding interface [6–12]. Our approach's ability to spot subtle variations on NA binding surfaces makes it applicable to a wide range of molecular recognition issues, with possible implications for drug design. It is necessary to find RNAs in specific genomic regions after identifying the genomic structures (RNA), because there is a direct connection between RNA molecules, which vary according to their specific shapes. Graphical profiles of RNA secondary structure and clustering analysis were used to predict a new type called RNA topologies, augmenting the number of specific RNA families shown in Fig. 1.

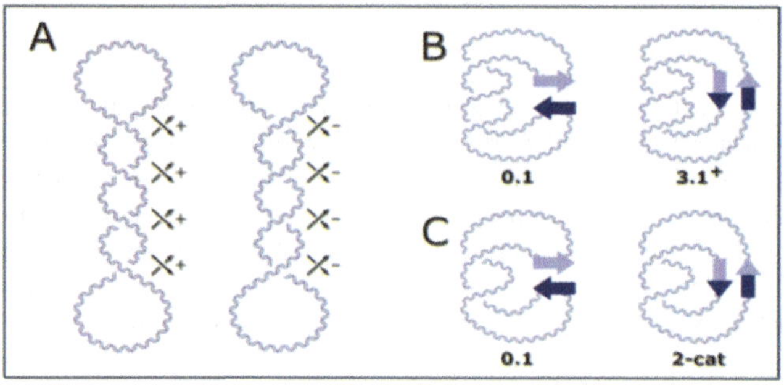

Fig. 1. [1] A) Supercoils in both positive and negative DNA. (B) Tyrosine recombinase activity at locations with inverted repeats DNA circle knotting aided by supercoils. (C) The catenation of DNA circles by supercoiling, aided by the same process seen in (B), on directly repeated sites, converting an unknot into two-crossing catenae, or 2-cat. The arrows in (A) in Fig. 1A indicate the direction of travel along the DNA defining the crossing locations, whereas the arrows in (B) and (C) indicate the direction of the recombination sites' DNA sequences.

In this paper is recognizing the Issue: The basic building blocks of life, DNA and RNA, are more than just straight strands; they frequently display intricate topological patterns like knots and supercoils. Their biological processes, such as transcription, replication, and protein synthesis, can be strongly impacted by these topological characteristics. Analyzing how topology impacts the rotational movement of DNA and RNA, which is essential for many cellular functions, is one important way to comprehend these consequences.

2 RNA Matrix: RNA 2D Topology to a Tree Graph Rag Using a Computer Programme

Definition 1.2 [7] . If a matrix A is of order $m \times n$, then the transpose of the matrix A^T is of order $n \times m$.

$$A = \begin{pmatrix} a_{11} & \cdots & a_{j1} \\ \vdots & \ddots & \vdots \\ a_{i1} & \cdots & a_{ij} \end{pmatrix}_{m\times n} \rightarrow A^T = \begin{pmatrix} a_{11} & \cdots & a_{i1} \\ \vdots & \ddots & \vdots \\ a_{j1} & \cdots & a_{ij} \end{pmatrix}_{n\times m}$$

$$\alpha_1 = \begin{bmatrix} 1 & -1 & 0 \\ -1 & 1 & 0 \end{bmatrix}$$

$$\alpha_2 = \begin{bmatrix} 1 & 0 & -1 \\ -1 & 0 & 1 \end{bmatrix}$$

$$\alpha_3 = \begin{bmatrix} 0 & 1 & -1 \\ 0 & -1 & 1 \end{bmatrix}$$

$$\alpha_4 = \begin{bmatrix} -1 & 1 & 0 \\ 1 & -1 & 0 \end{bmatrix}$$

$$\alpha_5 = \begin{bmatrix} -1 & 0 & 1 \\ 1 & 0 & -1 \end{bmatrix}$$

$$\alpha_6 = \begin{bmatrix} 0 & -1 & 1 \\ 0 & 1 & -1 \end{bmatrix}$$

Then $A^T = [A_{n\times m}]$ and therefore

$\alpha_1^T =$

1	-1
-1	1
0	0

$\alpha_2^T =$

1	−1
0	0
−1	1

$\alpha_3^T =$

0	0
1	−1
−1	1

$\alpha_4^T =$

−1	1
1	−1
0	0

$\alpha_5^T =$

-1	1
0	0
1	-1

$\alpha_6^T =$

0	0
−1	1
1	−1

2.1 Circular DNA Topology and Geometry

i. *Evolutionary Algorithm*

The following scheme is adopted to illustrate the work of the genetic algorithm in general:

Applying these steps to a type of cell in the liver in several detailed steps:

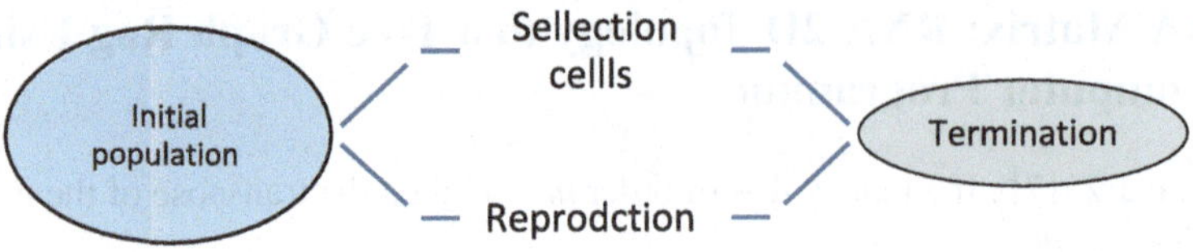

Fig. 2. Evolutionary Algorithm diagram.

- Config
- Determine which cells are being tested
- Changing locations by cutting or wounding
- Fusion of cut or wounded cells with normal cells
- Sort those cells
- Go back to step 2 if you're done.

ii. *Circular DNA Topology and Geometry*

This section discusses the fundamental physical topological properties of rotational motion. We must state right away that the hepatocyte molecules are in close proximity to the circular movement. We shall concentrate on the general structure of the molecule shape in order to analyses circular DNA and its characteristics. There will therefore be a number of important issues. It was barely touched or brushed aside to provide a more detailed analysis of how DNA's and RNA's topological conformations relate to one another. Circular dichroism spectrum (Fig. 14) in the near-ultraviolet (UV) band. The GC level of the DNA has a negligible impact on the transition, and it may be examined using the simplest basic statistical-mechanical model, which assumes that all base pairs are equivalent in this transition. The free energy of a base pair varies when it transitions from the B- to the A-form close to the transition range, as shown by Eq. 1.

$$\Delta y = (RT/Q)(x - x\_0) \tag{1}$$

Where the x concentration DNA of cells in the liver and x_0 the concentration at the point of uniform rotation and $1/Q$ is the proportionality factor. According to Ivanov et al. [8–13], Q = 20 and x_0 = 70% the solution of trifluoroethanol (the most convenient for monitoring reasons). For this transition, the parameter Δy is 1.8 kcal/mol. "Figure 5" provides an illustration of a $B - A$ curve of transition. Within the context of the aforementioned model, the breadth of the transition, Δx, is related to the parameters Q and Δy in Eq. 2 by a straightforward equation.

$$\Delta x = 4Q.exp(-y_i/RT) \tag{2}$$

It is currently unknown what Free energy for each pair B-A change is connected to the transition in solutions.

3 Topological Structures of RNA and DNA

When two strands of the macromolecule DNA twist around a common axis, they form a double helix. It seems as though the double helix is a bent ladder. In "Fig. 1", pairs of nitrogenous bases (base pairs) constitute the rungs of the ladder, while sugar molecules and phosphate groups alternately make up the sides (Fig. 4).

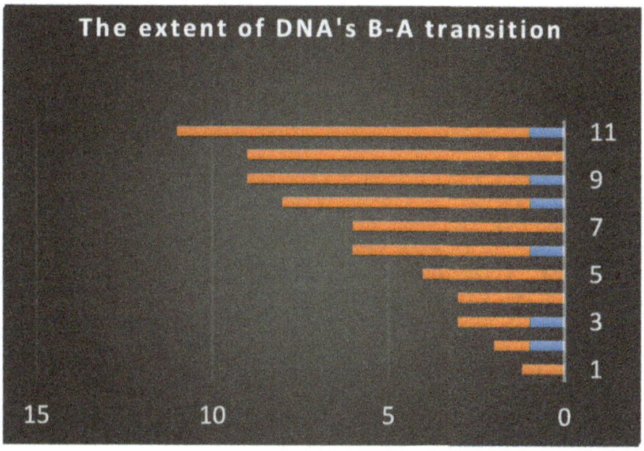

Fig. 3. The extent of DNA's B-A transition, is dependent on trifluoroethanol concentration. 32 The width of the transition, Δx, is displayed.

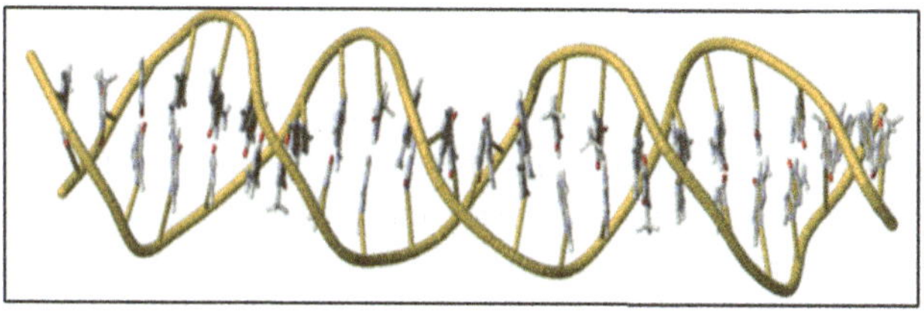

Fig. 4. Double Helix DNA

An organic compound with nitrogen and chemical characteristics like bases is called a nitrogenous base. The four nitrogenous bases (abbreviated C, G, A, and T) are all found in DNA molecules. The organic material's cytosine, guanine, and adenine, as well as nitrogenous bases—compounds containing nitrogen that share chemical properties with bases—make up RNA molecules. The four nitrogenous bases (abbreviated C, G, A, and T are all found in DNA molecules. RNA molecules with adenine, guanine, and cytosine are depicted in "Fig. 2," but uracil (U), a special nitrogenous base, is present in place of thymine.

Fig. 5. DNA's nitrogenous base.

There is a specific correlation between the nitrogenous base sequence on one strand and the sequence on the other strand of a double helix DNA molecule. Adenine is connected to thymine and cytosine to guanine in "Fig. 3."

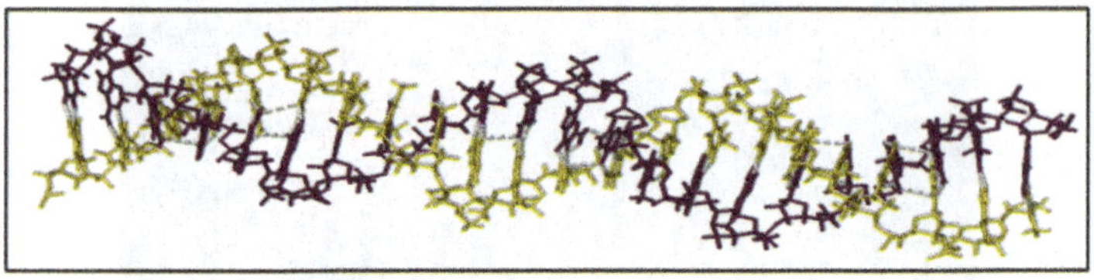

Fig. 6. Adenine and guanine pairs.

4 Topological Descriptors and Functional RNAs with Discussion

Despite our approaches not being able to precisely identify sequence characteristics, we find that most known RNAs are grouped together. Our topological descriptors are able to discriminate between topologies that are RNA-like and those that are not, as evidenced by our analysis of the significance patterns of putative, actual, and hypothetical RNAs. Furthermore, there are strong connections between functioning and hypothetical RNA topologies. Therefore, our research and the clustering results show that functional RNA structures' topologies encode significant traits. For example, our previous research demonstrated that some branching is present in functional RNA tree topologies.10 While our topological descriptors implicitly recognize the existence of functional motifs such as the GNRA and UNCG tetraloops, the A-stack, the biloop and triloop, the U-turn, the bulge-G, and the bulge-helix-bulge, it is clear that our level of description falls short in capturing the finer points of these motifs. Of course, tertiary interactions—such as those involving metal ions—which are significant in many RNA structures are absent from 2D RNA structures and their graphical representations [48, 49]. Finally, a solid grasp of 3D structures is necessary for a comprehensive knowledge of functional features.

Assume we have the following three DNA sequences that have been aligned:

$$\begin{array}{llllllllllll} Taxon1: & A & G & C & AU & U & G & A & A & A & GGCAAAAAU & G \\ Taxon2: & C & A & U & GA & U & G & C & C & C & AAUGGGGGU & G \\ Taxon3: & G & G & U & CG & U & C & G & G & G & GGUCCCCCU & C \end{array}$$

There are three taxon's for above DNA sequences with

$$|x| = \sum_{i \in I} x_i = L = 36$$

and six patterns ***AAA***, ***GGT***, ***CAT***, ***TTC***, ***GCA*** and ***ATG*** with frequencies **4, 3**... respectively.

Now, we input the patterns frequencies x_i of a bovo sequences in 2×3 – contingency table as follows (Table 1):

Table 1. Values of x of marginal and joint probability.

4	*3*	*7*
2	5	7
2	2	4
8	10	36

Then the table of marginal and joint probability is:

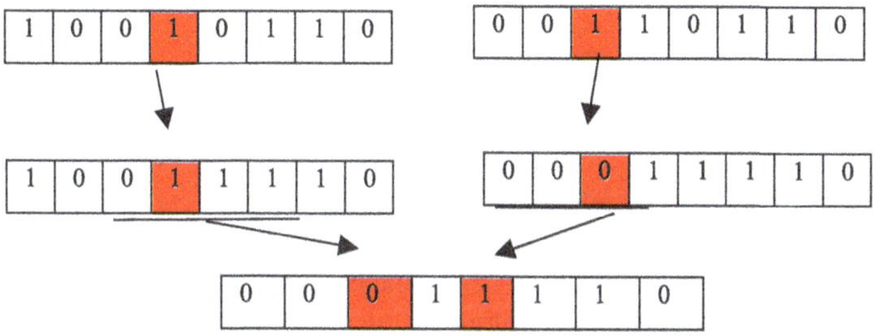

The polynucleotide known as DNA is composed of nucleotide units. Three distinct components make up a nucleotide: a nitrogenous base, a phosphate group, and a sugar molecule. Nucleotides, also known as deoxyribose sugar molecules, are the building blocks of DNA, which is an acronym for deoxyribonucleic acid. A particular kind of sugar called ribose is used by RNA molecules in "Fig. 6" (Fig. 7).

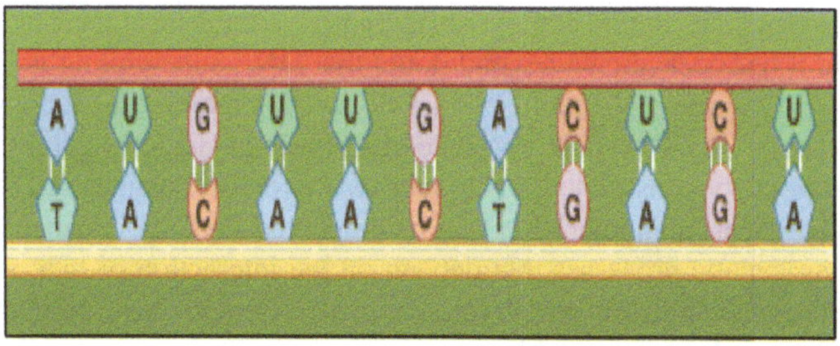

Fig. 7. A specific way in which the sequence on the other strand matches to the double helix RNA helix. Cytosine and guanine also form pairs, as do adenine and thymine.

5 Results and Discussion

There are topological domains in every DNA genome. These can be simple circular DNAs, like those found in bacteria, mitochondria, chloroplasts, and other cells, or massive DNA loops connected to the nuclear matrix, like those seen in eukaryotic chromosomes. Finally, some viruses have linear DNA terminals that can connect to the cell membrane. According to this, some topologically restricted DNA characteristics may have been advantageous during natural selection "Fig. 8".

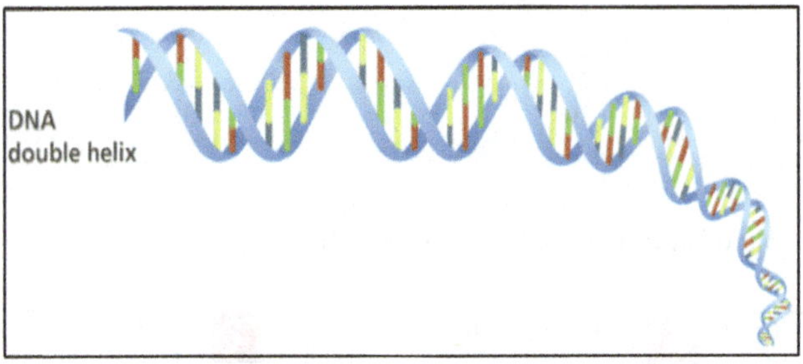

Fig. 8. Topological domains of DNA genomes.

Et al., [9] provides a crucial quality. It demonstrates that every change in the secondary DNA structure (Tw) of a topologically closed DNA domain is instantaneously represented by an alteration in the general shape of the domain (Wr). Consequently, a modification in the overall supercoiling of the molecule would promptly mirror an alteration in a local characteristic of a topologically closed DNA, like the unwinding resulting from protein interaction, and vice versa. There are two important biological benefits to this. First off, if cellular machinery detects a relationship between local and global DNA modifications, it may be assured that both DNA chains are intact, indicating that DNA has not been destroyed. The supercoiling of DNA adds to the evidence that it is undamaged. Naturally, knowing whether or not DNA has been damaged is useful when deciding how to replicate it. Replicants frequently have topological constraints, which is not surprising. In this context, prokaryotic systems have received the greatest attention, and it has been shown beyond doubt that DNA supercoiling is necessary for these systems to initiate replication. Further research is required to support the notion that this also applies to eukaryotes, despite certain findings suggesting this [10]. Second, because topologically constrained DNAs can swiftly detect changes in DNA structure in one DNA segment in another, they can be employed for long-distance communication. For many genetic processes that depend on the interaction of two or more DNA segments, this is essential. In prokaryotes as well as eukaryotes, proteins that interact with DNA segments separated by hundreds or thousands of base pairs commonly co-regulate transcriptional initiation. This is not possible for a linear DNA molecule's linear protein-protein interaction. Consequently, it is widely acknowledged that these distributed DNA segments are topologically interacting with one other. In "Figs. 9, 10", genetic recombination between extremely distant DNA regions happens often [11–15].

The *T* chromosome family consists of randomly selected individuals whose number of variants for each individual chromosome are equal dimensions in the morphological pattern of a two-dimensional phononic crystal unit cell, which is 10. Although it lacks a bandgap, the random structure helps it converge to approximately. In the case of a unit cell, the matrix is displayed for each chromosome that is represented by a distinct

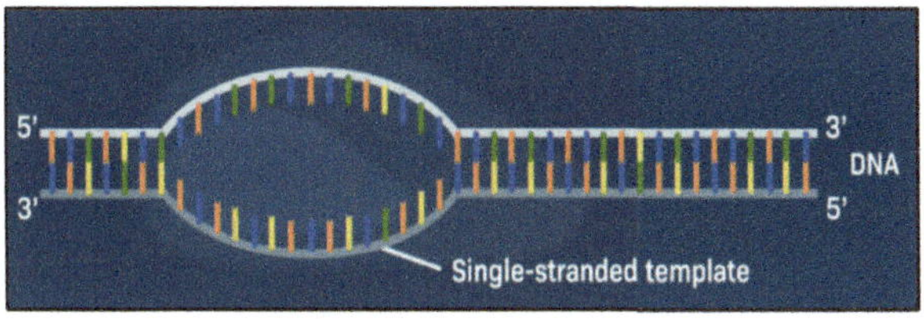

Fig. 9. DNA replication.

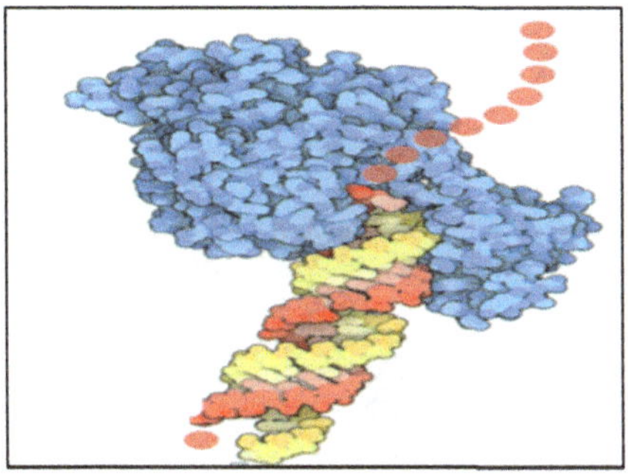

Fig. 10. DNA supercoiling.

chromosome in each row of the chromosomal matrix.

$$\begin{array}{lllllll} 0\ 0 & 0 & 1 & 1 & 1 & & 1 \\ A\ GC & AU & U & G & AA & & A \\ C\ AU & GA & U & G & CC & & C \\ G\ GU & CG & U & C & GG & & G \end{array}$$

Display the backup table. $x = \{x_i\}_{i \in I} \in \mathbb{N}^6$ as a 6–*dimensional* column vector of non-negative integers

$$x = (4, 3, 2, 5, 2, 2)'$$
$$= \left(i.e \quad x \in A^{-1}[t],\ where\, A^{-1}[t] = \left\{ x \in \mathbb{N}^6 : A \begin{bmatrix} x_0 \\ x_1 \\ x_2 \\ x_3 \\ x_4 \\ x_5 \end{bmatrix} = \begin{bmatrix} 7 \\ 7 \\ 4 \\ 8 \\ 10 \end{bmatrix} \right\}. \right)$$

where A^T is 6×5 matrix and

$$A^T = \begin{bmatrix} 1 & 1 & 0 & 0 & 0 & 0 \\ 0 & 0 & 1 & 1 & 0 & 0 \\ 0 & 0 & 0 & 0 & 1 & 1 \\ 1 & 0 & 1 & 0 & 1 & 0 \\ 0 & 1 & 0 & 1 & 0 & 1 \end{bmatrix}_{6 \times 5}$$

and $A^T x = t$ is written as

$$\begin{bmatrix} 1 & 1 & 0 & 0 & 0 & 0 \\ 0 & 0 & 1 & 1 & 0 & 0 \\ 0 & 0 & 0 & 0 & 1 & 1 \\ 1 & 0 & 1 & 0 & 1 & 0 \\ 0 & 1 & 0 & 1 & 0 & 1 \end{bmatrix}_{6\times5} \begin{bmatrix} 4 \\ 3 \\ 2 \\ 5 \\ 2 \\ 2 \end{bmatrix} = \begin{bmatrix} t_1 \\ t_2 \\ t_3 \\ t_4 \\ t_5 \end{bmatrix} = \begin{bmatrix} 7 \\ 7 \\ 4 \\ 8 \\ 10 \end{bmatrix}$$

Where the elements of the column vector x in the beneath diagram serve as the index for the columns of the matrix A^T:

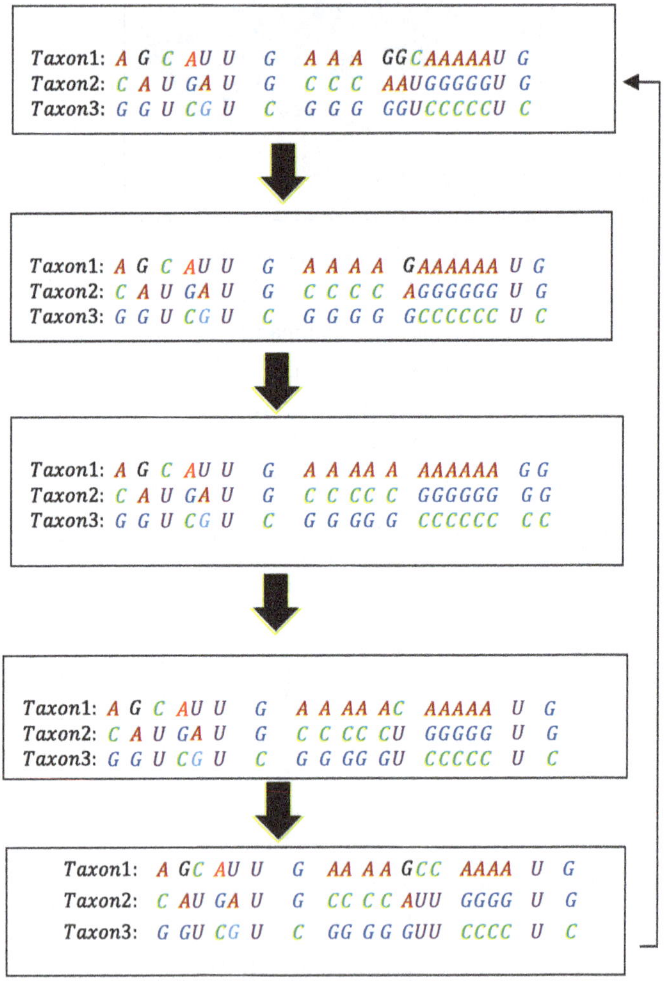

Then the table of marginal and joint probability in Table 2:

Table 2. Point of marginal and joint probability.

0.11	*0.08*	*0.19*
0.06	0.14	0.19
0.06	0.06	0.1

6 Conclusion

In this work some of the primary physical topological characteristics of the rotational motion that takes place in the vicinity of hepatocyte molecules were examined in this work. It turns out that the general structure of the shape of the molecule in the first place as it pertains to the examination of circular DNA and its qualities that regenerate on their own after a cut or wound or infection with various infectious diseases and viruses. Also, there will be numerous critical topics. The recurrent rumor that the molecule contains functional chromosomes has either ignored it or hardly impacted it. We employed the genetic algorithm to get a more thorough picture of the topological structure of DNA and RNA and to visualize it in a close-up and simulation system.

References

1. Gallier, J., Quaintance, J.: Differential Geometry and Lie Groups: A Computational Perspective, vol. 12. Springer, Cham (2020)
2. Wu, Q., Liu, H., Wang, S., Yu, B., Beck, R., Hinkel, K.: A localized contour tree method for deriving geometric and topological properties of complex surface depressions based on high-resolution topographical data. Int. J. Geogr. Inf. Sci. **29**(12), 2041–2060 (2015)
3. Shazman, S., Mandel-Gutfreund, Y.: Classifying RNA-binding proteins based on electrostatic properties. PLoS Comput. Biol. **4**(8), e1000146 (2008)
4. Liu, G.: The dynamic interplay between influenza a virus rna signatures and retinoic acid-inducible gene I. Doctoral dissertation, University of Saskatchewan Saskatoon (2018)
5. Shazman, S., Elber, G., Mandel-Gutfreund, Y.: From face to interface recognition: a differential geometric approach to distinguish DNA from RNA binding surfaces. Nucleic Acids Res. **39**(17), 7390–7399 (2011)
6. Kim, N., Fuhr, K.N., Schlick, T.: Graph applications to RNA structure and function. In: Biophysics of RNA Folding, pp. 23–51. Springer, New York (2013)
7. Radecki, P.: Accurate, Automated, and Scalable Identification of RNA Structure Motifs in Structurome Profiling Data. Doctoral dissertation, University of California, Davis (2021)
8. Ahuja, M.S.: Structure of DNA. Manual of Cytogenetics in Reproductive Biology (2014)
9. Rostron, C.: Drug Design and Development. Oxford University Press, Oxford (2020)
10. Saha, S.K., Pathak, N.N.: Proteins and other nitrogenous substances of nutritional significance. In: Fundamentals of Animal Nutrition, pp. 65–92. Springer, Singapore (2021)
11. Wu, R., Taylor, E.: Complete nucleotide sequence of cohesive ends of bacteriophage Î» DNA. J. Mol. Biol. **57**, 491 (1971)
12. Putz, M.V., Tudoran, A., Ori, O.: Topological organic chemistry: from distance matrix to Timisoara eccentricity. Curr. Org. Chem. **19**(3), 249–273 (2015)
13. Calladine, C.R., Drew, H.R.: Understanding DNA: The Molecule and How it Works, 2nd edn. Academic Press, Sandiego (1997)

14. Abbass, H.H., Aneed, M.W.: A new model of genetic algorithm using extended Markov basis. Eur. J. Sci. Res. **137**(4), 431–432 (2016)
15. Takemura, A., Aoki, S.: Some characterizations of minimal Markov basis for sampling from conditional distributions. Ann. Inst. Stat. Math. **56**(1), 1–17 (2004)

Reduction Specific Absorbing Rate on Human Brain for Rectangular Slot Patch Antenna in C Band

Hussien Y. Radhi[1](✉), Haraa Raheem Hatem[2], and Marwa Al-Sultani[2]

[1] Department of Computer Engineering, College of Engineering, University of Diyala, 32001 Ba'aqubah, Diyala, Iraq
hussienradhi_eng@uodiyala.edu.iq

[2] Department of Communications Engineering, College of Engineering, University of Diyala, 32001 Ba'aqubah, Diyala, Iraq
{haraa_altaie_eng,marwa_al-sultani_eng}@uodiyala.edu.iq

Abstract. These days, the utilization of Rectangular Patch Antennas (RPA) is widespread in wireless communication systems. The major source of electromagnetic radiation has come from wireless devices, which penetrate human tissues and cause health danger. This paper exam Rectangular Patch Antenna (RPA) and Rectangular Slot Patch Antenna (RSPA) that were designed, evaluated, and simulated, at c-band application using CST Microwave Studio 2020. The RSPA is a modified version of RPA via cutting the rectangular slot of the patch layer. The two antennas are designed with a copper material for ground and patch layers (thickness is 0.035 mm) whereas a FR-4 epoxy material for the substrate layer, (thickness is 1.5 mm). The study was examined and evaluated based on SAR to determine the humans' exposure to communication antenna radiation and its value was computed over 10 gm and 1 gm mass of tissues corresponding to (ICNIRP) and the IEEE standards for head safeness. The results revealed that the RSPA antenna produces the minimum SAR levels in head tissues of adult. The levels of SAR in the same tissues are recorded as high values for 10g compared to others.

Keywords: Head tissues · Rectangular patch antenna · Specific Absorption Rate · Gain and Slot

1 Introduction

In the last decades, the massive expansion of the wireless communication system and its application led to an increase in the requirement to create various antenna characteristics to address a particular application. A Rectangular Patch Antenna (RPA) is used widely in the application of wireless because of its lightweight and small size. However, it has many issues that need to be solved like narrow bandwidth and low efficiency (Abdalla & Ali, 2020) [1]. Many applications in wireless communication systems like WLAN, mobile phones, GPS, …etc. require an antenna with significant performance to decrease the size and cost (Raj & Dwivedi, 2023) [2].

S. Pal and Á. Rocha (Eds.): ICMMCS 2025, LNNS 1398, pp. 57–66, 2025.
https://doi.org/10.1007/978-3-031-90998-6_6

The multiband antenna witnessed significant attentiveness in terms of the reduction of the antennae numbers that combined multiple applications in one antenna. Therefore, researchers have been conducted to achieve the characteristics of multiband.

Therefore, there are many techniques used for achieving better performance of patch antennae like slot technique, the fractal-engineering technique, and notch technique (Gupta, Kundu, Rajawat, Sharma, & Gupta, 2022 [3]; Hediya, Attiya, & El-Deeb, 2022 [4]).

The use of wireless devices has been growing rapidly. The radiation of these devices results several health problems. The operating frequencies of these devices causes tissue heating (Tharakan, Deepthi, Sebastian, Gopika, & Krishna, 2015) [5].There is a relationship linking the radiation received by the body's tissues and their mass, this relationship is called the Specific Absorption Rate (SAR), which affects the entire body and sometimes part of it (Ali, Hamd, & Abdalla, 2018 [6]; Hamd, Al-Sultani, & Al-Rawi, 2021 [7]). It is possible to find the absorption rate from the following relationship (Ali et al., 2018 [6]; Mondal & Padmini, 2019 [8]):

$$SAR = \frac{\sigma}{2.\rho}|E|^2 \tag{1}$$

where E represents the magnitude of the electric field (V/m), while σ represents the conductivity (s/m), and ρ represents is the mass density of the tissue. As a result, many researchers have focused on the issue of safety when using these technologies that can be worn by humans for any purpose, by maintaining SAR levels at the lowest level without affecting the efficiency of the devices (Lu, Pang, Hu, & Jiang, 2021) [9]. SAR is known as one of the important concepts that can be used to examine the quality and safety of wireless devices emitting electromagnetic waves, as is the case with antennas (Akram & Jasmy, 2008) [10]. The factors and variables that affect SAR rates include the quality of the antenna, its size, its frequency range, and the energy coming out of it, in addition to its proximity to objects (Hediya et al., 2022) [3].

So that there are many searches have been focalized on the topic, a rectangular patch antenna on a precisely cut Hugo structure of 3mm x 3mm x 3mm is designed. It is noticed that by increasing the patch antenna area then reduces SAR value by 0.0005 to 0.00041, high band width frequencies as well as increasing gain (Khan, Ahmad, & Naeem, 2016) [11]. A flexible antenna is designed in the 2.4 GHz frequency range which can be used in human wearable applications. This proposed antenna achieves robust performance and a low specific absorption rate (Jiang, Cui, Yue, Zhu, & Werner, 2017) [12]. The researchers in (Ali et al., 2018) [6] design a half-wave type dipole antenna and Square patch antenna at 2.4 GHz,1.575 GHz, and 1.8 GHz to get a low SAR. The SAR was calculated on a mass of 1 g and 10 respectively for these two types of antennas. The obtained SAR by Square patch antenna is smaller than of a dipole antenna. While in (Kuther, Ali, & Ahmed, 2020) [13], the absorption effect of electromagnetic waves on human fingers was tested by calculating the specific absorption rate (SAR) level of the proposed Fractal Sierpinski Square Patch antenna, which was simulated over the 1.575 GHz frequency range using the CST studio Suite application. In (Allbadi, Hamd, & Qaddoori, 2022) [14], the M-slot antenna is designed to perform the necessary calculations for main 4 band frequencies, 8.3, 7.8, 4.5, and 2.4 GHz. Moreover, SAR is evaluated over figure tissues according to ICNIPR and IEEE standards. In (Gao et al., 2022) [15], a rectangular slot patch antenna is designed with cutting out of the patch for shifting the resonance of the unidirectional mode to get low frequency and low the antenna's size. The measured specific absorption rate (SAR) values meet the standards of Europe and the United States. This work is different from all mentioned works in designing to get the best results by cutting the rectangular slot of the patch layer in different places.

The paper is organized as follows: Sect. 1 presents the introduction with related work as well as the research objective. In Sect. 2, the design methodology of the proposed antenna is described. Additionally, the measurement and simulation results are discussed in Sect. 3. Finally, Sect. 4 concludes the research and suggests future work.

1.1 Research Objective

This paper presents two designs of patch antennae, which are designed to provide Wi-max and C-band resonant frequencies. The Rectangular Slot Patch Antenna is a modified version of the Rectangular Patch Antenna by cutting the rectangular slot of the patch layer. For the analysis and design of the proposed antenna, CST software is used. In addition, reduction of the specific absorbing rate on human brain is performed by designing rectangular slot patch antenna in c-band. The specific absorbing rate is recorded of head tissues for adult human, which is consist of (brain, skin, and skull).

2 Antenna Design

The tremendous development in wireless communications over the last two decades, especially in satellite and mobile Wi-Fi applications, has led to the search for an antenna with good properties s in terms of weight and size. The patch antenna is the most important antennas that used in wireless communications, especially for the bandwidth from 1–10 Gigahertz (Hamd, Al-Sultani, et al., 2021) [7]. The antenna contains three layers: the third and first layers are made of conductive materials, and the second layer is made of insulating material. In this paper, the material for ground and patch layers is copper while FR-4 is the material for substrate layer with 4.3 epsilon value. Thickness of layers must be chosen according to achieve better BW and good efficiency, so the thick of ground and patch is 0.035 mm and substrate is 1.5 mm. The fundamental Eqs. (2–7) are used to calculate the dimensions of proposed antenna at frequency 7.3 GHz (Ali et al., 2018 [6]; Hamd, Hatem, & Ali, 2021 [16]). It could be shown in Fig. 1 and Table 1 respectively. The width of rectangular patch is:

$$w_{patch} = \frac{u}{2f_{reson.}\sqrt{\frac{\varepsilon_{r+1}}{2}}} \quad (2)$$

where ε_r is the relative permittivity and u is speed of light in free space.

The dielectric substrate height can be evaluated using (3)

$$h_{sub.} = \frac{0.3u}{2\pi f_{res.}.\sqrt{\varepsilon_r}} h_{sub.} \leq 0.06\frac{\lambda d}{\sqrt{\varepsilon_r}} \quad (3)$$

where $\lambda d = \lambda_0/\sqrt{\varepsilon_{reff}}$, λd is the wavelength in dielectric medium, $\lambda 0$ is free space wavelength and ε_{reff} is the effective dielectric constant.

The actual length is obtained using (4):

$$L_{patch=}L_{peff} - 2\Delta L_{patch} \tag{4}$$

Equation (5) gives the length extension ($\Delta L_{pat.}$) as:

$$\Delta L_{patch} = 0.412 h_{sub.} \frac{(\epsilon_{reff} + 0.3)\left[\frac{w_{patch}}{h_{sub.}} + 0.264\right]}{(\epsilon_{reff} - 0.258)\left[\frac{w_{patch}}{h_{sub.}} + 0.8\right]} \tag{5}$$

Equation (6) determines ($\varepsilon reff$)

$$\epsilon_{reff} = \frac{\epsilon_r + 1}{2} + \frac{\epsilon_r - 1}{2}\left[1 + 12\frac{h_{sub.}}{w_{pat.}}\right]^{\frac{-1}{2}} \tag{6}$$

$$Lp_{eff} = \frac{c}{2f_r\sqrt{\epsilon_{reff}}} \tag{7}$$

Equations (8) and (9) give the length (LGD) and the width (WGD) of the ground plane & substrate (Ali et al., 2018 [6]; Hamd, Hatem, et al., 2021 [16]):

$$\mathrm{W_{GD}} = 6\, h_{sub} + w_{patch} \tag{8}$$

$$\mathrm{L_{GD}} = 6\, h_{sub} + L_{patch} \tag{9}$$

Fig. 1. Rectangular patch antenna

The designed antenna consists of two parts. The Rectangular Patch Antenna (RPA) represents the first one and the second part is a modifying Rectangular Slot Patch Antenna (RSPA) through cutting rectangular slot. Figures 1 and 2 represent the two parts of the designed antenna. While Table 1. Shows its dimensions.

Fig. 2. Modification antenna

Table 1. Proposed antenna dimensions

Parameters	Values (mm)	
L_{GD}	60	Length of ground layer
L_{GD}	60	width of ground layer
l_{patch}	38	Length pf patch
w_{patch}	18.2	Width of patch
h	1.5	substrate thickness
t	0.035	patch thickness
wf	3.7	Width of feeding
lf	9	Length of feeding

3 Simulation Results and Discussion

The simulation and design of the rectangular patch antenna were implemented utilizing CST software 2020. The results have been introduced in two parts: RPA and RSPA.

3.1 Rectangular Patch Antenna (RPA)

This section introduces the RPA results. Figure 3 presents the connection between the proposed antenna frequency and return loss (RL). It could be seen that the resonance frequencies are 7.31 GHz with regard to −25dB RL. In the results of program, the S11 refers to RL.

In addition, Fig. 4 reveals the radiation pattern for the proposed antenna where the directivity is 5.6 dBi, according to polar angle. Table 2. Lists the characteristics of the proposed antenna. It is obvious that the VSWR at resonant frequencies (7.31 GHz) is approaching one leading to a significant matching between the transmission line and the proposed antenna. We could see a reduction in the value of the impedance compared with the value utilized in the design because of the heat losses of the impedance.

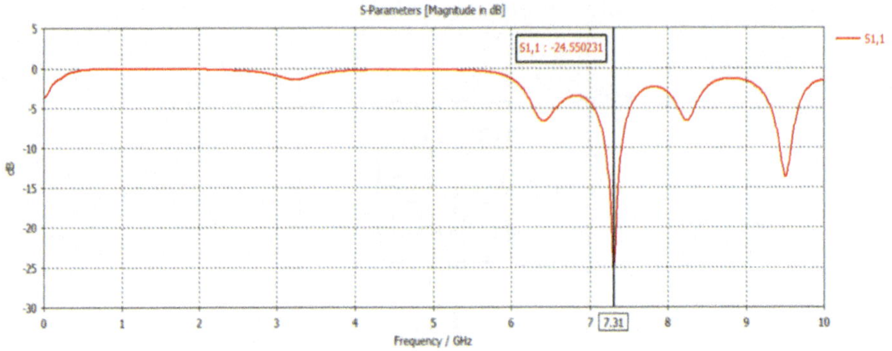

Fig. 3. Return loss for proposed antenna

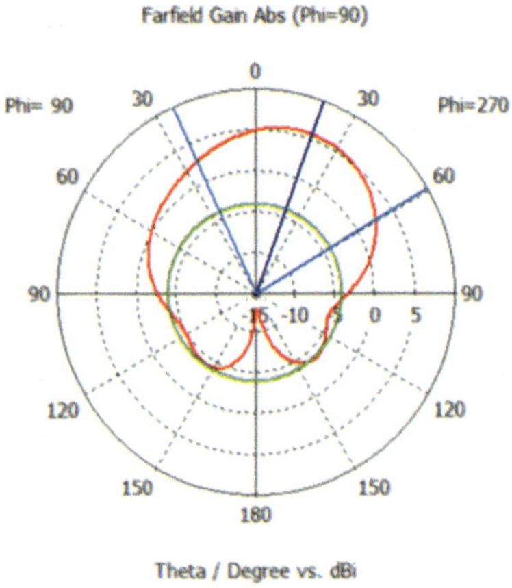

Fig. 4. Radiation pattern for proposed antenna

Table 2. Characteristics of proposed antenna

Ch/ch	7.31 GHz
RL (dB)	−25
VSWR	1.1259
Gain(dB)	5.2
Directivity(dB)	5.6
Z	40
BW(MHz)	220

3.2 Rectangular-Slot Patch Antenna (RSPA)

In this section, the results of the RSPA were presented. Figure 5 reveals that the operated frequency of RSPA is 5.71 GHz. Therefore, the new frequency could be utilized in various applications such as WLAN (A wireless local-area network) or Wi-Fi (Wireless Fidelity), Wi-max (Worldwide Interoperability for Microwave Access), and c-band. The RL is 25 dB at this frequency.

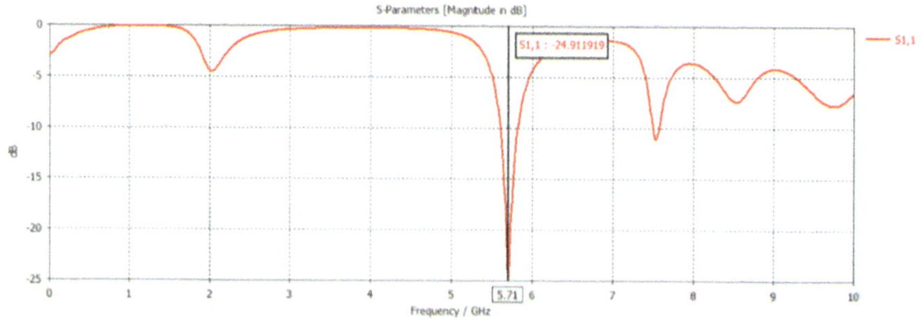

Fig. 5. Return loss for modification antenna

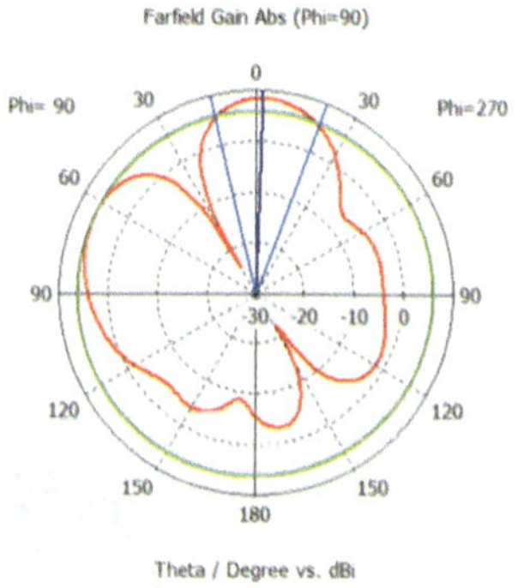

Fig. 6. Radiation pattern for modification antenna

Moreover, Fig. 6 represents the radiation pattern for the modification antenna. Table 3. Lists the characteristics of the modification antenna. Hence, the bandwidth was measured according to $(fH - fL)$ at (-10 dB) RL.

Depending on Tables 2 and 3, it could be noticed that the RSPA characteristics are better than the characteristics of RPA. Furthermore, the modified antenna has expanded in the working range to include the C-band for execution in Wi-max applications. In addition, it could be seen a significant improvement in directivity and gain while maintaining both SWR and Zo on acceptable values.

The research motivation is to secure a low-radiation antenna for protection living organisms from electromagnetic waves harmful effects. Figure 7 and Fig. 8 represent the SAR for adult heads at 1g and 10g for both RPA and RSPA respectively. According to

Table 3. Characteristics of modification antenna

Ch/ch	5.71 GHz
RL (dB)	−25
VSWR	1.120
Gain(dB)	7.48
Directivity(dB)	8.47
Z	46
BW(MHz)	220

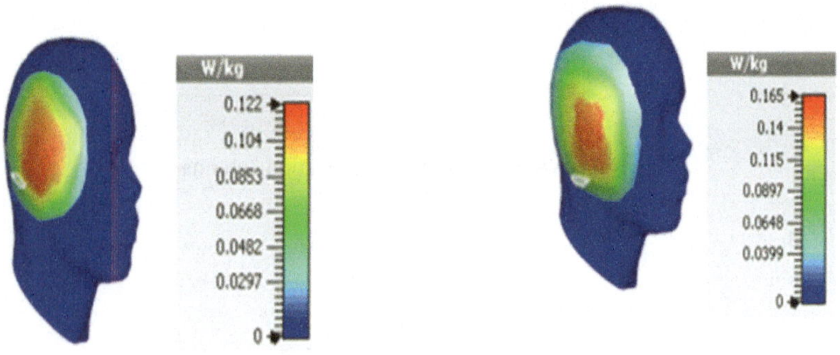

(a) (b)

Fig. 7. SAR for adult's head for RPA (a) 1 g, (b) 10 g

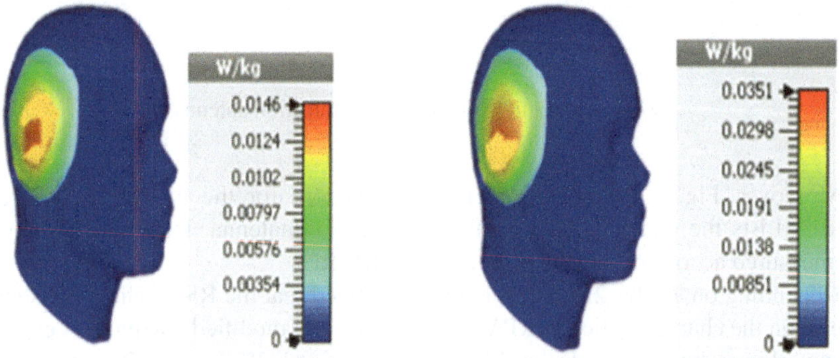

(a) (b)

Fig. 8. SAR for adult's head for RSPA (a) 1 g, (b) 10 g

the results from these figures, it could be noticed that SAR in 1g reduces with a decrease in the frequency. Accordingly, the standard limit for IEEE and USA, the highest value of SAR should equal to 0.0351 and 0.165 W/Kg respectively. Additionally, Table 4. Shows the SAR values for the proposed antenna and modified antenna, and all values of SAR are allow able levels with low radiation ratios for both frequencies (7.3 GHz and 5.7 GHz) as well as both standards.

Table 4. The SAR values for RPA and RSPA

Design	SAR value	
	1 g	10 g
RPA	0.122	0.165
RSPA	0.0146	0.0351

4 Conclusion

Because of the frequent utilization of antennas in daily life through included in electronic device components, they are considered a source of electromagnetic wave radiation. In addition, the increased health awareness in most countries has become an important criterion in the development of this country which looking at and approving standards. This paper presents a comparative study between two types of antenna RPA antenna and RSPA antenna in terms of measuring the SAR level. These two antennae are simulated at 7.3 GHz and 5.7 GHz of wireless communication applications under the exposition of RF frequency at a distance of 5 mm from the head to the wireless device. The results show that the modified antenna produces the minimum SAR levels in the head tissues of adults compared to the proposed antenna. The levels of SAR in the same tissues are highest in standard (ICNIRP) compared to (IEEE). The potential future research may focus on studying the effect of SAR levels in head tissues of different age groups.

References

1. Abdalla, A.I., Ali, I.H.: Design and modification of multiband M-slot patch antenna for wireless applications. In: Paper presented at the 2020 IEEE 17th International Conference on Smart Communities: Improving Quality of Life Using ICT, IoT and AI (HONET) (2020). https://doi.org/10.1109/HONET50430.2020.9322656
2. Raj, P., Dwivedi, R.P.: sar reduction techniques for wearable application (2023). ArXiv Preprint ArXiv:2304.07804. https://doi.org/10.48550/arXiv.2304.07804
3. Hediya, A., Attiya, A.M., El-Deeb, W.S.: Reduction of specific absorption rate: a review article. Egyptian Int. J. Eng. Sci. Technol. **39**(3), 80–96 (2022)
4. Gupta, S., Kundu, C., Rajawat, A., Sharma, R., Gupta, S.H.: Design and analysis of permitted specific absorption rate (SAR) antennas for IOT applications. Int. J. Syst. Assur. Eng. Manag. **13**(5), 2594–2603 (2022). https://doi.org/10.1007/s13198-022-01668-4

5. Tharakan, A., Deepthi, J., Sebastian, D.A., Gopika, J., Krishna, D.: Specific absorption rate reduced (SAR) mobile phone antenna designs. In: Paper presented at the 2015 Fifth International Conference on Advances in Computing and Communications (ICACC) (2015). https://doi.org/10.1109/ICACC.2015.89
6. Ali, I.H., Hamd, H.I., Abdalla, A.I.: Design and comparison of two types of antennas for SAR calculation in wireless applications. In: Paper presented at the 2018 Advances in Science and Engineering Technology International Conferences (ASET) (2018). https://doi.org/10.1109/ICASET.2018.8376891
7. Hamd, H.I., Al-Sultani, M., Al-Rawi, Z.N.A.: Influence different dielectric materials for square patch antenna on RCS. In: Paper presented at the 2021 5th International Symposium on Multidisciplinary Studies and Innovative Technologies (ISMSIT) (2021). https://doi.org/10.1109/ISMSIT52890.2021.9604564
8. Mondal, S.P., Padmini, T.: Specific absorption rate (SAR), ways to find in mobile devices, its measurement and side effects on human. In: Paper presented at the 2019 International Conference on Vision Towards Emerging Trends in Communication and Networking (ViTECoN) (2019). https://doi.org/10.1109/ViTECoN.2019.8899581
9. Lu, B., Pang, B., Hu, W., Jiang, W.: Low-SAR antenna design and implementation for mobile phone applications. IEEE Access **9**, 96444–96452 (2021). https://doi.org/10.1109/ACCESS.2021.3093720
10. Akram, G., Jasmy, Y.: Specific absorption rate (SAR) on the human head as function of orientation of plane wave radiation: FDTD-based analysis. In: Paper presented at the 2008 Second Asia International Conference on Modelling & Simulation (AMS) (2008). https://doi.org/10.1109/AMS.2008.193
11. Khan, S., Ahmad, N., Naeem, M.: The reduction of specific absorption rate at different frequencies. J. Eng. Appl. Sci. **34**(2) (2016)
12. Jiang, Z.H., Cui, Z., Yue, T., Zhu, Y., Werner, D.H.: Compact, highly efficient, and fully flexible circularly polarized antenna enabled by silver nanowires for wireless body-area networks. IEEE Trans. Biomed. Circuits Syst. **11**(4), 920–932 (2017). https://doi.org/10.1109/TBCAS.2017.2671841
13. Kuther, K.H., Ali, I.H., Ahmed, R.K.: Radiation effect of fractal sierpinski square patch antenna. Int. J. Electr. Comput. Eng. (IJECE) **10**(5), 5329–5334 (2020). https://doi.org/10.11591/ijece.v10i5.pp5329-5334
14. Allbadi, Y., Hamd, H.I., Qaddoori, I.H.: Radiation effect of M-slot patch antenna for wireless application. Bull. Electr. Eng. Inf. **11**(5), 2657–266 (2022). https://doi.org/10.11591/eei.v11i5.3801
15. Gao, G.-P., Zhang, B.-K., Dong, J.-H., Dou, Z.-H., Yu, Z.-Q., Hu, B.: A compact dual-mode pattern-reconfigurable wearable Antenna for the 2.4-GHz WBAN application. IEEE Trans. Antennas Propag. **71**(2), 1901–1906 (2022). https://doi.org/10.1109/TAP.2022.3225529
16. Hamd, H.I., Hatem, H.R., Ali, I.H.: Monostatic RCS of rectangular patch Antenna in C-Band frequency. J. Commun. **16**(4) (2021). https://doi.org/10.12720/jcm.16.4.150-154

Boosting Real-Time Intelligence: A Sturdy and Expandable Twitter Streaming Platform

V. Jeyajeev[1], D. Vignesh[1], S. P. Thirumukhil[1], and P. Shanmuga Sundari[2](✉)

[1] Computer Science and Engineering, SRM Institute of Science and Technology, Tiruchirappalli, Tamil Nadu, India

[2] School of Computing, SRM Institute of Science and Technology, Tiruchirappalli, Tamil Nadu, India

shanmugasundari.p@ist.srmtrichy.edu.in

Abstract. This study investigates real-time sentiment analysis of geolocated category-based tweets using a pre-trained language model called Dynamic Content Routing (DCR). Using Twitter's streaming API and keyword filtering, we collect and analyze sentiment during live category-based events. DCR eliminates the need for extensive training and provides an efficient and scalable approach. We evaluate the effectiveness of DCR to capture the nuances of location-based atmospheres without domain adaptation. Focusing on NLP techniques such as sentiment analysis and pre-trained language models, this research helps understand social media sentiment in real time. The framework highlights the potential of DCR and geolocation filtering for broader social media applications.

Keywords: Real-time Data Processing · Twitter streaming · Sentiment analysis · API interaction

1 Introduction

It is crucial to know what the public thinks, particularly when events are happening in person. Using natural language processing (NLP) techniques, this work explores a novel approach for real-time sentiment analysis of geolocated tweets based on category. Emotions and opinions in a certain area are what we want to capture. On the basis of labeled data (positive, negative, neutral, or irrelevant), sentiment analysis models are traditionally trained. That being said, it takes time to gather large amounts of category-based training data. Our suggested method is more effective and makes use of Dynamic Content Routing (DCR), a pre-trained language model. DCR analyzes linguistic subtleties without the need for domain adaptation since it has been trained on vast volumes of text data. The Twitter Streaming API allows us to collect live tweets with proper keywords. Geolocation filtering refines this further by ensuring that tweets originate from the location of the event. This combination provides a unique advantage. DCR's pre-trained capabilities capture emotion without any specific training, while geolocation filtering focuses on location-based emotion and captures the atmosphere of a live event. Focusing on NLP techniques such as sentiment analysis and pre-trained language models,

S. Pal and Á. Rocha (Eds.): ICMMCS 2025, LNNS 1398, pp. 67–76, 2025.
https://doi.org/10.1007/978-3-031-90998-6_7

this study aims to improve the understanding of real-time social media sentiment. The framework using DCR and geolocation filtering enables broader applications beyond sentiment analysis and provides a robust and adaptive approach to different social media contexts. Using pre-trained models and geolocation filtering, this study provides a new approach to real-time analysis of location-specific events. This approach overcomes the limitations of traditional sentiment analysis methods and paves the way for a more efficient and scalable social media analysis framework.

2 Literature Review

Study	Focus	Methods	Contributions	Limitations
Medhat et al. (2014)	Algorithms and uses for sentiment analysis	An overview of sentiment analysis methods	offered a thorough overview of sentiment analysis techniques	Conventional approaches call on domain-specific expertise
Pak & Paroubek (2010)	Sentiment analysis with data from Twitter	Examination of Twitter data	Twitter's sentiment analysis potential was demonstrated, and the noise limitations were emphasized	Managing irrelevant data and noise
Sakaki et al. (2010)	Using geotagged tweets to detect events	filtering geolocation data and detecting events	Used geotagged tweets to identify earthquakes and other real-time occurrences	Exclusive to event detection and not sentiment analysis in general
Devlin et al. (2018)	Models of pre-trained language (BERT)	Pre-training approach based on transformers	BERT was introduced; it works well for a variety of NLP tasks and requires little more training	Considerable computational resources are needed
Brown et al. (2020)	Short-term learning with GPT-3	Model based on transformers and few-shot learning	demonstrated GPT-3's proficiency with limited samples	Large model size and intensive computing needs
Radford et al. (2019)	Unsupervised multitask education	trained linguistic models beforehand (DCR)	shown how DCR can adjust to many situations	All-purpose model, not

3 System Architecture

Our real-time sentiment analysis system architecture uses three main components:

- **Twitter Streaming API and Keyword Filtering**: This allows us to collect tweets containing category-based keywords in real-time.
- **Geolocation filtering**: Tweets are refined based on geolocation information to ensure that they originate from the area around a specific category-based event.
- **Dynamic Content Routing (DCR)**: This pre-trained language model analyzes the sentiment of collected tweets. DCR's strength is its ability to understand the nuances of emotion without extensive category-specific training. This architecture facilitates real-time sentiment analysis of geo-targeted category based tweets and captures location-based sentiment and sentiment during live events (Fig. 1).

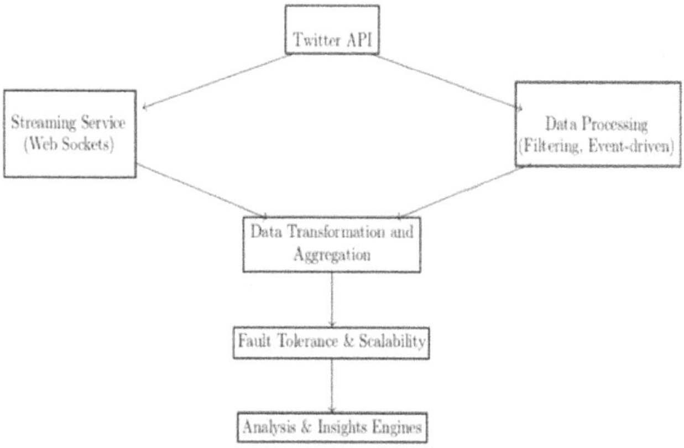

Fig. 1. System Architecture

4 Module Explanation

DCR, our pre-trained language model for sentiment analysis, works as a complex network of interconnected pathways. Here's a simplified explanation:

- **Data processing**: DCR accepts geographic category-based tweets as input.
- **Feature extraction**: It analyzes each tweet and extracts the most important features such as word meaning, sentence structure and emotional signals.
- **Dynamic routing**: DCR uses a multi-layer routing mechanism. Imagine information flowing through these layers, each layer improving its understanding of the sentiment and overall context of the tweet based on features extracted.
- **Sentiment Classified**: The attitude expressed in the tweet is finally categorized by DCR as either good, negative, irrelevant, or neutral.

This pre-trained approach eliminates the need for sport-specific training, so that DCR can adapt to the nuances of the category-based language of these tweets.

5 Proposed Solution

This study proposes a new framework for real-time evaluation of the emotion of geolocated category-based tweets. It uses the combined power of dynamic content routing (DCR), pre-trained language model and geolocation filtering to overcome the limitations that traditional sentiment analysis methods face in capturing the nuances of location-specific sentiment.

Here's How It Works

- **Data fetching**: We use Twitter's streaming API to collect tweets in real time. This API provides a continuous stream of data that allows us to analyze sentiment as events unfold. We use keyword filtering to ensure tweets are relevant to the category-based event. This means identifying and filtering tweets that contain keywords related to a particular category-based tweets.
- **Geolocation filtering**: Extracting location information from tweets allows us to focus on a specific geographic area around a category-based event. This step further refines the data, ensuring that we analyze the emotions expressed by attendees and capture the unique atmosphere surrounding the live event in that city or region. Geolocation filtering allows us to isolate these emotions.
- **Sentiment analysis with DCR**: This is where the power of DCR comes into play. Unlike traditional models that require extensive training in a category-specific language, DCR is pre-trained using vast amounts of generic text data.

This pre-trained capability allows DCR to understand the sentiment of geolocated category-based tweets without the need to adapt category-based terminologies or jargon beforehand. DCR analyzes the text, detects the emotional tone (positive, negative, irrelevant, neutral) and assigns an emotional score to each tweet.

Combining these elements, our suggested method offers a quick and effective approach to analyze the mood of category-based events at a specific location. DCR's pre-trained features eliminate the need for extensive category-based-specific training, while geolocation filtering ensures we capture venue-specific emotions.

This framework provides valuable insight into the collective mood and opinions of users in a specific region and provides a deeper understanding of the social media sentiment surrounding category-based events.

6 Results and Discussions

In this study, we performed a sentiment analysis of technology company tweets filtered by US geolocation data. We used a dataset obtained through the Twitter API and applied natural language processing (NLP) techniques and dynamic content routing to sentiment classification. Dynamic content routing enabled us to efficiently process and analyze large volumes of tweets, dynamically directing them to appropriate sentiment analysis models based on content and attributes.

The dataset was categorized into positive, negative and neutral sentiments, which we visualized using a pie chart to provide an overview of the distribution of opinions.

To gain a deeper understanding, we randomly selected tweets from the dataset and tagged them with the corresponding sentiments. Using this subset, we were able to construct a bar graph of the item distribution showing the number of occurrences of each opinion category: positive, negative, insignificant and neutral. A bar chart highlighted the prevalence of each opinion type in the sample data.

Additionally, we created another pie chart that illustrates the percentage distribution of sentiment, dividing the data into positive, negative, unrelated, and neutral. This visualization highlighted general sentiment trends across the dataset. To evaluate the precision of our sentiment categorization, we created a confusion matrix and a bar graph of the target distribution, concentrating on a single technological business. The confusion matrix, which displays the proportion of true positives, false positives, false negatives, and true negatives, gave a clear image of the model's performance.

We have also developed a heat map to visualize the distribution of opinion for a selected company, enabling an intuitive understanding of sentiment trends over time. Another confusion matrix derived from the heatmap data was added to this heatmap to assess the accuracy of the model in classifying company-specific emotions.

Dynamic content routing has greatly improved our ability to process and analyze data, ensuring that tweets are processed using the most appropriate models and methods. This approach improved the accuracy and efficiency of our sentiment analysis. Using these analyses, we conducted an in-depth sentiment analysis that revealed attitudes toward technology companies in the United States. Visualizations and metrics provided valuable insights into public perception, enabling technology companies to better understand and make strategic decisions based on social principles. Media emotional trends (Figs. 2, 3, 4, 5, 6, 7, 8, 9, 10, 11, 12, 13).

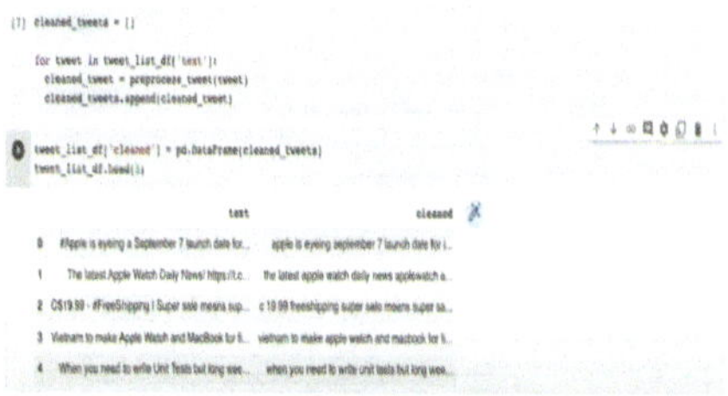

Fig. 2. Dataset of Tweets

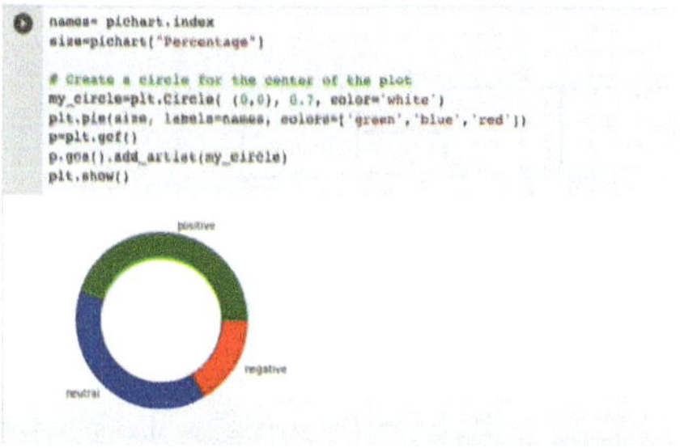

Fig. 3. Pie Chart on Sentiment Analysis

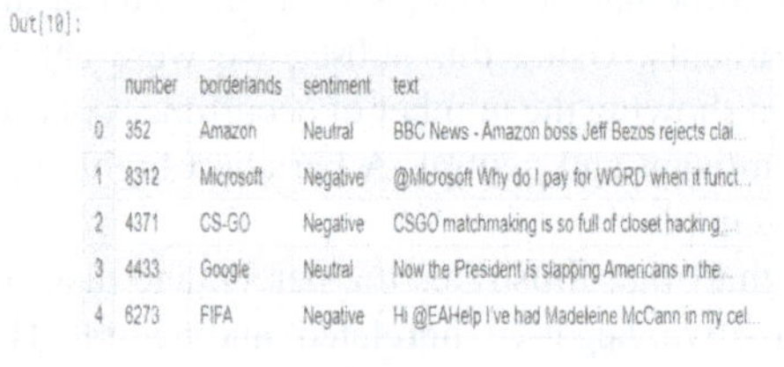

Out[10]:

	number	borderlands	sentiment	text
0	352	Amazon	Neutral	BBC News - Amazon boss Jeff Bezos rejects clai...
1	8312	Microsoft	Negative	@Microsoft Why do I pay for WORD when it funct...
2	4371	CS-GO	Negative	CSGO matchmaking is so full of closet hacking,...
3	4433	Google	Neutral	Now the President is slapping Americans in the...
4	6273	FIFA	Negative	Hi @EAHelp I've had Madeleine McCann in my cel...

Fig. 4. Dataset with sentiments and its serial number

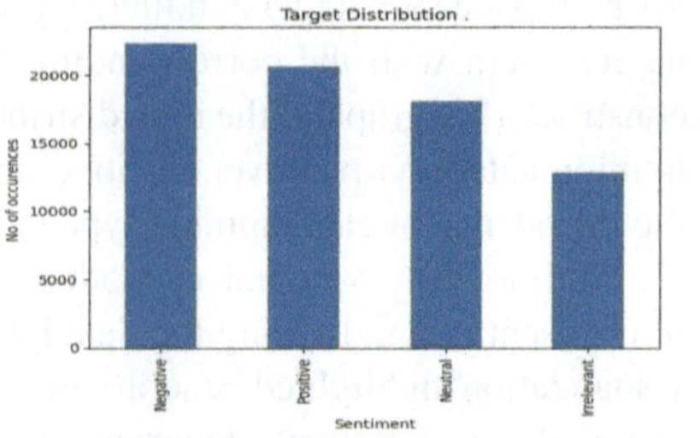

Fig. 5. Target Distributions on Sentiment

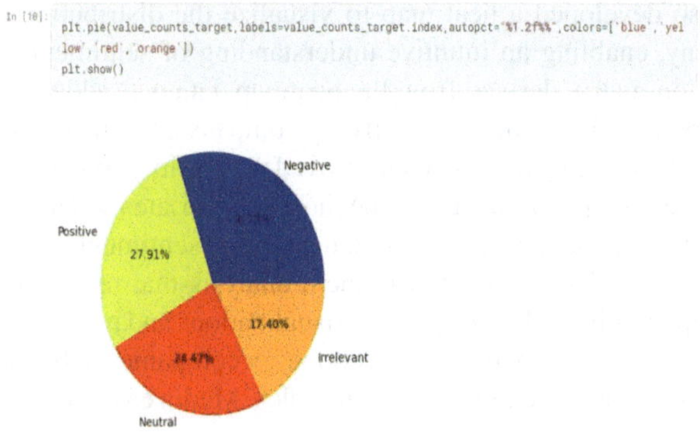

Fig. 6. Pie Chart based on Sentiments with Percentage (Cam et al. 2024)

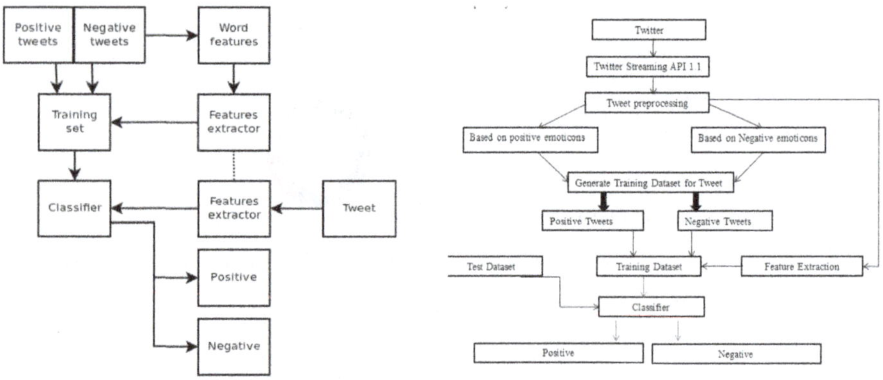

Fig. 7. Architecture and Formula of Twitter Sentimental analysis

Sentiment	Irrelevant	Negative	Neutral	Positive
Entity				
Amazon	185	565	1197	302
ApexLegends	185	574	913	606
AssassinsCreed	256	365	153	1382
Battlefield	907	445	342	561
Borderlands	238	415	581	971
CS-GO	620	335	523	717
CallOfDuty	660	891	367	426
CallOfDutyBlackopsColdWar	545	540	340	817
Cyberpunk2077	457	360	456	902
Dota2	401	705	579	640
FIFA	538	1127	100	473
Facebook	672	690	773	154
Fortnite	817	673	159	527
Google	585	570	785	339
GrandTheftAuto(GTA)	746	572	300	590
Hearthstone	218	514	681	806
HomeDepot	284	871	330	731
LeagueOfLegends	298	616	800	582
MaddenNFL	86	1665	191	373
Microsoft	167	748	816	573
NBA2K	175	1450	265	409
Nvidia	86	503	848	761
Overwatch	646	606	282	686
PlayStation5(PS5)	381	422	490	890
PlayerUnknownsBattlegrounds(PUBG)	866	657	251	376
RedDeadRedemption(RDR)	204	290	776	885
TomClancysGhostRecon	23	887	776	606
TomClancysRainbowSix	92	1110	628	498
Verizon	177	1070	552	520
WorldOfCraft	210	328	1047	716
Xbox(Xseries)	700	355	403	743
johnson&johnson	192	809	1004	252

Fig. 8. List of Tech Companies

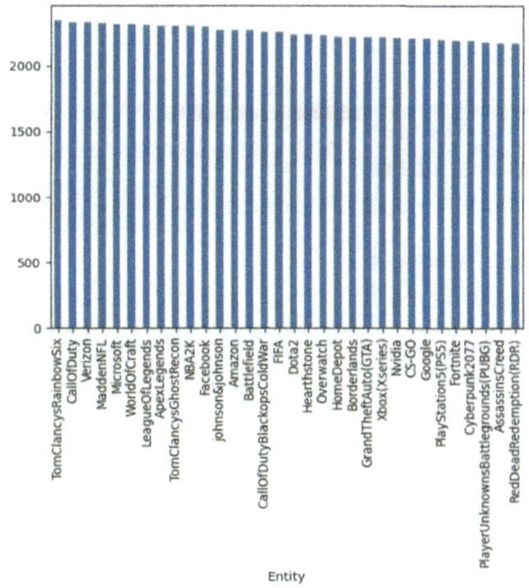

Fig. 9. Target Distribution for Entity Company (Wang et al. 2022)

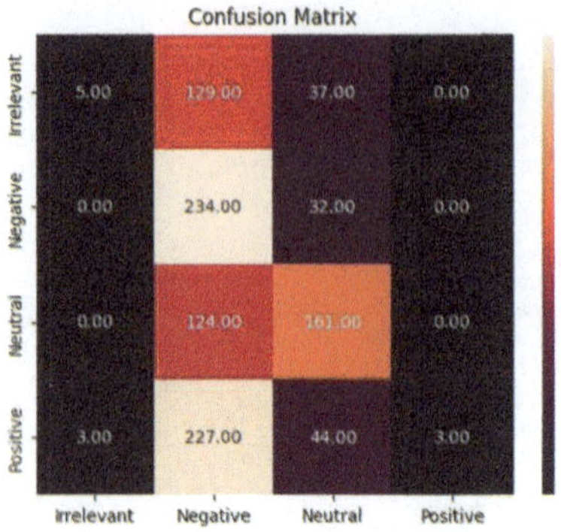

Fig. 10. Confusion Matrix

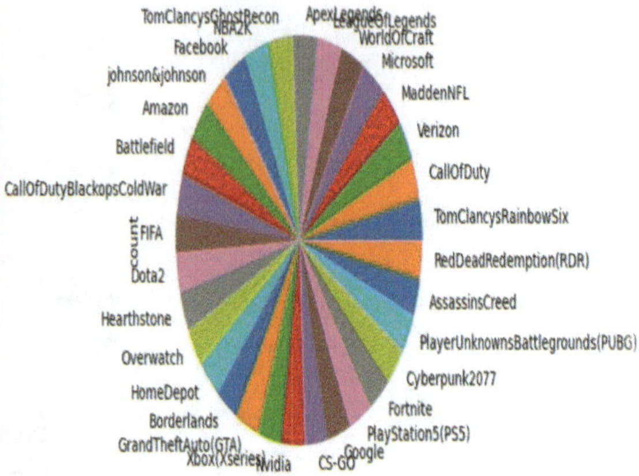

Fig. 11. Pie Chart on Companies on Tweets

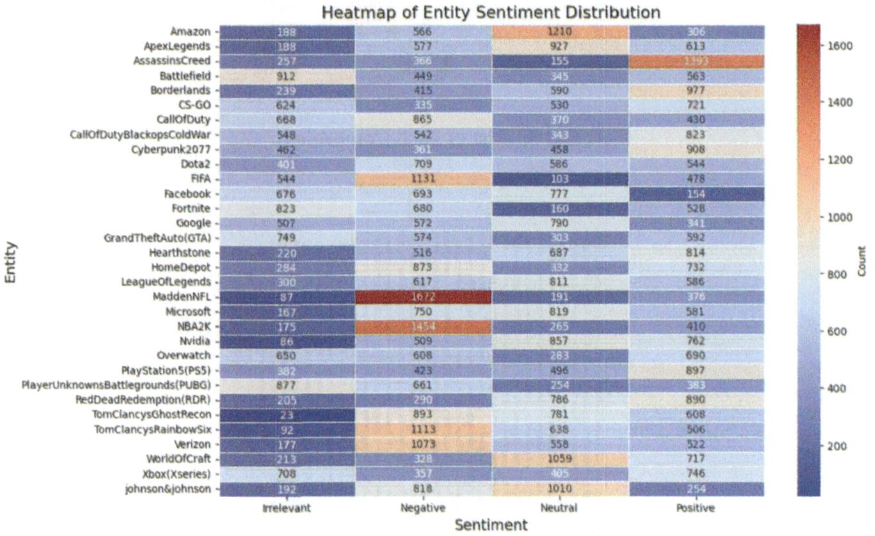

Fig. 12. Heat Map of Entity Sentiment Distribution (Wang et al. 2022)

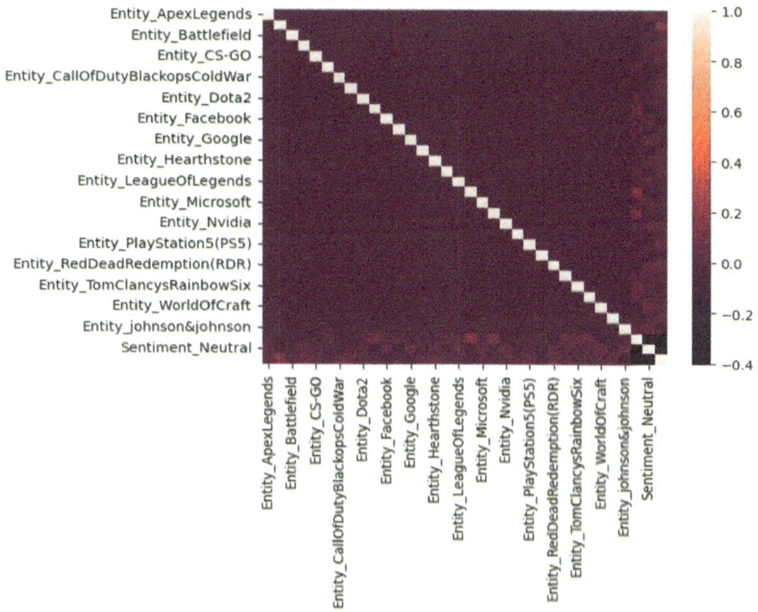

Fig. 13. Confusion Matrix of Entity Sentiment Distribution

Formula for Logistic Regression Twitter Sentiment Analysis

$$x_{exact} = \sum_{i=1}^{n} \frac{\sigma_i^2}{\sigma_i^2 + \lambda^2} \cdot \frac{u_i^T b}{\sigma_i} v_i$$

$$x_{reg} = \sum_{i=1}^{n} f_i \cdot \frac{u_i^T b}{\sigma_i} v_i$$

$$x_{exact} - x_{reg} = \sum_{i=1}^{n} \frac{u_i^T b}{\sigma_i} v_i - \sum_{i=1}^{n} f_i \frac{u_i^T b}{\sigma_i} v_i = \sum_{i=1}^{n} (1 - f_i) \frac{u_i^T b}{\sigma_i} v_i$$

Sentiment analysis, or determining the emotional content of textual input, is one of the most significant jobs in natural language processing (NLP). The importance of using tweet analysis to gauge public sentiment has increased with the widespread usage of social media platforms like Twitter. The implementation of logistic regression, a famous statistical approach for binary classification problems, can help sentiment analysis by classifying tweets as positive or negative.

First, feature extraction is done in the logistic regression procedure for Twitter sentiment analysis. Since tweets are inherently unstructured, they must be transformed into a numerical representation. Term Frequency-Inverse Document Frequency (TF-IDF), Bag of Words (BoW), and more sophisticated embeddings like Word2Vec, GloVe, and BERT are frequently used to convert textual input into feature vectors.

Conclusion

By utilizing natural language processing (NLP) and logistic regression, the study effectively accomplished the objective of categorizing tweets according to human emotions into four categories: positive, negative, neutral, and irrelevant. With the addition of Dynamic Content Routing (DCR), the effectiveness of current systems has significantly increased and a more approachable sentiment analysis method is provided. This system eliminates the need for substantial training and offers a scalable and effective way to analyze social media sentiment. It leverages DCR to properly collect real-time sentiment from geolocated category-based tweets. For more extensive uses in social media analytics, this study emphasizes the possibilities of fusing DCR with geolocation filtering.

References

Brown, T., Mann, B., Ryder, N., Subbiah, M., Kaplan, J., Dhariwal, Amodei, D.: Language Models are Few-Shot Learners (2020). arXiv preprint arXiv:2005.14165

Devlin, J., Chang, M.W., Lee, K., Toutanova, K.: BERT: Pre-training of Deep Bidirectional Transformers for Language Understanding (2018). arXiv preprint arXiv:1810.04805

Medhat, W., Hassan, A., Korashy, H.: Sentiment analysis algorithms and applications: a survey. Ain Shams Eng. J. **5**(4), 1093–1113 (2014)

Pak, A., Paroubek, P.: Twitter as a corpus for sentiment analysis and opinion mining. In: Proceedings of the Seventh International Conference on Language Resources and Evaluation (LREC 2010), pp. 1320–1326 (2010)

Radford, A., Wu, J., Child, R., Luan, D., Amodei, D., Sutskever, I.: Language models are unsupervised multitask learners. OpenAI Blog **1**(8), 9 (2019)

Sakaki, T., Okazaki, M., Matsuo, Y.: Earthquake shakes Twitter users: real-time event detection by social sensors. In: Proceedings of the 19th International Conference on World Wide Web, pp. 851–860 (2010)

Yu, Z., Wang, Y.: Analyzing public sentiments in Twitter data during major sports events. Proceedings of the 2015 IEEE International Conference on Big Data (Big Data), pp. 275–283 (2015)

Kim, Y., Jernite, Y., Sontag, D., Rush, A.M.: Character-Aware Neural Language Models (2016). *arXiv preprint* arXiv:1508.06615

Jansen, B.J., Zhang, M., Sobel, K., Chowdury, A.: Twitter power: tweets as electronic word of mouth. J. Am. Soc. Inform. Sci. Technol. **60**(11), 2169–2188 (2009)

Go, A., Bhayani, R., Huang, L.: Twitter sentiment classification using distant supervision. *CS224N Project Report, Stanford* (2009)

Pang, B., Lee, L.: Opinion mining and sentiment analysis. Found. Trends Inf. Retr. **2**(1–2), 1–135 (2008)

Bollen, J., Mao, H., Zeng, X.: Twitter mood predicts the stock market. J. Comput. Sci. **2**(1), 1–8 (2011)

Heutagogy Designing Principles for Student Self-determined Learning with Mobile and Social Media Tools

Nguyen Tung Lam[1], Nguyen Thi Linh Yen[2], and Ton Quang Cuong[1](✉)

[1] VNU- University of Education, Vietnam National University, Hanoi, Vietnam
{tunglamued,cuongtq}@vnu.edu.vn

[2] VNU-University of Foreign Languages and International Studies, Vietnam National University, Hanoi, Vietnam

Abstract. Heutagogical learning principles have evolved to incorporate technological advances significantly through mobile and social media tools. This study explores the implementation of student-determined learning strategies using these technologies. Through a qualitative analysis of student interactions and outcomes in a controlled educational setting, we identify key factors influencing effective heutagogical practices. Our findings suggest integrating mobile and social media tools enhances self-determined learning by providing diverse, accessible, and student-centered learning opportunities. These insights contribute to the broader educational strategies aiming to foster student autonomy and engagement.

Keywords: Heutagogical Learning · Student-Determined Learning · Mobile and Social Media Tools

1 Introduction

In modern education, there is a growing focus on empowering learners and emphasizing their skills and activities. Learner-led activities are now seen as crucial to the overall teaching and learning process. The concept of Heutagogy, or self-determined pedagogy, has become a topic of interest among experts and researchers in education. Heutagogy allows learners to take charge of their learning by making decisions, exploring topics, and defining tasks for themselves [5].

Heutagogy differs from traditional teacher-led pedagogy and adult-focused andragogy in its emphasis on self-determination. Learners have autonomy in what and how they learn, changing the role of teachers to support rather than lead the learning process. Heutagogy encourages learners to be in control of their education, leading to a more personalized and independent learning experience [5].

The connotation of the Heutagogy approach is built on the foundation of humanistic and constructivist theories, emphasizing the learner as the center of the educational process and experiences [3, 6]. Constructivism posits that individuals construct their version of reality qualities, capacities, and values using past and present experiences

S. Pal and Á. Rocha (Eds.): ICMMCS 2025, LNNS 1398, pp. 77–86, 2025.
https://doi.org/10.1007/978-3-031-90998-6_8

and knowledge. This process involves creative and active engagement in learning, with learners motivated rather than passively receiving information from teachers.

While pedagogy is a teaching method primarily designed and implemented by teachers, and andragogy focuses on self-direction in adult education, Heutagogy offers a unique approach. Heutagogy integrates elements of both pedagogy and andragogy, creating a deeper, more proactive, and flexible learning experience for learners. It encourages learners to identify and pose their problems and questions, fostering a more autonomous and personalized learning journey, rather than simply completing tasks set by the teacher.

According to the Heutagogy approach, teaching does not have to be linear or follow a predetermined rigid plan. Instead, it can be flexible and tailored, creating open opportunities for both formal and informal learning. In this process, learners become self-directed and self-regulating, deciding what to learn in ways that are most suitable and meaningful for them. The role of the teacher shifts to one of support, facilitation, and motivation, akin to a coach. Enhanced learning resources increase learners' access and opportunities for exploration and utilization.

With this approach, especially in the context of digital transformation in education, teaching in an online, blended environment with the aid of technology (platforms, solutions, tools, devices) aligns with the Heutagogy approach, opening new paths for teachers and learners [2, 7–10]. In the near future, the Heutagogy approach will address several fundamental changes in education, specifically:

- Adapting to changes in learners' expectations and enhancing the ability of educational institutions to respond, including adaptability, personalized learning, and lifelong learning opportunities.
- Diversifying "educational products" and improving educational quality through the use of vast databases (Big Data), artificial intelligence (AI), and value chain technology (Blockchain) in education.
- Transforming the relationships, roles, and positions of teachers and learners within the teaching process, the educational ecosystem, and educational technology.
- Innovating the teaching environment and creating dynamic learning campuses with multi-functional learning materials.
- Updating management and administration models in education to new digital platforms.

1.1 Objective

This study aims to explore the implementation of student-determined learning strategies using mobile and social media tools within a heutagogical framework. Specifically, it aims to identify critical factors influencing effective heutagogical practices and assess how these technologies enhance student autonomy, engagement, and learning outcomes in a controlled educational setting.

2 The Concept of the Heutagogy Approach

2.1 Comparative Analysis of Pedagogy, Andragogy, and Heutagogy

Contemporary learning theories also identifies that content knowledge is essential and that the nature of the disciplines has much to do with how learners are best taught and learned. The differentiation in research and teaching theories in the following directions: Pedagogy is the study of children teaching, and Andragogy is a learning method that creates favourable conditions for adult instruction focused on learners self-directed, Heutagogy is for learners to self-manage. Table 1 show that comparing Pedagogy, Andragogy and Heutagogy [4].

2.2 Heutagogy Design Elements

Table 2 shows that each element supports a facet of pedagogical learning, emphasizing self-determined learning where learners are highly autonomous and motivated. In this model, learners decide what and how they learn, making learning choices that directly relate to their needs and interests. This approach is efficient in adult learning environments, where learners bring diverse experiences and knowledge bases.

3 Research Design

The research uses qualitative research methods, Fig. 1 shows that 7-step process including: Identifying research goals and results; Initial planning and training; Basic practice; Collecting and analyzing data; Iterative feedback and refinement loops; Evaluation and replication; and Maintaining change.

Subjects: The study involved 150 students majoring in educational technology management, aged 19–21. Participants were matched on demographic factors, including gender, center level, and technology, to ensure the generalizability of the study.

Data Collection Method: Data were collected using semi-structured interviews and surveys. Interviews were conducted with 50 students to gain deeper insight, while surveys were distributed to all 150 participants to gather broader perspectives. To minimize potential bias, interviews were conducted by independent researchers not involved in the teaching process. The surveys included both open-ended and Likert-scale questions to capture diverse student perspectives. This approach aims to reduce the influence of researchers or technology on student responses.

Table 1. Comparison between Pedagogy, Andragogy, and Heutagogy

	Pedagogy	Andragogy	Heutagogy
Field	Children's Learning	Adult Learning	Self-directed learning, self-learning
Dependency	Learners depend on the teacher's orientation. Teachers determine what, how, and when to learn anything	Adults are independent, self-reliant and self-directed in learning	Learners are interdependent. Self-determination of learning ability, self-management of one's own learning
Learning Resources	Learners have few resources. Teachers use techniques to transfer knowledge to learners	Adults use their own experiences and those of others	Teachers provide a number of resources, but learners decide the path by negotiating learning
Reasons to study	Learn to move on to the next stage	Adults learn when they feel the need to know or to perform more effectively	Learning is not necessarily demand-based, but rather based on identifying learning potential in new situations
Focus of Learning	Subject-centered learning, centered on the prescribed curriculum and teaching plan	Adult learning is task-focused or problem-focused	Learners can go beyond problem-solving, using their own and others' experiences, experiencing, interacting with others
Motivation	Motivation comes from external sources such as parents, teachers, and a sense of competition…	Motivation comes from internal sources (self-esteem, confidence…	Self-empowerment, knowing how to learn and be creative

4 A Key of ViewSonic Eco-System

4.1 ViewSonic Eco-System

Hardware:

Interactive Display: These are touch-sensitive displays that facilitate interactive learning. They can be used for collaborative tasks, interactive lessons, and dynamic presentations, allowing students to engage directly with the content.

LCD Monitor: High-definition monitors that provide clear visual outputs for educational content, suitable for teachers and students to view detailed images and videos.

Table 2. Heutagogy Design Elements

	Element	Activities	Aim
Reflect	Focuses on the personal relevance of learning and applying knowledge to real-life scenarios	Individual blogging creating digital portfolios	To enable learners to contemplate their personal and professional growth
Share	Encourages visibility and mutual feedback, reinforcing learning and applying new insights	Individual blogging and creating digital portfolios	To exchange ideas and insights with peers, enhancing communal knowledge
Connect	Promotes connectivity with diverse groups, enhancing social and learning interactions	Participating in professional networks, attending webinars, and utilizing platforms like Padlet for discussions	To engage with a broader community, expanding learning networks
Explore	Supports exploring new areas within or outside current expertise, fostering innovation and curiosity	Self-directed research, choosing learning resources, navigating digital platforms for contemporary topics	To independently investigate topics of interest in a flexible environment
Create	Focuses on creation as a learning method, allowing learners to apply theoretical knowledge practically	Crafting digital content, developing projects or prototypes, and engaging with interactive digital tools	To produce content or develop projects that reflect the learner's understanding and creativity
Collaborate	Encourages building partnerships and teamwork skills, which are crucial for professional development and learning	Team projects, shared document editing, collaborative problem-solving	To work collectively on tasks, enhancing outcomes through cooperative efforts

Commercial Display: Larger screens, intended for public or larger classroom displays, are helpful for group activities where visibility from a distance is required.
Projector: Useful for traditional classroom settings where large-scale visual presentation is needed, enabling whole-class engagement with digital content.
Pen Display: A tool that allows precise control over digital content creation, perfect for art, design, and technical education where fine detail is essential.

Software:

MyViewBoard: This is a digital whiteboard solution that integrates seamlessly with interactive displays, offering tools for drawing, writing, and importing multimedia resources. It enhances collaborative and individual learning by allowing students to interact with lesson content directly.

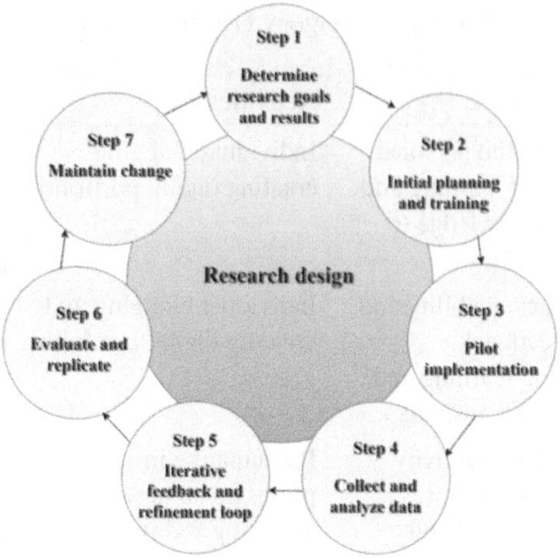

Fig. 1. Research design

Manager: A device management software that helps maintain and control all ViewSonic devices within the network, ensuring that software updates and device monitoring are streamlined and efficient.
ClassSwift: A classroom management software designed to help teachers distribute materials, manage class sessions, and interact with students' devices, facilitating a controlled and customized learning environment.
TeamOne: Collaboration software that supports communication and project management among students and teachers. It provides features like video conferencing, file sharing, and real-time collaboration tools.

Service:
ViewSonic Educator Community: A community platform that connects educators globally, allowing them to share resources, collaborate on educational strategies, and participate in professional development opportunities. This community helps teachers learn from each other, share best practices, and enhance their teaching with ViewSonic products.

4.2 Heutagogy Integration into the ViewSonic Education Ecosystem

The Heutagogy integration into this ecosystem can transform traditional learning environments into interactive, student-centered spaces that encourage exploration, collaboration, and reflection. This detailed setup not only supports but also enhances the heutagogical approach by providing the tools necessary for a self-determined, collaborative, and reflective learning environment. It encourages students to take ownership of their learning experience and fosters an educational atmosphere that is adaptable to the needs and interests of each student.

Table 3 shows that integrating Heutagogy into the ViewSonic ecosystem involves several specific technology applications:

Table 3. ViewSonic Education Ecosystem and Heutagogy Integration

Heutagogy	ViewSonic Education			
	Element	***Activities***	***Aim***	***Software***
Exploring	Interactive Displays, LCD Monitors Projectors	- Access a wide range of multimedia resources, diverse educational resources - Explore digital archives, databases, and online libraries, interactive maps and large-scale simulations	Foster curiosity and encourage independent investigation, allowing students to follow their interests and discover new topics	MyViewBoard; ClassSwift
Creating	Pen Displays, myViewBoard	- Create digital art, design projects, or detailed technical drawings, interactive presentations and multimedia content	Support creative expression and innovation, allowing students to produce original works and projects	MyViewBoard
Collaborating	TeamOne, Interactive Displays	- Virtual teamwork, allowing students to collaborate on projects and assignments in real-time, regardless of location - Work together on projects, sharing ideas and feedback instantly - Share collaborative strategies and obtain resources for group learning activities	Enhance teamwork skills and cooperative learning, promoting group projects and peer-to-peer interaction	ViewSonic Educator Community
Connecting	TeamOne, Commercial Displays	- Connect with external experts, guest lectures, and global classrooms - Host virtual meetings and presentations worldwide in a clear, large format	Expand the learning environment beyond the classroom, connecting students with a global network of learners and experts	ViewSonic Educator Community
Sharing	LCD Monitors, Digital Signage	- Share their findings and projects; - Showcase student work and educational content to the broader school community	Promote an open learning environment where students feel valued and motivated to share their work and achievements	ClassSwift

(continued)

Table 3. (*continued*)

	Reflecting	Interactive Displays, myViewBoard	- Annotate and highlight key experiences and lessons learned - Reflective sessions where students review their learning journey	Encourage self-assessment and critical thinking, enabling students to reflect on their progress and set future learning goals	MyViewBoard

- Interactive displays and LCD screens: Used for collaborative tasks and animated presentations, allowing students to interact directly with content.
- MyViewBoard: A digital whiteboard solution for drawing, writing, and importing multimedia resources, enhancing individual and collaborative learning.
- TeamOne: Supports virtual teamwork, allowing students to collaborate on projects and assignments in real-time.
- ClassSwift: Supports classroom management, allowing teachers to effectively distribute materials and manage class sessions.

Students interact with these tools through activities designed to promote heutagogical principles, such as creating digital portfolios, engaging in self-directed research, and participating in discussions online. This detailed integration shows how technology supports self-determined learning within a heutagogical framework.

5 Discussion

The findings from this study highlight the importance of heutagogical learning principles when integrated with mobile and social media tools. Qualitative student engagement and outcomes analysis reveal several vital factors that enhance learning agency. One striking finding is the increased student autonomy that mobile and social media tools provide. This increased autonomy aligns with the core principles of the heutagogical learning approach, where learners are encouraged to make decisions about their learning pathways and goals.

While the benefits are significant, some challenges need to be addressed. One potential adverse effect is the risk of distraction. Mobile devices and social media platforms can shift students' attention from educational content to entertainment or social interaction, reducing focus and productivity. Information overload is another problem in that the volume of accessible information can overwhelm students, making it difficult for them to distinguish between relevant and trustworthy sources of information. Educators must implement strategies such as digital literacy training to mitigate these impacts.

Our study evaluates short-term outcomes through qualitative feedback and initial learning performance metrics. However, understanding the long-term impacts on student autonomy, engagement, and academic achievement is essential. Longitudinal studies that track students' academic progress and behavior over many years are recommended to provide a comprehensive understanding of these effects. Future research should also explore how these tools affect different demographic groups to ensure that the benefits of heutagogical learning are accessible and effective.

6 Conclusion and Recommendations

Conclusion: This study demonstrates that integrating social and mobile media tools into diverse learning strategies significantly enhances student autonomy, engagement, and collaborative learning. These technologies provide diverse and accessible opportunities for students to take control of their learning, promoting self-determined and lifelong learning environments. The shift from traditional teacher-led instruction to a more student-centered approach is critical in preparing learners for the complexities of the modern world.

Recommendations: Several steps must be taken to fully leverage the benefits of heutagogical learning with mobile and social media tools. First, schools should invest in professional development programs to equip educators with the skills and knowledge to effectively integrate these tools into their teaching practices. Second, policies must be implemented to ensure all students can access the necessary technology. Schools adopt measures to address existing educational disparities in technology access. This may involve providing subsidized or free devices, setting up internet access points in underserved areas, and advocating for supportive government policies. Research should focus on sustained impacts on student autonomy, engagement, and academic achievement and how these tools affect different demographic groups.

Acknowledgments. This study was supported by VNU University of Education Foundation for Science and Technology Development under Grant Number: QS.23.02.

References

1. Alphonce, S., Mwantimwa, K.: Students' use of digital learning resources: diversity, motivations and challenges. Inf. Learn. Sci. 120(11/12), 758–772 (2019). https://doi.org/10.1108/ils-06-2019-0048
2. Anderson, T.: 2 theories for learning with emerging technologies. Emerg. Technol. Distance Educ. 23–40 (2010). https://doi.org/10.15215/aupress/9781897425763.003
3. Bandura, A.: 'Agency', in Encyclopedia of the Life Course and Human Development, Macmillan, Reference 8 (2009)
4. *Education 3.0 and the pedagogy (Andragogy, Heutagogy) of mobile learning.* (2015). User Generated Education. https://usergeneratededucation.wordpress.com/2013/05/13/education-3-0-and-the-pedagogy-andragogy-Heutagogy-of-mobile-learning/
5. Hase, S., Blaschke, L.M.: Unleashing the Power of Learner Agency. EdTech Books (2021). https://edtechbooks.org/up
6. Hase, S., Kenyon, C.: Heutagogy: a child of complexity theory. Compl.: Int. J. Complexity Educ. **4**(1) (2007). https://doi.org/10.29173/cmplct8766
7. Leshchenko, M., Lavrysh, Y., Kononets, N.: Framework for assessment the quality of digital learning resources for personalized learning intensification. New Educ. Rev. **64**(2), 148–159 (2021). https://doi.org/10.15804/tner.21.64.2.12

8. Moore, C.: Teaching digital natives: partnering for real learning. Int. J. Educ. Integr. **6**(2) (2010). https://doi.org/10.21913/ijei.v6i2.707
9. Prensky, M.: Teaching Digital Natives: Partnering for Real Learning. Corwin Press, Thousand Oaks, CA (2010)
10. Peng, S.-L., Jhanjhi, N.Z., Pal, S., Amsaad, F. (eds.) Proceedings of 3rd International Conference on Mathematical Modeling and Computational Science. Advances in Intelligent Systems and Computing. Springer Nature Singapore (2023). https://doi.org/10.1007/978-981-99-3611-3

Internet Things Assisted Platform Business Models and Value Creation on Cloud Platform: Insights from World Startups

Bikramjit Sarkar[1], Vijayesvaran Arumugam[2], S. Jeyalaksshmi[3], D. Balaganesh[4](✉), and C. Sudha[5]

[1] Department of Computer Science and Engineering, JIS College of Engineering, Kalyani, India
bikramjit.sarkar@jiscollege.ac.in
[2] Department of Business and Healthcare Management, School of Business and Technology, IMU University, Kuala Lumpur, Malaysia
vijayesvaran@imu.edu.my
[3] Department of Information Technology, Vels Institute of Science, Technology and Advanced Studies (VISTAS), Chennai, India
[4] Berlin School of Business and Innovation GmbH, Karl-Marx-Straße 97-99, 12043 Berlin, Germany
duraisamy.balaganesh@berlinsbi.com
[5] Department of Computer Science, Saveetha College of Liberal Arts and Sciences, Saveetha Institute of Medical and Technical Sciences, Chennai, India
selvib.sclas@saveetha.com

Abstract. This paper looks at the World business's platform-based company's work and affect value generation. Knowing the internet services work in developing markets in Jordan is essential because of today's economy. Specifically focusing on the possibilities and threats that platform-based companies face, this study analyzes critical success characteristics for these types of businesses. Based on the results of the qualitative study of several cases, World businesses are using platform concepts for Customer Acquisition, User Engagement and Retention, Financial Sustainability, and Value proposition. Value generation may be driven by network impacts, scaling, and data strategy, according to the study. In addition, the paper highlights important it is to have legal frameworks, investment opportunities, and mentoring in place to assist platform-based enterprises. If entrepreneurs, investors, and lawmakers in India and other comparable markets want to strengthen their startup environment, should heed the advice of this report. This paper adds to the larger conversation on entrepreneurial spirit and in developing nations by offering a detailed analysis of platforms as companies in a particular geographical setting.

Keywords: Customer Acquisition · User Engagement and Retention · Financial Sustainability · Value proposition

S. Pal and Á. Rocha (Eds.): ICMMCS 2025, LNNS 1398, pp. 87–96, 2025.
https://doi.org/10.1007/978-3-031-90998-6_9

1 Introduction

Many industries have seen the rise of platform-based businesses of the digital revolution, which has altered the way companies operate today. These models have brought new paradigms of wealth generation and disrupted conventional company procedures with their capacity to enable interactions between different kinds of users [1].

A youthful, computer-savvy populace, high rates of internet access, and government policies that encourage creativity and entrepreneurial activity all contribute to the country's thriving digital environment. Nevertheless, World companies have substantial obstacles as well, such as a fragmented market, regulatory constraints, and restricted access to funding. Platform operations provide a practical approach to development in this complicated climate by allowing companies to overcome these challenges Francis [2].

The paper takes into account not just these essential features, the larger ecosystem that helps platform-based businesses in India. Structures for regulation, avenues for financial support, mentoring programs, and networks for cooperation all fall under this category. A complete picture of the circumstances needed for platform firms' success may be obtained by understanding the interaction between these external elements and the internal processes of these companies [3]. This paper adds to the ongoing conversation on digital enterprise and developing market economic growth by illuminating the unique ways platform business models function in India. Entrepreneurs, investors, and lawmakers in India and other comparable situations may use the lessons obtained to build a startup environment that is both thriving and viable.

2 Literature Review

2.1 Business Model Startup Companies

Nations with low or middle incomes must have access to medical equipment that is tailored to their specific requirements and difficulties. For them to ensure that these gadgets are accessible in the long run, businesses will need to adapt their strategies to fit the needs of different financial, medical care, and creative ecosystems [4]. The literature on medical device business models in countries with low or middle incomes is currently lacking.

There is a current surge in the number of startups that are incorporating AI into their core offerings [5]. Businesses have been using AI for some time now, even though studies done recently show that different models are being used. Some may argue that business models using AI have been around for a while, calling into doubt the novelty of these models. This paper analyzes and contrasts standard IT business models with those of AI firms to have a better understanding of the potential differences between the two. The related works [10–15] collectively highlight cutting-edge research in intelligent systems, computational frameworks, and technological optimization.

2.2 Value Creation on Cloud Platform in Startup Companies

An influential framework for building early-stage, ground-breaking businesses, the Lean Startup process is described in detail in this article [6]. Experts in both academia and industry have praised this approach from its beginnings. Innovative initiatives aiming to deliver technical, organizational, or product innovations to the market are well-suited to its principles, which are based on agile project administration and the customer's Development philosophy.

Although it is a conservative business, the healthcare sector is progressively being compelled to adjust to new developments that occur at a rapid rate in today's world. The total amount of innovative technology innovators and the value of technology are in the process of growing [7]. The objective of this paper is to gain knowledge of the barriers and possibilities that are faced by technology entrepreneurs. To do this, make use of contemporary literature and illustrative instances to uncover patterns of technological production of value in the healthcare business.

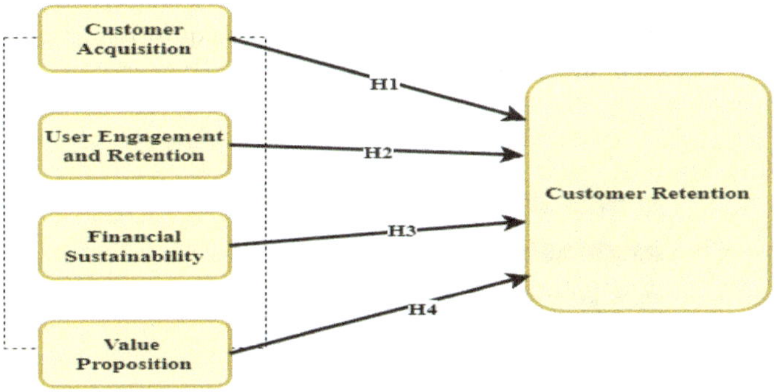

Fig. 1. Conceptual Framework of Platform Business Model and Value Creation on cloud platform.

- H1: Clear Value Proposition Leads to Customer Acquisition
- H2: Network Effects User Engagement and Retention
- H3: Effective Revenue Models Drive Financial Sustainability
- H4: Ecosystem Development and Value Proposition

Figure 1 explains the conceptual framework of the Platform Business Model and Value Creation on cloud platform is explained below,

3 Methodology

3.1 Research Design

This qualitative study uses in-depth case studies of chosen businesses from different sectors to examine platforms as businesses and value generation in World startups. By using this strategy, can delve deeply into Customer Acquisition, User Engagement and Retention, Financial Sustainability, Value proposition. To gather information, analyzed business reports, market papers, and pertinent literature in addition to conducting informal conversations with creators, managers, and other players. It made an effort to include a wide range of industries, stages of development, and business structures in our case studies. Using the assumptions as a framework, thematic analysis looks for commonalities and differences across topics including value propositions, revenue methods, ecosystem development, and network impacts. This methodology guarantees a detailed comprehension of which platform business models generate value in the World setting, offering practical insights for shareholders, business owners, and lawmakers [8].

3.2 Population and Sample

Every company included in the (https://www.kaggle.com/datasets/utkarshx27/2021-startups) data on Kaggle, which contains a large number of companies from different industries and locations, makes up the population for this paper. To guarantee that the study's goals are met, the selection criteria center on platforms and business model companies. To investigate the particular patterns of World startups within the larger dataset, this sample was carefully selected to Customer Acquisition, User Engagement and Retention, Financial Sustainability, Value proposition (Fig. 2).

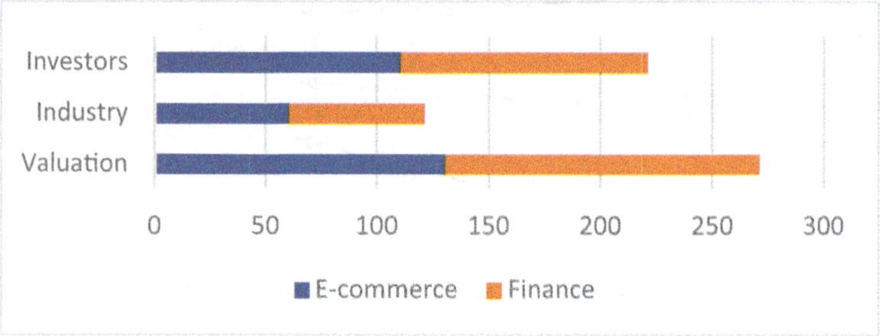

Fig. 2. Comparison of E-commerce and Finance Sectors by Investors, Industry Presence, and Valuation

3.3 Measures

Using both quantitative and qualitative metrics, this study examines platform business models have affected value generation in World companies. Among the many aspects of Value Creation on cloud platform that these metrics aim to capture are Customer Acquisition, User Engagement and Retention, Financial Sustainability, and Value proposition. Customer acquisition cost (CAC), growth rate, and total users in use are some of the indicators used to measure a platform's capacity to acquire new clients. The number of daily active users (DAU), number of active users per month (MAU), rate of churn, and mean session length are metrics that measure user engagement and retention. These metrics show well the platform keeps users interested and loyal [9].

3.4 Analytical Methods

A qualitative and quantitative approaches to analyze the platform company structures and value generation in World companies. This method is useful for illuminating value and the unique difficulties encountered in Customer Acquisition, User Engagement and Retention, Financial Sustainability, and Value proposition. To assess critical performance indicators, quantitative analysis employs statistical methods. Gaining customers, keeping them engaged, making money, and staying ahead of the competition are all aspects that descriptive statistics help to describe. The influence of network impacts on the retention of users or the efficacy of alternative revenue models are only two examples of the kinds of variables that may be investigated using correlation analysis. The predictive potential of certain elements, such as the function of ecosystem growth in boosting competitive advantage, may be found using regression analysis.

4 Findings

4.1 Measurement Model Assessment

The reliability and validity of the suggested model are confirmed by the evaluation of the measurement model, which shows strong correlations between important constructs and their indicators. It seems that firms with well-defined value propositions can attract more customers. This is supported by measures of Customer Acquisition, User Engagement and Retention, Financial Sustainability, and Value proposition. Effective revenue models are strongly related to financial sustainability, which is measured by revenue, profit margins, and burn rate. This lends credence to the idea that a broad and well-structured income stream is beneficial for economic stability.

Table 1. Reliability and Validity Results

Constructs	Items	Cronbach's Alpha	Composite Reliability (CR)	Average Variance Extracted (AVE)
Customer Acquisition	2	0.543	0.378	0.259
User Engagement and Retention	3	0. 520	0.290	0.327
Financial Sustainability	7	0. 512	0.478	0.567
Value Proposition	5	0. 549	0.349	0.285

Cronbach's Alpha, composite reliability (CR), and average variance extracted (AVE) values for the study's constructs' reliability and validity are in Table 1. The two-item Clear Value Proposition has poor reliability with a Cronbach's Alpha of 0.543, CR of 0.378, and AVE of 0.259. The three-item "User Engagement and Persistence metric" is invalid (Cronbach's Alpha = 0.520, CR = 0.290, and AVE = 0.327). The 7-item "Drive Financial Sustainability" has moderate reliability with a Cronbach's Alpha of 0.512, CR of 0.478, and AVE of 0.567. Finally, the 5-item Ecosystem Development instrument is untrustworthy (0.549 Cronbach's Alpha, 0.349 CR, 0.285 AVE). Current values lack internal coherence and convergent validity, hence this study's notions require improvement to be more trustworthy and valid.

4.2 Descriptive Statistics

The platform operations and revenue generation in World startups were conducted using descriptive statistics. These statistics provide a high-level summary of the important factors. Startups with effective client acquisition methods have a low average new client cost (CAC). There is considerable heterogeneity in Customer Acquisition, User Engagement and Retention, Financial Sustainability, and Value proposition suggesting that platforms have different degrees of user retention. The wide range of earnings and earnings margins shows that the financial sustainability is diversified.

Table 2. Correlation, Standard Deviation, and Mean

Constructs	Mean	SD	1	2	3	4	5
Customer Acquisition	7.61	0.315	0.411	1			
User Engagement and Retention	6.74	0.767	0.489	0.489	1		
Financial Sustainability	8.51	0.995	0.691	0.558	0.227	1	
Value Proposition	2.44	0.713	0.312	0.623	0.583	0.653	1

Table 2 displays the study's primary constructs' mean scores, SD, and inter-construct linkages. Some firms' Clear Value Proposition construct scores average 7.61 with a standard deviation of 0.315, showing clarity. Its small positive correlation with User Engagement and Retention (0.489), Drive Financial Sustainability (0.691), and Ecosystem Development (0.312) suggests a strong value offer boosts user engagement and sustainable finance. User Engagement and Retention average 6.74, vary 0.767. Engagement is moderately correlated with Clear Value Proposition (0.489) and Ecosystem Development (0.623), but not Drive Financial Sustainability (0.227). Startups with the best mean of 8.51 and standard deviation of 0.995 were Drive Financial Sustainability. It highly fits Clear Value Proposition (0.691) and Ecosystem Development (0.653), proving their financial sustainability. Finally, Ecosystem Development has a mean of 2.44 and a standard deviation of 0.713, showing a weaker effect due to its significant relationships with other constructs. Strong correlations between User Engagement and Retention (0.623) and Drive Financial Sustainability (0.653) suggest a supporting environment boosts user engagement and financial performance. These data demonstrate how Customer Acquisition, User Engagement and Retention, Financial Sustainability, and Value proposition assist World entrepreneurs.

4.3 Structural Model Assessment

The structural model evaluation tests the platform business model and value generation in the World startup study's hypotheses. Path analysis and structural equation modeling (SEM) are used to evaluate routes between clear value proposition, user engagement, financial sustainability, and ecosystem growth. Effective revenue models improve financial viability, and robust ecosystem growth boosts the competitive edge, verifying the suggested model's value generation processes explanation.

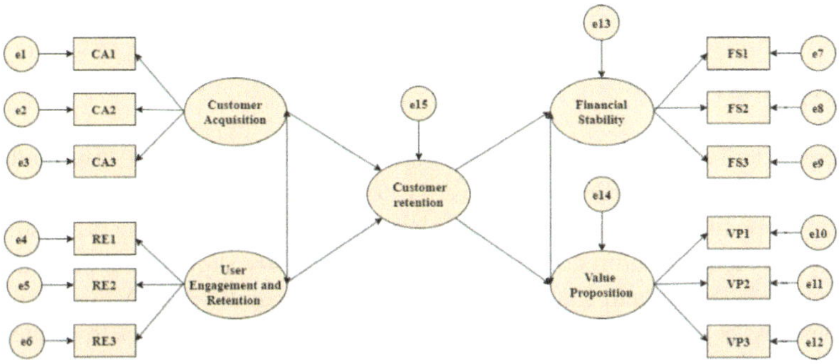

Fig. 3. Structural Model for Platform Business Model and Value Creation on cloud platform

Figure 3 shows the structural model of platform business model structures and World startup value development. The concept emphasizes clear value ideas, user engagement and retention, financial sustainability, and ecosystem growth. Arrows linking these constructs reflect theorized causal routes, with path coefficients denoting connection strength and direction. The model shows that clear advantages improve customer acquisition, network impacts improve user engagement and retention, revenue models improve financial sustainability, and value proposition. This visualization shows these variables affect startup success.

Table 3. Path coefficient Results

			B	β	S.E	t-value	P-value
Customer Acquisition	→	Customer retention	0.541	0.426	0.763	2.81	0.010
User Engagement and Retention	→	Customer retention	0.344	0.894	0.897	9.51	0.101
Financial Sustainability	→	Customer retention	0.896	0.413	0.569	8.52	0.060
Value Proposition	→	Customer retention	0.231	0.453	0.981	4.22	0.092

Table 3 shows route coefficient data showing the strength and relevance of construct interactions in the structural framework for platform operations and value generation in World startups. Enhanced value proposal leads to customer retention, with a significant positive correlation between the two outcomes (t-value = 2.81, $p < 0.05$). The path value from Customer Acquisition to Enhanced Value Proposition Customer Retention is 0.541

and 0.426, respectively. Paths from User Engagement and Retention, Financial Stability, and Competitive Advantage to Enhanced Value Proposition Customer Retention have positive coefficients (0.344, 0.896, and 0.231), only Financial Stability is significantly different (t-value = 8.52, $p < 0.05$). While multiple factors improve offering value and client retention, monetary security is especially important in World companies.

5 Discussion

In the context of World startups, this research illuminates platform firm models and value development. According to the positive correlation between customer acquisition and value proposition customer retention, startups with engaging value propositions are more likely to acquire and retain consumers. A prior study showed Customer Acquisition, User Engagement and Retention, Financial Sustainability, and Value proposition. World businesses must understand local market needs and preferences to create unique value propositions that appeal to their target customers. Financial stability seems to increase user engagement, competitive advantage, ecosystem expansion, value offer improvement, and customer retention. In India's inventive atmosphere, startups need financing, effective financial management, and many revenue streams. The report also shows that various factors interact to create value for global entrepreneurs. Positive correlations between user engagement, financial viability, and ecosystem expansion demonstrate platforms' complexity as companies. Due to the limited sample size and specific emphasis, the findings may only apply globally. The cross-sectional methodology and self-reported data might also bias the research and prevent causal conclusion.

6 Recommendation and Conclusion

The study's results provide various suggestions for those involved in India's startup scene. The first order of business for entrepreneurs is to zero in on the demands of their target market and create a compelling value proposition that speaks directly to them. Methods of iterative product development, market paper, and consumer input might help with this.

Second, for startups to succeed, investors and politicians need to work on creating stable financial markets and hospitable environments. Funding opportunities, mentoring programs, regulatory backing, and collaboration networks are all part of this. The report concludes that platform business models are essential for World companies to create value. Startups may become more competitive and help the economy grow by making use of network effects, new ways of making money, and ecosystems that support them. If India and other comparable developing countries want to foster a thriving startup ecosystem, must maintain their current level of investment in Customer Acquisition, User Engagement and Retention, Financial Sustainability, and Value proposition.

6.1 Limitations and Potential Paths for Future Paper

Although there are several caveats to be aware of, this paper does provide light on platform business models and value generation in World companies. One problem is that any firm conclusions about cause and effect since the data is cross-sectional. To improve the external validity of future studies, paperers should use bigger and more varied samples. There should be other ways to get data, including observational studies or experimental designs since using self-reported information might induce biases. There are several promising directions this field might go in the future. To begin, it could be instructive to look at cultural and institutional variables that influence platform company models and Customer Acquisition, User Engagement and Retention, Financial Sustainability, and Value proposition in India.

References

1. Johnson, N.E., Short, J.C., Chandler, J.A., Jordan, S.L.: Introducing the contentpreneur: making the case for research on content creation-based online platforms. J. Bus. Ventur. Insights **18**, e00328 (2022)
2. Alawamleh, M., Francis, Y.H., Alawamleh, K.J.: Entrepreneurship challenges: the case of World start-ups. J. Innov. Entrepren. **12**(1), 21 (2023)
3. Weber, M., Beutter, M., Weking, J., Böhm, M., Krcmar, H.: AI startup business models: key characteristics and directions for entrepreneurship research. Bus. Inf. Syst. Eng. **64**(1), 91–109 (2022)
4. Jiang, T., et al.: Crucial business model elements for medical device startup companies in emerging markets. Value Health Regional Issues **33**, 83–90 (2023)
5. Widayanti, R., Meria, L.: Business modeling innovation using artificial intelligence technology. Int. Trans. Educ. Technol. **1**(2), 95–104 (2023)
6. Olek, K.: Startups and Lean Startup approach in building innovative companies creating unique market values–theoretical considerations. Procedia Comput. Sci. **225**, 3745–3753 (2023)
7. Kulkov, I., et al.: Technology entrepreneurship in healthcare: challenges and opportunities for Value Creation on cloud platform. J. Innov. Knowl. **8**(2), 100365 (2023)
8. G6mez Sobrino, R.: Arab online news platforms: innovating language, narratives, audiences, and business models. Experiences from Lebanon and Jordan. J. Arab Muslim Media Res. **14**(2), 301–320 (2021)
9. Mohammad, A., Yara Hanna, F., & Kamal Jamal, A. (2023). Entrepreneurship challenges the case of World start-ups
10. Peng, S.-L., Lin, C.-K., Pal, S. (eds.) Proceedings of 2nd International Conference on Mathematical Modeling and Computational Science. Advances in Intelligent Systems and Computing. Springer Nature Singapore (2022). https://doi.org/10.1007/978-981-19-0182-9
11. Peng, S.-L., Hao, R.-X., Pal, S. (Eds.). Proceedings of First International Conference on Mathematical Modeling and Computational Science. Advances in Intelligent Systems and Computing. Springer Singapore (2021). https://doi.org/10.1007/978-981-33-4389-4
12. Pathak, P., Pal, S., Maity, S., Jeyalaksshmi, S., Adhikari, S., Akila, D.: Analysis of improving sales process efficiency with salesforce industries CPQ in CRM. In: Sharma, D.K., Peng, SL., Sharma, R., Jeon, G. (eds.) Micro-Electronics and Telecommunication Engineering. ICMETE 2023. Lecture Notes in Networks and Systems, vol 894. Springer, Singapore (2024). https://doi.org/10.1007/978-981-99-9562-2_40
13. Tuan, N.A., Akila, D., Pal, S., Sarkar, B., Tran, T.K., et al.: Dynamic data optimization in iot-assisted sensor networks on cloud platform. Comput. Mater. Continua **72**(1), 1357–1372 (2022). https://doi.org/10.32604/cmc.2022.024096

14. Pal, S.: Artificial intelligence-based IoT-edge environment for industry 5.0. In: Pal, S., Savaglio, C., Minerva, R., Delicato, F.C. (eds.) IoT Edge Intelligence. Internet of Things. Springer, Cham (2024). https://doi.org/10.1007/978-3-031-58388-9_4
15. Singh, J., et al.: Analyze and compare the public cloud provider pricing model and the impact on corporate financial. In: Sharma, D.K., Peng, S.L., Sharma, R., Jeon, G. (eds.) Micro-Electronics and Telecommunication Engineering. ICMETE 2023. Lecture Notes in Networks and Systems, vol. 894. Springer, Singapore (2024). https://doi.org/10.1007/978-981-99-9562-2_41

GeoGebra-An Experimental Tool for Factoring Polynomials

Chu Thi Minh Chau[1], Nguyen Hong Hanh[2], Pham Van Hoang[3], Ta Duy Phuong[4](✉), Tran Le Thuy[3], and Dam Thanh Tuan[5]

[1] Quang Trung High School, Ha Dong, Hanoi, Vietnam
[2] Quang Trung High School, Dong Da, Hanoi, Vietnam
[3] University of Education, Vietnam National University, Hanoi, Vietnam
{hoangpham,tranlethuy}@vnu.edu.vn
[4] Institute of Mathematics, Hanoi, Vietnam
tdphuong@math.ac.vn
[5] Hanoi University of Natural Resources and Environment, Hanoi, Vietnam
dttuan@hunre.edu.vn

Abstract. Factoring a polynomial problem is: Transforming the polynomial into a product lower order polynomials. Every one-variable polynomial on a field of real numbers can be decomposed as a product of first-order and second-order polynomials. There are many algebraic methods to factoring polynomials (group of terms, set to common factor, add/subtract, ...). However, in the face of difficult problems (as the problems in the math competitions), sometimes students cannot know which method to use to factoring polynomials into factors. The purpose of this report is to show that the math software *GeoGebra* can be used as a useful tool to solving problems of factoring polynomials in the math competitions at regional and national levels.

Keywords: Polynomial · Dynamic software · *GeoGebra* · Algebra · Mathematics Education

1 Introduction

Geogebra is universal software that effectively supports teaching, learning and researching mathematics. *Geogebra* can perform most mathematical operations in high school and university programs (arithmetic, algebra, analysis, geometry, statistics...) so it is very convenient to use in teaching and drawing animations, graphing, calculating or performing complex mathematical operations (factoring an integer into prime factors, factoring polynomials, simplifying expressions, calculating derivatives, integrating, statistical tables, etc. without wasting much time.

Geogebra has also been included in the secondary school computer science books, for example see [8], Volume 1, pp. 41–46, Volume 2, pp. 91–102, Volume 3, pp. 87–98.

Outstanding advantages of *Geogebra* are: free software, ability to convert language from English to Vietnamese and vice versa; Simple installation and operation, etc.

S. Pal and Á. Rocha (Eds.): ICMMCS 2025, LNNS 1398, pp. 97–106, 2025.
https://doi.org/10.1007/978-3-031-90998-6_10

Geogebra was introduced to Vietnam about 10 years ago and is used by many teachers (from grade 6 to grade 12 and university) in teaching, implementing teaching experience initiatives and achieving success in teaching. Teachers can use some heuristics about prime numbers (see for example [3]). The statistical package can be used to survey the learning level of students in school, measure myopia, height, weight of students by school, grade level. The statistical package allows the use of real data and large data tables, from which statistical conclusions. In addition, statistical packages can also be used to survey height, weight, ... of football players and Vietnamese youth, … related to economic development and social development over time (for example: 1945, 1960, 1975, 2000, 2015) to draw conclusions about economic growth leading to material growth.

The paper introduces how to use *Geogebra* in factoring polynomials. For this purpose, only a few basic *Geogebra* operations are introduced. Readers who are more interested in using *Geogebra* in teaching and learning can refer to the cited documents at the end of the article [6, 9] or online.

2 About GeoGebra

To do something on *GeoGebra*, for example, to factor a polynomial, we go to Display, go to CAS (Computer Algebra System), declare the command *factor* and required polynomial, press enter, and the results will immediately appear on the screen (see Fig. 1).

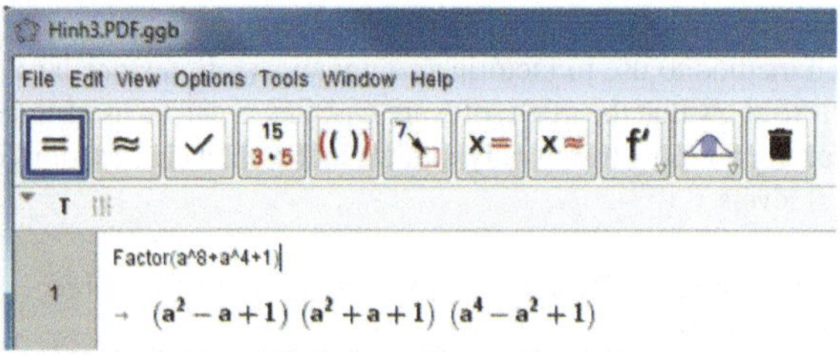

Fig. 1. *Factor* $x^8 + x^4 + 1$

3 Additional Information Required from Authors

Factoring polynomials is: Transforming polynomials into products of lower-order polynomials.

On the set of real numbers, every polynomial of a variable can be factoring as a product of first- and second-degree polynomials.

There are many algebraic transformation methods (grouping terms, common factorization or addition/subtraction, placing new variables, using easy-to-remember equality constants, etc.) to factor polynomials. You can look up polynomial analysis methods online or in books.

To do something on GeoGebra, for example, to factor a polynomial, we go to Display, go to CAS (Computer Algebra System), declare the command coefficient and polynomial to find, press enter, the results will appear immediately on the screen (see Figs. 1 and 2).

3.1 *Geogebra* - A Hint Towards Factoring Polynomials

Faced with the difficult problem of factoring a polynomial, we often don't know where to start. *Geogebra* helps with this, because it gives the answer right away.

Example 1 (Math Olympiad, 8 grad, Gia Bình District, Bac Ninh Province, School year 2012–2013). Factorize the polynomial $x^4 + 2008x^2 + 2007x + 2008$.

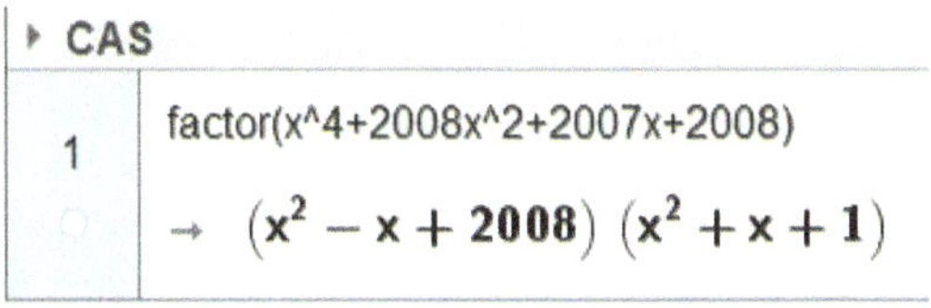

Fig. 2. Factorize $x^4 + 2008x^2 + 2007x + 2008$.

Thank to *GeoGebra*, we know the polynomial $x^4 + 2008x^2 + 2007x + 2008$ has one factor $x^2 + x + 1$, which suggests we solve by adding and subtracting, the group of terms is as follows:

$$\begin{aligned} x^4 + 2008x^2 + 2007x + 2008 &= x^4 + x^2 + 2007x^2 + 2007x + 2007 + 1 \\ &= x^4 + x^2 + 1 + 2007\left(x^2 + x + 1\right) \end{aligned}$$

We already know that there is a factor $x^2 + x + 1$, so we are sure $x^4 + x^2 + 1$ to be divisible by $x^2 + x + 1$.

Dividing $x^4 + x^2 + 1$ (by hand or by machine, see Fig. 4) for $x^2 + x + 1$ we have $(x^4 + x^2 + 1)/(x^2 + x + 1) = (x^2 - x + 1)$ (Fig. 3).

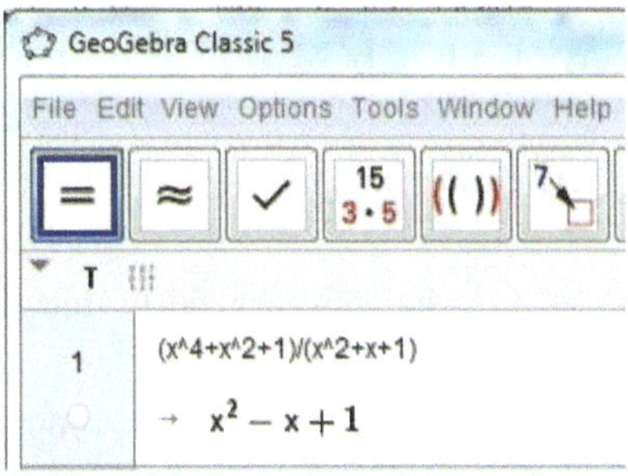

Fig. 3. Divide $x^4 + x^2 + 1$ for $x^2 + x + 1$.

So

$$\begin{aligned} x^4 + 2008x^2 + 2007x + 2008 &= x^4 + x^2 + 2007x^2 + 2007x + 2007 + 1 \\ = x^4 + x^2 + 1 + 2007\left(x^2 + x + 1\right) &= \left(x^2 + x + 1\right)\left(x^2 - x + 1\right) + 2007\left(x^2 + x + 1\right) \end{aligned}$$

$$= \left(x^2 + x + 1\right)\left(x^2 - x + 2008\right).$$

Geogebra can also help in factorizing the polynomial with multivariable polynomials.

Example 2 (Math Olympiad, 8 grad, Bac Giang Province, 2012–2013). Factorize

$$2a^3 + 7a^2b + 7ab^2 + 2b^3.$$

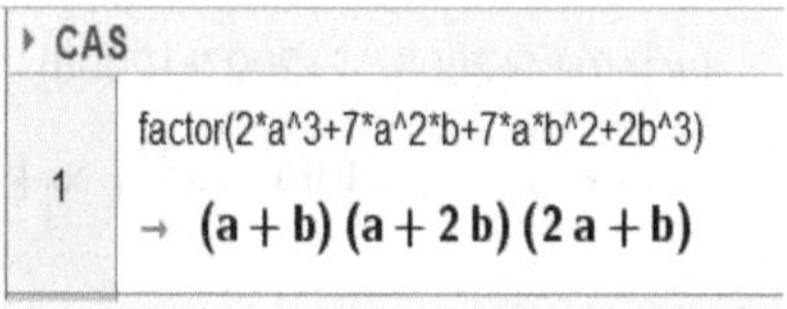

Fig. 4. Factorize $2a^3 + 7a^2b + 7ab^2 + 2b^3$.

Thus, the given polynomial has a factor $a + b$.
Leading to the solution (group of terms) as follow:

$$\begin{aligned} 2a^3 + 7a^2b + 7ab^2 + 2b^3 &= 2\left(a^3 + b^3\right) + 7ab(a + b) \\ &= 2(a + b)\left(a^2 - ab + b^2\right) + 7ab(a + b) \\ &= (a + b)\left(2a^2 + 2b^2 + 5ab\right) = (a + b)\left(2a^2 + 4ab + 2b^2 + ab\right) \\ &= (a + b)[2a(a + 2b) + b(a + 2b)] = (a + b)(a + 2b)(2a + b). \end{aligned}$$

Example 3 (6th Math Olympiad, Nord Vietnam, 1966–1967). Factorize $a^{16} + a^8b^8 + b^{16}$ (Fig. 5).

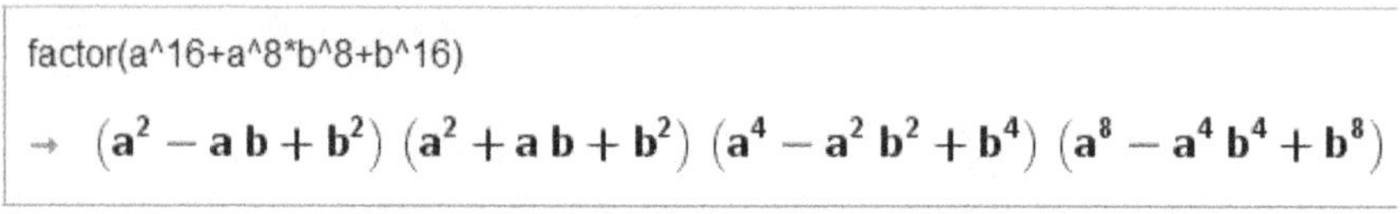

Fig. 5. Factorize $a^{16} + a^8b^8 + b^{16}$.

The above three examples are relatively easy, fairly students can do it. The example below is more difficult.

Example 4 (8th Math Olympiad, North Vietnam, 1968–1969). Factorize the polynomial (Fig. 6)

3.2 *Geogebra* - An Experimental Tool to Find Out the Rules of Factoring Polynomials

The exams on factoring polynomials are often quite diverse, for each problem, students often have to create a solution (by tinkering with groups of terms, setting the common factor, ...).

```
factor(b*c*(a+d)*(b-c)-a*c*(b+d)*(a-c)+a*b*(c+d)*(a-b))
```

$$\rightarrow \ (-a+c)\,(b-c)\,(b-a)\,d$$

$$bc(a+d)(b-c)-ac(b+d)(a-c)+ab(c+d)(a-b)$$

Fig. 6. Factorize the polynomial

Geogebra can be considered as an experimental tool to find out the rules of factoring for a class of seemingly very different problems.

Example 5 (Entrance Exam to Special class of Mathematics, Hanoi University of Education, 1992–1993, round 3). Factorize polynomial $x^{10}+x^5+1$ (Fig. 7).

```
factor(x^10+x^5+1)
```

$$\rightarrow \ \left(x^2+x+1\right)\left(x^8-x^7+x^5-x^4+x^3-x+1\right)$$

Fig. 7. Factorize $x^{10}+x^5+1$.

Example 6. Factorize x^7+x^5+1 (Fig. 8).

```
factor(x^7+x^5+1)
```

$$\rightarrow \ \left(x^2+x+1\right)\left(x^5-x^4+x^3-x+1\right)$$

Fig. 8. Factorize x^7+x^5+1.

Example 7. Factorize x^5+x^4+1 (Fig. 9).

```
ifactor(x^5+x^4+1)
```

$$\rightarrow \ \left(x^2+x+1\right)\left(x^3-x+1\right)$$

Fig. 9. Factorize x^5+x^4+1.

Solution. Adding/subtracting x^3+x^2+x to set the general factor

$$x^5+x^4+1=x^5+x^4+x^3-x^3-x^2-x+x^2+x+1$$
$$=x^3\left(x^2+x+1\right)-x\left(x^2+x+1\right)+\left(x^2+x+1\right)=\left(x^2+x+1\right)\left(x^3-x+1\right).$$

Example 8. Factorize $x^5 + x + 1$ (Fig. 10).

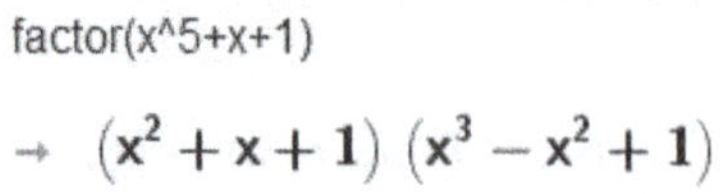

Fig. 10. Factorize $x^5 + x + 1$.

Solution. Adding/subtracting $x^4 + x^3 + x^2$ to set the general factor:

$$x^5 + x + 1 = x^5 + x + 1 + x^4 + x^3 + x^2 - x^4 - x^3 - x^2$$
$$= x^3\left(x^2 + x + 1\right) + \left(x^2 + x + 1\right) - x^2\left(x^2 + x + 1\right) = \left(x^2 + x + 1\right)\left(x^3 - x^2 + 1\right).$$

3.3 Factoring Polynomials and Some Related Problems

Factorizing polynomials is a great help in solving other problems: reducing expressions, solving equations, calculating the value of expressions. Here are some examples.

Example 9 (Entrance exam to 10 of Special School Nguyen Trai, Hai Duong Province, 2006–2007). Given two polynomials $f(x) = x^5 - 3x^4 + 7x^3 - 9x^2 + 8x - 2$ and $g(x) = x^2 - 2x + a$.

Find a to exist a polynomial $p(x)$ such that $f(x) = g(x) \cdot p(x)$ for all x.

Solution. Using *GeoGebra* (Fig. 11):

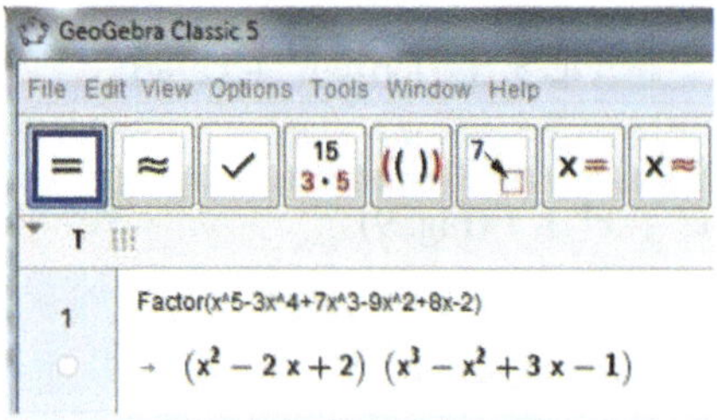

Fig. 11. Factorize $x^5 - 3x^4 + 7x^3 - 9x^2 + 8x - 2$.

Because for all a, Polynomial $x^3 - x^2 + 3x - 1$ cannot divisive by $x^2 - 2x + a$, then $a = 2$ is unique value the find.

Example 10 (Math Olympiad, 9 grad, Thai Nguyen Province, 2006–2007)

1) Factorize $f(x) = x^5 - 5x - 4$ into product of a quadratic polynomial and a cubic polynomial with integer coefficients
2) Simplify the expression $P = \frac{2}{\sqrt{4 - 3\sqrt[4]{5} + 2\sqrt{5} - \sqrt[4]{125}}}$.

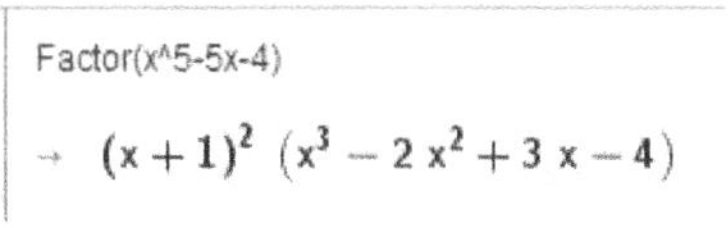

Fig. 12. Factorize $f(x) = x^5 - 5x - 4$.

Solution. Factorizing $f(x) = x^5 - 5x - 4$ on *GeoGebra*, we have (Fig. 12)

Then $f(x) = x^5 - 5x - 4 = (x + 1)^2 (x^3 - 2x^2 + 3x - 4)$. Put $x = \sqrt[4]{5}$, we have

$$P = \frac{2}{\sqrt{4 - 3\sqrt[4]{5} + 2\sqrt{5} - \sqrt[4]{125}}} = \frac{2}{\sqrt{4 - 3x + 2x^2 - x^3}} = \frac{2(x+1)}{\sqrt{4 + 5x - x^5}}$$

$$= \frac{2(\sqrt[4]{5}+1)}{\sqrt{4 + 5\sqrt[4]{5} - \sqrt[4]{5^5}}} = \sqrt[4]{5} + 1.$$

Example 11 (Math contest for girls, *Vietnam Children's Fun math Journal*, No. 134, April 2014). Solve the equation $(x + 1)^6 + (x + 2)^4 = x^2 + 3x + 3$.

Solution. Factorizing the polynomial $(x + 1)^6 + (x + 2)^4 - (x^2 + 3x + 3)$, we have (Fig. 13)

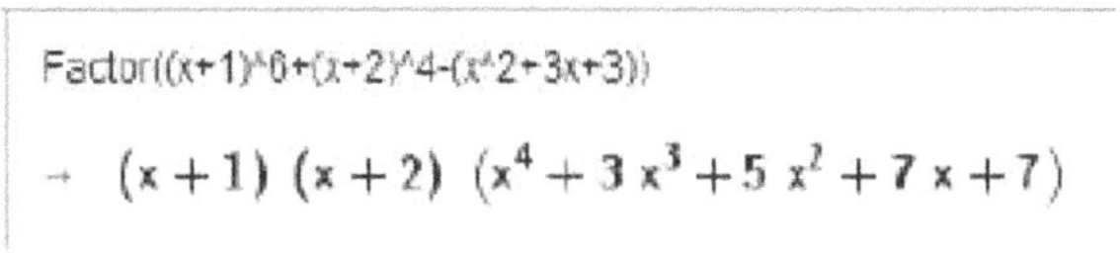

Fig. 13. Factorize $(x+1)^6 + (x+2)^4 - \left(x^2 + 3x + 3\right)$.

It can be prove that $x^4 + 3x^3 + 5x^2 + 7x + 7 > 0$ for all x. Then the given equation have to solutions $x_1 = -1$ and $x_2 = -2$.

3.4 *Geogebra* Does Not Replace Math Thinking and Techniques

Although *Geogebra* is universal, it is only a tool to help, not replace thinking and problem solving techniques.

Example 12 (Math Olympiad, Junior high school, Nord Vietnam, 1966–1967, Round 3). Factorize $a^8 + a^4 + 1$ (Fig. 14).

For the pupils of 8th graders, when they have not learned irrational numbers, the polynomial analysis is here to be achieved. However, since $a^4 - a^2 + 1$ is a quadratic polynomial, further analysis is possible. Try factorizing $a^4 - a^2 + 1$ on *Geogebra*, we get (Fig. 15)

Geogebra was powerless. Meanwhile, thanks to the method of uncertainty coefficients or setting auxiliary variables $x = a^2$, or adding/subtracting terms, we can easily factorize

$$a^4 - a^2 + 1 = a^4 + 2a^2 + 1 - 3a^2 = \left(a^2 + 1\right)^2 - (\sqrt{3}a)^2$$

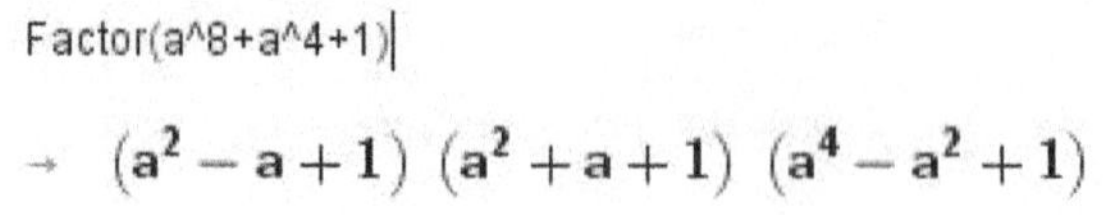

Fig. 14. Factorize $a^8 + a^4 + 1$.

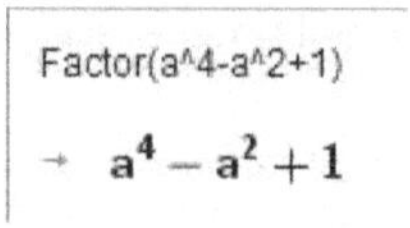

Fig. 15. Factorize $a^4 - a^2 + 1$.

$$= \left(a^2 - \sqrt{3}a + 1\right)\left(a^2 + \sqrt{3}a + 1\right).$$

Many of the examples above have the same phenomenon. For example, the polynomial $a^4 - a^2b^2 + b^4$ cannot be factored by the *Factor* command of *GeoGebra*, but can be easily factored by the method of uncertainty coefficients or by addition/subtraction:

$$a^4 - a^2b^2 + b^4 = a^4 + 2a^2b^2 + b^4 - 3a^2b^2 = \left(a^2 + b^2\right)^2 - (\sqrt{3}ab)^2$$
$$= \left(a^2 - \sqrt{3}ab + b^2\right)\left(a^2 + \sqrt{3}ab + b^2\right).$$

3.5 Some Other Examples and Exercises

Exercise 13. Factoring polynomial $x^4 - 6x^3 + 12x^2 - 14x + 3$.

Exercise 14. Factoring polynomial $3x^6 - 4x^5 + 2x^4 - 8x^3 - 4x + 3 + 2x^2$.

Exercise 15. Factorize polynomial $x^2(y - z) + y^2(z - x) + z^2(x - y)$.

Exercise 16. Factorize polynomial $P := (x - y)^3 + (y - z)^3 + (z - x)^3$.

Exercise 17 (Belarussian Math Olympiad, II round, 1958). Factorize the polynomial

$$\left[4abcd + \left(a^2 + b^2\right)\left(c^2 + d^2\right)\right]^2 - 4\left[cd\left(a^2 + b^2\right) + ab\left(c^2 + d^2\right)\right]^2.$$

Exercise 18 (Belorutssian Math Olympiad, 8 grad, I Round, 1952)

$$(x + y + z)^3 - x^3 - y^3 - z^3.$$

Exercise 19 (Belorutssian Math Olympiad, 8 grad, I Round, 1957)

$$\left(x^2 + y^2\right)^3 + \left(z^2 - x^2\right)^3 - \left(y^2 + z^2\right)^3.$$

Exercise 20 (Entrance Exam to 10 grad, Le Hong Phong special School, Ho Chi Minh City, 1988)

$$x^3 + y^3 + z^3 - 3xyz.$$

Exercise 21 (Math Olympiad, Ninh Thuan Province, 2014–2015; Lai Chau Province, 2016–2017)

$$xy(x + y) - yz(y + z) + xz(x - z).$$

Exercise 22 (Math Olympiad, District Tien Hai, Thai Binh Province, 2014–2015)

$$(x + 2)(x + 3)(x + 4)(x + 5) - 24.$$

4 Conclusion

Factoring a polynomial requires a lot of creative math skills. Mastering these skills takes time and practice. Therefore, the average students often have difficulty.

Geogebra is a pretty universal software that helps effectively in teaching and learning, including Factoring Polynomials.

The Examples and Exercises in the paper shows that, with just one command of *Geogebra*'s Factor (or factor), it is possible to factorize polynomials quite easily. First, students can use *Geogebra* to factor polynomials. From there, it is possible to give solutions, even different from the traditional ones.

Using math software can be considered as an experimental tool for finding and/or verifying mathematical conclusions or hypothesis (see, for example, [2, 3]).

It is also possible to consider using the tools of computer science as a new teaching method. Geogebra can also be used to improve the quality of lectures and create interest in learning mathematics.

Other fields (Factoring natural number into prime factors, see, for example, [3], drawing and calculating cross-sections [4], teaching and learning integral [5], geometry [3, 7, 10]).

Hopefully this article will provide suggestions for teachers and students interested in *GeoGebra* and using *Geogebra* in teaching and learning mathematics.

References

1. Ha, B.V.: Geogebra quick start guide, School@net. (in Vietnamese)
2. Hanh, N.T.H., Khanh, Đ.A., Mo, B.T.H., Phuong, T.D.: Use GeoGebra to factor the number of the form 100...001 into prime factors. Mathematics and Youth Magazine, No. 517, pp. 15–21, July 2020. (in Vietnamese)
3. Hanh, N.T.H., Phuong, T.D., Tam, P.T., Thuy, N.T.B., Le Thuy, T., Vu, N.H.: Using GeoGebra software for in geometry hypothesis testing. Mathematics and Youth Magazine, No. 529, pp. 10–15, July 2021. (in Vietnamese)

4. Hanh, N.T.H., Phuong, T.D., Thuy, N.T.B., Le Thuy, T.: Using GeoGebra software in teaching and learning the space geometry. In: Education of Things: Digital Pedagogy, pp. 73–102 (2021)
5. Hanh, N.T.H., Phuong, T.D., Thuy, N.T.B., Le Thuy, T., Vu, N.H.: Use GeoGebra in teaching definite integral. In: Proceedings of 2nd International Conference on Innovative Computing and Cutting-Edge Technologies (ICCT). Springer Series "Learning and Analytics in Intelligent Systems", pp. 327–335 (2021)
6. Hohenwarter, J., Hohenwarter, M.: Introduction to Geogebra. www.geogebra.org
7. Van Hoang, P., Phuong, T.D., Thuy, N.T.B., Le Thuy, T., Vu, N.H.: Modeling of teaching-learning process of geometrical LOCI in the plane with GeoGebra. In: Proceedings of 2nd International Conference on Mathematical Modeling and Computational Science, ICMMCS 2021, pp. 449–460. Springer, Cham (2022)
8. Trung, M. (ed), Thành, L.K., Thảo, T., Thuần, N.T., Vân, L.T.T.: Capacity Development for Middle School, Informatics. Vietnam Education Publisher, Hanoi (2018). Books 1, 2, 3 (in Vietnamese)
9. The official manual of GeoGebra. https://research.shu.ac.uk/geogebra
10. Nguyen, H.V., et al.: GeoGebra-assisted teaching of rotation in geometric problem solving. In: Proceedings of 3nd International Conference on Mathematical Modeling and Computational Science, ICMMCS 2022, pp. 451–460. Springer, Cham (2023)

Cubic Roots Cordial Labeling of Sunlet and Friendship Graphs

D. Ramani Bai[1(✉)] and J. Senbagamalar[2]

[1] Chevalier T. Thomas Elizabeth College for Women, Chennai, India
ramanibai1987@gmail.com

[2] Vel Tech Rangarajan Dr. Sagunthala R & D Institute of Science and Technology, Chennai, India

Abstract. A graph label involves allocating integers to the edges, vertices, or both, subject to certain restrictions. Assigning a value of 0 or 1 to a graph's vertices (or edges) in particular conditions is known as "cordial labeling". The absolute disparity between the total number of vertices marked with 0 and the total number of vertices marked with 1 must be either 0 or 1. A graph F with n vertices is said to have cubic roots cordial labeling if each vertex is marked either 1, δ or δ^2. The absolute disparity between the total number of vertices marked with 1 and the total number of vertices marked with δ must be either 0 or 1. The absolute disparity between the total number of vertices marked with δ and the total number of vertices marked with δ^2 must be either 0 or 1. The induced edge is from the set either 1, δ or δ^2. Also, the absolute disparity between the number of edges marked with 1 and the number of edges marked with δ must be either 0 or 1. The absolute disparity between the number of edges marked with δ and the number of edges marked with δ^2 must be either 0 or 1. In this paper, we prove the Sunlet graphs and Friendship graphs admit a cubic root cordial labeling.

Keywords: Cordial labeling · Cubic roots cordial labeling · Sunlet graphs · Friendship graphs and Network

1 Introduction

Graph theory is an important area of mathematics with a wide range of real-world applications. Let $F = (\eta, \xi)$ be the simple and undirected graph, then η and ξ stand for the vertices and the edges of the graph F. Graph labeling is one of the most actively studied topics in graph theory. The majority of graph labeling techniques stem from one Rosa [1] established in 1967. Cahit [2] first proposed the idea of cordial graph labeling in 1987, and S.W. Golomb established the term "graph numbering". According to Gallian [3], several scholars have looked into the cordialness of various graphs. Sundaram et al. [4] established the idea of prime cordial labeling, and in the same study, they looked into several prime cordial labeling findings. The pair difference cordial labeling technique was established by Ponraj et al. [5] and is used to study the pair difference cordial labeling behavior of graph unions such as path, cycle, star, and bistar graphs. The amicable

S. Pal and Á. Rocha (Eds.): ICMMCS 2025, LNNS 1398, pp. 107–115, 2025.
https://doi.org/10.1007/978-3-031-90998-6_11

behavior of the union of graphs as a k-product, as demonstrated by Daisey, K.J. et al. [6]. Cordial-related labeling was then defined and examined in [7–14]. In noisy communication channels and DNA code word design challenges, cordial labeling is beneficial. The cubic roots cordial labeling was first presented by Ashraf Elrokh et al. [15] they investigated the cubic roots cordial labeling in various graphs like Path, Cycle, Wheel, and union of two paths. In particular, this paper analyzed the cubic roots cordial labeling of the Sunlet and Friendship graphs.

2 Preliminaries

2.1 Definition

A graph $F = (\eta, \xi)$ is a Cordial labeling, each vertex(edge) labeled either 0 or 1 under certain function and satisfy the conditions $\mid \eta(0) - \eta(1) \mid \leq 1$ and $\mid \xi(0) - \xi(1) \mid \leq 1$, where $\eta(0)$ is equal to the total number of vertices of F assigned 0. $\eta(1)$ is equal to the total number of vertices of F assigned 1, $\xi(0)$ is equal to the number of edges of F assigned 0, $\xi(1)$ is equal to the number of edges of F assigned 1.

2.2 Definition

A graph $F = (\eta, \xi)$ have the vertex labeling of $f : \eta \to \{1, \delta, \delta^2\}$ and the induced edge labeling from the set $\{1, \delta, \delta^2\}$ is defined in Table 1.

Table 1. Conditions for Edge labeling

η	1	δ	δ^2
1	1	δ	δ^2
δ	δ	δ^2	1
δ^2	δ^2	1	δ

is called a cubic roots labeling if $\mid \eta\ (k) - \eta\ (w)\mid \leq 1$ & $\mid \xi\ (k) - \xi\ (w)\mid \leq 1$, $k \neq w$ and $k, w \in \{1, \delta, \delta^2\}$ where $\eta\ (k)$ and $\xi\ (k)$ are the number of vertices and edges assigned by $k \in \{1, \delta, \delta^2\}$. If a graph F allows for cubic roots cordial, then it is called cubic roots cordial labeling.

3 Main Results

3.1 Theorem

The cubic roots cordial labeling is acceptable in the Sunlet graph S_β, $\beta \geq 3$.

Proof: Consider, $\eta = \{\eta_1, \eta_2, \eta_3, \ldots\ldots \eta_{2\beta}\}$ and $|\eta| = 2\beta$ *ve* rtices obtained by attaching β pendent edges to β cycle graph C_β. The induced edge labeling from the set $\{1, \delta, \delta^2\}$ is defined in Table 2.

Table 2. Conditions for Edge labeling for sunlet graph

η	1	δ	δ^2
1	1	δ	δ^2
δ	δ	δ^2	1
δ^2	δ^2	1	δ

Also, the graph must be satisfying the conditions $|\ \eta\ (1) - \eta\ (\delta)| \le 1, |\ \eta\ (1) - \eta\ (\delta^2)| \le 1, |\ \eta\ (\delta^2) - \eta\ (\delta)| \le 1, |\ \xi\ (1) - \xi\ (\delta)| \le 1, |\ \xi\ (1) - \xi\ (\delta^2)| \le 1$ and $|\ \xi\ (\delta) - \xi\ (\delta^2)| \le 1$ where $\eta\ (1), \eta\ (\delta)$ and $\eta\ (\delta^2)$ are called the total number of vertices assigned with $1, \delta$ and δ^2 respectively.

Similarly, $\xi\ (1), \xi\ (\delta)$ & $\xi\ (\delta^2)$ are called the total number of edges assigned with $1, \delta$ & δ^2 respectively. Consider the following three cases.

Case (i)

$\beta \equiv 0(mod 3), \beta = 3\ \gamma, \gamma = 1, 2, 3, 4 \ldots\ldots.$

In this case consider, $\eta(1) = \eta(\delta) = \eta\ (\delta^2) = 2\gamma$ and the induced edges are

$\xi\ (1) = \xi\ (\ \delta) = \xi\ (\delta^2) = 2\gamma$

Also, it is satisfying the conditions $|\ \eta\ (k) - \eta\ (w)| \le 1$ & $|\ \xi\ (k) - \xi\ (w)| \le 1$, $k \ne w$ and $k, w \in \{1, \delta, \delta^2\}$.

Case (ii)

$\beta \equiv 1(mod 3), \beta = 3\ \gamma{+}1, \gamma = 1, 2, 3, 4 \ldots\ldots.$

In this case consider, $\eta 1) = \eta(\delta) = 2\gamma + 1$ and $\eta\ (\delta^2){=}2\gamma$.

Then the induced edges are $\xi\ (1) = \xi\ (\ \delta) = 2\gamma + 1, \xi\ (\delta^2) = 2\gamma$

and it is satisfying the condition $|\ \eta\ (k) - \eta\ (w)| \le 1$ & $|\ \xi\ (k) - \xi\ (w)| \le 1, k \ne w$ and $k, w \in \{1, \delta, \delta^2\}$

Case (iii)

$\beta \equiv 2(mod 3), \beta = 3\ \gamma{+}2, \gamma = 1{,}2, 3{,}4 \ldots\ldots.$

In this case consider, $\eta(1) = \eta\ (\delta^2){=}\ 2\gamma + 1, \eta(\delta) = 2\gamma + 2$

Then the induced edges are $\xi\ (1) = \xi\ (\delta^2) = 2\gamma + 1, \xi\ (\ \delta){=}2\gamma + 2$

It is satisfied the condition $|\ \eta\ (k) - \eta\ (w)| \le 1$ & $|\ \xi\ (k) - \xi\ (w)| \le 1, k \ne w$ and k, $w \in \{1, \delta, \delta^2\}$

Thus, we have seen in each case, the Sunlet graph $S_\beta, \beta \ge 3$ satisfies the conditions $|\ \eta\ (k) - \eta\ (w)| \le 1$ & $|\ \xi\ (k) - \xi\ (w)| \le 1, k \ne w$ and $k, w \in \{1, \delta, \delta^2\}$ Hence the Sunlet graph $S_\beta, \beta \ge 3$ is acceptable the cordial labeling of cubic roots. In general, the condition for vertex labelling and edge labeling of sunlet graph are as shown in Table 3.

Table 3. Conditions for vertex labeling and edge labeling of sunlet graph

β	$\eta(1)$	$\eta(\delta)$	$\eta\,(\delta^2)$	$\lvert\,\eta\,(k) - \eta\,(l)\rvert \le 1$ $k \ne l$ and $k,\, l \in \{1, \delta, \delta^2\}$
$\beta \equiv 0(mod3),\, \beta = 3\,\gamma$	2γ	2γ	2γ	(0, 0, 0)
$\beta \equiv 1(mod3),\, \beta = 3\,\gamma + 1$	$2\gamma + 1$	$2\gamma + 1$	2γ	(0, 1, 1)
$\beta \equiv 2(mod3),\, \beta = 3\,\gamma + 2$	$2\gamma + 1$	$2\gamma + 2$	$2\gamma + 1$	(1, 0, 1)
β	$\xi(1)$	$\xi(\delta)$	$\xi(\delta^2)$	$\lvert\,\xi\,(k) - \xi\,(l)\rvert \le 1,\, k \ne l$ and $k,\, l$ $\in \{1, \delta, \delta^2\}$
$\beta \equiv 0(mod3),\, \beta = 3\,\gamma$	2γ	2γ	2γ	(0, 0, 0)
$\beta \equiv 1(mod3),\, \beta = 3\,\gamma + 1$	$2\gamma + 1$	$2\gamma + 1$	2γ	(0, 1, 1)
$\beta \equiv 2(mod3),\, \beta = 3\,\gamma + 2$	$2\gamma + 1$	$2\gamma + 2$	$2\gamma + 1$	(1, 0, 1)

3.2 Example

Consider a Sunlet graph S_6 shown in Fig. 1, is acceptable the cordial labeling of cubic roots.

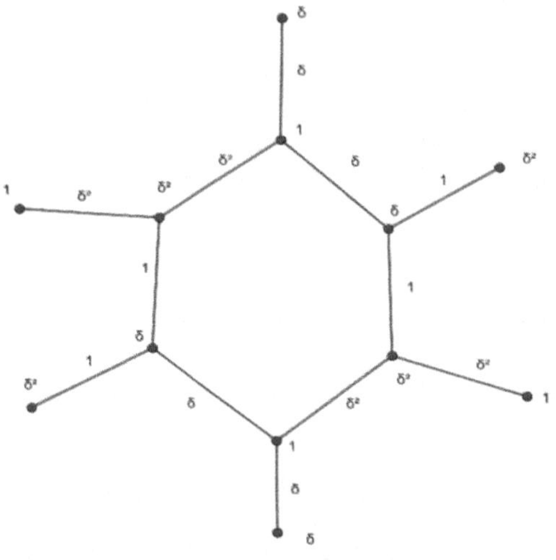

Fig. 1. The Sunlet graph S_6

Six pendent edges are added to cycle graph C6 to create the sunlet graph.
Here $|\eta| = 12$,
$\beta \equiv 0(mod3),\, \beta = 6,$

$\eta(1) = 4$
$\eta(\delta) = 4$
$\eta(\delta^2) = 4$

$\eta(1) = \eta(\delta) = \eta(\delta^2) = 4$

So that it satisfying the conditions $|\, \eta\ (k) - \eta\ (w)| \leq 1, k \neq w$ and $k,\ w \in \{1, \delta, \delta^2\}$. Also, the induced edge labeling,

$\xi\ (1) = 4$
$\xi\ (\delta) = 4$
$\xi\ (\delta^2) = 4$
$\xi\ (1) = \xi\ (\delta) = \xi\ (\delta^2) = 4$

So, it satisfying the conditions

$|\, \xi\ (k) - \xi\ (w)| \leq 1, k \neq w$ and $k,\ w \in \{1, \delta, \delta^2\}$

Hence, the Sunlet graph S_6 admits a cubic root cordial labeling

3.3 Example

Consider a sunlet graph S_8, as shown in Fig. 2, is acceptable the cordial labeling of cubic roots.

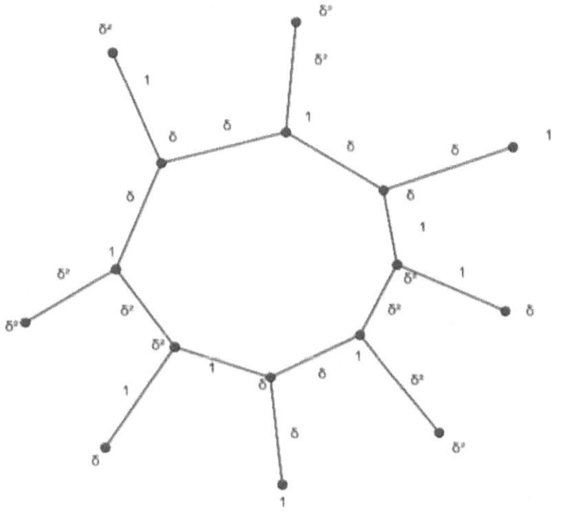

Fig. 2. The Sunlet graph S_8

Here, $\beta = 8,\ \beta \equiv 2(mod3)$,
$\eta(1) = 5,\ \eta\ (\ \delta) = 5,\ \eta\ (\delta^2)= 6$

$\therefore |\, \eta\ (1) - \eta\ (\delta)| = 0, |\, \eta\ (\delta^2) - \eta\ (1)| = 1, |\, \eta\ (\delta^2) - \eta\ (\delta)| = 1.$

$\xi\ (1) = 5,\ \xi\ (\ \delta) = 6,\ \xi\ (\delta^2) = 5$

$\therefore |\, \xi\ (1) - \xi\ (\delta)| = 1, |\, \xi\ (1) - \xi\ (\delta^2)| = 0, |\, \xi\ (\delta) - \xi\ (\delta^2)| = 1,$

It is satisfied the condition $|\, \eta\ (k) - \eta\ (w)| \leq 1$ & $|\, \xi\ (k) - \xi\ (w)| \leq 1, k \neq w$ and $k,\ w \in \{1, \delta, \delta^2\}$.
Hence, the Sunlet graph S_8 is acceptable a cubic root cordial labeling.

3.4 Theorem

The cubic roots cordial labeling is acceptable in the friendship graph $F_\beta, \beta \geq 3$.

Proof: Consider, $F = F_\beta, \beta \geq 3$ and $|\eta| = 2\beta + 1$, $|\xi| = 3\beta$

Let the vertex labeling of $f : \eta \rightarrow \{1, \delta, \delta^2\}$ with induced edge labeling from the set $\{1, \delta, \delta^2\}$ as shown in Table 4.

Table 4. Conditions for edge labeling of friendship graph

η	1	δ	δ^2
1	1	δ	δ^2
δ	δ	δ^2	1
δ^2	δ^2	1	δ

Also, the graph must be satisfying the conditions $|\eta(1) - \eta(\delta)| \leq 1$, $|\eta(1) - \eta(\delta^2)| \leq 1$, $|\eta(\delta^2) - \eta(\delta)| \leq 1$, $|\xi(1) - \xi(\delta)| \leq 1$, $|\xi(1) - \xi(\delta^2)| \leq 1$ and $|\xi(\delta) - \xi(\delta^2)| \leq 1$ where $\eta(1)$, $\eta(\delta)$ and $\eta(\delta^2)$ are stand for how many vertices are assigned by $1, \delta$ & δ^2 respectively.

Similarly, $\xi(1)$, $\xi(\delta)$ & $\xi(\delta^2)$ are stand for how many edges are assigned by $1, \delta$ & δ^2 respectively.

Consider the following three cases;

Case (i)

$\beta \equiv 0(mod 3)$, $\beta = 3\gamma$, $\gamma = 1, 2, 3, 4 \ldots\ldots.$

In this case consider, $\eta(1) = 2\gamma$, $\eta(\delta) = 2\gamma + 1$, $\eta(\delta^2) = 2\gamma$ and the induced edges are $\xi(1) = \xi(\delta) = \xi(\delta^2) = 2\gamma + 1$

Also, it is satisfying the conditions $|\eta(k) - \eta(w)| \leq 1$ & $|\xi(k) - \xi(w)| \leq 1, k \neq w$ and $k, w \in \{1, \delta, \delta^2\}$.

Case (ii)

$\beta \equiv 1(mod 3)$, $\beta = 3\gamma + 1$, $\gamma = 1, 2, 3, 4 \ldots\ldots.$

In this case consider, $\eta(1) = \eta(\delta) = \eta(\delta^2) = 2\gamma + 1$.

Then the induced edges are $\xi(1) = \xi(\delta) = \xi(\delta^2) = 3\gamma + 1$.

and it is satisfying the condition $|\eta(k) - \eta(w)| \leq 1$ & $|\xi(k) - \xi(w)| \leq 1, k \neq w$ and $k, w \in \{1, \delta, \delta^2\}$

Case (iii)

$\beta \equiv 2(mod 3)$, $\beta = 3\gamma + 2$, $\gamma = 1, 2, 3, 4 \ldots\ldots.$

In this case consider, $\eta(1) = \eta(\delta^2) = 2\gamma + 2$, $\eta(\delta) = 2\gamma + 1$

Then the induced edges are $\xi(1) = \xi(\delta) = \xi(\delta^2) = 3\gamma + 2$

It is satisfied the condition $|\eta(k) - \eta(w)| \leq 1$ & $|\xi(k) - \xi(w)| \leq 1, k \neq w$ and $k, w \in \{1, \delta, \delta^2\}$

Thus, we have seen in each case, the friendship graph $F_\beta, \beta \geq 3$ satisfies the $|\eta(k) - \eta(w)| \leq 1$ & $|\xi(k) - \xi(w)| \leq 1, k \neq w$ and $k, w \in \{1, \delta, \delta^2\}$. In general, the conditions for vertex labeling and edge labeling of friendship graph are as shown in Table 5.

Hence the cubic roots cordial labeling is acceptable in the friendship graph $F_\beta, \beta \geq 3$.

Table 5. Conditions for vertex and edge labeling for friendship graph

β	$\eta(1)$	$\eta(\delta)$	$\eta\ (\delta^2)$	$\mid \eta\ (k) - \eta\ (l) \mid \leq 1$ $k \neq l$ and $k,\ l \in \{1, \delta, \delta^2\}$
$\beta \equiv 0(mod3)$, $\beta = 3\ \gamma$	2γ	$2\gamma + 1$	2γ	(1, 0, 1)
$\beta \equiv 1(mod3), \beta = 3\ \gamma + 1$	$2\gamma + 1$	$2\gamma + 1$	$2\gamma + 1$	(0, 0, 0)
$\beta \equiv 2(mod3), \beta = 3\ \gamma + 2$	$2\gamma + 2$	$2\gamma + 1$	$2\gamma + 2$	(1, 0, 1)
β	$\xi(1)$	$\xi(\delta)$	$\xi(\delta^2)$	$\mid \xi\ (k) - \xi\ (l) \mid \leq 1, k \neq l$ and $k,\ l \in \{1, \delta, \delta^2\}$
$\beta \equiv 0(mod3)$, $\beta = 3\ \gamma$	$2\gamma + 1$	$2\gamma + 1$	$2\gamma + 1$	(0, 0, 0)
$\beta \equiv 1(mod3), \beta = 3\ \gamma + 1$	$3\gamma + 1$	$3\gamma + 1$	$3\gamma + 1$	(0, 0, 0)
$\beta \equiv 2(mod3), \beta = 3\ \gamma + 2$	$3\gamma + 2$	$3\gamma + 2$	$3\gamma + 2$	(0, 0, 0)

3.5 Example

Consider a friendship graph F_5 in Fig. 3, admits a cubic roots cordial labeling.

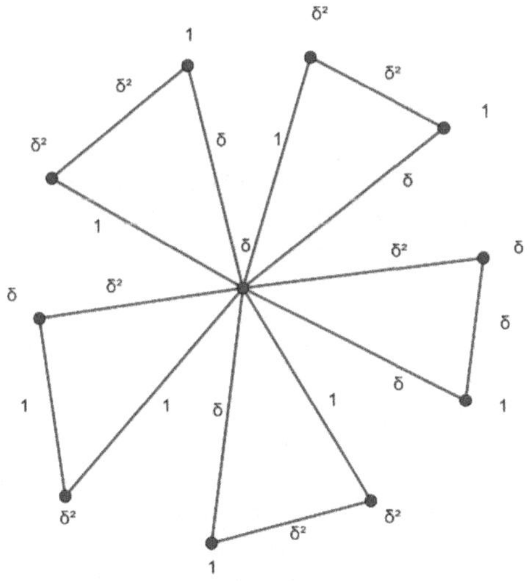

Fig. 3. The Friendship graph F_5

Here, $\beta = 5, \beta \equiv 2(mod3)$

$\eta(1) = 4$
$\eta(\delta) = 3$
$\eta\ (\delta^2) = 4$

$\therefore \eta(1) = \eta(\delta^2) = 4, \eta(\delta) = 3$
So that it satisfying the conditions $|\eta(k) - \eta(w)| \leq 1, k \neq w$ and $k, w \in \{1, \delta, \delta^2\}$.
Also, the induced edge labeling,

$\xi(1) = 5$
$\xi(\delta) = 5$
$\xi(\delta^2) = 5$

$\therefore \xi(1) = \xi(\delta) = \xi(\delta^2) = 5$
So, it satisfies the conditions
$|\xi(k) - \xi(w)| \leq 1, k \neq w$ and $k, w \in \{1, \delta, \delta^2\}$.
Hence, the cubic roots cordial labeling is acceptable in the Friendship graph F_5.

4 Conclusion

Cordial labeling has applications across various fields, including graph theory research, communication networks, error-correcting codes, optimization, cryptography, education, Coding theory, and the study of X-rays. In a variety of mathematical models, the labeled graphs perform admirably. Examples include cyber security and radar codes. Cordial labeling has been proven to be advantageous in noisy communication channels and DNA code word design difficulties. We have shown that cubic roots cordial labelling is admissible for both the Sunlet graph $S_\beta, \beta \geq 3$ and the Friendship graph $F_\beta, \beta \geq 3$. In future work, we investigate the operations, properties, and characteristics of some class of graphs using cubic roots cordial labeling.

References

1. Rosa, A.: On certain valuations of the vertices of a graph. In: Theory of Graphs, International Symposium, Rome, Gordon and Breach, N. Y. and Dunod Paris, pp. 349–355 (1967)
2. Cahit, I.: Cordial graphs: a weaker version of graceful and harmonious graphs. Ars combin. **23**, 201–207 (1987)
3. Gallian, J.A.: A dynamic survey of graph labeling. Electron. J. Combin. **19** (2016)
4. Sundaram, M., Ponraj, R., Somasundram, S.: Prime cordial labeling of graphs. J. Indian Acad. Math. **27**(2), 373–390 (2005)
5. Ponraj, R., Gayathri, A., Somasundaram, S.: Pair difference cordial labeling of graphs. J. Math. Comp. Sci. **11**(3), 2551–2567 (2021)
6. Daisey, K.J., Sabibha, R.S., Jeyanthi, P., Youssef, M.Z.: k-Product cordial behaviour Union of graphs. J. Indon. Math. Soc. 1–7 (2022). https://doi.org/10.22342/jims.28.1.1025.1-7
7. Peng, S.-L., Jhanjhi, N. Z., Pal, S., Amsaad, F. (eds.): Proceedings of 3rd International Conference on Mathematical Modeling and Computational Science. Advances in Intelligent Systems and Computing. Springer Nature Singapore (2023). https://doi.org/10.1007/978-981-99-3611-3
8. Peng, S.-L., Lin, C.-K., Pal, S.: (eds.): Proceedings of 2nd International Conference on Mathematical Modeling and Computational Science. Advances in Intelligent Systems and Computing. Springer Nature Singapore (2022). https://doi.org/10.1007/978-981-19-0182-9
9. Peng, S.-L., Hao, R.-X., Pal, S. (eds.): Proceedings of First International Conference on Mathematical Modeling and Computational Science. Advances in Intelligent Systems and Computing. Springer Singapore (2021). https://doi.org/10.1007/978-981-33-4389-4

10. AnanthaLakshmi, R., Jayalakshmi, K., Madhavi, T.: Prime labeling of jahangir graphs. Int. J. Eng. Technol. **7**(410), 389–392 (2018)
11. Francis, R.P.M., Geetha, V.V.: Signed product cordial labeling by double duplication of cycle related graphs. In: AIP Conference Proceedings, vol. 2852, p. 130016 (2023).https://doi.org/10.1063/5.0164861
12. Parthiban, A., Sharma, V.: A comprehensive survey on prime cordial and divisor cordial labeling of graphs. J. Phys.: Conf. Ser. **1531**(1), 012074 (2020)
13. Devakirubanithi, D., Jeba Jesintha, J.: Cordial labeling on few graphs of subdivided shell graphs. J. Algebraic Stat. **13**(2), 3455–3466 (2022)
14. Vaghela, U.V., Parmar, D.B.: Difference perfect square cordial labeling of some graphs. J. Xidian Univ. **14**(2), 87–98 (2020)
15. Elrokh, A., Ismail, R., El-hay, A.A.: On cubic roots cordial labelling for some graphs. Symmetry **15**(5), 990 (2023). https://doi.org/10.3390/sym15050990

Some Result of Approximate Serre Fibration

Ammar Qasim Naser(✉) and Daher Waly AL–Baydli

Department of Mathematics, College of Education for Pure Science, Wasit University, Kut, Iraq
{ammar.q,daheralbaydli}@uowasit.edu.iq

Abstract. In this paper we introduce, in studying the approximate space and approximate fibration space, we introduce a new concept, the approximate Serre fibration space (ASFS) by symbol. In addition, we advise new concepts of Mixed approximate Serre fibration space with some results and properties Among the results definition of Proximal Serre Fibrations (P.S.F), proximal homotopy covering properties (PHCP) and definition of the Mixed proximity homotopy covering property (M-PHCP).

Keywords: proximity space · proximal covering map · proximal Serre fibration · Mixed proximal Serre fibration

1 Introduction

The concept of topological perspective was initially introduced in 19th-century scholarly publications by Riemann and Poincare [4, 5]. The idea shows that topological space definitions arise either via Kuratoowski's closure operation [7] or by utilizing open sets. As with nearness theory, some other tactics and techniques look beneficial in various contexts and need further development given Kuratowski's closure. Riesz initially introduces the concept of near sets, Wallace revives it, and Efremovic axiomatically elaborates it [1]. This suggests that image analysis, human perception, and certain engineering difficulties are some near-set theoretical-related areas. Algebraic topology methods in proximity spaces, the primary topic of this article [8].

1.1 Research Objective

Introduce and study a new concept of the proximity Serre fibration space by symbol (ASFS) and mixed proximity Serre fibration (M-proximity Serre fibration) on CW-complex space.

2 Related Works

Definition 2.1 [6] **:** Let δ a binary relation that is defined on the set of A. It is referred to as proximity on A, provided that it satisfies the ensuing axioms:

1. $N\delta M$ implies $M\delta N$.

S. Pal and Á. Rocha (Eds.): ICMMCS 2025, LNNS 1398, pp. 116–123, 2025.
https://doi.org/10.1007/978-3-031-90998-6_12

2. $(N \cup M)\delta D$ if and only if $N\delta D \vee M\delta D$.
3. $N\delta M$ implies $N \neq \varnothing \wedge M \neq \varnothing$.
4. $(N \cap M) \neq \varnothing$ implies $N\delta M$.
5. $N\delta M \Rightarrow \exists D \subset A : N\underline{\delta}D \wedge (A - D)\underline{\delta}M$

$\bar{\delta}$ is a nearness relation and $N \ \delta \ M$ means that N is near M and $N\delta M$ is read N is far from M we refer to the pair (A, δ) as a proximity space (in short, we call A is P-spa ce), and Proximity is symbolized by the symbol (p).

Definition 2.6 [10] **:** When a mapping $q : (A, \delta_1) \rightarrow (B, \delta_2)$ is proximally continuous or pc- map if $N\delta_1 M$ then $q(N)\delta_2 q(M)$ for each $N, M \subseteq X$.

Definition 2.8 [3] **:** Given two proximally continuous maps $q_1, q_2 : A \rightarrow A^*$ if there is a proximity between continuous maps $F : A \times I \rightarrow A^*$ and the properties $F(a, 0) = q_1(a)$, $F(a, 1) = q_2(a)$, Thus, q_1 and q_2 are referred to as proximally homotopy maps.

Definition 2.9 [9] **:** A Hausdorff space A makes up a pair (A, ω) and a cell- composition w of A is referred to as a CW complex if the following are satisfies:

- For every m-cell, $e \in \omega$ a map is present $\varphi_e : D^m \rightarrow A$ restricted to a Homeomorphism $\varphi_{e|int(D^m)} : int(D^m) \rightarrow e$ and taking k^{m-1} into A^{m-1} It is referred to as Characteristic Maps.
- The closure $\underline{e}$ intersects For any cell $e \in \omega$ only a finite number of other cells in ω, is referred to as Closure Finiteness.
- A subset $Z \subseteq A$ is closed if $Z \cap \underline{e}$ is closed in A for every $e \in \omega$ It is referred to as Weak Topology.

The restrictions $\varphi_{e|\partial D_m}$ are referred to as attaching maps.

Notice that we can recover A (up to homeomorphism) from an awareness of A^0 and the attached maps. The following describes the recovery:

1. If we begin with a discrete collection A^0 whose points are thought of as 0-cells.
2. From A^{m-1} by attaching m-cells e_r^m by maps $\varphi_r : K^{m-1} \rightarrow A^{m-1}$, we get from the m-skeleton A^m. In the disjoint union, the quotient space $A^{m-1}\coprod_r D_r^m ofX^{m-1}$ a collection of m-disks D_r^m under the identification $A \backsim \varphi_r(X) forA \in \partial D_r^m isA^m$. Thus as a set $A^m = A^{m-1}\coprod_r e_r^m$ where each e_r^m is an open m-cell.
3. Setting $A = A^m$ for some$n < \infty$, setting $A = \bigcup_m A^m$ In the later case A The weak topology is assigned: A set $Z \subset A$ is closed (or open) iff $Z \cap A^m$ is closed (or open) in A^m for every m So we have,

$$A^0 \subset A^1 \subset A^2 \subset \cdots \subset A^m \subset \ldots$$

Provided that an integer exists m in a way that $A^m = A$ is then referred to as finite-dimensional.

Definition 2.10 [2] **:** Let $\{ E_i, f_i, A, \alpha\}$ be a mixed fibre space, where $i = 1,2$. A and B constitute a CW-complex space and a map $h_t : B \rightarrow A$, A continuous $k_1 : B \rightarrow E_1$, $k_2 : B \rightarrow E_2$ such that

1. $f_1 \circ k_1 = h_t$
2. $f_2 \circ k_2 = h_t$

Where $K_i = \{k_1, k_2\}$ is referred to as a mixed-covering or (M-covering of h_t).

3 Proximal Serre Fibrations (P.S.F)

It is possible to think of some topological issues as lifting or extension issues. From a homotopy theoretic perspective, they are dealt with by Serre fibrations, respectively. Important techniques based on homotopy theory, such as obstruction theory, spectral sequences, and Postnikov systems, involve fibrations. Conversely, we expand the concept of proximal Serre fibration of p-spaces with advantageous features in proximity contexts, which was initially introduced, and proximal Serre fibration is symbolized by the symbol (p.S.f).

Definition 3.1: A proximity continuous map $p : (E, \delta_1) \to (B, \delta_1^*)$ is said to have the approximal homotopy covering property (PHCP) concerning a p-space(A^{**}, δ_3^{**}), for every proximity continuous map.

$k : (A^{**}, \delta_3^{**}) \to (E, \delta_1)$, and proximity homotopy$h_t : (A^{**}, \delta_3^{**}) \to \left(B, \delta_1^*\right)$, then there exists a proximity homotopy $h_t^* : (A^{**}, \delta_3^{**}) \to (E, \delta_1)$ for which $h_0^*(A^{**}) = k(A^{**})$ and $p \circ h_t^*(A^{**}) = h_t(A^{**})$ (Fig. 1).

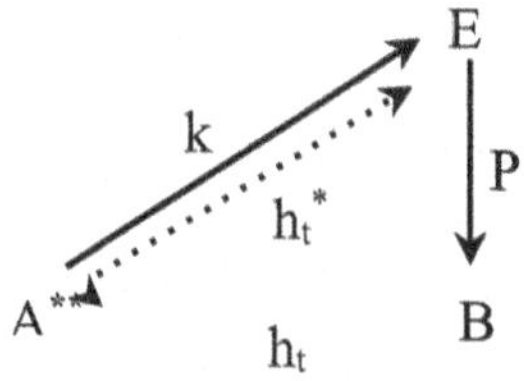

Fig. 1. Proximal Homotopy Covering Property (P.H.C.P)

Definition 3.2:

1 - The map p is a ***proximity Hurewicz fibration*** if it has the proximal homotopy covering property with respect to all spaces A.
2 - The map p is a ***proximity Serre fibration*** if it has the proximal homotopy covering property with respect to all CW-complexes a p-space A**.

Example 3.3: Let A^{**} be CW – complex and For any constant map, $c : (X, \delta) \to (\{x_0\}, \delta_1)$ of proximity space Considering the diagram,

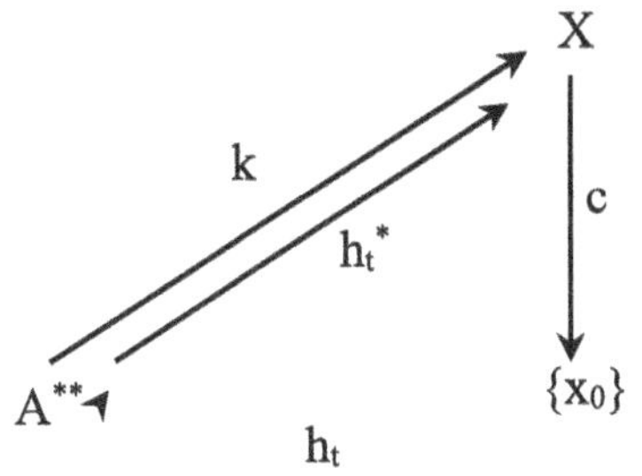

With the condition $c \circ h_t^* = h_t(A^{**}) = \{x_0\}$ then a constant is present proximal homotopies $h_t^* : (A^{**}, \delta_3^{**}) \to (X, \delta)$ define $h_t^*(A^{**}) = k(A^{**})$ then

$$c \circ h_t^*(A^{**}) = c(h_t^*(A^{**})) = \{x_0\} = h(A^{**})h_t^*(A^{**}) = h_0^*(A^{**}) = k(A^{**})$$

This indicates that **c** is is approximate Serre fibration (P,S.F).

Theorem 3.4: (i) Approximal Serre fibration (P. S. F) is also the composition of two approximal Serre fibrations (P. S. F).

(ii) Approximal Serre fibration (P.S.F) is also the cartesian product of two approximal Serre fibrations (P.S.F).

Proof (i): Consider $P_1 : (S_1, \delta_1) \to (Y_1, \delta_1^*)$, $P_2 : (Y_1, \delta_1^*) \to (Y_2, \delta_2^*)$ be any P-Serre fibration and proximity continuous maps.

$k_1 : (A^{**}, \delta_3^{**}) \to (S_1, \delta_1)$,$k_2 : (A^{**}, \delta_3^{**}) \to \left(Y_1, \delta_1^*\right)$ and p-homotopies $g_t : (A^{**}, \delta_3^{**}) \to \left(Y_1, \delta_1^*\right)$ and $G_t : (A^{**}, \delta_3^{**}) \to (Y_2, \delta_2^*)$ there exist two p-homotopies

$g_t^* : (A^{**}, \delta_3^{**}) \to (S_1, \delta_1)$ and $G_t^* : (A^{**}, \delta_3^{**}) \to \left(Y_1, \delta_1^*\right)$ then

$g_0^* = k_1, P_1 \circ g_t^* = g_t$ and $G_0^* = k_2, P_2 \circ G_t^* = G_t$

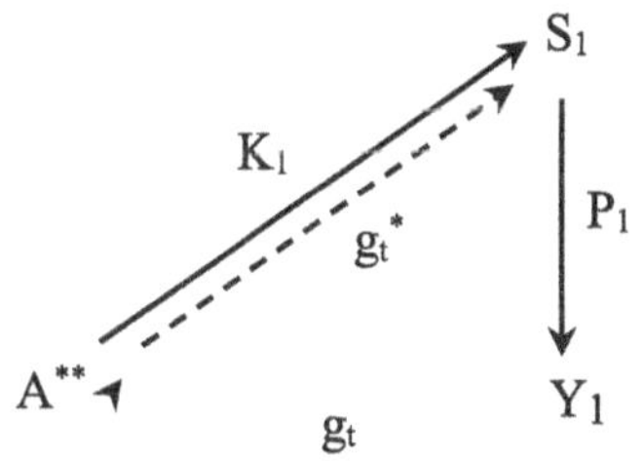

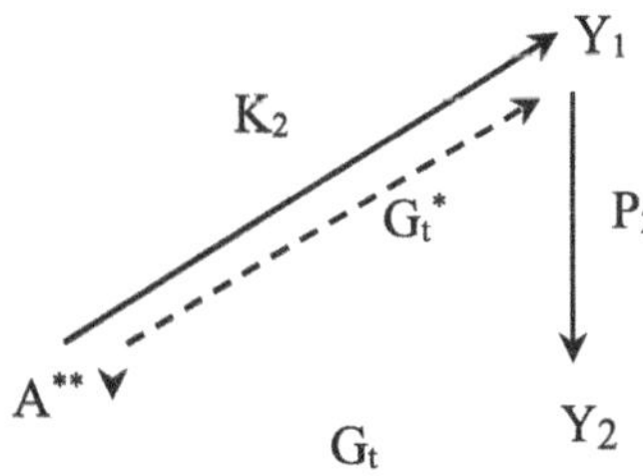

If we take $G_t^* = g_t$ and $k = k_1 = k_2$, then the commutative diagram that follows:

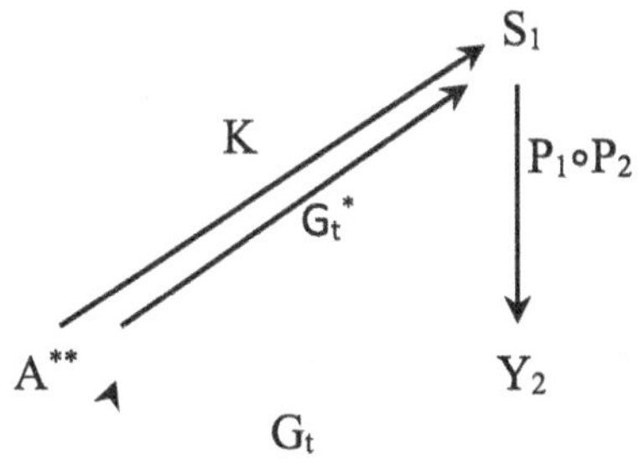

Thus we get $g_t^* = g_t$ and $(P_1 \circ P_2) \circ g_t^* = G_t$.
This indicates that the composition $P_1 \circ P_2$ is a (P.S.F).

Proof (ii): Consider $P_1 : (S_1, \delta_1) \to (Y_1, \delta_1^*)$ and $P_2 : (S_2, \delta_2) \to (Y_2, \delta_2^*)$ be any approximal Serre fibration and proximity continuous maps.

$k_1 : (A^{**}, \delta_3^{**}) \to (S_1, \delta_1)$ and $k_2 : (A^{**}, \delta_3^{**}) \to (S_2, \delta_2)$ and p-homotopies $G_1 : \left(A^{**}, \delta_3^{**}\right) \to \left(Y_1, \delta_1^*\right), G_2 : (A^{**}, \delta_3^{**}) \to (Y_2, \delta_2^*)$ There exist two proximal homotopies.

$G_1^* : (A^{**}, \delta_3^{**}) \to (S_1, \delta_1)$ and $G_2^* : (A^{**}, \delta_3^{**}) \to (S_2, \delta_2)$.
satisfying that:
$G_1^*(A^{**}) = k_1(A^{**})$ and $P_1 \circ G_1^* = G_1$ then $G_2^*(A^{**}) = k_2(A^{**})$ and $P_2 \circ G_2^* = G_2$.

Let the map $G_3^* = (G_1^*, G_2^*)$ Then G_3^* be a p -homotopy and This is the commutative diagram that we have

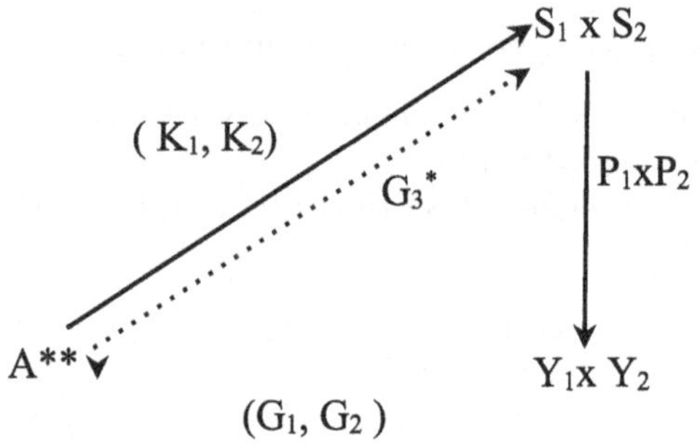

Thus we get,

$$G_3^* = (k_1, k_2)(p_1 \times p_2) \circ G_3^* = (G_1, G_2)$$

This indicates that $(P_1 \times P_2)$ is a (P.S.F).

4 Mixed Proximity Covering Homotopy

In this section, we introduce and study a new concept which is Mixed proximal Serre fibration (M-proximal Serre fibration). We start with the following definition:

Definition 4.1: (1) Let (E_1, δ_1), (E_2, δ_2) and (A^{**}, δ_3^{**}) be three proximity topological spaces, let $E_i = \{E_1, E_2\}, f_i = \{f_1, f_2\}$ where

$f_1 : (E_1, \delta_1) \to (A^{**}, \delta_3^{**})$ and $f_2 : (E_2, \delta_2) \to (A^{**}, \delta_3^{**})$ are two proximity continuous maps and $\alpha : (E_2, \delta_2) \to (E_1, \delta_1)$ such that $f_1 \circ \alpha = f_2$ then (E_i, f_i, X, α) is a Mixed proximal fibre space (M. P. F. S) (Fig. 2).

1) If $E_1 = E_2 = E_i$, $\alpha =$ identity, $f_1 = f_2 = f_i$ then (E_i, f_i, A^{**}) is the usual proximal fibre space
2) Let $\{E_i, f_i, X, \alpha\}$ be a M- proxmal fiber space, Let $x_0 \in X$ then $f = \{ f_i^{-1} (x_0)\}$ is the M- proximal fibre over x_0

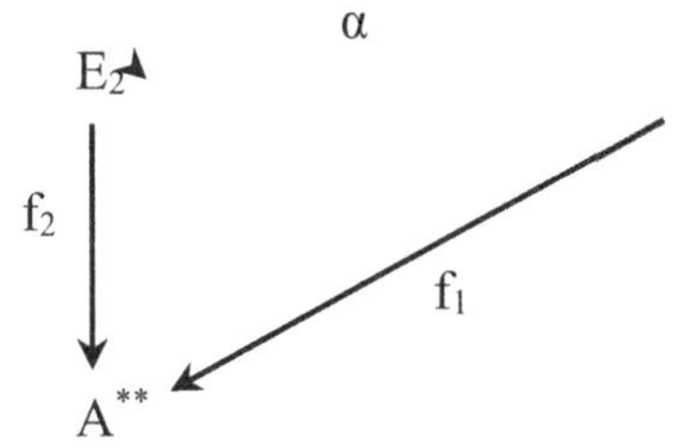

Fig. 2. Mixed Proxmal Fber Space

Definition 4.2: Let $\{(E_i, \delta_i), f_i, (A^{**}, \delta_3^{**}), \alpha\}$ be a M- proximal fibre space, where i = 1, 2. Let (A^{**}, δ_3^{**}), (B, δ_1^*) be a CW-complex proximity spaces and $h_t : (B, \delta_1^*) \to (A^{**}, \delta_3^{**})$ be aproximity maps. A proximity continuous map, $k_1 : (B, \delta_1^*) \to (E_1, \delta_1)$ and $k_2 : (B, \delta_1^*) \to (E_2, \delta_2)$ such that $f_1 \circ k_1 = h_t$ and $f_2 \circ k_2 = h_0$ where $k_i = \{ k_1, k_2\}$ is called a Mixed proximity covering (M.P.C of h_t) (Fig. 3).

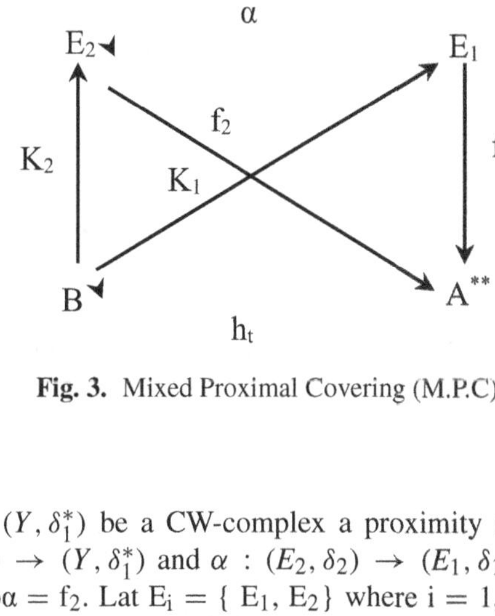

Fig. 3. Mixed Proximal Covering (M.P.C)

Definition 4.3: Let (Y, δ_1^*) be a CW-complex a proximity space, $f_1 : (E_1, \delta_1) \to (Y, \delta_1^*)$, $f_2 : (E_2, \delta_2) \to (Y, \delta_1^*)$ and $\alpha : (E_2, \delta_2) \to (E_1, \delta_1)$ are proximity map of a space such that $f_1 o\alpha = f_2$. Lat $E_i = \{ E_1, E_2\}$ where i = 1, 2. Let $f_i = \{f_1, f_2\}$ the quartic $\{E_i, f_i, Y, \alpha\}$ has the Mixed proximity homotopy covering property (M-PHCP) with respect to a CW-complex a p - space A^{**} if given a proximity map $k_1 : (A^{**}, \delta_3^{**}) \to (E_1, \delta_1)$, $k_2 : (A^{**}, \delta_3^{**}) \to (E_2, \delta_2)$ and proximity homotopy $h_t : (A^{**}, \delta_3^{**}) \to (Y, \delta_1^*)$ such that $f_i o k_i = h_0$ then exists a proximity homotopy $g_t : (A^{**}, \delta_3^{**}) \to (E_1, \delta_1)$ and $g_t^* : (A^{**}, \delta_3^{**}) \to (E_2, \delta_2)$ such that

(1)$g_0(A^{**}) = k_1(A^{**})$ and $f_1 \circ g_t = h_t$ (2)$g_t^*(A^{**}) = k_2(A^{**})$ and $\alpha \circ g_t^* = g_0$ (Fig. 4).

Mixed proximity fiber space is called mixed proximity Serre fibration, is it has the (M-PHCP) with respect to all CW-complex a proximity space Y.

Theorem 4.4: Every proximity Serre fibration is a Mixed proximity Serre fibration.

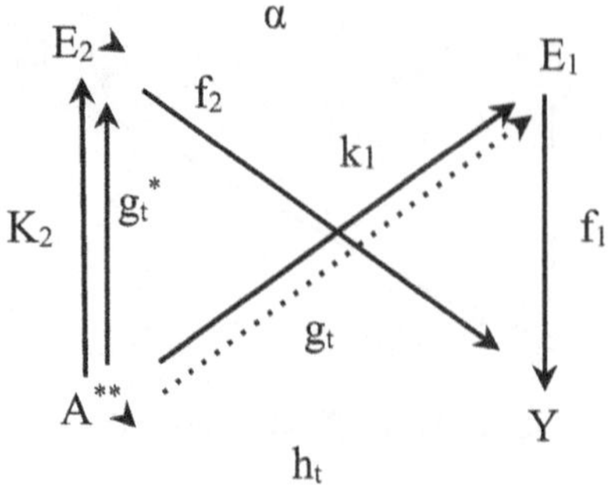

Fig. 4. Mixed Proximity Serre Fibration (M.P.S.F)

Proof: Let $\{(E_i, \delta_i), f_i, (Y, \delta_1{}^*), \alpha\}$ is proximity fibre space such that $E_i = E_1 = E_2$, $\alpha = I_d$ (identity), $f_i = f_1 = f_2$

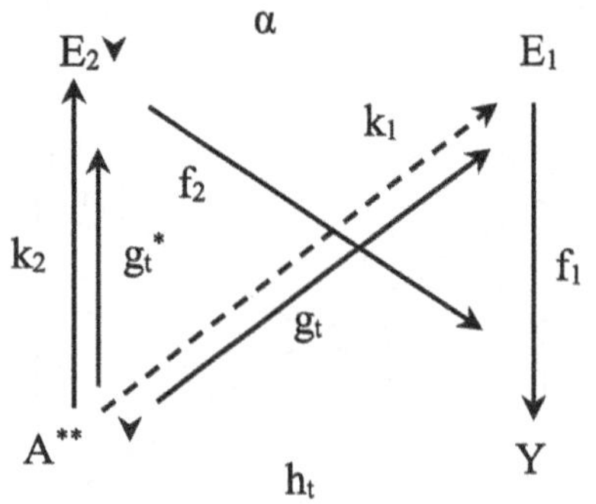

Let $g_t^* : (A^{**}, \delta_3^{**}) \to (E_2, \delta_2)$ be proximity map and proximity homotopy $h_t : (A^{**}, \delta_3^{**}) \to (Y, \delta_1^*)$ such that $f_i \circ k_i = h_0$ then there exist $g_t : (A^{**}, \delta_3^{**}) \to (E_1, \delta_1)$ such that

(1) $g_t = k_1$ and $f_1 \circ g_t = h_t$

(2) $g_t^* = k_2$ and $\alpha \circ g_t^* = g_0$

for all $a \in A^{**}$, $t \in I$ Then f has Mixed proximal covering property with respect to space CW-complex a proximity space Therefore f has Mixed proximity Serre fibration.

Proposition 4.5: Mixed proximity Serre fibration may not be a proximity Serre fibration As shown by the following example.

Example: Let $E_1 \neq E_2$ and $(E_1, \delta_1), (E_2, \delta_2), (B, \delta_1^*)$ are CW-complex proximity spaces, $f_1 : (E_1, \delta_1) \times (B, \delta_1^*) \to (B, \delta_1^*)$ be the projection defined as $f_1(e_1, b) = b, f_2 : (E_2, \delta_2) \times (B, \delta_1^*) \to (B, \delta_1^*)$ be the projection defined as $f_2(e_2, b) = b, \alpha : (E_2, \delta_2) \to (E_1, \delta_1)$ be any proximal map, $E_i = \{ E_1, E_2\}$, $f_i = \{ f_1, f_2\}$, $\delta_i = \{ \delta_1, \delta_2\}$

Let $p : (E_1, \delta_1) \times (B, \delta_1^*) \to (E_1, \delta_1)$ be the projection defined as $p(e_1, b) = e_1$ for all $(e_1, b) \in (E_1, \delta_1) \times (B, \delta_1^*)$ and $q : (E_2, \delta_2) \times (B, \delta_1^*) \to (E_2, \delta_2)$ be the projection defined as $q(e_2, b) = e_2$ for all$(e_2, b) \in (E_2, \delta_2) \times (B, \delta_1^*)$.

Let $k_1 : (A^{**}, \delta_3^{**}) \to (E_1, \delta_1) \times (B, \delta_1^*)$ and $k_2 : (A^{**}, \delta_3^{**}) \to (E_2, \delta_2) \times (B, \delta_1^*)$ be any proximal map and $h_t : (A^{**}, \delta_3^{**}) \to (B, \delta_1^*)$ be any proximal homotopy such that $f_I \circ k_I = h_0$.

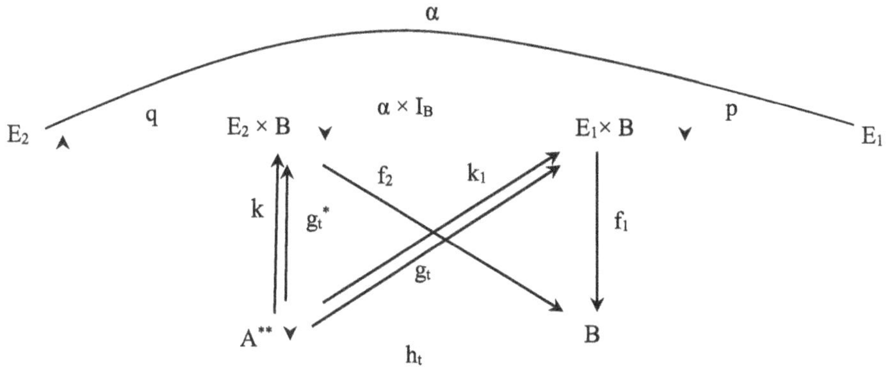

Define $g_t : (A^{**}, \delta_3^{**}) \to (E_1, \delta_1) \times (B, \delta_1^*)$
and $g_t^* : (A^{**}, \delta_3^{**}) \to (E_2, \delta_2) \times (B, \delta_1^*)$ as follows $g_t(x) = \{\alpha \circ q \circ g_t^*(X), ht\}$ the g_t satisfy (1) $g_t = k_1$ and $f_1 \circ g_t = h_t \forall t \in I$ (2) $g_t^* = k_2$ and $(\alpha \times I_B) \circ k = g_0$.

Therefore f_i: (E_i, δ_i) × (B, $\delta_1{}^*$) → (B, $\delta_1{}^*$) is M- proximity Serre fibration, which is not proximity Serre fibration.

5 Conclusion

In this paper we the results as the following:

The composition of two proximal Serre fibrations is also a proximal Serre fibration. Also, The cartesian product of two proximal Serre fibrations is a proximal Serre fibration. Additionally, Every proximity Serre fibration is a Mixed proximity Serre fibration. Mixed proximity Serre fibration may not be a proximity Serre fibration as shown by the following Example.

References

1. Efremovich, V.A.E.: The geometry of proximity. I. Mat. Sbornik **73**(1), 189–200 (1952)
2. AlBaydli, D.: The mixed slicing structure property. Results Nonlinear Anal. **7**(1), 1–7 (2024)
3. Peters, J.F., Vergili, T.: Good coverings of proximal Alexandrov spaces. Homotopic cycles in jordan curve theorem extension. arXiv preprint arXiv:2108.10113 (2021)
4. Mawhin, J.: Poincaré's early use of analysis situs in nonlinear differential equation: variations around the theme of Kronecker's integral. Philos. Sci. **4**(1), 103–143 (2000)
5. Riemann, B.: Grundlagen fur eine allgemeine Theorie der Functionen einer veränderlichen complexen Grösse. Adalbert Rente (1867)
6. Smirnov, J.M.: On proximity spaces. Math. Sb.(NS) 31(1952), no. 73, pp. 543–574 (1964). English translation: Amer. Math. Soc. Trans. Ser. **2**, 38.
7. Kuratowski, C.: Topologie. I, Panstwowe Wydawnictwo Naukowe, Warsaw, p. xiii+494 (1958)
8. Pei-Ren, F.: Proximity on function spaces. Tsukuba J. Math. **9**(2), 289–297 (1985)
9. Soren Hansen. Lecture note on algebraic topology (2013). http://www.math.ksu.edu/hansen/CWcomplex.pdf
10. Naimpally, S.A.: Proximity spaces. Cambridge Tracts Math. Math. Phys. (1970)
11. Is, M., Karaca, I.: Some properties of proximal homotopy theory. arXiv preprint arXiv:2306.07558 (2023)

A Proposed Item-Based Collaborative Filtering Model for e-Book Recommendation with a Weighted KNN

Abdullah Mohammed Saleh[1](✉) and Alaa Yaseen Taqa[2]

[1] College of Computer Sciences and Mathematics, University of Mosul, Nineveh, Iraq
abdullah.20csp39@student.uomosul.edu.iq
[2] College of Education for Pure Sciences, University of Mosul, Nineveh, Iraq
Alaa.taqa@uomosul.edu.iq

Abstract. Recommender systems are crucial in modern software, aiding users by suggesting items aligned with their preferences within a large amount of items. This paper proposes an item-based e-book recommendation model utilizing a Weighted KNN algorithm. The data set is subjected to cleaning, pre-processing and segmentation using cross-validation into 3 folds, and an item similarity matrix is created that is used in model evaluation and item recommendation. Weighted KNN is presented as an alternative to KNN-Means, KNN-Baseline, and KNN-ZScore for rating prediction. In this model, weights were adopted for the elements proportional to the number of evaluations for them, as elements that have a large number of evaluations have higher weights in addition to the similarity value used in the basic nearest neighbor algorithm. Testing on the book-crossing dataset reveals that the proposed model achieves the lowest error rates, particularly with the Pearson-Baseline methodology, showing a mean absolute error of 1.252 and a root mean square error of 1.650. With the cosine metric, the errors are 1.271 and 1.693 respectively. This study provides an overview of these methods and highlights the potential improvements of the Item-based Collaborative Filtering methodology through further research.

Keywords: E-Books · Collaborative Filtering · Book-Crossing · Recommender System · Item-Based · KNN-Weight

1 Introduction

E-commerce businesses utilize recommendation systems, which are technologies that offer suggestions on different products on a website to increase sales. From the standpoint of the user, it is advantageous because it lets users keep a personalized read-list [1, 2]. Machine learning and data mining techniques are applied to massive volumes of past data in order to forecast customer requirements. Collaboration filtering technology is an integral component of the recommendation system, which generates suggestions for the target user based on the preferences of users whose demands are comparable to those of the current users. A portion of the issues that the item-based collaborative filtering algorithm aims to address are data sparsity, inadequate scalability, and inaccurate recommendations [2, 3].

S. Pal and Á. Rocha (Eds.): ICMMCS 2025, LNNS 1398, pp. 124–136, 2025.
https://doi.org/10.1007/978-3-031-90998-6_13

Collaborative filtering technologies are frequently employed, with Amazon serving as a successful example. However, data scarcity and "cold start" issues have had a significant influence on the application. When providing recommendations to users, the algorithm does not take into account the current user's subjective assessment of the item and determines the similarity between products based only on user ratings. Furthermore, multiple people may rate the same thing with the same score from various perspectives. With the emergence of Web2.0, tag technology is frequently employed to help users better explore and retrieve data [4, 5].

1.1 Objective

The primary objective of this paper is to enhance the performance of recommendation systems, specifically for e-books, by proposing a weighted K-Nearest Neighbors (KNN) approach. The study aims to address common issues in recommendation systems such as data sparsity, inadequate scalability, and inaccurate recommendations by leveraging collaborative filtering techniques. It evaluates the effectiveness of different KNN models (KNNBaseline, KNNMeans, and KNNZ-Score) and compares them against the newly proposed weighted KNN model. The ultimate goal is to identify the most accurate and efficient model for predicting user ratings and improving the overall quality of recommendations.

2 Related Works

It is difficult for users to choose a suitable book from a big number of candidate books when the library provides a reasonably large number of books to choose from. In this case [6] Develop a personalised recommendation system utilising a hybrid recommendation algorithm for implementation in college libraries. An additional focus is the transfer of knowledge from source domains to target domains; for instance, incorporating insights from the book domain to enhance movie recommendations would constitute transfer learning methods. It encourages us to utilize supplementary and primary sources of information in order to make suggestions in the article. To achieve this [7] Present a Hybrid Transfer Meeting framework for unstructured text-based recommendation. [8] Offer a novel set based recommendation method that allows the user model to be clearly joined to users in regular language, allowing users to comprehend suggestions and improve them dynamically. While functioning similarly to existing collaborative filtering algorithms in a static situation, the method enables users to efficiently enhance recommendations. Collaborative filtering predicts a user's rating on an item by identifying similar users who rated the item or comparable items that the user rated, and then utilising the selected similar neighbours to "collaboratively filter" the Recommendation. The research indicates that item centrality negatively affects the accuracy of rating predictions; this must be taken into account in order to improve algorithm performance [9]. [10] presents a new transfer learning model for cross-domain recommendation, based on the existing Graph Collaborative Filtering Network (BiTGCF). Unlike previous methods of cross domain collaborative filtering, BiTGCF incorporates shared and domain specific features of users during knowledge transfer. Collaborative filtering and content-based filtering are among numerous recommendations made by various filtering algorithms. The paper presents a new transfer learning model for cross-domain recommendation, based on the existing Graph Collaborative Filtering Network (BiTGCF). Unlike previous methods of cross domain collaborative filtering, BiTGCF incorporates shared and domain specific features of users during knowledge transfer. Collaborative filtering and content-based filtering are among numerous recommendations made by various filtering algorithms. [11] Present a hybrid book recommendation system with recommendation anticipation. The task of top-N recommendation is a formidable challenge due to the need to effectively handle intricate and sparse user-item interactions in order to produce recommendations of superior quality.

– Collaborative Recommender System

A collaborative system generates recommendations based on other users' ratings and reviews [12].

- User-Based CF: When using collaborative filtering, people that share similar traits with the current user are taken into account for ratings and preferences. Afterwards, this information is used to suggest goods that similar customers have liked in the past [13].
- Item-based CF: These are product recommendations relying on collaborative filtering that factor in the individual's previous likes. These were suggestions which were given based on other customers' tastes and preferences who had also used these types of items [14].
- Data collection

It is important because the machine learning model uses it as input, and it predicts the output according to the information in the dataset. The dataset a BRMS uses will typically vary depending on the kind of system being used. This dataset is widely used in the collaborative filtering literature. It was obtained in 2004 via the Book-Crossing website in less than four weeks. The files are in the ".csv" format and are named Books.csv, Ratings.csv, and Users.csv [15]. The next is a dataset description:

- Book file (csv): includes the following attributes: Author of Book, Publisher Publication Year, ISBN, Picture-URL-L,Picture-URL-S, Book-Title, Picture-URL-M)
- Rating file (csv): Includes the characteristics (Rating of Book, ID of User, ISBN)
- Users (csv): Included are the attributes (User-ID, Age, Location) 1,149,780 ratings for 271,379 books have been submitted by a total of 278,858 users. A scale of 1 to 10 is used. The highest number is 10. Details about books can be added to the dataset by researchers. Because it offers a wealth of information that may be utilized to produce precise suggestions, BRMS is helpful for collaborative filtering when applied to the Book-Crossing dataset [16]. By looking at how others rate books, RMs can learn about other users' interests and give recommendations.

– Dataset Description

  A big part of the literature on collaborative filtering is the Book-Crossing dataset. It was found on the Book-Crossing website in 2004 in less than four weeks. It is in ".csv" format and is called Books.csv, description Ratings.csv, and Users.csv. It gives information about the books that can be used to find books that are similar and make personalized suggestions based on the user's tastes and how similar they are to other users. The dataset description is shown in Table 1.

Table 1. Book-Crossing Dataset Description [13, 14]

Features Name	Values
Rating values range	1–10
Demographic	Age and Location
Metadata	Publisher, authors, title, year, and cover image
Description	N / A
Number of Item	271, 379
Number of User	278, 858
Number of Rating	1, 149, 780
Rating/Item	4. 236
Rating/User	4. 123

– Similarity Techniques

The calculation of user similarities is a critical factor that has a substantial impact on the performance of CF. The consumers' historical tastes or favorites are therefore denoted as rating vectors. The module includes functions that calculate the degree of similarity between two individuals or items through the utilization of diverse metrics. These metrics comprise mean square dissimilarity, Pearson correlation and cosine similarity [17, 18].

1. Cosine similarity:

A specific user is represented by a unique rating vector as opposed to each item being represented by a vector of ratings from the collection of user ratings. The cosine measure shows how similar two separate vectors which represent different users (or items) [19]. There is a high correlation between two variables when the similarity value is close to one. A value close to 0 indicates that there is no association (independent variables). Something similar is usually used by the CF recommender system. The similarity of user u and v defined in Eq. (1) [17, 20].

$$COSSim(i,j) = \frac{\overrightarrow{R_u} \times \overrightarrow{R_v}}{\left\|\overrightarrow{R_u}\right\| \times \left\|\overrightarrow{R_u}\right\|} = \frac{\sum_{i \in I_u \cap I_v} R_{ui} \times R_{vi}}{\sqrt{\sum_{i \in I_u \cap I_v} R^2 u_i \times \sum_{i \in I_u \cap I_v} R^2 v_i}} \quad (1)$$

where Iu and Iv stand for the sets of items rated by users u and v, respectively, and IuIv stands for the set of items rated by both u and v. Item I received ratings from users V and U with the values Rvi and Rui [20].

2. Mean squared dissimilarity:

In some cases, the similarity metric can be erroneous. Unlike Jaccard, Mean Squared Dissimilarity (MSD) places more weight on absolute ratings. Another popular CF statistic is Jaccard, which counts the number of things assessed by two users instead of the ratings; the more co-rated items there are, the more similar they are. Equations (2) and (3) display the Jaccard and MSD formulae, respectively [20].

$$JaccardSim(u, v) = \frac{|I_u \cap I_v|}{|I_u \cup I_v|} \tag{2}$$

$$MSDSim(u, v) = 1 - \frac{\sum_{i \in I|_u \cap I_v|} (R_{ui} \times R_{vi})^2}{\sqrt{\sum_{i \in I_u \cap I_v} R_{ui}^2} \times \sqrt{\sum_{i \in I_u \cap I_v} R_{vi}^2}} \tag{3}$$

3. Pearson correlation similarity:

This metric was created by Karl Pearson to measure linear correlations, and it quickly became popular in the field of statistics. The PCC formula gives a number between −1 and 1, where 1 means a strong positive correlation, −1 means a strong negative correlation, and 0 means there is no connection. The similarity between two users, u and v, is found by Eq. (4) [21]:

$$C_{Sim(i,j)} = \frac{\sum_{i \in I_{uv}} \left(R_{u,i} - \underline{R_u}\right) \times \left(R_{v,i} - \underline{R_v}\right)}{\sqrt{\sum_{i \in I_{uv}} \left(R_{u,i} - \underline{R_u}\right)^2} \times \sqrt{\sum_{i \in I_{uv}} \left(R_{v,i} - \underline{R_v}\right)^2}} \tag{4}$$

The collection of objects that are regularly assessed by both u and v is indicated by the symbol Iuv. The means of user u and v's ratings for item I in Iuv are represented by the symbols (Ru) and (Rv). Rating of User for the same item I from users u and v are denoted by R(u,i) and R(v,i) [21].

4. Correlation similarity using Pearson's baseline:

This assessment determines the Pearson correlation coefficient between each pair of users by concentrating on baselines rather than means (or objects). When there are few ratings, the shrinkage parameter keeps overfitting from happening. The pearson-baseline correlation coefficient formula is displayed in Eq. (5) [17, 20]:

$$PCC\_BaselineSim(u, v) = \frac{\sum_{i \in I_{uv}} \left(R_{u,i} - BL_{u,i}\right) \times \left(R_{v,i} - BL_{v,i}\right)}{\sqrt{\sum_{i \in I_{uv}} \left(R_{u,i} - BL_{u,i}\right)^2} \times \sqrt{\sum_{i \in I_{uv}} \left(R_{v,i} - BL_{v,i}\right)^2}} \tag{5}$$

BL(u,i) is a estimate of baseline that accounts for item and user influences for an unidentified rating R(u,i) (6), as shown by:

$$\text{BLui} = \mu + \text{BLu} + \text{BLi} \tag{6}$$

The BLu and BLi parameters denote the deviations from the mean that are observed and item-specific, respectively.

- Evaluation Model

It is an essential part of machine learning and are used to evaluate learning algorithms. The measurements of error are much different in the recommendation because we are anticipating production. As is customary, the error metrics are generated by averaging the differences between the target variable values as predicted by the model and their actual values. Two often used metrics to assess the correctness of a proposed system are mean absolute error (MAE) and root mean square error (RMSE) [22]. Equation (7) for the mean deviation is as follows:

$$MAE = \frac{\sum_{i=1}^{n} |p_i - q_i|}{n} \tag{7}$$

What is the difference between the anticipated and realised user ratings? The closer the projected score of the recommendation system is to the actual score, the less the deviation from the mean value. Equation (8) calculates the root mean square error:

$$RMSE = \frac{\sum_{i=1}^{n} (p_i - q_i)^2}{n} \tag{8}$$

Which displays the following information: the user, the book, the testing data, the size of the test dataset, the user's actual rating on the book, and the user's expected score on the book [23].

3 Methodology

The plan has been to make a model that is based on the item for e-book recommendation as shown in Fig. 3. The data used was the book crossing dataset which went through pre-processing. Then, came the idea of Cross-validation in which it folds the data into three, one test and other two for item-item similarity matrix using cosine, MSD, Pearson and Pearson baseline metrics. Three nearest neighbour techniques were employed to predict ratings: KNNBaseline, KNNMeans, and KNNZ-Score. Worth to note all these three algorithms are direct derivatives of basic nearest neighbors approach. It uses average rating of each user to forecast ratings while the latter predicts ratings using normalization of every users' z-score. This shows that Eq. 9 is the core formula for KNN. [19].

$$\widehat{R_{ui}} = \frac{\sum_{v \in N_i^k(u)} R_{vi}}{k} \tag{9}$$

This study proposes the use of weighted KNN to predict ratings. Similar to the necessary user, the weights were determined for each item based on the distance with the other items as shown in Eq. 10. The basic equation mentioned above served as the basis for the rating prediction Eq. (11) (9). This is how the distance is calculated: Dist (u,v) = 1-Sim(u,v)

$$Weights(v) = \frac{1}{Dist(u, v)^2} \tag{10}$$

$$\widehat{R_{ui}} = \frac{\sum_{v \in N_i^k(u)} W_v.R_{vi}}{\sum_{v \in N_i^k(u)} W_v} \tag{11}$$

Active user is represented as u, similar user as v, the rating of user v for item i is Rvi, the set of similar users is represented as N, k stands for the number of users in N and Wv measures how important user v is to user u. Figure 1 shows that This matrix on similarity is updated based on customer behavior (interaction between a customer and an item) every time a recommendation process takes place so as to identify its closest neighbors.

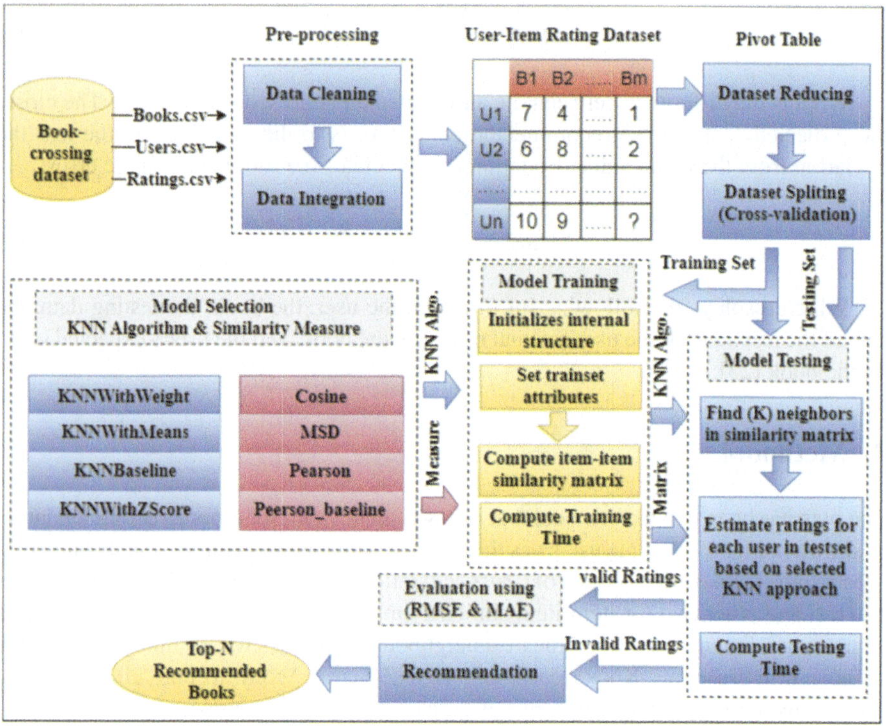

Fig. 1. The Proposed Item-based CF Model

However, the evaluation approach mainly relies on the information from two main sources: the user-user matrix for prediction and the matrix of user-item (actual or available ratings/testing set). The following algorithm outlines the main steps involved in implementing this collaborative filtering model based on items.

Proposed Algorithm:

```
1. Let k be the number of use  rs and n be the number of
books; read datasets (csv) files (Users info, Books info,
Ratings count) as Uk, Bn and Rs correspondingly.
2. Removal of null values, handling low rated records,
redundant characters during processing and replacing
invalid data are some of the pre-processing steps that
are often done to clean the data.
3. This is achieved by creating a user-book matrix or
UBknn using those three files mentioned above.
4. Separate the data set into three parts by cross
validation approach.
5. For each items bi in UBkn do
6. For each items bj in UB do
7. Compute CosSim(bi,bj) similarity of bi - bj
8. Compute MSDSim(ui,uj) similarity between bi - bj
9. Compute PearsonSim(ui,uj) similarity between bi - bj
10.Compute PearsonBLSim(ui,uj) similarity between bi - bj
11.End
12.Find the (k) Nearest Neighbours for item bi.
13.Use KNNwithMeans, KNNWithZscore ,KNNBaseline and
KNNwithweight to find Rating Prediction for itembi based
on ratings of similar items rated by user U .
14.Using MAE  and RMSE to evaluate the model based on
actual rating and predicted rating of item bi
15.The top-N books  with large prediction have been
recommended to item bi
```

4 Result Analysis and Discussion

An investigation into the effectiveness of a number of different similarity metrics and data normalization approaches for KNN recommendation algorithms was carried out in this paper. A summary of the results is provided in Table 2, which includes the Root Mean Squared Error (RMSE), Mean Absolute Error (MAE), fit time (Fig. 2), and test time (in Fig. 3) for each model setup.

Based on Cosine Similarity and MSD, KNN with Weight seems to be the best performing model overall, obtaining the lowest RMSE as illustrated in Fig. 4 and MAE (in Fig. 5) Nonetheless, Pearson baseline is a competitive substitute, particularly in cases where computational efficiency is crucial or where MAE is more significant. The KNN baseline model has the lowest overall performance but is the fastest to train and test. The impact of data normalization techniques is mixed. For Cosine Similarity and

Table 2. Accuracy results of the proposed approach with traditional models

Similarity Techniques	Evaluation Metric	KNN Baseline	KNN With Means	KNN With ZScore	KNN with Weight
Cosine	RMSE	1.788	1.827	1.830	1.693
	MAE	1.331	1.384	1.387	1.271
	Fit_time	4.633	6.815	6.134	5.973
	Test_time	3.015	3.190	3.244	3.406
MSD	RMSE	1.794	1.834	1.847	1.708
	MAE	1.332	1.387	1.398	1.280
	Fit_time	2.975	3.187	4.941	4.199
	Test_time	2.962	3.056	3.252	3.423
Pearson	RMSE	1.783	1.840	1.847	1.647
	MAE	1.367	1.412	1.418	1.254
	Fit_time	9.356	9.860	10.916	9.723
	Test_time	3.109	3.166	3.201	3.252
Pearson-Baseline	RMSE	1.787	1.842	1.843	1.650
	MAE	1.362	1.410	1.413	1.252
	Fit_time	5.507	5.682	6.054	5.834
	Test_time	2.936	2.918	2.863	3.009

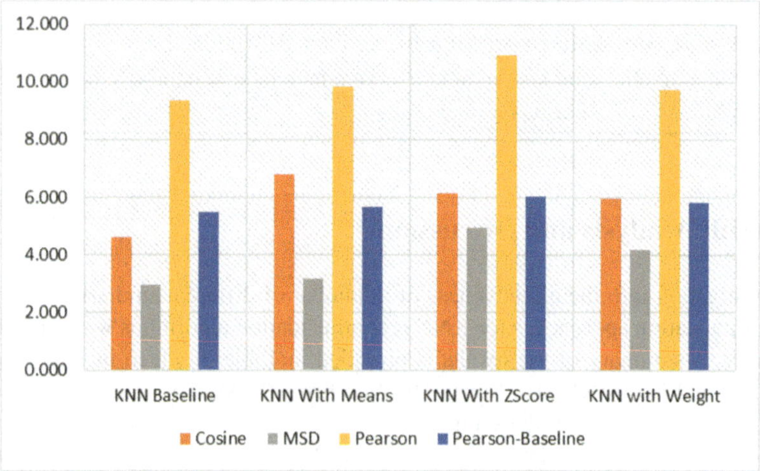

Fig. 2. The comparison of models based on fit-time

MSD, applying ZScore and Means did not significantly improve performance. In some cases, it even slightly degraded performance. However, for Pearson Correlation, Pearson baseline, which applies ZScore normalization, outperforms KNN with Weight, which does not apply any normalization. This suggests that Pearson Correlation may be more sensitive to the data distribution.

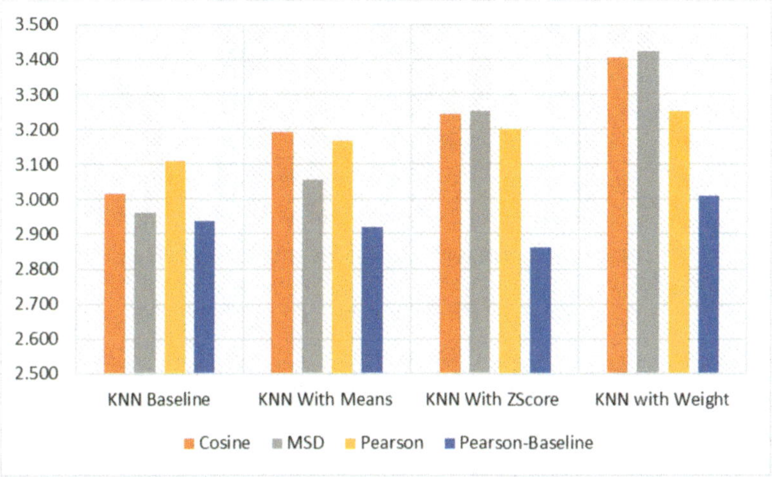

Fig. 3. The comparison of models based on test-time

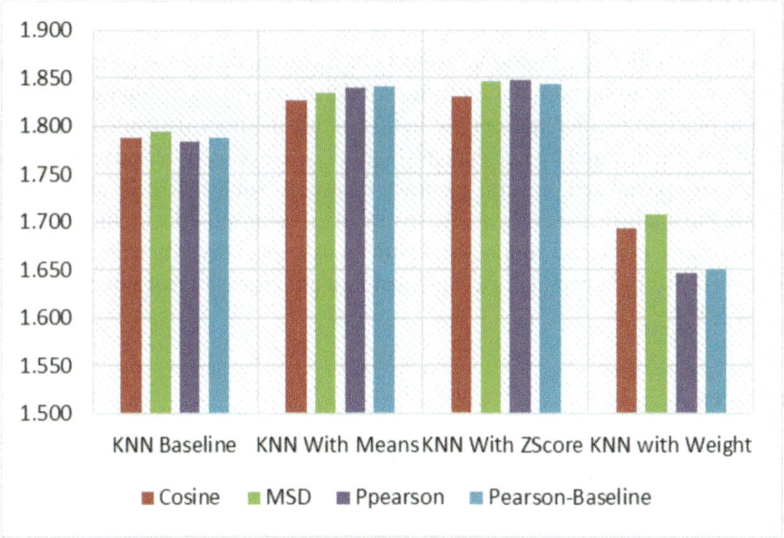

Fig. 4. Results evaluation of KNN models using RMSE

Based on Cosine Similarity and MSD, KNN with Weight seems to be the best performing model overall, obtaining the lowest RMSE in Fig. 3 and MAE. Nonetheless, Pearson baseline is a competitive substitute, particularly in cases where computational

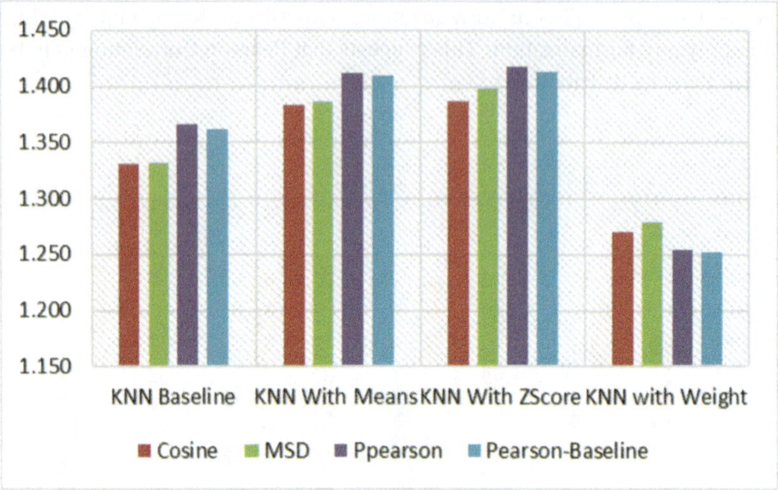

Fig. 5. Results evaluation of KNN models using MAE

efficiency is crucial or where MAE is more significant. The KNN baseline model has the lowest overall performance but is the fastest to train and test.

5 Conclusion

This research revealed the important aspects of recommendation systems, particularly for e-books where strong nomination form, similarity metric and evaluation metric need to be put in place to predict appropriate items. In this memory-dependent collaborative filtering algorithm inspired by K-Nearest Neighbor, some content-based technologies are no longer over-specialized. Several accuracy measures, such as RMSE and MAE, were used in order to analyze and compare these models. The proposed weighted KNN model shows the lowest error rate among other methods especially when Pearson-Baseline technique is utilized. Whereas for cosine technique its lowest error rate was 1.693 and MAE was 1.271.

References

1. Sreenivasa Raju, K.: Recommendation system in social media: a case study. J. Res. Lepidoptera, **51**(1), 574–579 (2020). https://doi.org/10.36872/lepi/v51i1/301051
2. Dr, V.D.: Review of recommendation system methodologies. Int. J. Psychosoc. Rehabil. **23**(1), 524–531 (2019). https://doi.org/10.37200/ijpr/v23i1/pr190495
3. Wang, Q.: Design and implementation of the web-based book recommendation system. Appl. Mech. Mater. **427–429**, 2143–2146 (2013). https://doi.org/10.4028/www.scientific.net/amm.427-429.2143
4. Sharma, P., Yadav, L.: Movie recommendation system using item based collaborative filtering. Int. J. Innovative Res. Comput. Sci. Technol. **8**(4) (2020). https://doi.org/10.21276/ijircst.2020.8.4.2

5. Gharat, J., Dixit, H., Dhurat, S., Shinde, S.: Enhanced book recommendation system using collaborative filtering. SSRN Electron. J. (2021).https://doi.org/10.2139/ssrn.3867376
6. Tian, Y., Zheng, B., Wang, Y., Zhang, Y., Wu, Q.: College library personalized recommendation system based on hybrid recommendation algorithm. Procedia CIRP **83**, 490–494 (2019). https://doi.org/10.1016/j.procir.2019.04.126
7. Hu, G., Zhang, Y., Yang, Q.: Transfer meets hybrid: a synthetic approach for cross-domain collaborative filtering with text. In: The World Wide Web Conference (2019)
8. Balog, K., Radlinski, F., Arakelyan, S.: Transparent, scrutable and explainable user models for personalized recommendation. In: Proceedings of the 42nd International ACM SIGIR Conference on Research and Development in Information Retrieval, Paris, France, pp. 265–274 (2019). https://doi.org/10.1145/3331184.3331211
9. Ai, J., Su, Z., Wang, K., Wu, C., Peng, D.: Decentralized collaborative filtering algorithms based on complex network modeling and degree centrality. IEEE Access **8**, 151242–151249 (2020). https://doi.org/10.1109/access.2020.3017701
10. Liu, M., Li, J., Li, G., Pan, P.: Cross domain recommendation via bi-directional transfer graph collaborative filtering networks. In: Proceedings of the 29th ACM International Conference on Information & Knowledge Management, Virtual Event, Ireland, pp. 885–894 (2020). https://doi.org/10.1145/3340531.3412012
11. Sharma, S., Vijay, R., Malhotra, M.: Automatic recommendation system based on hybrid filtering algorithm. Educ. Inf. Technol. **27**(2), 1523–1538 (2021). https://doi.org/10.1007/s10639-021-10643-8
12. Arezoumandan, M., Ghannadrad, A., Candela L., Castelli, D.: Recommender systems for science: a basic taxonomy. Zenodo (2022). https://doi.org/10.5281/zenodo.6006906
13. Hikmatyar and Ruuhwan, M.: Book recommendation system development using user-based collaborative filtering. J. Phys. Conf. Ser. **1477**(3), 032024 (2020). https://doi.org/10.1088/1742-6596/1477/3/032024
14. Tewari, S.: Generating items recommendations by fusing content and user-item based collaborative filtering. Procedia Comput. Sci. **167**, 1934–1940 (2020). https://doi.org/10.1016/j.procs.2020.03.215
15. Ziegler, C.-N., McNee, S.M., Konstan, J.A., Lausen, G.: Improving recommendation lists through topic diversification. In: Proceedings of the 14th International Conference on World Wide Web, Chiba, Japan, pp. 22–32 (2005).https://doi.org/10.1145/1060745.1060754
16. Pujahari, A., Sisodia, D.S.: Modeling side information in preference relation based restricted Boltzmann machine for recommender systems. Inf. Sci. **490**, 126–145 (2019). https://doi.org/10.1016/j.ins.2019.03.064
17. Suryakant, Mahara, T.: A new similarity measure based on mean measure of divergence for collaborative filtering in sparse environment. Procedia Comput. Sci. **89**, 450–456 (2016). https://doi.org/10.1016/j.procs.2016.06.099
18. Feng, J., Fengs, X., Zhang, N., Peng, J.: An improved collaborative filtering method based on similarity. PLoS ONE **13**(9), e0204003 (2018). https://doi.org/10.1371/journal.pone.0204003
19. Saeed, A.A.M., Taqa, A.Y.: A proposed approach for plagiarism detection in article documents. SinkrOn **7**(2), 568–578 (2022). https://doi.org/10.33395/sinkron.v7i2.11381
20. Fkih, F.: Similarity measures for collaborative filtering-based recommender systems: review and experimental comparison. J. King Saud Univ.-Comput. Inf. Sci. **34**(9), 7645–7669 (2022). https://doi.org/10.1016/j.jksuci.2021.09.014
21. Logesh, R., Subramaniyaswamy, V., Malathi, D., Sivaramakrishnan, N., Vijayakumar, V.: Enhancing recommendation stability of collaborative filtering recommender system through bio-inspired clustering ensemble method. Neural Comput. Appl. **32**(7), 2141–2164 (2018). https://doi.org/10.1007/s00521-018-3891-5

22. Alita, D., Putra, A.D., Darwis, D.: Analysis of classic assumption test and multiple linear regression coefficient test for employee structural office recommendation. IJCCS (Indonesian J. Comput. Cybern. Syst.) **15**(3), 295 (2021). https://doi.org/10.22146/ijccs.65586
23. Xia, H., Li, J.J., Liu, Y.: Collaborative filtering recommendation algorithm based on attention GRU and adversarial learning. IEEE Access **8**, 208149–208157 (2020). https://doi.org/10.1109/access.2020.3038770

Computing Sombor Indices on Special Networks

M. S. Ramani(✉) and J. Senbagamalar

Vel Tech Rangarajan Dr Sagunthala R&D Institute of Science and Technology, Chennai 600 062, India
ramanibalaji73@gmail.com

Abstract. A molecular network or structure can be linked to different physical properties, chemical reactivity, or biological activity using a topological descriptor, also known as an index, which is a numerical number that correlates to chemical composition. Because they help distinguish and comprehend physicochemical features, uncover structure-activity connections, and design pharmaceutical medications, topological descriptors are essential in the field of chemistry. A molecular structure can be represented as a network that can be analyzed using mathematical techniques called molecular descriptors. The degree-based index examines mathematical figures to gain insight into specific molecular physicochemical characteristics. This study focuses on the degree based topological indices, specifically the Sombor index of chemical networks. Numerous topological indices with diverse applications in chemistry have been reported in literature. For instance, vertex degrees in a graph are utilized in the calculation of the Sombor index. In this research we computed the Sombor indices of networks, such as flabellum graphs, rhenium trioxide lattices, silicate, chain silicate, and star silicate, as molecular descriptors to aid in comprehending the physicochemical properties of molecules.

Keywords: Sombor index · Flabellum graphs · Edge partition · Complex network

1 Introduction

A subfield of mathematical chemistry called "chemical graph theory" models chemical processes using graph theory. These mathematical structures are composed of nodes and edges that represent the chemical structures and their properties are studied. A graph $\mathcal{G}$ is defined as a nonempty two sets of vertices and edges. A set of lines that represent the covalent bonds between molecules and a collection of points that represent the atoms in the molecule make up a molecular graph., which in turn helps better visualize the structure of any molecule. Topological descriptors have contributed to the comparison of boiling points of different alkane isomers and have facilitated the discovery of numerous topological graph indices. Silicate minerals are rock-forming minerals that are composed of silicate groups. They comprise 90% of the Earth's crust and are the most important class of minerals. Silicate is extremely important in the cement, ceramic, and glass industries. A silicate network can be represented by SL_η, where η is the number of

S. Pal and Á. Rocha (Eds.): ICMMCS 2025, LNNS 1398, pp. 137–145, 2025.
https://doi.org/10.1007/978-3-031-90998-6_14

hexagons between the center and the boundary of the network. Figure 1 depicts a two-dimensional silicate network. Figure 2 represents a family of chain silicate networks, assigned by the symbol CS_η which are created by linearly arranging n tetrahedra. In [11], the unit cell of rhenium trioxide consisted of a cube with rhenium atoms at the corners and an oxygen atom on each of the 12 edges. The atoms touch the edges of the unit cell. Rhenium trioxide, or rhenium (VI) oxide, is an inorganic compound with the formula ReO3. It is a red solid with a metallic luster that resembles copper in appearance, and only the stable trioxide of Group 7 elements. From [11], Flabellum graphs are a specific type of graph obtained by duplicating a cycle graph *Cn* with a central vertex at least twice ($k \geq 2$) and adding additional edges then in j of these duplicates, connecting the central vertex to all non-adjacent vertices. Flabellum graphs are utilized in chemical graph theory to represent and analyze the structure and properties of chemical compounds, providing insights into molecular characteristics. This paper [11] study the first Zagreb index that serves as a crucial metric for assessing the significance and impact of individual vertices within the cyber flabellum graph, aiding in the identification of key players in cybercrime networks.

Motivated by previous studies [1, 3–6, 14], we investigated the sombor index of certain unique graphs. In [10], the authors calculated the Randić, Zagreb, and ABC indices of silicate, honeycomb, and hexagonal networks, and used these indices to predict the bioactivity of chemical compounds. From [7–9], Banhatti, F-sombor, Modified sombor, and Nirmala indices of various chemically interesting networks that focus on correlating topological indices with the chemical properties of molecules. In [12], the authors presented algorithms to find perfect packings of the Silicate Network SL(η) and Oxide Network OX(η) with constraints K1, K2, and C3. It also aims to determine the packing numbers of these networks, highlighting the significance of H-packing in scheduling theory and its applications in biotechnology and nanotechnology. The [13] various applications of silicon oxide (SiO2) in numerous technological advancements and the relevance of the Winer index in comprehending its molecular structure were examined in this study.

1.1 Objective

Aim of this paper is to compute the Sombor index for certain special type of graphs.

2 Definitions

Many graph invariants have been applied and are being utilised in pharmacology, chemistry and etc. The introduction of Zagreb indices dates back over thirty years. Let $\mathcal{g}$ be a graph, and its vertices are u and v; then the Zagreb indices of $\mathcal{g}$ are defined as follows:

$$M_1\mathcal{g} = \sum_{uv \in E(\mathcal{g})} d(u) + d(v)$$

$$M_2\mathcal{g} = \sum_{uv \in E(\mathcal{g})} d(u) + d(v)$$

In the late 1990's Estrada et al. [16] defined Atom–Bond Connectivity index in short ABC index; which is based on degrees of the vertices used to examine the stability of alkanes and etc., ABC index of any graph g is defined as

$$ABC(g) = \sum_{u \in V(g)} \sqrt{\frac{d_u + d_v - 2}{d_u d_v}}$$

In 2021, Gutman et al. introduced the Sombor index which is the most recent of the indices and is defined as:

$$SO(g) = \sum_{e=uv \in E(g)} \sqrt{d_u^2 + d_v^2}$$

3 Results on Silicate Networks

Theorem 1: The Sombor index of SL_η silicate network is

$$SO(SL_\eta) = 54\eta(2\eta - 1)\sqrt{2} + 18\eta(3\eta + 1)\sqrt{5}$$

Proof. From the molecular graph (Fig. 1) the vertex and edge sets of the silicate network SL_η has $|V(SL_\eta)| = 15\eta^2 + 3\eta$, $|E(SL_\eta)| = 36\eta^2$. In SL_η, depending on the degree of each edge's vertices, there are three different types of edges. The silicate network's [10] edge partition using the algebraic technique is presented in Table 1.

Table 1. SL_η Graph's edge partition

(d_u, d_v) where $uv \in E(SL_\eta)$	Total count of the edges
(3, 3)	6η
(3, 6)	$18\eta^2 + \eta$
(6, 6)	$18\eta^2 - 12\eta$

$$SO(SL_\eta) = \sum_{e=uv \in E(SL_\eta)} \sqrt{d_u^2 + d_v^2}$$

$$SO(SL_\eta) = 6\eta\sqrt{3^2 + 3^2} + (18\eta^2 + 6\eta)\sqrt{3^2 + 6^2} + (18\eta^2 - 12\eta)\sqrt{6^2 + 6^2} + 6$$
$$+(18\eta^2 + 6\eta)\sqrt{45} + (18\eta^2 - 12\eta)\sqrt{72}$$
$$SO(SL_\eta) = 54\eta(2\eta - 1)\sqrt{2} + 18\eta(3\eta + 1)\sqrt{5}$$

Sombor index of the silicate network of Fig. 2, $SO(SL_1) = 54\sqrt{2} + 72\sqrt{5}$.

Theorem 2: The Sombor index for chain silicate CS_η: $SO(CS_\eta) = 9\eta\sqrt{2} + (12\eta - 6)\sqrt{5}$.

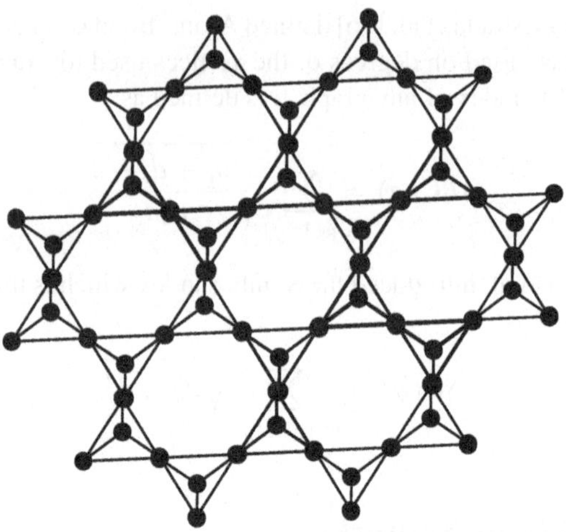

Fig. 1. The Silicate network of dimension two SL_2

Proof: The chain silicate network CS_η (Fig. 2) of the vertex set and edge set is $|V(CS_\eta)| = 3\eta + 1, |E(CS_\eta)| = 6\eta$. Based on the degree of each edge's vertices, CS_η has three different kinds of edges. Table 2 displays the silicate network's edge partitions that were created using the algebraic approach.

Table 2. CS_η Graph's edge partition

(d_u, d_v) where $uv \in E(CS_\eta)$	Total count of the edges
(3, 3)	$\eta + 4$
(3, 6)	$4\eta - 2$
(6, 6)	$\eta - 2$

$$SO(CS_\eta) = \sum_{e=uv\in E(CS_\eta)} \sqrt{d_u^2 + d_v^2}$$
$$SO(CS_\eta) = (\eta + 4)\sqrt{3^2 + 3^2} + (4\eta - 2)\sqrt{3^2 + 6^2} + (\eta - 2)\sqrt{6^2 + 6^2}$$
$$SO(CS_\eta) = (\eta + 4)3\sqrt{2} + (4\eta - 2)3\sqrt{5} + (\eta - 2)6\sqrt{2}$$
$$SO(CS_\eta) = 9\eta\sqrt{2} + (12\eta - 6)\sqrt{5}$$

Theorem 3: The Sombor index for star silicate SS*Ln*.

$$SO(SSL\eta) = 6\sqrt{2}(3 + 16\eta) + \left(\eta - \sqrt{2}\right)120 - 60 + \left(108\eta^2 - 180\eta + 108\right)\sqrt{5} + 48(\eta - 1)\sqrt{13} + \left(72\eta^2 - 144\eta + 96\right)\sqrt{6}$$

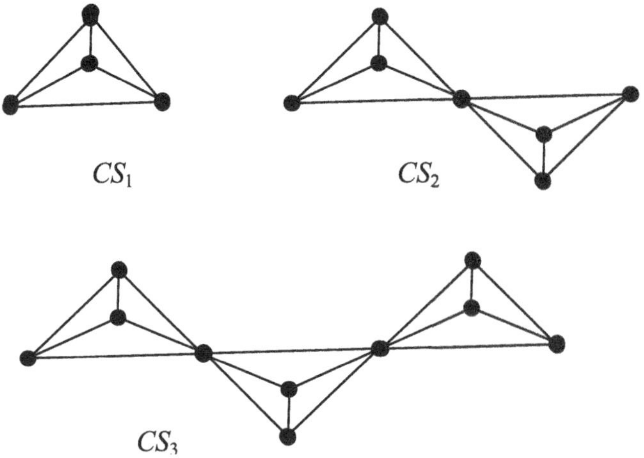

Fig. 2. Chain silicate network CS_η

Proof: The star of the silicate network $SSL\eta$ (Fig. 3) with vertex set and edge set is $|V(SSL\eta)| = 30\eta^2 - 18\eta + 6$, $|E(SSL\eta)| = 42\eta^2 - 10\eta + 4$. In $SSL\eta$ depending on the degree of each edge's vertices, there are six different kinds of edges. Moreover, Table 3 presents the star silicate network's edge divisions using the algebraic method.

Table 3. $SSL\eta$ Graph's edge partition

(d_u, d_v) where $uv \in E(SSL\eta)$	Total count of the edges
(3, 3)	6
(3, 4)	$12(2\eta - 1)$
(3, 6)	$36\eta^2 - 60\eta + 36$
(4, 4)	$24\eta - 30$
(4, 6)	$24(\eta - 1)$
(6, 6)	$36\eta^2 - 72\eta + 48$

$$SO(SSL\eta) = \sum_{e=uv\in E(SSL\eta)} \sqrt{d_u^2 + d_v^2}$$

$$SO(SSL\eta) = 6\sqrt{3^2+3^2} + 12(2\eta-1)\sqrt{3^2+4^2} + \left(36\eta^2 - 60\eta + 36\right)$$
$$\sqrt{3^2+6^2} + (24\eta - 30)\sqrt{4^2+4^2} + 24(\eta-1)\sqrt{4^2+6^2}$$
$$+\left(36\eta^2 - 72\eta + 48\right)\sqrt{6^2+6^2}$$

On simplification we get the required result.

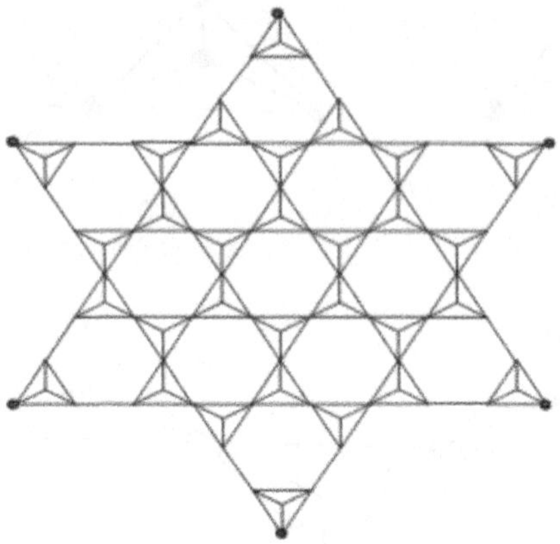

Fig. 3. Star silicate network SSL_2

4 Results on Some Special Graphs

Theorem 4: The Sombor index for Rhenium trioxide lattice ReO_3(Fig. 4).

$$SO(ReO_3) = 24\sqrt{13} + 16(\alpha + \beta + \gamma - 6)\sqrt{20}$$
$$+10(\alpha\beta + \beta\gamma + \gamma\alpha - 4\alpha - 4\beta - 4\gamma + 12)\sqrt{29}$$
$$+6(\alpha\beta\gamma - 2\alpha\beta - 2\beta\gamma - 2\gamma\alpha + 4\alpha + 4\beta + 4\gamma - 8)\sqrt{40}$$

Proof: The rhenium trioxide lattice (Fig. 4), we represent $\alpha x\beta x\gamma ReO_3$ lattice, which is a three-dimensional sequence of α unit cell rows along the X-axis, β unit cell columns along the Y-axis, and γ unit cell pages along the Z-axis of the vertex set and edge set to be $|V(ReO_3)| = 4\alpha\beta\gamma - \alpha\beta - \beta\gamma - \gamma\alpha$, $|E(ReO_3)| = 6\alpha\beta\gamma - 2\alpha\beta - 2\gamma\alpha - 2\beta\gamma$. In ReO_3 depending on the degree of each edge's vertices, there are four different types of edges. Table 4 presents the edge partition of the rhenium trioxide lattice using algebraic approach.

Table 4. By [2] ReO_3 Graph's edge partition

(d_u, d_v) where $uv \in E(ReO_3)$	Total count of the edges
(2, 3)	24
(2, 4)	$16(\alpha + \beta + \gamma - 6)$
(2, 5)	$10\ (\alpha\beta + \beta\gamma + \alpha\gamma - 4\alpha - 4\beta - 4\gamma + 12)$
(2, 6)	$6\ (\alpha q\gamma - 2\alpha\beta - 2\beta\gamma - 2\gamma\alpha + 4\alpha + 4\beta + 4\gamma - 8)$

$$SO(ReO_3) = \sum_{e=uv\in E(ReO_3)} \sqrt{d_u^2 + d_v^2}$$

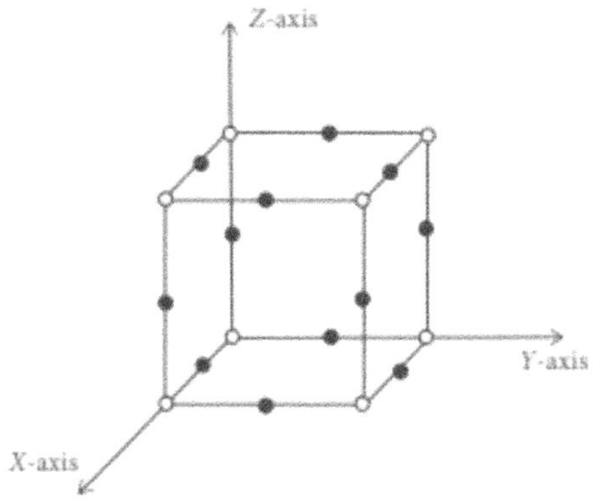

Fig. 4. Rhenium trioxide lattice ReO_3

$$SO(ReO_3) = 24\sqrt{2^2+3^2} + 16(\alpha+\beta+\gamma-6)\sqrt{2^2+4^2}$$
$$+ (10(\alpha\beta+\beta\gamma+\gamma\alpha-4\alpha-4\beta-4\gamma+12))\sqrt{2^2+5^2}$$
$$+ 6(\alpha\beta\gamma - 2\alpha\beta-2\beta\gamma-2\gamma\alpha+4\alpha+4\beta+4\gamma-8)$$
$$\sqrt{2^2+6^2}$$

On simplification we get the required result.

Theorem 5: The Sombor index for Flabellum graph $F_\eta^{(\hbar,\jmath)}$ (Fig. 5).

$$SO\left(F_\eta^{(\hbar,\jmath)}\right) = 6\sqrt{2}(3+16\eta) + \left(\eta-\sqrt{2}\right)120 - 60 + \left(108\eta^2 - 180\eta + 108\right)$$
$$\sqrt{5} + 48(\eta-1)\sqrt{13} + \left(72\eta^2 - 144\eta + 96\right)\sqrt{6}$$

Proof: The flabellum graph (Fig. 5) is obtained by considering $\hbar \geq 2$ duplicates of cycles C_η, with each copy sharing a common central vertex and then in '$\jmath$' duplicates of the cycle graph letting an additional edge between the central vertex and each non-adjacent vertex, for $0 \leq \jmath \leq \hbar$. The cardinality of the vertex and edge sets of the flabellum graph is $\hbar$ (η -1) + 1 and $k\eta + \eta\,\jmath - 3\,\jmath$, respectively. Hence using the algebraic technique, the edge partition [11] of $F_\eta^{(\hbar,\jmath)}$ is presented in Table 5.

$$SO\left(F_\eta^{(\hbar,\jmath)}\right) = \sum_{e=uv\in E\left(F_\eta^{(\hbar,\jmath)}\right)} \sqrt{d_u^2+d_v^2}$$

$$SO\left(F_\eta^{(\hbar,\jmath)}\right) = (h\eta-\eta j - 2\hbar + 2\jmath)\sqrt{8} + 2\jmath\sqrt{13} + 2\hbar\sqrt{(2\jmath+\eta j-3\jmath)^2+2^2}$$
$$+(\eta j - 3\jmath)\sqrt{(2\hbar+\eta j-3\jmath)^2+3^2}$$

On simplification we obtain the required result.

Table 5. $F_{\eta}^{(k,j)}$ Graph's edge partition

(d_u, d_v) where $uv \in E(F_{\eta}^{(k,j)})$	Total count of edges
(2, 2)	$k\eta - \eta j - 2k + 2j$
(2, 3)	$2j$
(3, 3)	$j(\eta - 4)$
$((2k + \eta j - 3j), 2)$	$2k$
$((2k + \eta j - 3j), 3)$	$\eta j - 3j$

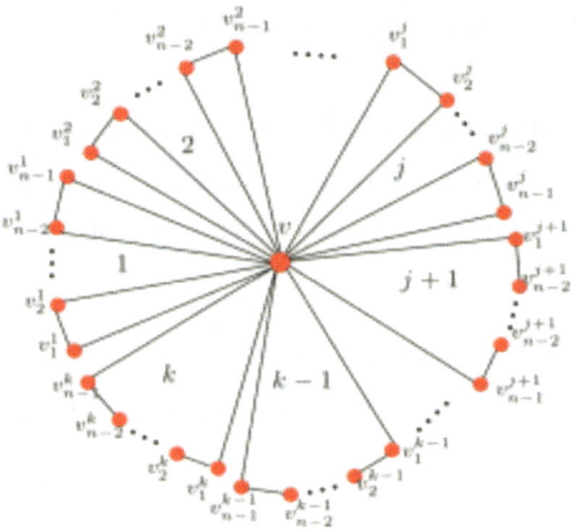

Fig. 5. Flabellum graph $F_{\eta}^{(,)}$

5 Conclusion

Computational techniques, exemplified by topological indices, have proven valuable in the modelling and prediction of the properties of nanomaterials. Computational methods are commonly used to examine the behavior of nanomaterials before actual synthesis experiments. The polynomials of these graphs can be analyzed, and a comparison between the indices can be conducted to improve the prediction of the properties of these molecules.

References

1. Abdelgader, M.S., Wang, C., Mohammed, S.A.: Computation of topological indices of some special graphs. Mathematics, **6**(3) (2018). https://doi.org/10.3390/math6030033
2. Ding, L., Ul Haq Bokhary, S.A., Rehman, M.U., Ali, U., Mubeen, H., Iqbal, Q., Liu, J.B.: Degree-based indices of some complex networks. J. Math. **2021** (2021).https://doi.org/10.1155/2021/5531357

3. Ghods, M., Rostami, Z.: Sombor indices of Silicon Carbide Graphs
4. Hui, W., Sherin, L., Javed, S., Khalid, S., Asghar, W., Fufa, S.A.: On topological indices for complex indium phosphate network and their applications. Complexity, **2022** (2022). https://doi.org/10.1155/2022/9707749
5. Imran, M., Iqbal, M.A., Liu, Y., Baig, A.Q., Khalid, W., Zaighum, M.A.: Computing eccentricity-based topological indices of 2-power interconnection networks. J. Chem. **2020** (2020). https://doi.org/10.1155/2020/3794592
6. Imran, M., Siddiqui, M.K., Naeem, M., Iqbal, M.A.: On topological properties of symmetric chemical structures. Symmetry, **10**(5) (2018). https://doi.org/10.3390/sym10050173
7. Kulli, V.R.: Computing banhatti indices of networks. Int. J. Adv. Math. **2018**(1) (2018). https://www.researchgate.net/publication/322602387
8. Kulli, V.R.: F-Sombor and modified F-Sombor indices of certain nanotubes. Ann. Pure Appl. Math. **27**(01), 13–17 (2023). https://doi.org/10.22457/apam.v27n1a03900
9. Raghu, S., Kulli, V.R., Niranjan, K.M., Goudar, V.M.: Nirmala indices of certain antiviral drugs. Int. J. Math. Comput. Res. **11** (2023). https://doi.org/10.47191/ijmcr/v11i11.05
10. Rajan, B., William, M.A., Grigorious, C., Stephen, S., William, A.: On certain topological indices of silicate, honeycomb and hexagonal networks. J. Comput. Math. Sci. **3**(5) (2012). https://www.researchgate.net/publication/257297085
11. Shi, X., Kosari, S., Ahmad, U., Hameed, S., Akhter, S.: Evaluation of various topological indices of flabellum graphs. Mathematics, **11**(19) (2023). https://doi.org/10.3390/math11194167
12. Sindiya Therese, S., Jude Annie Cynthia, V.: Packing of silicate and oxide networks. Mater. Today: Proc. **45**, 6864–6869 (2020).https://doi.org/10.1016/j.matpr.2020.12.1039
13. Thayamathy, P., Elango, P., Koneswaran, M.: M-polynomial and degree based topological indices for silicon oxide. Int. Res. J. Pure Appl. Chem. **16**(4), 1–9 (2018). https://doi.org/10.9734/irjpac/2018/42645
14. Peng, S.-L., Lin, C.-K., Pal, S.: Proceedings of 2nd International Conference on Mathematical Modeling and Computational Science. Advances in Intelligent Systems and Computing. Springer, Singapore (2022). https://doi.org/10.1007/978-981-19-0182-9
15. Peng, S.-L., Hao, R.-X., Pal, S.: Proceedings of First International Conference on Mathematical Modeling and Computational Science. Advances in Intelligent Systems and Computing. Springer Singapore (2021). https://doi.org/10.1007/978-981-33-4389-4
16. Estrada, E., Torres, L., Rodríguez, L., Gutman, I.: An atom-bond connectivity index: modelling the enthalpy of formation of alkanes. Indian J. Chem. *37A*, 849–855 (1998)

Using Tri Uan Game in Teaching Math in Primary School to Develop Students' Mathematical Competence

Tran Le Thuy(✉)

University of Education, Vietnam National University, Hanoi, Vietnam
tranlethuy@vnu.edu.vn

Abstract. Using games in teaching mathematics is an engaging and effective method that fosters students' interest in the learning process and helps them develop their mathematical skills through play. Vietnamese folklore boasts numerous entertaining games that also serve as exercises for cognitive development. Incorporating these traditional games into mathematics education not only supplements theoretical and practical knowledge about using games in teaching but also enhances students' understanding and pride in the intellectual prowess of the Vietnamese people, thus deepening their cultural appreciation. The article introduces Tri Uan, a game invented by revolutionary Nguyen Tri Uan in the 1940s, which has gained widespread popularity in Vietnam and internationally. This game can be applied to teaching natural numbers and geometric concepts. The study proposes a methodology for utilizing Tri Uan in teaching and illustrates its implementation in teaching natural number concepts and calculating the area of shapes in elementary mathematics. Integrating Tri Uan into mathematics education creates an environment conducive to autonomous learning, fostering creative thinking and cognitive development in students.

Keywords: Tri Uan game · Primary school students · Teaching mathematics · Competency development

1 Introduction

From an educational perspective, teaching to develop competencies for primary school students is always a crucial objective, opens up opportunities for students to form logical thinking, problem-solving skills and creativity from an early age. In an age where education needs to create a more engaging and interactive learning environment, the utilization of games became an interesting method in helping students develop both general and core competencies outlined in the 2018 General Education Program (Ministry of Education and Training, 2018) that is being implemented in Vietnam. Instead of feeling pressured and bored while learning math, students often feel intrigued and actively participate in activities when they are combined with games. This excitement not only keeps students more focused on the lesson but also encourages them to learn proactively and creatively. In addition to commonly used games in teaching mathematics, we can also

S. Pal and Á. Rocha (Eds.): ICMMCS 2025, LNNS 1398, pp. 146–159, 2025.
https://doi.org/10.1007/978-3-031-90998-6_15

exploit and develop traditional folk games to apply in education, enriching the resource pool of educational games.

Jigsaw puzzles have origins dating back thousands of years, when humans had known how to use wood and stone to build houses or famous structures such as the pyramids of Egypt. In 1940, a puzzle game similar to that of an older German one but reflecting many Vietnamese habits and cultural characteristics emerged in Vietnam. It was developed by the revolutionary Nguyen Tri Uan (1916–1995), who devised a way to cut cardboard pieces in order to form shapes according to his imagination and reasoning. After much experimentation and imagination, he finally managed to use seven pieces that were methodically cut from a rectangular sheet measuring 8 cm x 10 cm to produce: one pentagon, two identical right-angled isosceles triangles, and four right-angled trapezoids (two of which were identical small trapezoids, and two larger trapezoids differing by exactly one square). He was able to create over 1000 different shapes. This set of seven pieces and various arrangements was published in a book in 1957 and named the "Tri Uan" game [9–11]. The book was later translated into English, French, Russian and various other languages (Fig. 2).

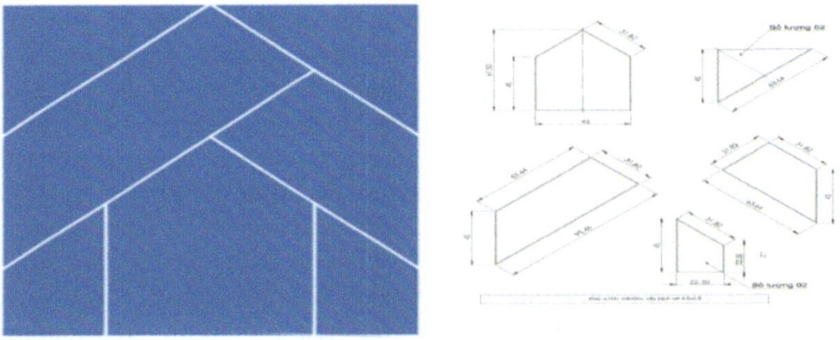

Fig. 1. Of the Tri Uan Puzzle Set

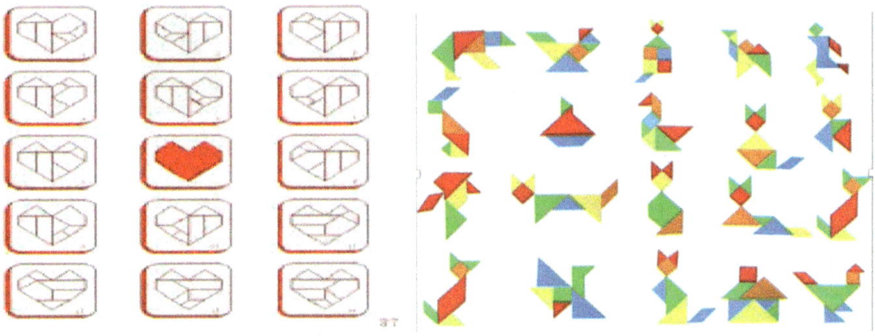

Fig. 2. Some shapes that can be assembled from the seven puzzle pieces of the Tri Uan game.

This game can be made of paper, cardboard, or wood. Players in later generations have created numerous shapes from these seven wooden pieces, covering themes such as people, labor, structures, animals, plants, letters, and numbers. With flexible means of assembling, the game is highly effective in generating interest, developing creative thinking skills and promoting intellectual growth. However, it is often used as an entertaining game to challenge the wits and creativity of the player, as well as their peers. We have conducted research, experiments and found that the Tri Uan game could be used to teach various math topics in primary schools. Here, we will introduce its applications and propose a process for organizing this game in teaching. Then, the proposed process is illustrated through teaching natural number content and calculating the area of a shape in elementary mathematics.

2 Application of the Tri Uan Game in Teaching Primary Education Mathematics

2.1 Teaching Geometry Through the Tri Uan Game

Chu Cam Tho and colleagues (2020) used the Puzzle game to help students: - Perform operations: observe, imagine; - Assemble, create shapes; - Familiarize themselves with basic shapes and their geometric properties; - Develop geometric thinking. To assess the level of implementation of these operations for each student, the study presented a table of quality criteria for each behavior on a scale consisting of three levels, with evaluation elements, indicators of student behavior, and quality criteria according to the scale, so that teachers can evaluate the abilities of students formed through the implementation of this game, such as the ability to observe and reproduce the image of an object based on its geometric characteristics, the ability to orient in space and imagine the replacement or displacement of parts of an object. In addition, the authors also exploited another aspect of the puzzle game, using the relationships of perpendicularity, parallelism, equality, and symmetry appearing in this game to form and reinforce these mathematical relationships for students, and presented a table describing these relationships (see Fig. 3) [6].

Illustrated pieces		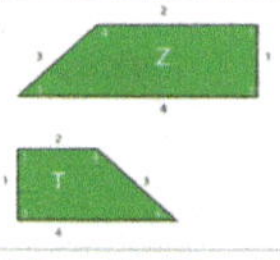	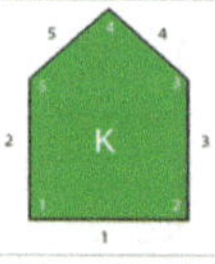
Equal relations	Equal relations of side	$SZ_1 = ST_1; SZ_3 = ST_3$	$SK_2 = SK_3; SK_4 = SK_5$
	Equal relations of angle	$AZ_1 = AZ_2 = AT_1 = AT_2$ $AZ_3 = AT_3; AZ_4 = AT_4$	$AK_1 = AK_2; AK_4 = AK_5$
Parallel relations		$SZ_1 \parallel SZ_4; ST_2 \parallel ST_4; SK_2 \parallel SK_3$	
Perpendicular relations		$SZ_1 \perp SZ_2; ST_1 \perp SZ_4; ST_1 \perp ST_2; ST_1 \perp ST_4;$ $SK_1 \perp SK_2; SK_1 \perp SK_3$	
Symmetric relations		In Tri Uan: + X and Y are symmetric through axis d. + U and V are symmetric through axis d.	d is the symmetric axis of K.

Fig. 3. Table Describing Equality Relationships, Perpendicular Relationships, Parallel Relationships, and Symmetry Relationships in the Tri Uan Game

Through research and experimentation, we have utilized the concept of jigsaw thinking to teach lessons on the area of flat geometric shapes in primary school. The construction of formulas for calculating geometric areas is challenging for primary school students and many teachers also find it difficult to teach this. In reality, we do not always encounter situations where the areas of familiar shapes need to be calculated, such as areas of land not always being a square or a rectangle.

An interesting approach to this problem is to use the jigsaw thinking of the Tri Uan game to calculate the area of a shape by cutting and assembling it into shapes with known area formulas. In this way, the area of the original shape will equal the area of the assembled shape (as long as no pieces are added or removed). After assembling, students can calculate the area of the newly formed shape using the learned formulas, thereby determining the area and the formula for the original shape.

This teaching method helps students independently derive area formulas, understand the essence of area calculation formulas, and develop logical thinking, problem analysis skills, and the ability to create unique problem-solving strategies. Through this, students develop the competence to address practical problems as they arise.

2.2 Using the Tri Uan Game to Teach Natural Numbers to Students

When it comes to the puzzle game Tri Uan, we often think of the field of geometry. However, this set of 7 puzzle pieces can be assembled into natural numbers. Each set of puzzles can be assembled into a single-digit natural number. Therefore, during the process of teaching natural numbers, teachers can have students arrange these puzzle sets into the numbers they have learned in practice to deepen their knowledge and develop their puzzle-solving skills. Teachers can organize individual play for students to develop their puzzle-solving skills, form autonomous and self-developing groups. In addition, teachers can also organize team games in groups. Teachers can divide the class into pairs, then give a two-digit natural number and assign the task for each group to use the puzzle set to assemble the number requested by the teacher. The group that assembles correctly and quickly will be the winning group. Students will need to assign each other a digit and then combine the two digits to get the number requested by the teacher. Through this activity, in addition to knowledge and puzzle-solving skills, students also develop their communication and cooperation skills with their group members to complete a common task. In the following rounds, teachers can change the number of students in each group and at the same time, give a number with more digits to increase the challenge for students. Below, I will propose a teaching organization process to apply the above ideas to the implementation of teaching for elementary school students.

2.3 Organizing the Tri Uan Game in Teaching Primary School Mathematics

2.3.1 Process of Applying the Tri Uan Game in Teaching Primary School Mathematics

Referencing the research of Nguyen Anh Tuyet (2000) [7] and Nguyen Thi Bich Hong (2014) [4], as well as the learning characteristics of primary school students, we propose the following process for organizing teaching using the Tri Uan game:

Phase 1: Preparation

Step 1: Conceptualization. In this step, the teacher needs to research the lesson content and resources to determine which objectives and content the Tri Uan game will serve. This helps to establish the game's goals by answering questions such as: What is the purpose of using this game? What will students learn and practice through this game? What competencies will be developed? The teacher needs to understand the students' familiarity with the Tri Uan game, whether they have played it before, and their proficiency level.

Step 2: Preparation of conditions and playing tools. In this step, the teacher guides students to prepare learning tools, specifically the 7 puzzle pieces of the Tri Uan game, which can be purchased as a learning set at bookstores, or students can use paper or cardboard to cut out the puzzle pieces according to the provided sizes (see Fig. 3). Next, the teacher needs to establish the rules and regulations of the game, thoroughly study the gameplay and organization of the game, determine the process of the game, and the necessary conditions and tools to implement the game. In this step, the fundamental factors that need to be prepared include: the objectives of the educational game; equipment and tools for organizing the educational game; the number of participants in the game; gameplay rules; assessment methods; and regulations for rewards and penalties.

Phase 2: Organizing the game activities

This phase is conducted in class with the following steps for organizing the game:

Step 1: Introduction and explanation of the game. The teacher introduces the name of the game, asks students about their knowledge of the Tri Uan game, and provides a brief history of the game, integrating lessons on Vietnamese intelligence and cultural pride through the game's development. Based on the lesson content and specific context, the teacher explains the game rules and regulations, and sets out the rewards and penalties for participating teams. The effectiveness of the educational activity depends on the teacher's pedagogical skills in organizing the game. If not done well, students may lose interest, misunderstand the rules, break the rules, or become discouraged by failure. The teacher can let students try out the game first, reiterating the rules.

Step 2: Organizing game activities. During the game, the teacher should monitor students' activities, encourage and motivate them, and create an environment for students to self-experience and acquire knowledge and skills. The teacher must also manage time and provide timely support for those who need help.

Step 3: Evaluation and summary of the game. After the game ends, the teacher should allow students to present and explain their methods. Besides evaluating the results, the teacher should comment on the students' spirit, attitude, and problem-solving abilities, helping them to draw knowledge, skills, and problem-solving methods from the game. The teacher should conduct evaluations and clearly define rewards and penalties to ensure fairness. After summarizing and reviewing the tasks performed, the teacher can adjust any areas that did not meet expectations to achieve better results in the future. Below are specific examples of applying the above process to teaching specific content in mathematics in elementary schools.

2.3.2 Illustration of the Process of Applying the Tri Uan Game in Teaching Mathematics in Primary School

Example 1: Using the Tri Uan game to teach natural numbers to students.

Phase 1: Preparation
Step 1: Conceptualization. For the content on natural numbers and developing students' puzzle-solving skills, the activity aims to help students:

- Recognize natural numbers.
- Match natural numbers from 7 puzzle pieces.
- Develop puzzle-solving skills and thinking to prepare for subsequent content in the primary school math curriculum.

Once students form an understanding of natural numbers, the teacher can organize the game during practice sessions to reinforce knowledge and create a positive learning environment.
Step 2: Preparing conditions and materials for the game. In this step, the teacher instructs students to prepare learning tools, specifically the seven pieces of the Tri Uan game, which can be purchased or made using paper and cardboard cut according to instructions (see Fig. 1). The game should incorporate learned knowledge, ensuring all students participate.

Phase 2: Organizing the game activities
This phase is conducted in class with the following steps for organizing the game:
Step 1: Introduction and explanation of the game. The teacher introduces the "Tri Uan" game and its history, allowing students to share their insights and lessons learned about the game's development. The teacher integrates lessons on Vietnamese intelligence and cultural pride.

The game explanation is: "From the seven pieces of the Tri Uan game, we can form any natural number from 0 to 9". The teacher demonstrates using the game set to form a natural number while students observe, then lets students practice to familiarize themselves with the game. Afterward, the teacher explains the rules: The game will be organized into multiple rounds until the activity time ends. In the first round, students play individually to practice puzzle-solving skills, fostering self-reliance and personal learning. The teacher presents a single-digit number, and students use the Tri Uan game set to form that number. The game lasts a maximum of 5 min; students who complete the task correctly within the time limit earn 1 star, those who finish 1 min earlier earn 2 stars, and those finishing 2 min or more earlier earn 3 stars.

In the second round, the teacher divides students into pairs. The teacher presents a two-digit number, and each student in the pair assembles one digit, then they combine the digits to form the required number. This activity involves both individual and group work, developing communication and teamwork skills. The star reward system for groups will be similar to the first round. The total stars accumulated by both members of the group contribute to individual achievements for all rounds. The teacher can organize additional rounds, change the number of group members, introduce numbers with more digits corresponding to group size, and follow similar steps.
Step 2: Organizing game activities.

In the first round, the teacher lets students play individually, ensuring they understand the rules and time limits. The teacher reminds students to raise their hands when they finish to mark the completion time. Once ready, the teacher presents a single-digit number and starts the timer for students to assemble the number using the Tri Uan game set. During the game, the teacher monitors student activities, encourages them, and creates a supportive environment for self-experience and knowledge acquisition, providing timely assistance when needed. The teacher prepares a class list to record completion times when receiving signals from students (Fig. 4).

Fig. 4. Tri Uan Game Forming Natural Numbers

Evaluation and Summary of the First Round: After the first round ends, the teacher displays the answers for students to check their results and cross-check the results of the student sitting next to them. The teacher evaluates the results, comments on the students' spirit, attitude, and problem-solving abilities, and awards stars based on the initial game rules, considering the products and the time taken to complete them.

In the second round, the teacher reiterates the rules to ensure the groups understand how to play, then divides students into pairs. Students, holding their Tri Uan game sets, move around the classroom while singing along to a piece of music. When the music stops, the teacher says, "Form groups, form groups," and the students ask, "How many, how many?" The teacher responds, "Pairs, pairs." Students then find the nearest peer to form a pair and find a convenient location to work together. This method of group formation helps students change their state, preventing fatigue and lack of concentration from sitting too long. Random pairing allows students to interact and work with different classmates, exposing them to various personalities and thought processes, and learning more from their peers. Once the groups are settled, the teacher writes their names on the board to track their results. When the groups are ready, the teacher presents a two-digit number, and each pair divides the task of forming one digit each, then combining them to form the required number. This activity involves both individual work and collaboration, fostering communication and teamwork skills. During the game, the teacher monitors student activities, encourages and motivates them, creates an environment for self-experience, and provides timely support when needed. A list is prepared to record the completion times as students signal they are done.

Evaluation and Summary of the Second Round: After the second round ends, the teacher displays the answers for students to check their results, and the groups cross-check the results of the nearest group. The teacher evaluates the results, comments on the students' spirit, attitude, and problem-solving abilities, and awards stars based on the initial game rules, considering the products and the time taken to complete them.
Step 3: Evaluation and summary of the game. After the game ends, the teacher evaluates and comments on the students' spirit, attitude, and problem-solving abilities throughout the rounds. The teacher asks students to share their strategies for completing the game quickly or the mistakes they made that caused delays, helping them derive knowledge, skills, and problem-solving methods through the game.

Example 2: Designing a Cut-and-Paste activity to teach students how to derive the formula for the area of a rhombus.

Phase 1: Preparation
Step 1: Conceptualization. For the lesson on the area of a rhombus, we will expand the Tri Uan game. Instead of using the pre-made pieces in the game set, students will role-play as the game's creator. They will cut a given rhombus into pieces according to their imagination and assemble it into a new shape with a known area. This requirement is more challenging but will enhance students' positivity, initiative, and creativity. The activity aims to help students:

- Derive the formula for the area of a rhombus based on a shape with a known area formula;
- Apply learned content to solve real-life situations;
- Develop puzzle-solving skills and thinking to prepare for subsequent primary school math curriculum content.

In this lesson, the puzzle game will be used to form new knowledge.

Step 2: Preparing conditions and materials for the game. The teacher instructs students to prepare study tools, specifically a rhombus made of paper or cardboard, scissors, and a ruler.

Phase 2: Organizing the game activities
This phase is conducted in class with the following steps for organizing the game:
Step 1: Dividing teams. The teacher divides the class into four teams.

Step 2: Introducing and explaining the game. The teacher prepares worksheets with three tasks. Students will discuss to complete the tasks within 15 min. Task 1: From the prepared rhombus, cut and assemble it into a shape with a known area formula. In this task, the teacher will have students recall shapes with known area formulas but will not guide them on how to cut and assemble, leaving them to figure it out independently. Task 2: Compare the area of the rhombus with the area of the newly assembled shape. Task 3: Derive the formula for the area of a rhombus from the area formula of the newly assembled shape. Evaluation: Task 1 (3 points), Task 2 (2 points), Task 3 (3 points), presentation (1 point), and fastest completion (1 point). Teams will be ranked based on their total points.
Step 3: Organizing the activities. The teacher instructs students to take out their prepared study tools, distributes the worksheets to the teams, signals the start, and starts the timer.

During the game, the teacher monitors students' activities, encourages and motivates them, creates an environment for self-experience, and provides timely support when needed.

Task 1: Students discuss how to cut and assemble the prepared rhombus into a shape with a known area formula (students have previously learned the area formulas for rectangles and parallelograms). Students may come up with various ways to cut and assemble the rhombus into a rectangle or parallelogram, as shown in Fig. 5. Remind students not to add or remove any pieces during the process.
Task 2: Students compare the area of the original rhombus with the area of the newly assembled rectangle (or parallelogram). The area of the rhombus is equal to the area of the newly assembled rectangle (or parallelogram).
Task 3:

- Case 1: Students assemble the rhombus ABCD into the rectangle MNCA as shown in Fig. 7. Students have learned the formula for the area of a rectangle as length multiplied by width. The area of rectangle SMNCA = AC × AM = AC × (BD / 2) = ½ (AC × BD) (since AM = BO = ½ BD). Therefore, the area of the rhombus equals half the product of its diagonals.
- Case 2: Students assemble the rhombus ABCD into the parallelogram ABHC as shown in Fig. 5. Students have learned the formula for the area of a parallelogram as base multiplied by height. The area of the parallelogram SABHC = AC × BO = AC × (BD / 2) = ½ (AC × BD). Therefore, the area of the rhombus equals half the product of its diagonals.
- Case 3: Students assemble the rhombus ABCD into the rectangle PBDQ as shown in Fig. 5. Students have learned the formula for the area of a rectangle as length multiplied by width. The area of rectangle SPBDQ = QD × BD = BD × (AC / 2) = ½ (AC × BD). Therefore, the area of the rhombus equals half the product of its diagonals.
- Case 4: Students assemble the rhombus ABCD into the parallelogram KBDA as shown in Fig. 5. Students have learned the formula for the area of a parallelogram as base multiplied by height. The area of the parallelogram SKBDA = BD × AO = BD × (AC / 2) = ½ (AC × BD). Therefore, the area of the rhombus equals half the product of its diagonals.
- Step 4: Evaluation and Summary of the Educational Game. After the game ends, the teacher reviews the answers with the students, comments, announces the scores, and awards the winning team. The teacher summarizes the knowledge and has students recite the formula several times to remember it. Some exercises are given for students to practice applying the newly learned formula.

After the game, the teacher allows the groups to present how they cut and assembled the rhombus into a rectangle (or parallelogram). By cutting and assembling without adding or removing any pieces, students will recognize that the two shapes have equal areas. Students then explain their reasoning process to derive the formula for the area of a rhombus from the area formula of the rectangle (or parallelogram). The teacher evaluates the cutting and assembling methods, the reasoning process of each team and awards points for each task based on the established criteria, as well as giving extra points to the fastest team. The teacher totals the points and announces the winning team. The teacher reinforces the knowledge of the rhombus area formula and has students recite it several times to memorize it. Some exercises are assigned for students to practice applying the newly learned formula.

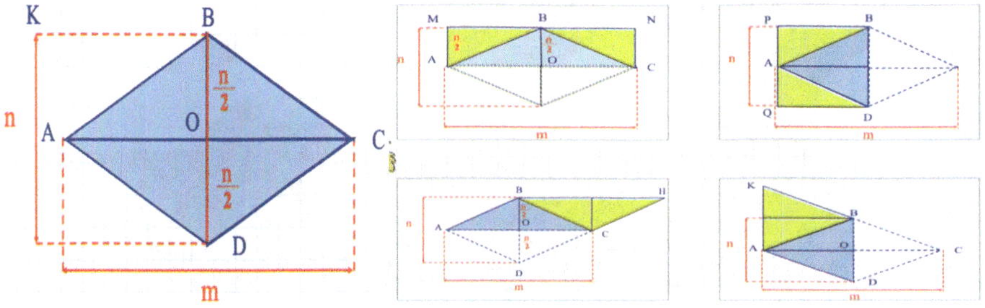

Fig. 5. Students cut the rhombus into four pieces to assemble into a rectangle or a parallelogram.

Note: For the rhombus, students can cut and assemble it into a parallelogram or a rectangle in four different ways as described above. Teachers need to be flexible and well-prepared to handle various situations that students may present. When teaching lessons on the area of other flat geometric shapes, teachers can also apply a similar process to form the area calculation formulas for students.

3 Practice Instructions for the Puzzle Game on the App

The game is designed with levels ranging from easy to difficult to gradually increase the challenge for players. Level 1 presents a complete model and an incomplete shape that is missing only one piece. The player needs to move the piece to the required position, rotate the piece in the appropriate direction to match the required shape, and they will win the game (Fig. 6).

On next levels, the number of puzzle pieces that the player needs to assemble will increase to 3 pieces. The player will move each piece to the appropriate position to match the required shape as specified by the game.

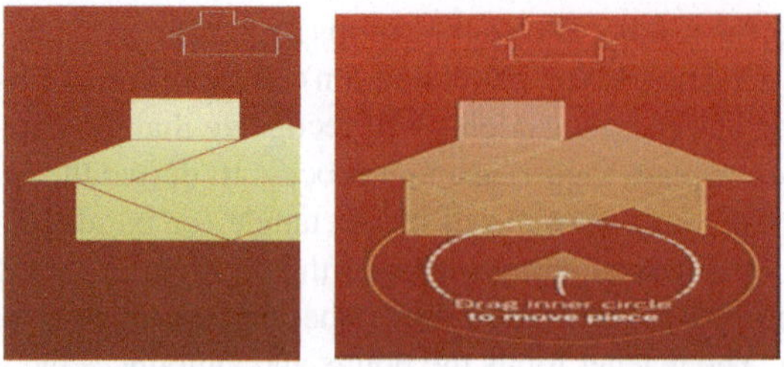

Fig. 6. Image of puzzle game on the app at level 1

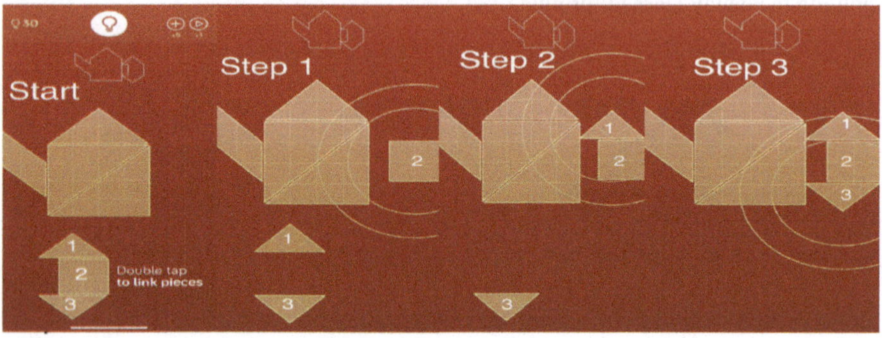

Fig. 7. Image of puzzle game on the app at level 2

As levels go up, hints will gradually decrease until there are no pre-assembled pieces left, increasing the challenge for the player. Using the app in the teaching process will change from making puzzles with real objects to practicing moving puzzle pieces on the screen of an electronic device (smart board, phone, iPad,...). This form of teaching will help teachers and students not have to prepare too many teaching aids that are puzzle pieces, but requires students to be equipped with appropriate electronic devices. This may be appropriate when teachers model in class for students to observe, have students study in the computer room, or require parent support when students practice at home. In order for teaching to take place effectively, teachers need to determine the lesson objectives, from which they will select jigsaw tasks related to the lesson content and consistent with their lesson plan ideas. Teachers need to determine the level of students to choose the puzzle level and set appropriate requirements for students, instructing students how to operate the software. Through that practice process, students can form new knowledge or review what they have learned. The software only replaces the jigsaw activity with real puzzle pieces and the teaching process will follow the steps stated in Sect. 2.2.

4 Evaluating the Level of Students' Puzzle-Solving Skills

To evaluate the level of performing the following actions: Observation, imagination; Familiarizing with basic shapes and their geometric properties; Developing geometric thinking during the process of puzzle-solving for each student, we can refer to the table presenting quality criteria for each behavior on a scale of three levels (one being the lowest and three being the highest) [6] (Table 1).

Table 1. Quality criteria of behavior indicators elements competency F1; F2.

Elements	Behavior indicators	Quality criteria
F1. The ability to observe and reproduce the pattern based on geometric characteristics	1. Observe and select pieces based on geometric characteristics	1. Choose some pieces to fit into the pattern without any geometric characteristics (edges, angles, dimensions, etc.). 2. Choose some pieces to fit into the pattern based on one geometric characteristic (edges, angles, dimensions, etc.). 3. Choose some pieces to fit into the pattern based on up to two geometric characteristics (edges, angles, dimensions, etc.)
	2. Observe and select pieces in the given time	1. For a given time, remember and position 3 pieces correctly in the pattern. 2. For a given time, remember and position 5 pieces correctly in the pattern. 3. For a given time, remember and position 7 pieces correctly in the pattern
	3. Create a new one, name it and explain the reasons	1. Use 3 pieces to create a new one, name the picture, and explain the reasons. 2. Use 5 pieces to create a new one, name the picture, and explain the reasons. 3. Use 7 pieces to create a new one, name the picture, and explain the reasons
F2. The ability to spatially orientate and imagine the substitution or displacement of the part of the pattern	1. Determine the direction of the puzzle	1. Fit a given piece at corresponding shapes in two different directions. 2. Fit two given pieces at corresponding shapes in two different directions. 3. Fit three given pieces at corresponding shapes in two different directions

(*continued*)

Table 1. (*continued*)

Elements	Behavior indicators	Quality criteria
	2. Locate the missing pieces in the picture	1. Placing the 7th piece is missing that has related to other pieces in the pattern. 2. Placing the 6th and 7th pieces are missing that have related to other pieces in the pattern. 3. Placing the 5th, 6th, and 7th pieces are missing that have related to other pieces in the pattern

Conclusion

The research results have shown that the game Tri Uan is suitable for application in teaching natural numbers and some basic geometry content in elementary mathematics curriculum. Specifically, this game has been proven to help form symbols of geometry and develop skills in constructing formulas for calculating the area of plane geometry for students. In addition, through participating in the game, exploring the puzzle pieces of the game set, students also have the opportunity to deepen their understanding of mathematical concepts and relationships such as parallel, perpendicular, equal, and symmetrical. These results highlight the important role of the game in creating a positive and creative learning environment, helping students develop not only mathematical abilities but also logical thinking skills and promoting knowledge and interest in folk culture. Research on the use of the game Tri Uan to develop the capacity of elementary school students through teaching geometry and mathematics can open up new opportunities to improve the quality of education and lay the foundation for the comprehensive development of students.

Acknowledgments. This research is funded by VNU- University of Education under Research Project number QS.24.03.

References

1. Ministry of Education and Training: General Education Program in Mathematics (issued with Circular No. 32/2018/TT-BGDĐT dated December 26, 2018 by the Minister of Education and Training) (2018)
2. Chung, V.Q., (ed.): *Teaching Methods in Elementary Mathematics*. Education Publishing House (2007). [3] Tran Nam Dung (General Editor, 2022). *Mathematics 3*. Vietnam Education Publishing House
3. Dung, T.N. (ed.): Mathematics 3. Vietnam Education Publishing House, Hanoi (2022)
4. Hong, N.T.B.: Methods of using games in teaching. J. Sci. Ho Chi Minh City University of Education **54**, 174–179 (2014)
5. Hien, T.D., Khue, N.D., Lai, D.T., Oanh, N.T.K., Tam, P.T., Van, N.T. (eds.): Mathematics 3. Vinh University Press, Vinh (2021)

6. Chu, C.T., Vu, A.T., Nguyen, T.D.: Potential development of geometric thinking for children through Tri Uan. Vietnam J. Educ. Sci. **1**, 52–57 (2020)
7. Tuyet, N.A.: Children's Games. Women's Publishing House, Goa (2000)
8. Uan, N.T.: 7 Pieces make 1000 Shapes. Kim Dong Publishing House, Hanoi (2010)
9. Uan, N.T.: Tri Uan: A Game of 7 Pieces made 1000 Images: A Puzzle Book. National Library of Vietnam, Hanoi (1957)
10. Uan, N.T.: Tri Uan Game. National Library of Vietnam, Hanoi (1959)
11. Uan, N.T.: Tan Tri Uan. National Library of Vietnam, Hanoi (1960)

Fuzzy-CNN Based Framework for Handwritten Signature Verification

Nisha Chaurasia(✉) and Ishu Priya(✉)

Dr. B R, Ambedkar National Institute of Technology, Jalandhar, Punjab, India
{chaurasian,ishup.cs.21}@nitj.ac.in

Abstract. Every individual possesses a unique signature, serving as a cornerstone for personal authentication and the validation of critical documents or formal procedures. Signatures, classified as biometric identifiers, play a pivotal role in confirming one's identity, thereby bolstering security measures. Handwritten signatures are crucial in underprivileged areas where digital authentication is scarce. They serve as primary identity verification for banking, government paperwork, and legal agreements. However, traditional offline verification is slow, especially with high paperwork volumes. To overcome the inefficiencies of traditional methods, automated signature verification (ASV) systems have been developed, leveraging advanced technologies to differentiate between genuine and forged signatures. Advanced technologies are utilized by these systems for determining the authenticity of signature images, discriminating between forged and genuine signatures. We present a novel approach in our work that combines the benefits of Fuzzy Logic (FL) and CNN thus enhancing accuracy of signature verification. In our methodology, within the fuzzy logic framework, we employ the Fuzzy c-Means Clustering method. BHSig260 dataset comprising Hindi signatures is used to conduct the experiments. Our hybrid model achieved a significant 89.29% accuracy, validating the potential of our hybridized approach in accurately classifying real and forged Hindi signatures, thus, in return depicting its effectiveness in advancing ASV technology. Moreover, our approach builds upon the accuracy & efficiency of verifying handwritten signatures in case of critical real-world applications.

Keywords: Handwritten signatures · deep learning · Convolutional neural networks · Automated system verification · Fuzzy logic

1 Introduction

Handwritten signature verification (HSV) is essential for security & privacy, persuading surging research focus on offline forgery detection efforts [1]. A reliable or dependable forgery detection system must be able to discern genuine signatures from the fake ones ensuring transaction integrity. Low-quality images and contextual influences impose noteworthy challenges in accurately analyzing and authenticating signatures. Deciphering the genuineness of such secluded signatures becomes a complex venture, as outlying background elements may complicate the actual properties of the signature itself [2]. Identifying counterfeit signatures that closely resemble the real ones is quite challenging

S. Pal and Á. Rocha (Eds.): ICMMCS 2025, LNNS 1398, pp. 160–169, 2025.
https://doi.org/10.1007/978-3-031-90998-6_16

because of sophisticated forgery approaches. Hence, enhancing the signature verification methodologies becomes crucial with evolution of forgery methods so as to maintain trust & security.

1.1 Key Objectives of Our Work

The main aim of our work is to explore and assess pre-processing techniques enhancing data quality of signatures for FL and CNN. These techniques include normalization, noise reduction, as well as feature extraction. It also aims in building a FL-based verification module amalgamated with a CNN architecture to serve signature verification. Overall extensive experiments are carried out to gauge and collate the performance of our proposed hybrid model with other existing models, using several performance metrics.

1.2 Significance of Automated Signature Verification (ASV) System

In today's digital scenario, the commonness of e-transactions & documentation outlines the critical need of robust methods to support authenticity of handwritten signatures. Standard manual verification processes are often time-consuming and also susceptible to human error, thus making then inadequate in handling large volumes of transactions skillfully [3]. ASV systems facilitate faster, reliable & scalable solutions to verify signatures. These systems build upon security, mitigating fraud risk. Additionally, they ensure accountability & regulatory assent offering transaction audit trails [4].

1.3 Deep Neural Networks (DNNs)

In pattern recognition, DNNs are intimidating in learning complex features and patterns from data. In HSV, these excel in capturing signatures' intricate details such as stroke variations, curvature, spatial relationships etc. [5] Using CNNs, designed for visual data analysis, hierarchical features can be retrieved from signature images efficiently, enabling accurate differentiation between fake and real. CNNs learn relevant details & patterns from raw pixel data directly, eradicating the need of manual feature engineering, also enhancing adaptability to different signature styles [6]. DNNs, by enabling end-to-end systems, simplify signature verification & eradicating intermediate pre-processing steps. Overall, DNNs enhance efficiency and robustness of the system [7].

1.4 Fuzzy Logic (FL) Systems

FL systems incorporate human-like reasoning, complementing capabilities of DNNs. These facilitate uncertainty handling into verification process [8]. Unlike traditional binary logic systems that classify inputs as either true or false, FL systems permit degrees of truth, ensuring more nuanced decision-making in case of ambiguity. In HSV, fuzzy logic handles variability and uncertainty effectively capturing nuances and discrepancies in signature data. Combining DNNs and FL, a Fuzzy-CNN framework offers an exhaustive solution for signature verification [9, 21]. FL offers a captivating direction for exploration despite crisp (boolean) logic being in existence. This is because it can lodge imprecise data without demanding explicit inputs, a capability missing in crisp systems. This flexibility of FL is mainly advantageous in signature verification, where they can extract features resilient to variations among signatures, including size, rotation, & slant [10, 22].

2 Literature Review

Both FL and NNs facilitate numerous advantages and drawbacks in signature verification. Neural networks (NNs) excel at efficiently spotting intricate patterns within data, but they also demand significant volume of training data. Also, NNs sometimes lack transparency in the decision-making process. However, FL systems, being capable enough to handle uncertainty well, are often complex and require fine-tuning and certain domain knowledge. Despite these advancements, challenges like data requirements, scalability and interpretability prevail, demanding ongoing research to enhance effectiveness and applicability of systems. Ultimately, a combining both these methodologies can leverage constructing an overall resilient and better signature verification system capable of identifying forgery attempts.

Table 1 elucidates a summary of various methodologies refined and developed by many researchers over the period.

Table 1. Comparison of Neuro-Fuzzy Techniques: A Concise Overview.

Ref	Methodology	Dataset	Observations
[11]	Image classification, CNN, RMSprop optimizer	MNIST	99.06% accuracy obtained
[12]	Compared VGG16, VGG19 & ResNet50	SigComp2009	VGG19 performed better than other models
[13]	Hybrid Dual-Fuzzy CNN, Probabilistic Hesitant fuzzy elements	MNIST	Optimized algorithm, performs better
[14]	Haar cascade classifier, SVM, CNN, fuzzy logic	Raw Data, FERC-2013	95% accuracy obtained
[15]	Fuzzy logic, VGG16	HAR dataset from Kaggle	The model outperformed the existing methodologies
[16]	Fuzzy logic, DNN	CIFAR-10	65.20% accuracy on CIFAR-10
[17]	CNN, fuzzy logic	BH-Sig260	86.75% accuracy obtained on the hybrid model
[18]	Fuzzy logic with ML algorithms (SVM, KNN, CNN)	Raw data	The model outperformed existing models, yielding better accuracy

3 Proposed Methodology

In decision-making, especially in case of situations where ambiguity may arise, FL offers flexibility, while CNNs excel in pattern recognition. Our proposed hybrid model integrates FL and CNNs aiding robust decision-making as well as accurate pattern recognition, significantly beneficial in many applications such as healthcare [19]. Our methodology uses Fuzzy c-means clustering (FCM) to extract signature region from an image. This region of interest (RoI) is then fed into a CNN architecture for peculiar feature learning and determining authenticity of signature.

- **Preprocessing:** The signature images taken from the BHSig260 dataset, require preprocessing to reduce noise. This is achieved by image enhancement, normalization, and segmenting as required.
- **RoI Extraction:** These preprocessed images are then processed using FCM in order to isolate the RoI consisting the signature. FCM basically clusters pixels based on intensity values, assigning each pixel a certain degree of membership. This helps identifying and clustering these pixels within the RoI representing signature. Unlike K-Means, that uses hard assignments, FCM facilitates soft clustering, in which each pixel may belong to more than one cluster with varying membership degrees. This provides greater flexibility in image segmentation.

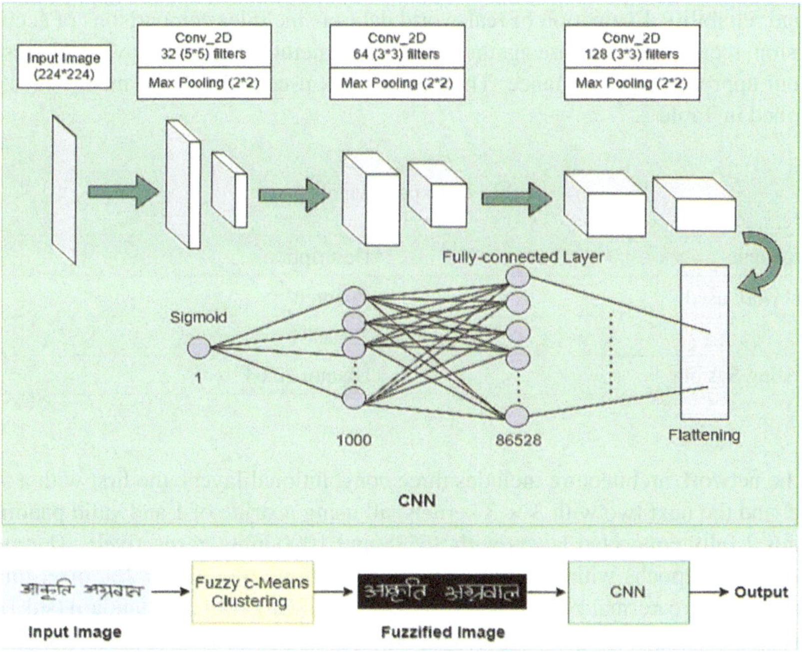

Fig. 1. Proposed architecture for our hybrid Fuzzy-CNN model

- **Feature Extraction:** This ROI is then inputted to the CNN architecture to extract features from the signature image. This process helps the CNN to learn a range of features encapsulating essential characteristics of the signature such as its shape, texture, and curvature. A CNN is a deep learning model that excels at image recognition, often used for processing grid-like data, e.g. images. It employs convolutional layers to apply filters for detecting patterns, and creating feature maps. These feature maps are then down-sampled by pooling layers facilitating dimensionality reduction. Then these flattened feature maps are processed by fully connected layers for predicting the final outcome.
- **Classification:** Finally, the images are classified into real or forged signature by the CNN.

Figure 1 elucidates the proposed methodology. This figure visually outlines the steps employed in our methodology, illustrating how FL and CNN are integrated together for signature verification in our work.

4 Experimental Results

This section showcases a comprehensive analysis of the deduced findings of our work, highlighting the effectiveness of our proposed system. Our results depict the potential to enhance signature verification, particularly in critical applications emphasizing security and reliability. Evaluation of real-world datasets includes comparisons of accuracy, precision, recall, and f1-Score against established methods, providing valuable insights into our approach's performance. The components used to build the model have been described in Table 2.

Table 2. Experimental Setup

Component	Description
Framework used	PyTorch
GPU	Tesla T4
Operating System	Ubuntu 20.04

The network architecture includes three convolutional layers: the first with a 5×5 kernel and the next two with 3×3 kernels, all using a stride of 1 and valid padding. It features 2 fully connected layers with 86528 and 1000 units, respectively. The model, trained for 10 epochs with a sigmoid output activation, uses 224×224 pixel images. These images are normalized using mean [0.485, 0.456, 0.406] and standard (std.) deviation [0.229, 0.224, 0.225]. The training dataset consists of 8640 samples, with 80% for training, a batch size of 64, and the optimizer being used is Adam optimizer (learning rate 0.001). Binary cross-entropy loss evaluates the model's performance.

4.1 Dataset

Our work utilizes the BHSig260 [20] dataset of Hindi signatures, containing 260 genuine and 2400 forged signatures from 100 individuals, to address challenges in Hindi signature verification. Leveraging this dataset, we aim to develop a robust verification system tailored to the nuances of the Hindi script, promoting broader adoption in diverse sectors.

4.2 Performance Results and Evaluation

This sub-section aims to offer a comprehensive analysis of the system's performance, emphasizing its efficacy in accurately verifying handwritten signatures. The evaluation of the system's performance encompasses several key performance metrics detailed below in Table 3.

Table 3. Results Obtained on our Hybrid Model

Parameter	Result
Accuracy	89.29%
Precision	88.16%
Recall	88.98%
f1-Score	88.17%

4.2.1 Accuracy

Accuracy graph has been plotted to evaluate our model. The accuracy is depicted against the number of training epochs on both training and test datasets. It was observed after 10 epochs that our fuzzy-CNN hybrid model attained 99.78% training accuracy, showing excellent learning performance on training data. Also, the test accuracy was recorded at 89.29%, indicating the model's strong ability to generalize to unseen data.

$$Accuracy = \frac{TP + TN}{TP + TN + FP + FN} \tag{1}$$

True Positive (TP): A TP represents the case where the model correctly recognizes a positive instance.
False Positive (FP): It occurs when the model identifies a positive instance incorrectly as a negative instance.
True Negative (TN): A TN indicates that the model has correctly identified a negative instance.
False Negative (FN): A false negative occurs when the model identifies a negative instance incorrectly as a positive one.

4.2.2 Precision

Along with accuracy, we have assessed our model's performance on various other metrics also. For instance, the precision, value was recorded at 0.8816, gauging the effectiveness of our model in correctly identifying genuine signatures.

$$Precision = \frac{TP}{TP + FP} \tag{2}$$

4.2.3 Recall

Recall, which reflects the model's ability in identifying all positive instances, was 0.8898. This high recall value indicates that our model can successfully detect 88.98% of genuine signatures, minimizing missed positive instances.

$$Recall = \frac{TP}{TP + FN} \tag{3}$$

4.2.4 f1-Score

The f1-score draws a balance between the precision and recall. The f1-score for our model was recorded at 0.8817, demonstrating that it effectively identified positive instances maintaining a strong balance between the precision and recall.

$$f1 - Score = \frac{2 * Precision * Recall}{TP + FP\ Precision + Recall} \tag{4}$$

4.3 Model Prediction

To evaluate our HSV model, we generated predictions for real as well as fake signature images, visually comparing the model's performance on real v/s fake image data, offering deeper insights into its effectiveness. Figure 2 depicts sample images of predictions drawn by our model.

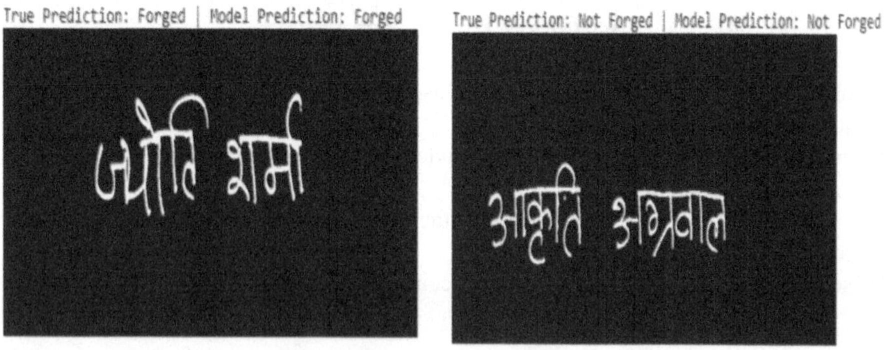

Fig. 2. Sample images illustrating predictions made by the model for Hindi signatures

5 Conclusion and Future Work

This paper presents a novel approach in verifying signatures amalgamating FCM clustering and CNN, also showcasing significant progress in handwritten signature authentication. Our hybrid approach effectively addresses the challenges imposed by variations in signatures due to factors such as fatigue, age, or illness, yielding an impressive 89.29% accuracy. The use of FCM clustering in our work handles uncertainty in signature data using soft assignment to multiple clusters, improvising on adaptability to different writing styles. Moreover, integrating CNNs helps capturing intricate patterns, improving the system's reliability in segregating genuine images from forged signatures. The key contributions of our work include the development of a hybrid approach combining FCM clustering with CNN for robust signature verification. Furthermore, our work embarks achievement of high accuracy in this challenging domain with significant variations in signature styles, as well as ensures enhanced adaptability & reliability through soft clustering & deep learning (DL) techniques. In future, we aim on expanding the dataset encompassing a wider range of handwriting styles & languages and refining the model's architecture, making it more robust and scalable for broader real-world applications.

References

1. Zhang, T.P., Fang, B., Xu, B., Chen, H.X., Chen, M., Tang, Y.Y.: Signature envelope curvature descriptor for offline signature verification. Proceedings of the 2007 International Conference on Wavelet Analysis and Pattern Recognition, vol. 3, pp. 1262–1266. IEEE, Beijing, China. 2–4 Nov 2007. https://doi.org/10.1109/ICWAPR.2007.4421628
2. Akhundjanov, U., et al.: Off-line handwritten signature verification based on machine learning. E3S Web of Conferences **508**, 03011 (2024). https://doi.org/10.1051/e3sconf/202450803011
3. Batuwita, K.B.M.R., Bandara, G.E.M.D.C.: Fuzzy recognition of offline handwritten numeric characters. In: Proceedings of the 2006 IEEE Conference on Cybernetics and Intelligent Systems, pp. 1–5. IEEE, Bangkok, Thailand (2006). https://doi.org/10.1109/ICCIS.2006.252356
4. Prakash, G.S., Sharma, S.: Computer vision & fuzzy logic based offline signature verification and forgery detection. In: Proceedings of the 2014 IEEE International Conference on Computational Intelligence and Computing Research, pp. 1–6. IEEE, Coimbatore, India (2014). https://doi.org/10.1109/ICCIC.2014.7238363
5. Zhang, C., Zhou, Z., Lin, L.: Handwritten digit recognition based on convolutional neural network. In: Proceedings of the 2020 Chinese Automation Congress (CAC), pp. 7384–7388. IEEE, Shanghai, China, 6–8 Nov 2020. https://doi.org/10.1109/CAC51589.2020.9326781
6. He, K.: Handwritten digit recognition based on convolutional neural network. In: Proceedings of the 2023 IEEE 3rd International Conference on Electronic Technology, Communication and Information (ICETCI), pp. 16–19. IEEE, Changchun, China. 26–28 May 2023. https://doi.org/10.1109/ICETCI57876.2023.10176680
7. Zheng, J., Zhang, Y., Wang, X., Liu, X., Li, J.: DWSCNN online signature verification algorithm based on CAE-MV feature dimensionality reduction. IEEE Access **12**, 22144–22157 (2024). https://doi.org/10.1109/ACCESS.2024.3355449

8. Gader, P.D., Keller, J.M. (eds.) Fuzzy methods in handwriting recognition: an overview. In: Proceedings of the North American Fuzzy Information Processing, pp. 137–141. IEEE, Berkeley, CA, USA. 19–22 June 1996. https://doi.org/10.1109/NAFIPS.1996.534718
9. Yazdanbakhsh, O., Dick, S.: A deep neuro-fuzzy network for image classification. Neural Evol. Comput. (2019). https://doi.org/10.48550/arXiv.2001.01686
10. Anikin, I.V., Anisimova, E.S. (eds.) Handwritten signature recognition method based on fuzzy logic. In: Proceedings of the 2016 Dynamics of Systems, Mechanisms and Machines (Dynamics), pp. 1–5. IEEE, Omsk, Russia. 15–17 Nov 2016. https://doi.org/10.1109/Dynamics.2016.7818968
11. Agrawal, A.K., Shrivas, A.K., Awasthi, V.K. (eds.): A robust model for handwritten digit recognition using machine and deep learning techniques. In: Proceedings of the 2021 2nd International Conference for Emerging Technology (INCET), pp. 1–4. IEEE, Belagavi, India. 21–23 May 2021. https://doi.org/10.1109/INCET51464.2021.9456118
12. Sudharshan, D.P., Vismaya, R.N. (eds.): Handwritten signature verification system using deep learning. In: Proceedings of the 2022 IEEE International Conference on Data Science and Information System (ICDSIS), pp. 1–5. IEEE, Hassan, India. 29–30 July 2022. https://doi.org/10.1109/ICDSIS55133.2022.9915833
13. Zhou, W., Liu, M., Xu, Z.: The dual-fuzzy convolutional neural network to deal with handwritten image recognition. IEEE Trans. Fuzzy Syst. **30**(12), 5225–5236 (2022). https://doi.org/10.1109/TFUZZ.2022.3170657
14. Abhishek, K., Appuhamy, E.J.G.S. (eds.): Emotion classification comparison in convolutional neural networks and fuzzy logics for service robotic applications. In: Proceedings of the 2023 3rd International Conference on Advanced Research in Computing (ICARC), pp. 36–41. IEEE, Belihuloya, Sri Lanka. 23–24 Feb 2023. https://doi.org/10.1109/ICARC57651.2023.10145753
15. Nithin, P.S., Alekhya, J.L., Enosh, P., Devika, Y. (eds.): Human action recognition using fuzzy logic and VGG16 classifier. In: Proceedings of the 2023 Second International Conference on Advances in Computational Intelligence and Communication (ICACIC), pp. 1–5. IEEE, Puducherry, India. 7–8 Dec 2023. https://doi.org/10.1109/ICACIC59454.2023.10435332
16. Mammadli, A., Ismayilov, E. (eds.): Application of fuzzy deep learning to classify images. In: Proceedings of the 2023 IEEE 17th International Conference on Application of Information and Communication Technologies (AICT), pp. 1–5. Baku, Azerbaijan. IEEE. 18–20 Oct 2023. https://doi.org/10.1109/AICT59525.2023.10313155
17. Priya, I., Chaurasia, N. (eds.): Estimating the effectiveness of CNN and fuzzy logic for signature verification. In: Proceedings of the 2023 Third International Conference on Secure Cyber Computing and Communication (ICSCCC), pp. 01–06. 26–28 May 2023. IEEE, Jalandhar, India. https://doi.org/10.1109/ICSCCC58608.2023.10176476
18. Yadav, R.S., Singh, A., Yadav, S.P. (eds.): The smart fuzzy based image recognition model for modern healthcare. In: Proceedings of the 2024 2nd International Conference on Disruptive Technologies (ICDT), pp. 250–254. IEEE, Greater Noida, India. 15–16 March 2024. https://doi.org/10.1109/ICDT61202.2024.10489059
19. Priya, I., Chaurasia, N., Singh, A.K., Mehta, N., Kilak, A.S., Alkhayyat, A.: Fuzzified deep learning based forgery detection of signatures in the healthcare mission records. ACM Trans. Asian Low-Resour. Lang. Inform. Process. (2023). https://doi.org/10.1145/3641818
20. Handwritten Signature Datasets (2021)

21. Nagarajan, D., Chourashia, K., Udhayakumar, A.: Neuro-fuzzy logic application in speech recognition. In: Peng, S.-L., Jhanjhi, N.Z., Pal, S., Amsaad, F. (eds.) Proceedings of 3rd International Conference on Mathematical Modeling and Computational Science: ICMMCS 2023, pp. 1–9. Springer Nature Singapore, Singapore (2023). https://doi.org/10.1007/978-981-99-3611-3_1
22. Narayan, L.K., Vishwakarma, V.P.: A Study on different fuzzy image enhancement techniques. In: Peng, S.-L., Jhanjhi, N.Z., Pal, S., Amsaad, F. (eds.) Proceedings of 3rd International Conference on Mathematical Modeling and Computational Science: ICMMCS 2023, pp. 135–142. Springer Nature Singapore, Singapore (2023). https://doi.org/10.1007/978-981-99-3611-3_11

An Extension of Marshall Olken Weibull Inverse Weibull Distribution with Application

Hemin A. Mohammed(✉) and Qasim N. Hussien

Department of Mathematics, College of Education for Pure Sciences, University of Tikrit, Tikrit, Iraq

ha230024pep@st.tu.edu.iq

Abstract. By combining the Inverse Weibull Distribution with The Marshal Olkin Weibull H-Family, a novel continuous distribution was introduced in this study. The Marshall Olkin Weibull Inverse Weibull Distribution is the name of the distribution that has been suggested. Then, the probability density function (PDF) and cumulative distribution function (CDF) of the new distribution are determined. Also extracted are a number of crucial functions, including the new distribution's survival function and hazard function. After that, the probability density function and the cumulative distribution function were both increased.

Keywords: Marshall Olken Weibull Family · Probability function · Distribution function · Moment. Quantitative function · inverse Weibull

1 Introduction

In 1997, Marshall and Olkin presented a new family of continuous distributions that are more flexible and meet the analysis requirements, called the Marshall-Olkin family. In 2010, Inverse Weibull distribution was expanded using the beta function by Khan et al. In this research, a new class of continuous distributions called the Marshall-Olkin-Weibull inverse-Weibull family will be defined by combining the inverse-Weibull distribution with the Marshall-Olkin Weibull suite.

Then by simulation method with the statistical software R the result has been compared using some statistical criteria like AIC, AICc, BIC, and HQIC to verify the efficiency of the suggested distribution in terms of flexibility and fitting for different kinds of data sets.

2 Methods

Inverse Weibull Distribution

The inverse Weibull distribution is a simple modification of the Weibull distribution, and is used in many fields such as medical, reliability and environment. Let X be a random

S. Pal and Á. Rocha (Eds.): ICMMCS 2025, LNNS 1398, pp. 170–179, 2025.
https://doi.org/10.1007/978-3-031-90998-6_17

variable having an inverse Weibull distribution with cdf and pdf are given respectively as follows (Alkarni, S. et al. 2020) [1]:

$$F(X) = e^{-b(X)^{-a}} \qquad X > 0, a, b > 0 \tag{1}$$

$$f(X) = abe^{-b(X)^{-a}}(X)^{-a-1} \tag{2}$$

where $a, b > 0$ and $X > 0$, a represents the shape parameter while b is the scale parameter.

Marshal Olkin Weibull H-Family

The cdf and pdf for Marshal Olkin Weibull H-Family can be given by:

$$G(X;\theta) = \frac{1 - Exp\left\{-\left[\frac{H(X)}{\overline{H}(X)}\right]^{\beta}\right\}}{1-(1-\theta)Exp\left\{-\left[\frac{H(X)}{\overline{H}(X)}\right]^{\beta}\right\}} X \epsilon R \tag{3}$$

$$g(X;\theta) = \frac{\theta\beta h(x)\left(H(x)^{\beta-1}\right)\left(1 - Exp\left\{-\left[\frac{H(X)}{\overline{H}(X)}\right]^{\beta}\right\}\right)}{\left(\overline{H}(X)\right)^{\beta+1}\left[1 - Exp\left\{-\left[\frac{H(X)}{\overline{H}(X)}\right]^{\beta}\right\}\right]^2} \tag{4}$$

where $\overline{H}(x) = 1 - H(x)$, and $\overline{\theta} = 1 - \theta$, $\theta > 0$ (Afify et al., 2022) [2].

In this paragraph, a new distribution was presented resulting from the composition of The Marshal Olkin Weibull H-Family with the Inverse Weibull Distribution. In order to derive a completely new distribution, namely the Marshal Olkin Weibull Inverse Weibull Distribution. Code may be used to encode the new distribution. (MOWIW) This new distribution consists of five features.

Distribution Function (CDF) of MOWIW Distribution

For the purpose of obtaining the cumulative distribution function CDF for the new distribution, we substitute Eq. (1) into Eq. (3) and obtain a new formula with relation to the upgraded distribution, which has the following format:

$$G(X) = \frac{1 - Exp\left\{-\left[\frac{e^{-b(X)^{-a}})}{1-e^{-b(X)^{-a}}}\right]^{\beta}\right\}}{1-(1-\theta)Exp\left\{-\left[\frac{e^{-b(X)^{-a}}}{1-e^{-b(X)^{-a}}}\right]^{\beta}\right\}} \tag{5}$$

where $\theta, b, a,$ and β are the parameter of MOWIW $\theta, b, a,$ and $\beta > 0.X > 0$. (Fig. 1)

For the purpose of obtaining the probability density function PDF for the new distribution, we substitute Eq. (2) with Eq. (4) and obtain a formula about the revised distribution, which has the following shape:

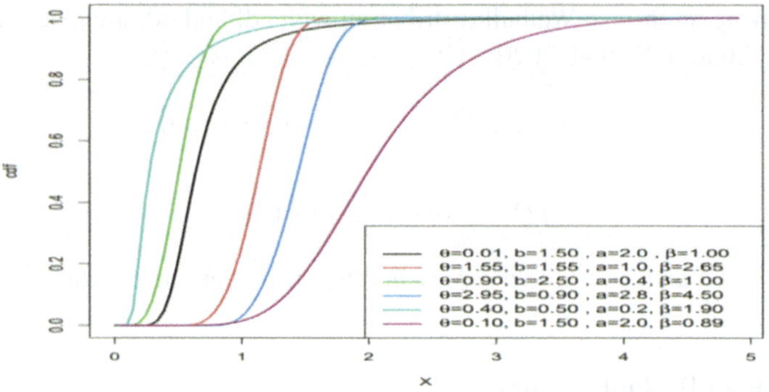

Fig. 1. For varying parameter values, the MOWIW distribution's CDF distribution function is shown.

$$g(X) = \frac{\theta\beta abe^{-b(X)^{-a}}(X)^{-a-1}\left(\left[e^{-b(X)^{-a}}\right]^{\beta-1}\right)\left(1-e^{\left\{-\left[\frac{e^{-b(X)^{-a}}}{1-e^{-b(X)^{-a}}}\right]^{\beta}\right\}}\right)}{\left(1-e^{-b(X)^{-a}}\right)^{\beta+1}\left[1-e^{\left\{-\left[\frac{e^{-b(X)^{-a}}}{1-e^{-b(X)^{-a}}}\right]^{\beta}\right\}}\right]^{2}}$$

$$g(X) = \theta\beta abe^{-b(X)^{-a}}(X)^{-a-1}\left(\left[e^{-b(X)^{-a}}\right]^{\beta-1}\right)\left(1-e^{-b(X)^{-a}}\right)^{-(\beta+1)}\left(1-e^{\left\{-\left[\frac{e^{-b(X)^{-a}}}{1-e^{-b(X)^{-a}}}\right]^{\beta}\right\}}\right)^{-1} \tag{6}$$

where $\theta, b, a,$ and β are the parameter of MOWIW $\theta, b, a,$ and $\beta > 0$ $.X > 0$ (Fig. 2).

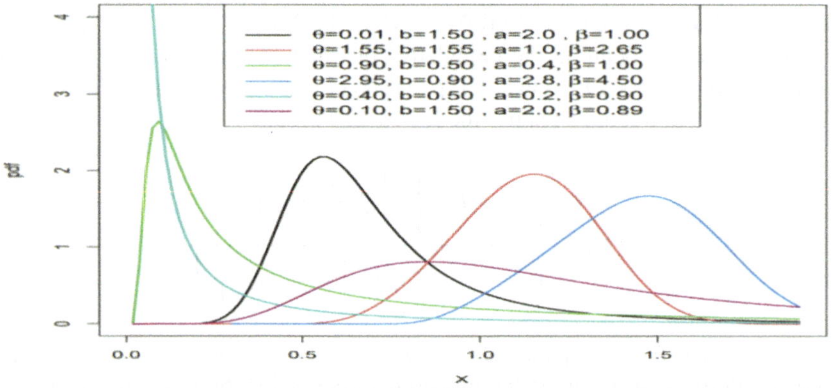

Fig. 2. A plot of the MOWIW distribution's PDF for various parameter values.

Survival Function

We can obtain the formula for the survival function through the following relationship: (Fox et al., 2002) [3].

$$S(X) = 1 - g(X)$$

$$S(X; \theta, b, a, \beta)_{\mathrm{MOWIW}} = 1 - \theta\beta a b e^{-b(X)^{-a}} (X)^{-a-1} \left(\left[e^{-b(X)^{-a}} \right]^{\beta-1} \right) \left(1 - e^{-b(X)^{-a}}\right)^{-(\beta+1)} \left(1 - \mathrm{e}^{\left\{ -\left[\frac{e^{-b(X)^{-a}}}{1-e^{-b(X)^{-a}}} \right]^{\beta} \right\}} \right)^{-1} \tag{7}$$

where $0 \leq S(X; \theta, b, a, \beta)_{\mathrm{MOWIW}} \leq 1$ and $\theta, b, a,$ and $\beta > 0$ $.X > 0$ (Fig. 3).

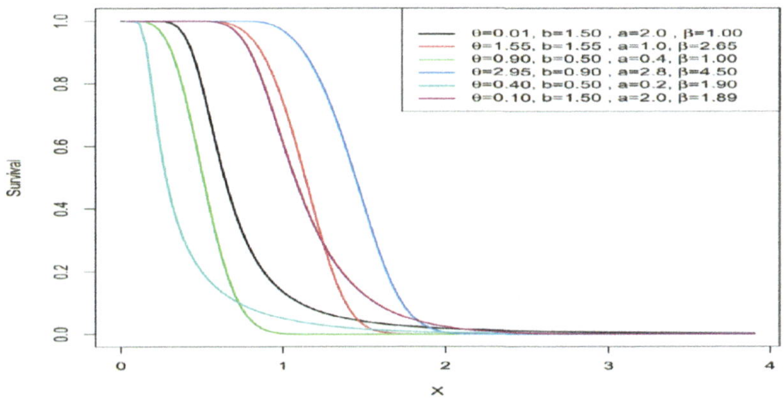

Fig. 3. Plot of the $S(X; \theta, b, a, \beta)_{\mathrm{MOWIW}}$ for different values of the parameters.

The hazard function can be formed as: (Johnson et al., 1975) [4]

$$h(x; \theta, b, a, \beta)_{MOWIW} = \frac{f(x; \theta, b, a, \beta)_{MOWIW}}{S(x; \theta, b, a, \beta)_{MOWIW}} \tag{8}$$

$$\left[\frac{\theta\beta a b e^{-b(x)^{-a}} (x)^{-a-1} \left(\left[e^{-b(x)^{-a}} \right]^{\beta-1} \right) \left(1 - e^{-b(x)^{-a}}\right)^{-(\beta+1)} \left(1 - \mathrm{e}^{\left\{ -\left[\frac{e^{-b(x)^{-a}}}{1-e^{-b(x)^{-a}}} \right]^{\beta} \right\}} \right)^{-1}}{1 - \theta\beta a b e^{-b(x)^{-a}} (x)^{-a-1} \left(\left[e^{-b(x)^{-a}} \right]^{\beta-1} \right) \left(1 - e^{-b(x)^{-a}}\right)^{-(\beta+1)} \left(1 - \mathrm{e}^{\left\{ -\left[\frac{e^{-b(x)^{-a}}}{1-e^{-b(x)^{-a}}} \right]^{\beta} \right\}} \right)^{-1}} \right] \tag{9}$$

where $x > 0$ and $\theta, b, a, \beta > 0$ (Fig. 4).

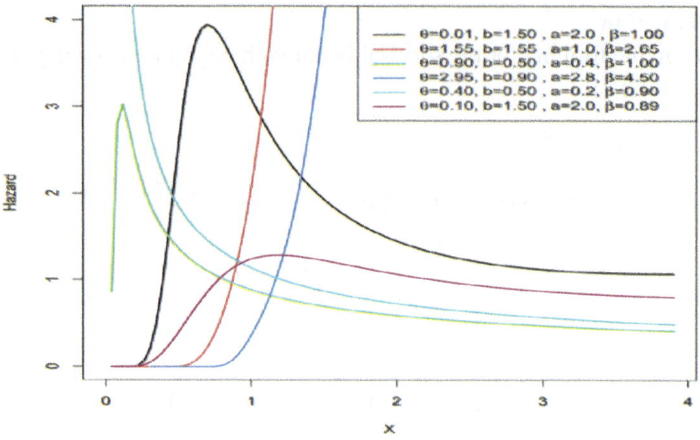

Fig. 4. Plot of the $h(X; \theta, b, a, \beta)_{\text{MOWIW}}$ For various parameter values.

2.1 Some Statistical Properties of the New Distribution

Moments

The mathematical expectation is one of the statistical properties that are used in moments. It is divided into two parts. The first part is called the moments about the origin and is symbolized by the symbol $\mu_r = E(x^r)$, and the second part is called the moments about the mean or central moments.

We can obtain the moments about the origin of degree r of the EChEW distribution from the following relationship by relying on the probability density function after expanding it according to Eq, (5).

$$\mu_r = E\left(X^r\right)_{MOWIW} = \mho_{p,q,k,v} \int_0^\infty X^r a\delta b(X)^{-a-1} e^{-b\delta(X)^{-a}} dX \tag{10}$$

Now assume that:

$$u = b\delta(X)^{-a} \Rightarrow du = -ab\delta(X)^{-a-1} dX$$

$$uX^a = b\delta \Rightarrow X^a = b\delta u^{-1} \Rightarrow X = (b\delta)^{\frac{1}{a}} u^{\frac{-1}{a}}$$

When $u = 0$, $X \Rightarrow \infty$ and $u \Rightarrow \infty$ then $X = 0$. (Kersey 2010) [5]

$$\mu_r = E\left(X^r\right)_{MOWIW} = \mho_{p,q,k,v} \int_0^\infty \left[(b\delta)^{\frac{1}{a}} u^{\frac{-1}{a}}\right]^r e^{-u} du$$

We integrate this type using gamma integration, and we obtain the final form of the equation, which is as follows. (Artin, 2015) [6]

$$\mu_r = E\left(X^r\right)_{MOWIW} = \mho_{p,q,k,v} \left[(b\delta)^{\frac{1}{a}}\right]^r \Gamma\left(1 - \frac{r}{a}\right) \tag{11}$$

Now, Eq. (11) is for the moment function of order r in its simplest form, as:

Through Eq. (11), we can find the first moment μ_1, which represents the mean at $r = 1$, the second moment μ_2, the third moment μ_3, and the fourth moment μ_4 for a new Marshall Olkin Weibull inverse Weibull Distribution, which It shall be as follows:

$$\mu_1 = E\left(X^1\right)_{MOWIW} = \mho_{p,q,k,v}\left[(b\delta)^{\frac{1}{a}}\right]^1 \Gamma\left(1 - \frac{1}{a}\right) \tag{12}$$

$$\mu_2 = E\left(X^2\right)_{MOWIW} = \mho_{p,q,k,v}\left[(b\delta)^{\frac{1}{a}}\right]^2 \Gamma\left(1 - \frac{2}{a}\right) \tag{13}$$

$$\mu_3 = E\left(X^3\right)_{MOWIW} = \mho_{p,q,k,v}\left[(b\delta)^{\frac{1}{a}}\right]^3 \Gamma\left(1 - \frac{3}{a}\right) \tag{14}$$

$$\mu_4 = E\left(X^4\right)_{MOWIW} = \mho_{p,q,k,v}\left[(b\delta)^{\frac{1}{a}}\right]^4 \Gamma\left(1 - \frac{4}{a}\right) \tag{15}$$

Characteristic Function

The characteristic function of the new Marshall Olkin Weibull inverse Weibull distribution can be found as follows.

$$Q_{(X)}(t) = E\left(e^{itX}\right) = \int_0^\infty e^{itX} f(X)dx \tag{16}$$

$$Q_{(X)}(t)_{MOWIW} = E\left(e^{itX}\right) = \mho_{k,L,M,p,q} \int_0^\infty e^{itX}(q+1)g(X;\zeta).G(X;\zeta)^q dX$$

$$Q_{(X)}(t)_{MOWIW} = E\left(e^{itX}\right) = \mho_{p,q,k,v}\sum_{z=0}^{\infty}\frac{i^z.t^z}{z!}\int_0^\infty X^z a\delta b(X)^{-a-1}e^{-b\delta(X)^{-a}}dX \tag{17}$$

$$Q_{(X)}(t)_{MOWIW} = E\left(e^{itX}\right) = (\Psi_z)\mho_{p,q,k,v}[b\delta]^{\frac{z}{a}}\Gamma\left(1 - \frac{z}{a}\right)$$

The Quantile Function

The partition function was studied by the researcher Hyndman in 1996 [7]. We can find the quantity function through the inverse of the new distribution, the Gombertztube-Lyon distribution and the inverse Weibull, relying on the cumulative distribution function of the new distribution MOWIW. Therefore, it can be represented mathematically as follows.

$$u = G^{-1}(X), u \in (0, 1) \tag{18}$$

$$u = 1 - \frac{1 - \mathrm{Exp}\left\{-\left[\frac{e^{-b(X)^{-a}})}{1-e^{-b(X)^{-a}}}\right]^\beta\right\}}{1 - (1-\theta)\mathrm{Exp}\left\{-\left[\frac{e^{-b(X)^{-a}}}{1-e^{-b(X)^{-a}}}\right]^\beta\right\}} \tag{19}$$

After a series of operations, the expression and the root a are simplified to both sides as follows:

$$X = \sqrt[a]{\frac{-\ln\left[\frac{\left[-\ln\left(\frac{1}{[[\theta]-u[\theta]+1]}\right)\right]^{\frac{1}{\beta}}}{\left(1+\left[-\ln\left(\frac{1}{[[\theta]-u[\theta]+1]}\right)\right]^{\frac{1}{\beta}}\right)}\right]}{b}} \quad b \neq 0 \; a > 0 \tag{20}$$

Through the last equation, the Median can be obtained by setting u=0.5, and the skewness, oblateness, quartiles, octaves, decimals, etc. can be found.

Kurtosis

The scientist Karl Pearson proposed a formula for the kurtosis of the new MOWIW distribution, which has the following formula:

$$Ku = \frac{\mu_4}{\mu_2^2} - 3$$

Since μ_4 is the fourth central moment and μ_2 is the second central moment, it depends on the value of Ku.

$$Ku_{\mathrm{MOWIW}} = \frac{\mho_{p,q,k,v}\left[(b\delta)^{\frac{1}{a}}\right]^4 \Gamma\left(1-\frac{4}{a}\right)}{\left(\mho_{p,q,k,v}\left[(b\delta)^{\frac{1}{a}}\right]^2 \Gamma\left(1-\frac{2}{a}\right)\right)^2} - 3 \tag{21}$$

If $Ku_{\mathrm{MOWIW}} > 3$, this means that the new MOWIW distribution is tapered (leptokurtic) and its degree increases with increasing degree Ku_{MOWIW}.

If $Ku_{\mathrm{MOWIW}} < 3$, this means that the new MOWIW distribution is flat (platykurtic).

If $Ku_{\mathrm{MOWIW}} = 3$, this means that the new MOWIW distribution is neither pointy nor flat, but rather has a moderate shape and is sometimes called average kurtosis.

Application

This data represents engineering data, which represents a data set for the tensile strength of glass fibers with a diameter of 1.5 cm, taken from European laboratories. This data has received extensive analysis from many researches, and we mention some of them, such as (Oguntunde et al., 2018) [8], here is where the glass fiber engineering info is:

(0.55, 0.74, 0.77, 0.81, 0.84, 1.24, 0.93, 1.04, 1.11, 1.13, 1.30, 1.25, 1.27, 1.28, 1.29, 1.48,

1.36, 1.39, 1.42, 1.48, 1.51, 1.49, 1.49, 1.50, 1.50, 1.55, 1.52, 1.53, 1.54, 1.55, 1.61, 1.58,

1.59, 1.60, 1.61, 1.63, 1.61, 1.61, 1.62, 1.62, 1.67, 1.64, 1.66, 1.66, 1.66, 1.70, 1.68, 1.68,

1.69, 1.70, 1.78, 1.73, 1.76, 1.76, 1.77, 1.89, 1.81, 1.82, 1.84, 1.84, 2.00, 2.01,

2.241.518).

Some measures for this sample of data were also extracted using the describe function provided by the R statistical program. Table 1 shows the descriptive statistics for the data (Fig. 5).

Table 1. The parameter values for the real data set with the MOWIW distribution and other competing distributions.

Distributions	Statistical criteria					
	AIC	CAIC	BIC	HQIC	W^*	A^*
MOWIW	32.140	32.829	40.712	35.511	0.109	0.612
GamIW	93.548	93.954	99.977	96.076	1.144	6.089
BeIW	50.89 6	51.585	59.468	54.267	0.481	2.620
KuIW	49.134	49.823	57.706	52.505	0.451	2.468
EGIW	56.435	57.124	65.007	59.806	0.572	3.107
WeIW	39.941	40.630	48.513	43.312	0.273	1.498
TEESIW	48.801	49.208	55.230	51.330	0.480	2.628
MOIW	75.509	75.916	81.938	78.038	0.8585	4.695
IW	97.706	97.906	101.992	99.392	1.225	6.486

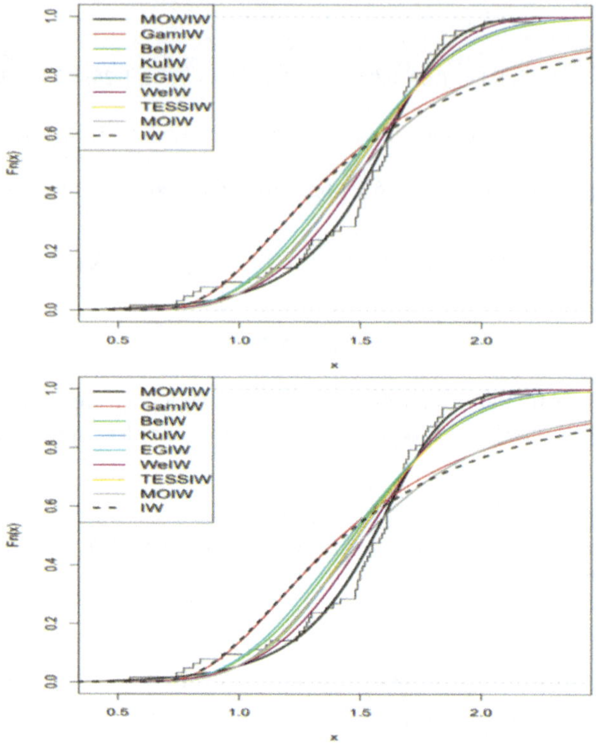

Fig. 5. The histogram and curves for the MOWIW distribution and some competing distributions.

3 Conclusion

Statistical distributions are crucial in inferences and applications to fit advanced real-world phenomena. In this study, a new distribution called MOWIW is proposed and applied to real data taken from the site. Several other distributions were used in the research to determine the suitability of the proposed distribution with these distributions. Through the results shown in the experiment, it is clear that the new distribution is preferable to the distributions compared to it and is distributable. There is great flexibility in modeling data of this type. We recommend that researchers experiment with the distribution of other data and see that the results will be positive or that they conduct an expansion that improves the accuracy of the model's match with real data.

References

1. Alkarni, S., Afify, A.Z., Elbatal, I., Elgarhy, M.: The extended inverse Weibull distribution: properties and applications. Complexity **2020**, 1–11 (2020)
2. Afify, A.Z., Al-Mofleh, H., Aljohani, H.M., Cordeiro, G.M.: The Marshall–Olkin–Weibull-H family: estimation, simulations, and applications to COVID-19 data. J. King Saud Univ.-Sci. **34**(5), 102115 (2022)

3. Fox, E., Russo, R., Dutton, K.: Attentional bias for threat: Evidence for delayed disengagement from emotional faces. Cognit. Emot. **16**(3), 355–379 (2002)
4. Johnson, N.L., Kotz, S.: A vector multivariate hazard rate. J. Multivariate Anal. **5**(1), 53–66 (1975)
5. Kersey, J.X.: Weighted inverse Weibull and beta-inverse Weibull distribution (2010)
6. Artin, E.: The gamma functions. Courier Dover Publications (2015)
7. Hyndman, R.J., Fan, Y.: Sample quantiles in statistical packages. The Am. Stat. **50**(4), 361–365 (1996)
8. Oguntunde, P.E., Adejumo, A.O., Khaleel, M.A., Okagbue, H.I., Odetunmibi, O.A.: The logistic inverse exponential distribution: Basic structural properties and application. World Congress on Engineering (2018)

Exploring the Role of Augmented Reality (AR) Advertising on Brand Engagement in Jordanian FMCG Sector

Bikramjit Sarkar[1], Souvik Pal[2,3(✉)], Vijayesvaran Arumugam[4], S. R. K. Juhi Reashma[5,6], and Saurabh Adhikari[7]

[1] Department of Computer Science and Engineering, JIS College of Engineering, Kalyani, India
bikramjit.sarkar@jiscollege.ac.in

[2] Department of Management Information Systems, Saveetha College of Liberal Arts and Sciences, Saveetha Institute of Medical and Technical Sciences, Chennai, India
Souvikpal22@gmail.com

[3] Department of Computer Science and Engineering, Sister Nivedita University (Techno India Group), Kolkata, India

[4] Department of Business and Healthcare Management, School of Business and Technology, IMU University, Kuala Lumpur, Malaysia
vijayesvaran@imu.edu.my

[5] Department of Computer Science, Faculty of Science and Humanities, SRM Institute of Science and Technology, Kattankulathur, Chengalpattu, India

[6] Department of Computer Science, SSKV College of Arts and Science for Women, Kanchipuram, India

[7] School of Engineering, Swami Vivekananda University, Barrackpore, India
saurabhadhikari@svu.ac.in

Abstract. Advertisements help establish the reputation of brands and provoke consumer engagement by conveying brand message deliberately and creating memorable experiences. Well-executed advertising campaigns can create a strong bond between consumers and inspire loyalty that will last ages. Influences of augmented reality (AR) ads on customer interaction with Jordan's fast-moving consumer goods (FMCG) brands are examined in this paper, which highlights both positive and negative aspects of this marketing gimmick. The approach to brand engagement in the FMCG sector has made customers less interested in FMCG brands due to sterile forms of brand interactions that the industry usually employs. To address these problems, a different way of engaging called brand engagement through augmented reality technology (BE-ART) is recommended by the authors. This would enable FMCG firms to establish meaningful connections with their customers through immersive brand experiences. Employ BE-AR in advertising campaigns for Jordanian FMCG companies with a view to enhancing consumer interaction, perceived usefulness, Consumer attitudes towards AR and purchase intention. Moreover, it increases research evidence towards such applications as BE-AR for fostering efficient brand relationships within marketing literature investigations concerning augmented reality use in marketing perspective.

S. Pal and Á. Rocha (Eds.): ICMMCS 2025, LNNS 1398, pp. 180–189, 2025.
https://doi.org/10.1007/978-3-031-90998-6_18

Keywords: brand engagement · Augmented Reality · Fast-Moving Consumer Goods

1 Introduction

The rise of digital technology has completely transformed the landscape of advertising, leading to new forms of consumption experience never seen before. One such development is AR which offers more interactive and immersive experiences than conventional media by Niedermeier, A., Mergel, C (2024) [1]. For an industry like FMCG where creative ways are always sought after to enchant them into participation this change is notable.

Augmented reality (AR) advertisements may contain engaging tales to make them more enticing to buyers. This is in contrast to the experience of uninteresting product interactions, in which customers may only view characteristics of the product (Tanudiharjo, R., Yun, F (2021)) [2].

Customers in Jordan are always looking for new methods to engage with companies, regardless of where they are located for the most part. An example of something that may satisfy this requirement is commercials that use augmented reality, which provide something that is distinct from traditional advertisements George, A. S., & George, A. H. (2023) [3].

FMCGs Augmented reality marketing may help Jordanian firms engage customers. Augmented reality gives consumers more immersive and engaging experiences, helping firms stand out in a competitive market. Relationships between vendors and purchasers typically boost sales. The fast development and widespread availability of technology has led to numerous augmented reality ads on internet sites. However, it implies better news for buyers as corporations may now engage them creatively instead of just using billboards. This article examines how AR commercials affect Jordanian brand-customer interactions. Customers respect augmented reality technologies for improving their shopping experience. When helpful, AR technologies like virtual try-ons and product representations increase customer contact, perceived utility, AR attitudes, and purchase intention. Because helpful and entertaining AR experiences may make purchasing easier, more educational, and more enjoyable, enhancing the brand-consumer bond. Augmented reality lets consumers imagine furniture in their living space, making them more inclined to buy from that firm.

2 Literature Review

2.1 AR Advertising on Brand Engagement

Advertisements in AR smartphone apps provide users more ways to engage with brands, which may make ads more effective and get people to like them. Sung, E., Han (2022) [4] examines escapism's role by investigating its causes and consequences. By analyzing two mental imaging processes, this research aims to provide insight into the potential of augmented reality mobile app advertising to enhance escapist experiences. Two of these mechanisms—reactions to narrative advertisements—are spatial immersion and narrative transportation.

Yang, S., Carlson, J. R (2020) [5], suggested several consumer-oriented firms have adopted AR technology and apps to improve customer shopping experiences, increase brand engagement, and increase profits. Conversely, there has been a dearth of research on AR potential in the advertising industry. An augmented reality ad improves consumers' attitude toward the commercial by increasing their interest in and focus on the ad (as measured physiologically by eye-tracking).

Ham, J., Li, S, (2023) [6] research on the consequences of social media's hybrid reality/virtuality on anthropomorphism, authenticity, advertising perception, and implicit behaviour is expanding, and this study adds to that body of knowledge. Virtual human influences were the setting for this paper.

Incorporating the pleasure factor and innovation diffusion and flow theories, Lin, K. Y., & Huang, T. K. (2024) [7] examine the factors influencing customer involvement with augmented reality commercial systems. An integrated research approach is introduced to study what makes augmented reality technology so special that consumers have a good time and are satisfied with AR services. Customer engagement intentions will be developed as a result. Various digital policies and pricing models [15–19] have been examined in the pertinent research literature, which underpins Brand Engagement in this article and associated prior procedures.

2.2 Fast-Moving Consumer Goods Sector

Lacy and Long (2020) [8] researched the possible functions of ten different industries within the circular economy and how these businesses might benefit greatly from embracing circular principles. Some of the businesses under this category are metals and mining, oil and gas, chemicals, electricity, machinery and industrial equipment, personal mobility, fashion, home goods, and FMCG.

Derqui and Fayos's (2022) [9] research aims to clarify the dynamics of downstream open innovation (OI) in the fast-moving consumer goods industry and identify successful innovations' traits. The paper also delves into the causes of the decline in innovations with disruptive potential in this industry inside the European Union (EU).

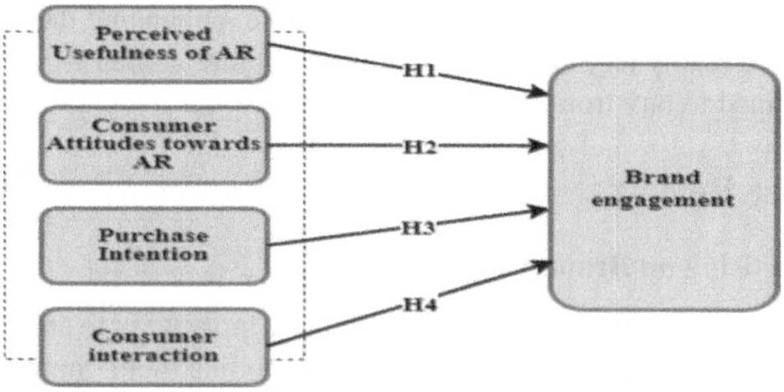

Fig. 1. Conceptual Framework of AR Advertising and Brand Engagement

- H1: Consumers perceive AR as useful so they tend to engage with it more often.

- H2: Positive attitudes towards AR lead to more brand engagement.
- H3: Higher purchase intentions will result in an increased level of involvement with a particular brand.
- H4: Active consumer interaction with AR features increases brand engagement.

Figure 1 explains the conceptual framework of AR advertising and brand engagement is explained below,

3 Methodology

3.1 Research Design

The purpose of this study is to provide a comprehensive understanding of the impact that AR advertising has on brand engagement in the Jordanian FMCG industry. Accordingly, this research makes use of a mixed-methods research methodology, which combines qualitative and quantitative research approaches Asgharizadeh, E., Daneshvar, A. (2023) [10]. To gather information on the viewpoints, attitudes, and levels of involvement of customers in relation to augmented reality advertising, a survey is used as part of the quantitative component. Included in the qualitative component are in-depth interviews with marketing professionals as well as focus groups with consumers (Paul, D., Malik, S (2023)) [11]. To get a deeper understanding of the factors that contribute to the success of augmented reality advertising, the objective of these interviews and groups is to study more fundamental insights and contextual factors of consumer interaction.

3.2 Population and Sample

Within the Jordanian fast-moving consumer goods (FMCG) industry, the population for this research consists of both customers and marketing experts. Individuals between the ages of 18 and 55 who are residents of Jordan's metropolitan regions and who have access to internet services and cellphones make up the customers in this market group. This group has been selected because they are more likely to come into contact with augmented reality advertising and because they constitute a significant section of the fast-moving consumer goods industry. To guarantee that the sample is representative of the population, a multi-stage sampling procedure is used (https://www.kaggle.com/datasets/karnikakapoor/video-ads-engagement-dataset). The collection of 3 million annotated advertising auction lines aims to further Machine Learning, notably for predicting video commercial user involvement. Cyrille Dubarry worked hard on this dataset for an École Polytechnique Machine Learning class tournament (Fig. 2).

3.3 Measures

Using established measures, several dimensions are examined to evaluate the effect of augmented reality advertising on brand engagement. Engagement with a brand is evaluated using a multi-dimensional scale that considers cognitive, emotional, and behavioural engagement aspects. Existing scales, such as the Brand Engagement in Self-Concept (BESC) scale, are used as a basis for developing these items. The perceived usefulness

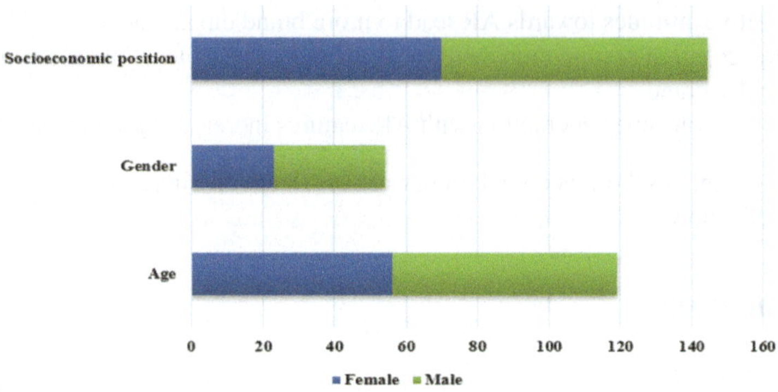

Fig. 2. Demographic profile of the sample

of augmented reality (AR) was evaluated using questions derived from the Technology Acceptance Model (TAM). These items examine how customers feel that augmented reality advertising improves their shopping enjoyment Sarangi, S., Chakraborty, A., & Triantis, K. P. (2021) [12].

3.4 Analytical Methods

The statistical methods are used to examine the quantitative data obtained from the survey. Descriptive statistics provide a concise summary of the demographic traits and important factors (consumer interaction, perceived usefulness, Consumer attitudes towards AR and purchase intention). To test hypotheses and investigate the connections between AR advertising and brand engagement, and A method known as theme analysis is used to examine qualitative data obtained from interviews and focus groups Mahajan, Y. (2020) [13].

4 Findings

4.1 Measurement Model Assessment

Evaluation of the validity and reliability of the measuring technique was carried out as part of our inquiry into the influence that augmented reality advertising has on the level of brand engagement in the Jordanian fast-moving consumer goods sector Prashar, A., & Sunder M, V. (2024) [14]. Evaluation was conducted on a variety of different variables, some of which were as follows: consumer interaction, Perceived Usefulness of Augmented Reality (AR), Consumer Attitudes Towards Augmented Reality Advertising, and Purchase Intention. Convergent and discriminant validity were used to assess the validity of the study, while Cronbach's alpha and composite reliability (CR) were utilized to determine the dependability of the findings. This assessment procedure included both of these approaches within its overall framework. The results are broken out in further depth in Table 1, which can be seen down below.

Table 1. Result of Validity and Reliability

Constructs	Items	Cronbach's Alpha	Composite Reliability (CR)	Average Variance Extracted (AVE)
Perceived Usefulness of AR	4	0.846	0.861	0.605
Consumer Attitudes towards AR	5	0.857	0.874	0.598
Purchase Intention	6	0.892	0.905	0.612
Consumer interaction	3	0.822	0.603	0.643

The Cronbach's alpha and composite reliability (CR) scores for each construct are more than 0.7, which indicates that the constructs have a high degree of internal consistency. It has been shown that the Average Variance Extracted (AVE) values are more than the threshold of 0.5, which is evidence that each construct has convergent validity. Based on these findings, it can be concluded that the constructs that were used to evaluate the influence of augmented reality advertising on brand engagement are highly trustworthy and valid.

4.2 Descriptive Statistics

The data is provided by descriptive statistics, which include the mean, the standard deviation, and correlations between various constructs. Some examples of these constructions are brand engagement, perceived usefulness of augmented reality, consumer attitudes toward augmented reality advertising, and purchase intention. A summary of the findings may be seen in the Table 2 that follows.

Table 2. Mean, Standard Deviation, and correlation

Constructs	Mean	SD	1	2	3	4	5
Perceived Usefulness of AR	3.61	0.915	0.441	1			
Consumer Attitudes towards AR	3.74	0.867	0.482	0.459	1		
Purchase Intention	3.51	0.925	0.624	0.598	0.627	1	
Consumer interaction	3.44	0.743	0.322	0.673	0.563	0.673	1

The concept mean values ranged from 3.51 to 3.75, indicating that respondents generally liked and engaged with augmented reality advertising. Overall, mean values are positive. The standard deviations range from 0.867 to 0.925, indicating enough response diversity. The constructs' correlation coefficients range from 0.441 to 0.627, indicating moderate to strong positive associations. Brand Engagement and Purchase Intention are most correlated (0.624), whereas Brand Engagement and Augmented Reality Perceived Usefulness are least correlated (0.441). 0.322 and 0.482 are consumer interaction and AR attitude. These findings support the measurement model's discriminant validity by showing that the constructs are distinct yet related. The positive correlation between the two variables shows that as consumers develop favorable opinions of augmented reality advertising and feel it to be useful, their involvement with the brand and buy intentions will increase.

4.3 Structural Model Assessment

The structural model assessment indicates that AR advertising significantly enhances brand awareness, brand attitude, and brand engagement in the Jordanian FMCG sector. The high R-squared values suggest that AR experiences are strong predictors of these constructs. Additionally, the model fit indices confirm a good fit between the hypothesized model and the observed data, underscoring the potential of AR advertising to engage consumers effectively in this market (Fig. 3).

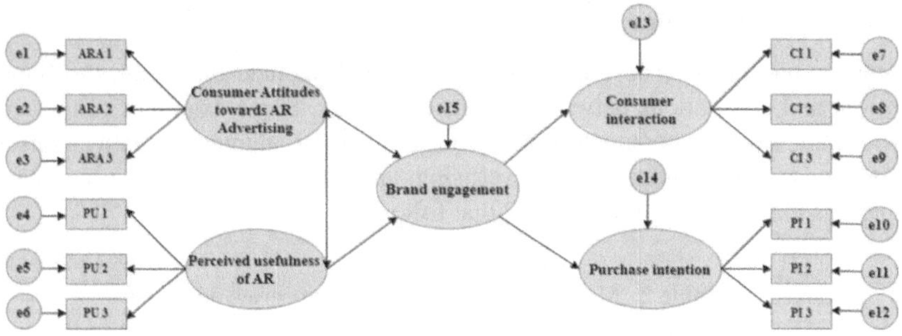

Fig. 3. Structural Model for AR Advertising on Brand Engagement

The structural model for examining the impact of AR advertising on brand engagement in the Jordanian FMCG sector involves four key constructs: Perceived Usefulness of AR (PUAR), Consumer Attitudes towards AR Advertising (CAAR), Brand Engagement (BE), and Purchase Intention (PI). The model hypothesizes that PUAR influences CAAR, BE, and PI directly. Additionally, CAAR affects BE and PI, and BE influences PI. This model visually represents how AR advertising can enhance consumer engagement and drive purchase intentions, highlighting the interconnected pathways between consumer perceptions, attitudes, interaction and behaviors in response to AR advertising.

Table 3 shows route coefficients from a research on brand engagement and several characteristics. A row represents a distinct predictor variable's impact on brand engagement. The table displays the unstandardized coefficient (B), standardized coefficient (β), standard error (S.E.), t-value, and p-value for each predictor. As shown by their positive

Table 3. Result of path coefficients

			B	β	S.E	t-value	P-value
Perceived Usefulness of AR	□	Brand engagement	0.121	0.136	0.173	4.81	0.01
Consumer Attitudes towards AR	□	Brand engagement	0.124	0.134	0.167	5.51	0.001
Purchase Intention	□	Brand engagement	0.126	0.143	0.169	6.52	0.006
Consumer interaction	□	Brand engagement	0.321	0.193	0.171	7.22	0.002

standardized coefficients and significant p-values, perceived utility of augmented reality (AR), customer attitudes about AR, consumer interaction and purchase intention all favorably affect brand engagement.

5 Discussion

The use of AR in advertising is a significant factor that contributes to the level of customer interaction with brands in the Jordanian FMCG sector, as shown by the data set. An illustration of the manner in which the different components interact with one another is provided by the structure of the representation. It is feasible to make the remark that the consumer's perceptions of AR and the value that they believe AR to provide are factors that determine the degree to which one gets engaged in augmented reality advertising. It is impossible to exaggerate the value of augmented reality advertising when it comes to increasing the number of interactions that consumers have with a company consumer interaction, perceived usefulness, Consumer attitudes towards AR and purchase intention and cultivating brand loyalty, as shown by this particular occurrence. Despite this, some difficulties still need to be surmounted, such as the limitations of the technology and the tolerance of the consumers. It has been shown that advertisements that make use of augmented reality are able to boost brand engagement. This finding underscores the value of this kind of media for marketers of fast-moving consumer products, especially those working in dynamic countries like Jordan.

6 Recommendation and Conclusion

It is recommended that FMCG companies operating in Jordan use AR advertising tactics since doing so would boost consumer interaction, perceived usefulness, Consumer attitudes towards AR and purchase intention. It is necessary for marketers to create brand experiences with a focus on ensuring not only beauty but also meaning, taking into account individual preferences and lifestyle, as well as other values held by consumers during the selection process. This is done to ensure that the brand experiences not only appear visually appealing but also provide genuine utility to the consumers. Furthermore, the cultivation of trust among consumers may result in the facilitation of greater effectiveness linked with the enhancement of advertising efforts, while simultaneously resolving concerns about the privacy and security of data. Augmented reality advertising has fantastic opportunities for FMCG companies in Jordan to differentiate themselves from the competition, provide experiences that are unforgettable, and cultivate brand loyalty.

7 Limitations and Potential Paths for Future Research

This article has a number of shortcomings, one of which is that it relied on self-reported data, which may be susceptible to inaccuracies and biases. During the course of future study, experimental designs or longitudinal studies may be used to get more convincing information about the causal connections that exist between augmented reality advertising and brand engagement. However, the scope of the current study was restricted to FMCGs in Jordan; hence, it is challenging to generalize the results to other industries or geographies. What would be advantageous would be to research how successful augmented reality advertising is across a broad variety of product categories and diverse cultural backgrounds. This would allow one to acquire relevant insights that relate to different product categories and different cultural backgrounds. Nevertheless, more study needs to investigate the ways in which individual characteristics, such as technology savvy and personality, impact the reactions of consumers to advertisements that are presented in augmented reality. This will assist in gaining a better understanding of the processes that contribute to the success of augmented reality advertising, as well as the implications these processes have for the engagement of businesses in forums of this kind.

References

1. Niedermeier, A., Mergel, C., Emberger-Klein, A., Menrad, K.: A predictive model approach to forecast consumers' cluster membership in the green fast moving consumer goods sector. EFB Bioecon. J. **4**, 100064 (2024)
2. Tanudiharjo, R., Yun, F., Joo, J., Arokiam, I.: Investigation of factors impacting lean implementation in the Indonesian fast-moving consumer goods industry. Operat. Supply Chain Manage. An Int. J. **14**(2), 162–172 (2021)
3. George, A.S., George, A.H.: FMCG's digital dilemma: the consequences of insufficient IT expertise in the fast-moving consumer goods industry. Partners Univ. Int. Innov. J. **1**(3), 46–69 (2023)
4. Sung, E., Han, D.I.D., Choi, Y.K.: Augmented reality advertising via a mobile app. Psychol. Mark. **39**(3), 543–558 (2022)
5. Yang, S., Carlson, J.R., Chen, S.: How augmented reality affects advertising effectiveness: the mediating effects of curiosity and attention toward the ad. J. Retail. Consum. Serv.Consum. Serv. **54**, 102020 (2020)
6. Ham, J., Li, S., Shah, P., Eastin, M.S.: The "mixed" reality of virtual brand endorsers: understanding the effect of brand engagement and social cues on technological perceptions and advertising effectiveness. J. Interact. Advert. **23**(2), 98–113 (2023)

7. Lin, K.Y., Huang, T.K.: Shopping in the digital world: How augmented reality mobile applications trigger customer engagement. Technol. Soc. **77**, 102540 (2024)
8. Lacy, P., Long, J., Spindler, W., Lacy, P., Long, J., Spindler, W.: Fast-moving consumer goods (fmcg) industry profile. The Circular Economy Handbook: Realizing the Circular Advantage, pp. 173–184 (2020)
9. Derqui, B., Fayos, T., Occhiocupo, N.: The virtuous cycle of trust. Unveiling clues to successful innovation in the Fast Moving Consumer Goods industry. Europ. J. Innov. Manag. **25**(6), 1036–1056 (2022)
10. Asgharizadeh, E., Daneshvar, A., Homayounfar, M., Salahi, F., Amini Khouzani, M.: Modeling the supply chain network in the fast-moving consumer goods industry during COVID-19 pandemic. Oper. Res. Int. J. **23**(1), 14 (2023)
11. Paul, D., Malik, S., Mishra, D.K., Teotia, A.: A Study on effectiveness of ecolabels in the fast-moving consumer goods sector. IOP Conf/ Ser.: Earth Environ. Sci. **1161**(1), 012004 (2023). https://doi.org/10.1088/1755-1315/1161/1/012004
12. Sarangi, S., Chakraborty, A., Triantis, K.P.: Multimarket competition effects on product line decisions–A multi-objective decision model in fast moving consumer goods industry. J. Bus. Res. **133**, 388–398 (2021)
13. Mahajan, Y.: Impact of Coronavirus pandemic on fast moving consumer goods (FMCG) sector in India. J. Xi'an Univ. Architect. Technol. 12 (2020)
14. Prashar, A., Sunder M.V.: Sustainability drivers of fast-moving consumer goods supply chains: a multi-method study. Benchmark. An Int. J. (2024)
15. Peng, S.-L., Jhanjhi, N. Z., Pal, S., Amsaad, F. (eds.): Proceedings of 3rd International Conference on Mathematical Modeling and Computational Science. Advances in Intelligent Systems and Computing. Springer Nature Singapore (2023). https://doi.org/10.1007/978-981-99-3611-3
16. Peng, S.-L., Hao, R.-X., Pal, S. (eds.):Proceedings of First International Conference on Mathematical Modeling and Computational Science. Advances in Intelligent Systems and Computing. Springer Singapore (2021). https://doi.org/10.1007/978-981-33-4389-4
17. Pathak, P., Pal, S., Maity, S., Jeyalaksshmi, S., Adhikari, S., Akila, D.: Analysis of improving sales process efficiency with salesforce industries CPQ in CRM. In: Sharma, D.K., Peng, S.-L., Sharma, R., Jeon, G. (eds.) Micro-Electronics and Telecommunication Engineering: Proceedings of 7th ICMETE 2023, pp. 481–495. Springer Nature Singapore, Singapore (2024). https://doi.org/10.1007/978-981-99-9562-2_40
18. Rana, A., et al.: A Data-Driven Analytical Approach on Digital Adoption and Digital Policy for Pharmaceutical Industry in India. In: Sharma, D.K., Peng, S.L., Sharma, R., Jeon, G. (eds.) Micro-Electronics and Telecommunication Engineering. ICMETE 2023. Lecture Notes in Networks and Systems, vol 894. Springer, Singapore (2024). https://doi.org/10.1007/978-981-99-9562-2_42
19. Singh, J., et al.: Analyze and compare the public cloud provier pricing model and the impact on corporate financial. In: Sharma, D.K., Peng, S.L., Sharma, R., Jeon, G. (eds.) Micro-Electronics and Telecommunication Engineering. ICMETE 2023. Lecture Notes in Networks and Systems, vol 894. Springer, Singapore (2024). https://doi.org/10.1007/978-981-99-9562-2_41

Enhancing Stress Detection: A Comprehensive Approach Using Wearable Sensors and Sentiment Analysis with CNN

Faizan Mirza[1], Nischay Tushir[1], Nitant Anand[1], Deepali Vora[1](✉), Shubhangi Deokar[1], and Aditi Sharma[2]

[1] Computer Science and Engineering, Symbiosis Institute of Technology, Pune Campus, Symbiosis International (Deemed University) (SIU), Pune, India
{mirza.baige.btech2021,nishchay.tushir.btech2021, nitant.anand.btech2021,deepali.vora, shubhangi.deokar}@sitpune.edu.in

[2] Department of Electrical and Computer Engineering, Nazarbayev University, Astana, Kazakhstan
aditi.sharma@sitpune.edu.in

Abstract. Stress substantially impacts both intellectual and physical health, growing the likelihood of mental troubles and chronic infection. This study indicates a completely unique combinational method for strain detection that combines mental and physical facts from wearable sensors (watches). In order to improve the accuracy of strain identity, emotional analysis is carried out to voice-transformed textual content information in the suggested observe. The method analyses textual content information and extracts the sentiment or emotional tone of a sentence the use of the Convolutional Neural Network (CNN) model for Natural Language Processing (NLP). To get more specific and beneficial findings, the bodily traits of the user are integrated with the sentimental analysis output generated by using a gadget learning model. After then, this information is stored on document for a totally long time. In order to higher classify the person's pressure stage, this statistic is then collected over an extended time period and shown as graphs. The aim of the undertaking is to increase a novel and green method of stress detection as a way to allow human beings to control their strain ranges.

Keywords: CNN · Sentimental Analysis · Stress · Physical Sensors · Psychological analysis · NLP

1 Introduction

In cutting-edge society, pressure has grown to be a vast problem [1], affecting humans' bodily and emotional properly-being. It has been proven that long-time period stress has destructive effects, inclusive of multiplied danger of developing some of continual ailments and mental issues. Accurate and well-timed pressure detection is crucial for each powerful strain control and average health. Conventional techniques of measuring

S. Pal and Á. Rocha (Eds.): ICMMCS 2025, LNNS 1398, pp. 190–201, 2025.
https://doi.org/10.1007/978-3-031-90998-6_19

pressure have depended on self-reporting, that is subjective and liable to bias. As a result, there is an increasing want to research current procedures that exactly stumble on and track stress stages the use of goal metrics. Because wearable sensor technology [2, 3] can acquire physiological data [5, 14, 15] associated with stress responses, they have got received reputation currently. Through non-stop tracking of variables inclusive of frame temperature, skin conductance, and heart rate, those gadgets offer critical facts approximately the physiological outcomes of stress. But concentrating best on clinical issues may not address how the thoughts and frame respond under strain. Emotional tone and other mental factors are crucial to comprehending a man or woman's strain. The accuracy and functionality of pressure sensors may be better by means of merging psychological cues with the physical information gathered by way of wearable technology. This look at offers a unique integration of wearable era, mainly smart watches, with the idea of speech-to-textual content statistics. The apparatus uses Convolutional neural network (CNN) fashions for natural language processing (NLP) are used as a cognitive evaluation of speech-to-speech data that describes the emotional tone in the sentence. The physical parameters from the apparatus are then integrated with the consequences of the analytical model to create a extra effective and efficient research. The amassed facts is tracked over the years and gives a specific depiction to decorate the consumer's stress class. This observes predominant intention is to indicate a novel, practical method to stress studies so one can assist humans control their stress stages. This have a look at attempts to establish a basic framework for gaining an objective information of the consequences of social contact and physical and psychological signs of stress via the usage of technology and theoretical analysis. The results of this may resource inside the advent of individualized pressure-discount plans and enhance well-known health.

1.1 Use of Machine Learning in Stress Detection

Machine gaining knowledge of (ML) performs a crucial role in stress detection based on the given context. Here are some methods device studying can be useful in this context:

1. **Testing Hypothesis**: Machine mastering techniques [4, 20], mainly Convolutional Neural Network (CNN) models, are used to perform notion evaluation of speech-to-speech facts. This leads to the exploration and evaluation of the emotional or emotional nation in the sentence. Machine studying algorithms examine patterns and associations in reading cloth to appropriately identify emotional stressors.
2. **Physical and Emotional Integration**: Machine gaining knowledge of algorithms facilitate the integration of the bodily frame captured via the device with the emotional outcomes. By combining those two records streams, device learning models can identify patterns and relationships that could indicate extra pressure than conventional techniques. Machine getting to know algorithms [20] excel at processing complicated and various statistics that could offer insights from an aggregate of physical and intellectual procedures.
3. **Improved Accuracy**: Machine gaining knowledge of algorithms [4] can improve the accuracy of pressure detection through mastering from big datasets and uncovering hidden patterns or functions that may not be apparent to people. Machine getting to know models trained on recorded data can clarify strain with the aid of identifying relationships among diverse physical and mental sensations.

4. **Personalization and Adaptation**: The ML version can change and adjust itself to understand stress relying at the man or woman and the version. By continuously mastering from user information, device mastering algorithms can tailor stress assessments [5] to people, along with their bodily and mental traits. This customization improves the accuracy and performance of strain detection and management strategies.
5. **Mod Presentation**: The use of system model to visualize strain levels may be facilitated by using system studying strategies. By changing gathered information into comprehensible visuals, system getting to know algorithms (ML) assist humans understand and analyse stress styles greater efficaciously [6]. By enabling humans to tune their pressure ranges and take proactive measures to manage their fitness, these insights contribute to the development of self-cognizance. In end, emotional evaluation, the fusion of intellectual and bodily techniques, multiplied accuracy, the ability to in my opinion apprehend tension algorithms, and the provision of visual imagery are only a few ways that device studying has substantially superior the research of tension. By enabling people to control their pressure, those gadget mastering algorithms offer a thorough, statistics-driven technique to strain studies, ultimately leading to upgrades in fitness, easy ingesting, and preferred properly-being.

1.2 Ease of Use

The plan is to integrate wearable sensors for stress tracking into each day existence, developing smooth-to-wear clever watches that can come across body issues. This non-invasive approach guarantees continuity and smooth monitoring of stress tiers. A person-pleasant voice-converted textual content records evaluation method is being evolved for sentiment detection [10, 16] converting audio documents to text and the usage of convolutional neural networks for natural language processing. This approach simplifies the technique, supplying users with a easy and intuitive experience.

The proposed method aims to create an intuitive and accessible stress detection device [7] for individuals. It features a consumer-centric interface, imparting actual-time remarks and notifications, and personalised stress management strategies [8]. This system is designed to be effortlessly integrated into daily routines, selling a proactive method to stress management and normal nicely-being.

Effortless management of strain ranges using the proposed combinational method: The most important goal of the concept is to offer an accessible, user-friendly, and intuitive pressure test. People can sincerely discover and engage with the gadget way to its intuitive design, clear commands, and consumer-pleasant functions. By giving customers immediately remarks and updates on their pressure stages, the stress detection gadget [9] permits users to take their time evaluating stress. Furthermore, the gadget ought to contain customized strain management strategies or advice primarily based at the requirements and possibilities of the consumer. Stress checking out is turning into a helpful tool that people can also use in their daily lives to sell stress management and improve universal health due to its simplicity of use and the significance of handy availability.

2 Related Works

A novel, integrated approach for strain detection is provided within the summary. It makes use of photograph representation of speech-to-text information, wearable sensors, and emotional sensors. Numerous studies on diverse sides of this strategy exist, which each show the development of strain studies and remedy and created the muse for its creation.

Studies like [1] have investigated the usage of physiological indicators which include heart rate, electro dermal pastime, and body temperature to screen the strain reaction inside the area of wearable device-based totally stress detection. This take a look at suggests that the tool's effectiveness may be utilised to acquire bodily facts in real-time for the purpose of gaining knowledge of stress. Growing emphasis on processing interventions and properly phrase calculation bureaucracy the inspiration for integrating sentiment analysis [21] and speech-to-records analysis for pressure evaluation. Furthermore, graphical representations are frequently hired in statistics visualisation and analysis for pressure control. Studies that have already been achieved, like [5], have regarded into how to carry strain and physical statistics through the years using charts and graphs. These research show how beneficial visualisation is for assisting to interpret and understand strain patterns. The purpose of integration is to enhance pressure consciousness and control by means of creating and incorporating content from applicable initiatives. It provides thorough and approachable facts for pressure evaluation and personal care with the aid of combining the functionality of wearable electronics, voice-to-text information belief [12], and image display. It affords thorough and approachable statistics for stress assessment and private care by means of combining the capability of wearable electronics, voice-to-text statistics belief and image shown. As shown in Table.1, various methodologies used previously along with the key findings are as follows:

Table 1. Key Findings with Different Methodologies

Authors	Methodology	Key Findings
Sheikh M. A. et al. [11]	Machine Learning (Random Forest)	Developed a Voltage Based Measure to assess motor senses with 85% accuracy
Gers et al. [13]	Deep Learning (Long Short-Term Memory Networks)	Using Smartphone Based Sensors and NLP Techniques to Detect Conversational Stress with 78% Accuracy
Chaoran Liu et al. [14]	Ensemble Learning (Support Vector Machine, K-NN, RF)	Combined with various Machine Learning Algorithms, the wearable sensor stress with an accuracy of 92%

(*continued*)

Table 1. (*continued*)

Authors	Methodology	Key Findings
Chen et al. [15]	Deep Learning (CNN)	Proposed a method to detect Anxiety using Faces and Achieved 81% Accuracy
Wang et al. [17]	Transfer Learning (Pre-Trained CNN)	Used Electro dermal activity and obtained accuracy of 87% by optimizing pre-trained models
Lee et al. [18]	Reinforcement Learning (Q-Learning)	Developed a Mobile Application that Combines Practical Tools and learning Techniques for Real Time Stress Research

3 Problem Statement

Stress is a serious trouble which can affect the body and thoughts, growing the chance of continual diseases and mental problems. Current practices for stress analysis and management regularly depend upon self-reporting, leading to bias and inaccuracy. A goal and consumer-centred method is wanted to accumulate more records to measure and manipulate stress stages. Additionally, cutting-edge methods often ignore the relationship between mind and frame throughout strain, impeding a higher understanding of stress patterns.

Targeting the above troubles, the proposed have a look at proposes a new combination for pressure detection. The proposed approach uses the power of portable devices, the detection of speech-to-text records, and visual representations to provide complete and user-friendly techniques for dimension, pressure, and control. First, the proposed studies have contributed by way of seamlessly integrating wearable sensors [7] into someone's daily existence. The proposed approach permits the improvement of wearable gadgets together with smart watches that are simple, unobtrusive, and capable of detecting bodily signs of stress. This mixture makes it smooth and comfortable to cope with anxiety without strain or soreness. Second, right here efforts to analyze the emotions of speech-to-textual content statistics [19] to perceive strain-related emotions. By converting speech records to textual content and the use of convolutional neural networks (CNNs) for herbal language processing (NLP), you'll be able to perceive and describe emotions in textual content. This application improves know-how of mood all through pressure and supports the evaluation of strain ranges. Third, the contribution by means of providing the stored statistics in a graphical illustration has been made. These visualizations offer remedy from depression by allowing humans to without problems interpret strain styles through the years.

4 Proposed Methodology

The proposed methodology revolves around a combinational method for strain detection, incorporating wearable sensors, sentiment evaluation on voice-converted textual content statistics, and system model display. The following steps define the method right here in the proposed methodology as follows:

Data Collection: Collecting statistics from individuals the use of wearable sensors, including smart watches has been performed inside the beginning. These sensors seize diverse physiological signals related to stress responses, such as heart price, pores and skin conductance, and body temperature. Simultaneously, voice data is accumulated and transformed into text the usage of suitable speech reputation strategies.

Sentiment Analysis [10, 21] on Voice-Converted Text Data: The voice-transformed text facts is subjected to sentiment evaluation using a Convolutional Neural Network (CNN) version for Natural Language Processing (NLP). The CNN version analyzes the textual content facts to discover emotional tones, sentiments, and underlying strain signs present inside the user's speech.

Integration of Physiological and Psychological Signals: The effects acquired from sentiment analysis are combined with the physiological signals [1, 5, 14, 15] captured by using the wearable sensors. By correlating the emotional tones expressed inside the voice-transformed textual content information with the physiological responses, a complete know-how of the person's stress degrees is completed. This integration enhances the accuracy and effectiveness of pressure detection.

Long-Term Data Recording and Analysis: The combined statistics, comprising physiological signals [22] and sentiment analysis effects, are recorded over an extended period. This long-time period statistics recording enables the identity of styles, trends, and triggers related to strain. By studying the statistics, won insights into the consumer's pressure patterns and their variations over the years.

System Model with OLED Display: The gathered statistics is displayed in the form of OLED display so that it's miles without difficulty comprehensible. These visualizations offer a simplified type of pressure degrees, enabling customers to visualise their stress patterns and fluctuations at a glance. Such representations facilitate easy interpretation and understanding of strain ranges, empowering people to music their stress over the years and make knowledgeable selections concerning pressure control.

Stress Level Awareness: Based at the analysed records and graphical representations, stress levels are categorized into distinctive categories, inclusive of low, moderate, or high. Users can screen their strain tiers and take appropriate movements to manipulate and reduce strain. The combinational approach affords insights into customized strain management strategies, empowering individuals to make proactive selections for his or her nicely-being.

5 Working of the System

Overall, the proposed method combines the energy of wearable sensors, sentiment evaluation on voice-transformed textual content statistics, and visual representations to provide an effective and person-pleasant approach to stress detection and control. By integrating both [22, 23] physiological and mental indicators, it offers a holistic know-how of stress degrees and enables personalized pressure control strategies for individuals. A machine that objectives to remedy stress troubles with the aid of combining realistic and emotional tools. The gadget makes use of a smart watch prepared with frame sensors to accumulate real-time physical statistics consisting of coronary heart charge, pores and skin conductivity, and frame temperature, imparting a better information of the frame's stress. To locate mental issues, the device plays sentiment analysis on changed speech statistics. A convolutional neural community (CNN) model [16] is used for natural language processing (NLP) to become aware of sounds or thoughts provided in sentences.

Combine emotional analysis effects [20] with physical samples gathered from wearable devices to growth the accuracy and effectiveness of pressure detection. The blended facts are recorded over a long time and provided in a graph to higher categorize the user's tension.

The proposed approach combines bodily and mental indicators, presenting an effective method for strain detection. This new aggregate allows greater correct and speedy analysis of strain levels, allowing human beings to manage pressure efficaciously.

The proposed technique helped broaden a new pressure research approach that allows control stress in people. It gives a framework for know-how the impact of social interactions and the bodily and psychological measures of strain. The findings of this take a look at have the capability to help the development of private pressure management strategies and improve average health and nicely-being.

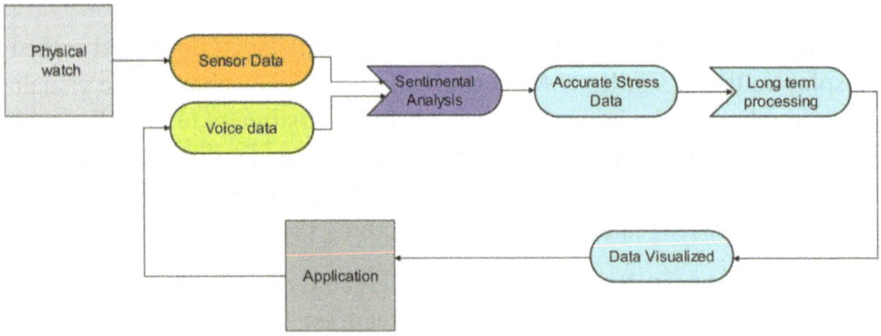

Fig. 1. Block Diagram

1. **Input**: As in the Fig. 1 of Block Diagram, the machine takes the person's input as a log and converts it into data for emotional evaluation as well as physiological records accumulated from gadgets inclusive of coronary heart charge, skin conductivity, and frame temperature.

2. **Concept Analysis**: Data processing for speech transformation is carried out the use of the Convolutional Neural Network (CNN) model for Natural Language Processing (NLP), which describes the noise inside the sentence or the present-day view.
3. **Physiological records**: Physiological information accrued by way of wearable devices presents real-time facts about the wearer's physiological reaction to strain, consisting of coronary heart rate, pores and skin conductivity, and frame temperature.
4. **Integration**: Combining emotional evaluation results with physical measurement facts to create a more efficient and effective stress perception system that consists of both intellectual and physical.
5. **Data output**: Integrated information is recorded and organized over the years, resulting in a higher fit among consumer strain and a higher know-how of control Respect pressure, and standard health.

Details of Sensors Used

The MAX30102 is a coronary heart charge and temperature sensor that can be used in devices together with watches. The sensor uses two LEDs (one infrared and one purple) to degree the amount of mild absorbed by way of the blood. The distinction between the absorption of the two LEDs is used to calculate the heart rate and oxygen inside the blood. The MAX30102 also has a temperature sensor that can be used to measure body temperature. The sensor may be used to screen pressure degrees via monitoring adjustments in coronary heart price, blood oxygen stages, and frame temperature. When someone is pressured, their coronary heart price and blood oxygen stages upward thrust and their body temperature may upward thrust. The MAX30102 can be used to monitor these modifications to provide early caution signs of strain. This have a look at has the ability to provide a new measure of pressure to assist people manage pressure. The method is still in its infancy; however, it has the capacity to alternate the manner we reflect on consideration on and manipulate stress.

Some blessings of the usage of the MAX30102 to monitor pressure stages are:

- Early warning of pressure may be detected.
- Stress degrees can be monitored over time.
- Can music the effects of strain management.
- People can recognize the motive of their strain and how to keep away from or avoid it.

Node MCU and it's Usage:

The ESP8266 NodeMCU is a low-value, open-source improvement board that may be used to build numerous IoT gadgets, which include smart watches. The NodeMCU is based on the ESP8266 Wi-Fi chip that may connect with the internet and speak with other devices. This is good for applications together with fitness tracking, weather tracking, and notifications. Here are a few applications ESP8266 NodeMCU may be used in a smart watch.

a. ***Fitness Tracker***: NodeMCU may be used to monitor heart charge, steps, and calories burned. This data can be used to reveal fitness dreams and improve overall fitness.

Weather tracking: NodeMCU can be used to get real-time climate updates. These statistics may be used to plot outdoor activities, dress as it should be, and live safe in inclement climate.

b. ***Notification:*** NodeMCU may be used to obtain notifications including text messages, emails, and new media announcements from smartphones. This allows users to stay linked even if far from their telephones.
c. ***Controlling Home Appliances***: NodeMCU may be used to govern clever domestic home equipment such as lighting, thermostats, and locks. This can be accomplished remotely, making it easy to control the home even when the user is away.

The ESP8266 NodeMCU is a powerful and versatile device that can be used to build a variety of smart watch applications (Fig. 2).

Fig. 2. System Model

6 Results and Discussions

The findings of this take a look at have the capacity to help increase techniques for managing non-public strain and enhancing general fitness and well-being. By expertise pressure styles and offering human beings with timely and accurate pressure tests, we are hoping to help humans efficaciously control their stress tiers. It is critical to acknowledge the limitations of this have a look at. The pattern length and length of the look at have been small, and more studies is needed with larger, extra numerous populations. In addition, while the proposed common approach shows promise, non-stop development and validity of the method are guaranteed. From the comparison in Table.2, these observations based on health parameters taken with the stress detection parameters which gives a beneficial approach to the look at of strain with the aid of combining electronic, mental, and experimental concerns. By integrating the bodily and emotional factors, we have established the capability to dissipate pressure more successfully and efficiently.

We desire proposed research will contribute to the development of stress management gear and in the long run help human beings lead healthy, for evaluation of the parameters

Table 2. Health Parameters and Comparison

Make of the Watch	Health Parameters	Stress Detection Parameters Used	Cost (in Rupees)	Target Market	App Features	Direct Competitor	Battery Capacity	Display Size
APPLE WATCH	3	No	Rs. 27900/-	Upper-class	Health Condition	Fit bit	245 mAh	44 mm
FITBIT SENSE	2	No	RS.18500/-	Upper-class	Fitness and Calorie	Apple watch SE	266 mAh	40 mm
SAMSUNG GALAXY WATCH- 4	3	Yes	Rs.13000/-	Upper-class	Fashion, Basic Health Features	Fit bit	361 mAh	44 mm
BOAT FLASH EDITION	5	Yes	RS.1500/-	Middle class	Low Price Value	Noise	2000 mAh	33 mm
NOISE COLOR FIT	2	Yes	Rs.2500/-	Middle class	Health Aspects	Boat	210 mAh	44.45 mm
VITA-HEALTH WATCH	3 to 7			Middle and lower class	Stress Detection Mental Health	Boat Flash Edition	450 mAh	33.02 mm

of the competitors, the proposed method of watch founds pleasant being a differentiator as follows:

- Health Parameters: Three (increasable up to 7)
- Stress Detection Parameter: Yes
- App Feature Centralized on: Coast: Rs.800 /- (For Prototype)
- Target Market: Middle and Lower class
- Battery Capacity: 420 mAh
- Display Size: 33.02 mm

7 Conclusion

In conclusion, this studies paper offers a new included method to the study of anxiety using practical and psychological markers. By combining physical facts amassed from wearable gadgets with the perception of speech-to-text data, we intention to increase the accuracy of pressure detection. Technology lets in us to display bodily parameters consisting of heart charge, pores and skin conductivity, and body temperature, allowing us to better apprehend the frame's resilience to pressure. However, we realize that focusing entirely on physical weakness can forget the vital effects of thoughts and body in stressful conditions. To cope with this, we combine the consequences of psychology, along with sound perception and emotion notion, within the proposed pressure studies. Using a convolutional neural network (CNN) model for herbal language processing (NLP), we analyse speech-to-text facts to decide the pronunciation of a sentence. The results of this evaluation are mixed with bodily parameters to create a greater green and powerful strain detection machine. The proposed studies is to provide desires and methods for identifying and controlling strain ranges. Using research and theoretical understanding, we intent to create a framework which will explain the effects of social, physical, and emotional measures of stress and can be considered for future research.

Future studies may investigate several directions to improve and expand on the current framework. For example, improving the CNN model's ability to grasp subtle emotional cues in speech could increase the accuracy of detection. Furthermore, using the system in various real-world contexts, such offices or classrooms, may offer insightful information on its usefulness. To further our understanding of stress, interdisciplinary collaboration combining knowledge in data science, physiology, psychology, and other fields is also possible. By pursuing these research avenues further, we intend to stimulate ongoing innovation in stress detection and management, thereby improving people's wellbeing in a variety of settings.

References

1. Gedam, S., Paul, S.: Towards mental stress detection using wearable physiological sensors. In: Proceedings of the 2021 IEEE International Conference on Power Electronics, Computer Science and Intelligent Control Systems (PECCS), pp. 1–6. IEEE (2021)
2. Elzeiny, S., Abdelrahman, M.: Mental stress detection using data from wearable and non-wearable sensors: a review. Sensors **21**(5), 1490 (2021)
3. Liu, Zhang, M., Zhang, L.: Continuous stress detection using wearable sensors in real life: Algorithmic programming contest case study. Sensors 21(11), 3891 (2021)
4. Sadruddin, S., Khairnar, V.D., Vora, D.R.: Issues and challenges in detecting mental stress from multimodal data using machine intelligence. SN Comput. Sci. **5**, 358 (2024). https://doi.org/10.1007/s42979-024-02730-7
5. Yadav, K., Shukla, A., Tiwari, R., Singh, Y.P.: Combining physiological and psychological signals to improve stress detection accuracy. IEEE Trans. Biomed. Eng. **69**(1), 1–12 (2022)
6. Gjoreski, M., Lustrek, M.G., Mlakar, M.: A wearable device for stress detection and management using physiological and psychological signals. IEEE Trans. Biomed. Eng. **65**(10), 2174–2183 (2018)
7. Al-Osaimi, M.H., Khalil, M., Al-Shehri, S., AlDossari, M.: Deep learning for stress detection using wearable sensors and smartphone data. IEEE Trans. Consum. Electron. **66**(4), 440–449 (2020)
8. Yang, Zhao, S., Li, Y., Jiang, X.: Stress detection using wearable sensors and multimodal data fusion. IEEE Sensors J. 22(1), 45–55 (2022)
9. Hossain, M.S., Muhammad, G.: A personalized stress detection system using wearable sensors and smartphone data. IEEE Trans. Biomed. Eng. **67**(12), 3003–3012 (2020)
10. Li, Y., Yang, H., Zhao, S., Jiang, X.: A wearable stress detection system using physiological signals and sentiment analysis. IEEE Trans. Biomed. Eng. **69**(4), 1394–1404 (2022)
11. Sheikh, M.A,. Iqbal, Mahgoub, I., Du, E., Leavitt, M.A., Asghar, W.: Advances in healthcare wearable devices. **9** (2021). https://doi.org/10.1016/j.matdes.2021.110164 0264-1275
12. Hossain, M.S., Islam, M.S., Muhammad, G.: Stress detection using wearable sensors and artificial intelligence: a review. IEEE Trans. Biomed. Eng. **68**(11), 3018–3030 (2021)
13. Gers, F., Schmidhuber, J., Cummins, F.: Learning to forget: continual prediction with LSTM. In: 1999 Ninth International Conference on Artificial Neural Networks ICANN 99. Conf. Publ. No. 470. https://doi.org/10.1049/cp:19991218

14. Li, Y., et al.: Integrated wearable smart sensor system for real-time multi-parameter respiration health monitoring **4**
15. Chen, Z., et al.: Network analysis of depression and anxiety symptoms and their associations with mobile phone addiction among Chinese medical students during the late stage of the COVID-19 pandemic. SSM - Population Health **25**, 101567 (2024). https://doi.org/10.1016/j.ssmph.2023.101567
16. Kumar, A.S., Tiwari R.: Detecting stress using wearable sensors and voice analysis, IEEE Access **9**, 133531133540 (2021)
17. Wu, D., Han, X., Yang, Z., Wang, R.: Exploiting transfer learning for emotion recognition under cloud-edge-client collaborations. IEEE J. Sel. Areas Commun. **39**(2), 479–490 (2020)
18. Chen, Q., Lee, B.G.: Deep learning models for stress analysis in university students: a sudoku-based study. Sensors **23**(13), 6099 (2023). https://doi.org/10.3390/s23136099
19. Yadav, K., Shukla, A., Tiwari, R., Singh, Y.P.: Combining physiological and psychological signals for stress detection using a hybrid deep learning model. IEEE Trans. Biomed. Eng. **69**(7), 2246–2259 (2022)
20. Ashwin, V.H., Jegan, R., Rajalakshmy, P.: Stress detection using wearable physiological sensors and machine learning algorithm. In: 2022 6th International Conference on Electronics, Communication and Aerospace Technology (ICECA). https://doi.org/10.1109/ICECA55336.2022.10009326
21. Cambria, E.: Affective computing and sentiment analysis. IEEE Intell. Syst. **31**(2), (2016). https://doi.org/10.1109/MIS.2016.31
22. Peng, S.-L., Lin, C.-K., Pal, S. (eds.): Analysis and classification of physiological signals for emotion detection. In: Proceedings of 2nd International Conference on Mathematical Modeling and Computational Science. Advances in Intelligent Systems and Computing. Springer Nature Singapore (2022). https://doi.org/10.1007/978-981-19-0182-9
23. Peng, S.-L., Hao, R.-X., Pal, S. (eds.): Proceedings of first international conference on mathematical modeling and computational science. In: Advances in Intelligent Systems and Computing. Springer Singapore (2021). https://doi.org/10.1007/978-981-33-4389-4

A Suggested Formula of Exponentiated Weibull Distribution with Application

Diyar J. Mohammed(✉) and Qasim N. Hussien

Department of Mathematics, College of Education for Pure Sciences, University of Tikrit, Tikrit, Iraq
dj230008pep@st.tu.edu.iq

Abstract. Via the literature of statistical studies, many continuous distributions have been presented that can help in many scientific and applied areas. on the other hand, variant properties of those distributions are derived and studied. However, there is a need to obtain highly flexible distributions to match different types of real data. For that purpose, new families of distributions have been established to have the best flexibility when dealing with variant kinds of data.

In this study, a proposed family of continuous distributions is presented by fitting the Exponentiated Weibull distribution with the Exponentiated Chen G-family to obtain the new distribution namely the Exponentiated Chen Exponentiated Weibull distribution. The cumulative distribution function (CDF) and the probability density function (PDF) have been derived. Also, some important functions such as the characteristic function, the quantile, the skewness, and kurtosis have been derived. Using real engineering data and data generated by the statistical software R and comparing the results with eight competing distributions, the suggested distribution showed highly flexible, efficient, and suitable for different real data sets using some statistical criteria such as AIC, AICC, BIC, and HQIC and more.

Keywords: Generated Exponential distribution · Chen-family · Moment · Quantile function · Order Statistics

1 Introduction

Numerous generalizations of continuous families of distributions have been introduced and covered over the last recent years for modeling data in different applied fields like economics, engineering, biology, environment, and financial affairs. Thus, many kinds of distributions have been established by matching common families of continuous distributions (Awodutire,2022) [1], presented the combined properties of the extended family are introduced. Some statistical properties of the family were considered, including moments, moment generating function, order statistics, and quantile function. The proposed generalized distributions gave more flexibility using additive one or more shape parameters. In this paper, a suggested combination of Chen G-family with weibull distribution has been presented to have a satisfied flexibility property to fit many Kinds of

S. Pal and Á. Rocha (Eds.): ICMMCS 2025, LNNS 1398, pp. 202–215, 2025.
https://doi.org/10.1007/978-3-031-90998-6_20

models or data sets. The results have been compared with some previous distributions using some statistical measuring tools such as AIC, AICC, BIC, and HQIC as shown in the application part.

2 Methodology

Exponentiated Chen Exponentiated Weibull Distribution

In this section, he presents a new distribution resulting from the combination of the Exponentiated Chen G family with the Exponentiated Weibull Distribution to obtain a new distribution, which is the Exponentiated Chen Exponentiated Weibull distribution. The new distribution can be encoded with a symbol. (EChEW) This new distribution consists of five parameters.

Let X be a random variable and let $G(x)$ be the probability distribution function (CDF) and g(x) is the probability density function PDF corresponding to the exponential Weibull distribution. The formula of CDF for the exponential Weibull distribution with two parameters can be written respectively as:

$$G(X)_{EW} = \left[1 - e^{-X^{\delta}}\right]^{a} \quad X > 0, a, \delta > 0 \tag{1}$$

$$g(X)_{EW} = a\delta(X)^{\delta-1}e^{-X^{\delta}}\left[1 - e^{-X^{\delta}}\right]^{a-1} \tag{2}$$

where a, $\delta > 0, X > 0,$ a represents the shape parameter, and δ is the scale parameter (Refaie et al., 2023) [2]

The distribution function (CDF) and probability density function (PDF) for the random variable X can be formulated as:

$$F(X)_{ECh-G} = \left[A\left(1 - e^{\tau(1-e^{G(X)^{\theta}}}\right)\right]^{\gamma} \tag{3}$$

$$f(X)_{ECh-G} = \gamma A^{\gamma}\tau\theta g(X)G(X)^{\theta-1}e^{G(X)^{\theta}}e^{G(X)^{\theta}}e^{\tau(1-e^{G(X)^{\theta}}}\left[1 - e^{\tau(1-e^{G(X)^{\theta}}}\right]^{\gamma-1} \tag{4}$$

Exponentiated-G family of distributions (Gupta et al. 1998) [3].

CDF for EChEW Distribution

For the purpose of obtaining the CDF for the new distribution, we substitute Eq. (1) into Eq. (3), then a formula for CDF of the new distribution ECHEW is obtained as follows:

$$\mathrm{F(X)_{EChEW}} = [\mathrm{A}(1 - \mathrm{e}^{\{\tau[(1-\mathrm{e}^{(1-\mathrm{e}^{-\mathrm{X}^{\delta}})^{\mathrm{a}}})^{\theta}]\}})]^{\gamma} \tag{5}$$

Such that γ, τ, θ, a and δ are the parameters of the new distribution ECHEW and $\gamma, \tau, \theta, a, \delta > 0$ (Fig. 1).

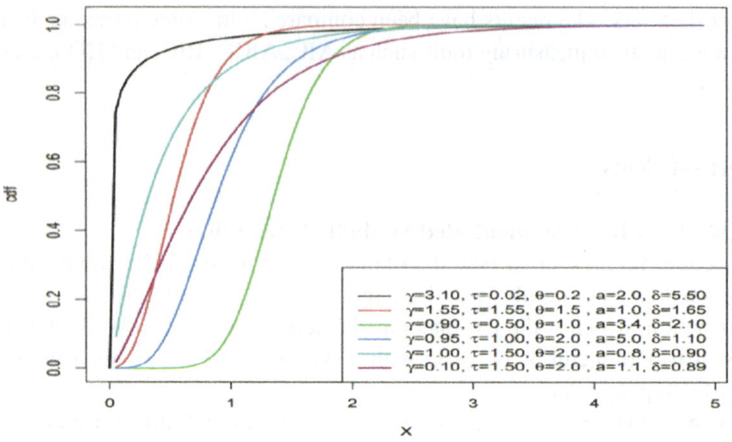

Fig. 1. CDF distribution function plot of the ECHEW distribution for different parameter values.

Probability Density Function (PDF)
To obtain the probability density function PDF for the new distribution, we substitute Eq. (2) and (1) into Eq. (4) and obtain a formula for the new distribution, which is in the following form:

$$f(x)_{\text{EChEW}} = \gamma A^{\gamma}\tau\theta a\delta(x)^{\delta-1}e^{-x^{\delta}}\left[1-e^{-x^{\delta}}\right]^{a-1}e^{\left[1-e^{-x^{\delta}}\right]^{a\theta}}e^{\tau\left(1-e^{\left(\left[1-e^{-x^{\delta}}\right]^{a}\right)^{\theta}}\right)}$$

$$\left[1-e^{\tau\left(1-e^{\left(\left[1-e^{-x^{\delta}}\right]^{a}\right)^{\theta}}\right)}\right]^{\gamma-1}\left(\left[1-e^{-x^{\delta}}\right]^{a}\right)^{\theta-1} \qquad (6)$$

where $\gamma, \tau, \theta, a, \delta > 0$.$x > 0$, $0 \leq f(x)_{\text{EChEW}} \leq 1$ (Fig. 2)

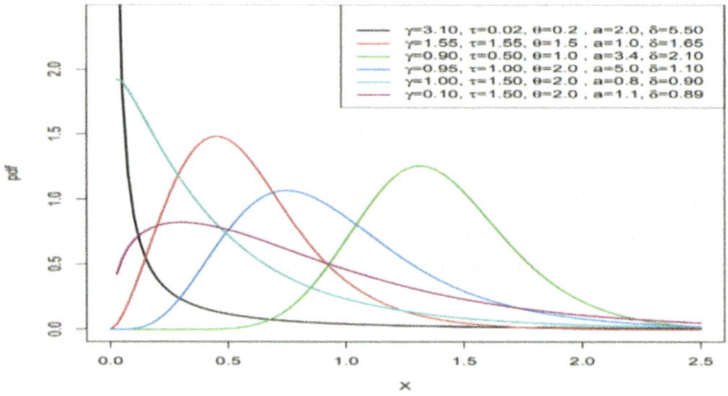

Fig. 2. PDF plot of the ECHEW distribution for different parameter values.

Survival Function

The survival function is one of the complementary functions to the distributional function of the new distribution in the possibility of preserving the organism after t of time. It is clear that the survival function decreases as the failure time increases. This is evidence that there is an inverse relationship between the failure time and the survival function, while the risk factor increases with the increase in failure time, so the relationship that unites them is a direct relationship. We can obtain it through the following relationship: (Fox et al., 2002) [4].

$$S(x) = 1 - F(x) \tag{7}$$

$$S(x)_{\text{EChEW}} = 1 - F(x; \gamma, \tau, \theta, a, \delta)_{\text{EChEW}}$$

$$S(x)_{\text{EChEW}} = 1 - \left[A\left(1 - e^{\left\{ \tau \left[\left(1-e^{(1-e^{-x^{\delta}})^{a}} \right)^{\theta} \right] \right\}} \right) \right]^{\gamma} \tag{8}$$

where $x > 0$ and $\gamma, \tau, \theta, a, \delta > 0$, $0 \leq S(x)_{\text{EChEW}} \leq 1$ (Fig. 3)

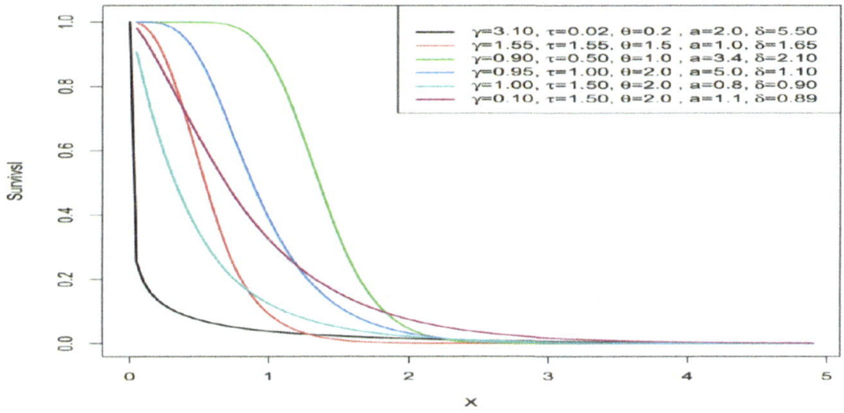

Fig. 3. Plot of the survival function $S(x)_{\text{EChEW}}$ for the new EChEW distribution for different values of the parameters.

Hazard Function

The hazard function is an important function that provides a mathematical model for survival function data and is also important in engineering applications. Therefore, some researchers worked to find distributions of different shapes that may be increasing at times or may be decreasing or constant, which are positive functions. The risk function may have a specific formula for the duration of failure or failure, and this period is limited between two times $t-1$ and $t-2$.. The probability function may be a single conditional function at time, and the risk function can be found using the following relationship: (Johnson et al., 1975) [5] (Fig. 4)

$$h(X)_{\text{EChEW}} = \frac{f(X;\gamma,\tau,\theta,a,\delta)_{\text{EChEW}}}{S(X;\gamma,\tau,\theta,a,\delta)_{\text{EChEW}}}$$

$$\left[\frac{\gamma A^{\gamma}\tau\theta a\delta(X)^{\delta-1}e^{-x^{\delta}}[1-e^{-x^{\delta}}]^{a-1}([1-e^{-x^{\delta}}]^{a})^{\theta-1}e^{[1-e^{-x^{\delta}}]^{a\theta}}e^{\tau(1-e^{([1-e^{-x^{\delta}}]^{a})^{\theta}}}\left[1-e^{\tau(1-e^{([1-e^{-x^{\delta}}]^{a})^{\theta}}}\right]^{\gamma-1}}{1-\left[A\left(1-e^{\{\tau[(1-e^{(1-e^{-x^{\delta}})^{a}})^{\theta}]\}}\right)\right]^{\gamma}}\right] \tag{9}$$

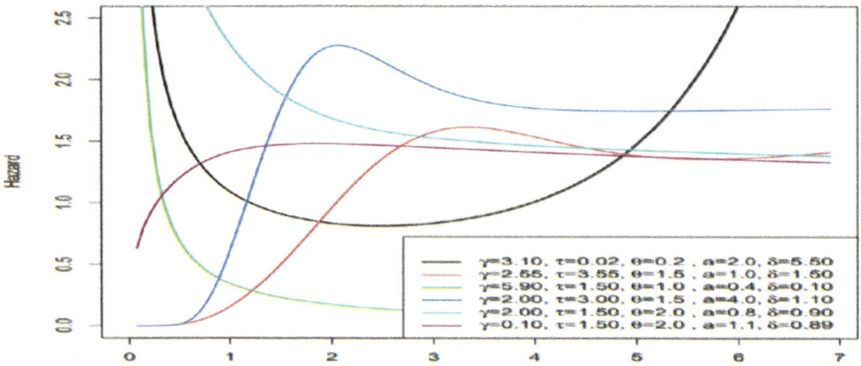

Fig. 4. Plot of the hazard function $h(X)_{\text{EChEW}}$ for the new distribution EChEW for different values of the parameters.

Expansion of PDF

In order for the probability density function to be in an understandable and simplified form, we have in this part to simplify the probability density function for a new distribution (ECHEW) for the purpose of studying some of the statistical properties of the new distribution in an easy and without complexity because merging some distributions with others may cause difficulty in dealing with or dealing with some distributions, and for this reason we simplify the functions. According to the expansion of some ordinary functions, whether it is a binomial expansion or a Tyler expansion for exponential functions. In Eq. (4) we expand the family, then after that we substitute Eqs. (1), (2), and then we perform some expansions again:

$$f(X) = \gamma A^{\gamma}\tau\theta g(X)G(X)^{\theta-1}e^{G(X)^{\theta}}e^{G(X)^{\theta}}e^{\tau(1-e^{G(X)^{\theta}}}\left[1-e^{\tau(1-e^{G(X)^{\theta}}}\right]^{\gamma-1}$$

Using the Taylor series equation for exponential functions, which is:

$$\left[1-e^{\tau(1-e^{G(X)^{\theta}}}\right]^{\gamma-1} = \sum_{j=0}^{\infty}(-1)^{j}\binom{\gamma-1}{j}e^{j\tau(1-e^{G(X)^{\theta}}} \tag{10}$$

The equation will be:

$$f(X) = \gamma A^{\gamma} \tau \theta g(X) G(X)^{\theta-1} e^{G(X)^{\theta}} e^{G(X)^{\theta}} e^{\tau(1-e^{G(X)^{\theta}}} \sum_{j=0}^{\infty} (-1)^j \binom{\gamma-1}{j} e^{j\tau(1-e^{G(X)^{\theta}}}$$

$$f(X) = \gamma A^{\gamma} \tau \theta g(X) G(X)^{\theta-1} e^{G(X)^{\theta}} e^{G(X)^{\theta}} \sum_{j=0}^{\infty} (-1)^j \binom{\gamma-1}{j} e^{(1+j)\tau(1-e^{G(X)^{\theta}}} \tag{11}$$

Then after a series of procedures, we obtain the final form of the new distribution, and thus the new form of the probability density function for a new distribution is in the form:

$$f(X) = \mathcal{M}_{j,k,l,m,q} (X)^{\delta-1} \delta(q+1) e^{-(q+1)X^{\delta}} \tag{12}$$

Properties of the New Distribution

This section will discuss some of the mathematical properties of the new proposed distribution, represented by the moments, the function generating the moments, the quantity function, and the Rhine entropy.

Moments

We can obtain the moments about the origin of degree r of the EChEW distribution from the following relationship by relying on the probability density function after expanding it according to Eq. (12).

$$\mu_r = \mathrm{E}(X) = \int_0^{\infty} X^r f(X) dX \tag{13}$$

After substituting Eq. (12) into Eq. (13), we obtain the new equation:

$$\mu_r = \mathrm{E}(X^r)_{\mathrm{EChEW}} = \mathcal{M}_{j,k,l,m,q} \int_0^{\infty} x^r (X)^{\delta-1} \delta(q+1) e^{-(q+1)X^{\delta}} dX$$

$$Y = (q+1)X^{\delta} \Rightarrow dY = (X)^{\delta-1} \delta(q+1) dX$$

$$Y = (q+1)X^{\delta} \Rightarrow X = Y^{\frac{1}{\delta}} (q+1)^{\frac{-1}{\delta}}$$

$$\mu_r = \mathrm{E}(X^r)_{\mathrm{EChEW}} = \mathcal{M}_{j,k,l,m,q} \int_0^{\infty} \left(Y^{\frac{1}{\delta}} (q+1)^{\frac{-1}{\delta}} \right)^r e^{-Y} dY$$

$$\mu_r = \mathrm{E}(X^r)_{\mathrm{EChEW}} = \mathcal{M}_{j,k,l,m,q} \left((q+1)^{\frac{-r}{\delta}} \right) \int_0^{\infty} Y^{\frac{r}{\delta}} e^{-Y} dY$$

$$\mu_r = \mathrm{E}(X^r)_{\mathrm{EChEW}} = \mathcal{M}_{j,k,l,m,q} \left((q+1)^{\frac{-r}{\delta}} \right) \int_0^{\infty} Y^{(\frac{r}{\delta}+1)-1} e^{-Y} dY$$

We integrate this type of equation using gamma integration to obtain the final formula for the equation, which is as follows. (Artin, 2015) [6]

$$\mu_r = E(X^r)_{\text{EChEW}} = \mathcal{M}_{j,k,l,m,q}\left((q+1)^{\frac{-r}{\delta}}\right)\Gamma\left(\frac{r}{\delta}+1\right) \tag{14}$$

Now, equation No. (14) for the r-order moment function in its simplest form is:

Through Eq. (14), we can find the first moment μ_1, , which represents the arithmetic mean (mean) at r = 1, the second moment μ_2, the third moment μ_3 and the fourth moment μ_4 for the new EChEW distribution, which is as follows:

$$\mu_r = E\left(X^1\right)_{\text{EChEW}} = \mathcal{M}_{j,k,l,m,q}\left((q+1)^{\frac{-1}{\delta}}\right)\Gamma\left(\frac{1}{\delta}+1\right) \tag{15}$$

$$\mu_r = E\left(X^2\right)_{\text{EChEW}} = \mathcal{M}_{j,k,l,m,q}\left((q+1)^{\frac{-2}{\delta}}\right)\Gamma\left(\frac{2}{\delta}+1\right) \tag{16}$$

$$\mu_r = E\left(X^3\right)_{\text{EChEW}} = \mathcal{M}_{j,k,l,m,q}\left((q+1)^{\frac{-3}{\delta}}\right)\Gamma\left(\frac{3}{\delta}+1\right) \tag{17}$$

$$\mu_r = E\left(X^4\right)_{\text{EChEW}} = \mathcal{M}_{j,k,l,m,q}\left((q+1)^{\frac{-4}{\delta}}\right)\Gamma\left(\frac{4}{\delta}+1\right) \tag{18}$$

Through these moments, we can obtain the variance of the new distribution, EChEW, as follows.

$$\text{Var}(X) = E\left(X^2\right) - (E(X))^2 \tag{19}$$

$$Var(X) = \left[\mathcal{M}_{j,k,l,m,q}\left((q+1)^{\frac{-2}{\delta}}\right)\Gamma\left(\frac{2}{\delta}+1\right)\right] - \left(\mathcal{M}_{j,k,l,m,q}\left((q+1)^{\frac{-1}{\delta}}\right)\Gamma\left(\frac{1}{\delta}+1\right)\right)^2 \tag{20}$$

The Moments Generating Function

This moment-generating function (mgf) can be defined as follows. Let X be a random variable with a probability density function $g(X)$. If the expectation of E(X) exists in the interval (-h,h), then mgf will found as:

$$M_X(t) = E\left(e^{tX}\right)$$

And mgf for the proposed distribution can be given as follows:

$$M_x(t) = E\left(e^{tx}\right) = \int_{-\infty}^{\infty} e^{tx} f(x)dx$$

$$M_x(t)_{\text{EChEW}} = E\left(e^{tx}\right) = \mathcal{M}_{j,k,l,m,q} = \int_{0}^{\infty} e^{tx}(x)^{\delta-1}\delta(q+1)e^{-(q+1)x^{\delta}}dx \tag{21}$$

$$e^{t.g(x)} = \sum_{k=0}^{\infty} \frac{[t.g(x)]^k}{k!}$$

$$e^{tx} = \sum_{v=0}^{\infty} \frac{t^v (x^v)}{v!}$$

$$M_x(t)_{\text{EChEW}} = E\left(e^{tx}\right) = \mathcal{M}_{j,k,l,m,q} \Phi_v \int_0^{\infty} x^v (x)^{\delta-1} \delta(q+1) e^{-(q+1)x^{\delta}} dx$$

Then by Eq. (14) the new mgf for EChEW will be:

$$M_X(t)_{\text{EChEW}} = E\left(e^{tX}\right) = \Phi_v \mathcal{M}_{j,k,l,m,q}\left((q+1)^{\frac{-v}{\delta}}\right)\Gamma\left(\frac{v}{\delta}+1\right) \tag{22}$$

The Quantile Function

The quantile function was studied by Hyndman in (1996) [7], and the function for the new distribution can be found based on the distribution function of EChEW, and therefore it can be represented mathematically as follows:

$$u = F^{-1}(X), u \in (0, 1) \tag{23}$$

It is also used to find the median, skewness, and kurtosis for distributions that do not contain moments or in which the skew value is very high. It is also necessary for generating data, where $Q(u)$ is the quantile function of the distribution function $F(X)$.

For every $0 < u < 1$ to obtain the quantitative function of the EChEW distribution, we take the inverse of the distribution function of the new distribution as we mentioned in Eq. (5) and it is as follows.

$$u = 1 - \left[A\left(1 - e^{\left\{\tau\left[\left(1-e^{(1-e^{-X^{\delta}})^{a}}\right)^{\theta}\right]\right\}}\right)\right]^{\gamma} \Rightarrow 1 - u = \left[A\left(1 - e^{\left\{\tau\left[\left(1-e^{(1-e^{-X^{\delta}})^{a}}\right)^{\theta}\right]\right\}}\right)\right]^{\gamma}$$

By taking the root γ of both sides, it becomes as follows:

After taking the root of both sides and simplifying, we obtain the final form of the new EChEW distribution

$$X = \sqrt[\delta]{-log(1 - \left(log\left[1 - \left(\frac{1}{\tau}log\left(1 - \frac{1}{A}(1-u)^{1/\gamma}\right)\right)^{1/\theta}\right]\right)^{1/a}} \tag{24}$$

Through the last equation, the median can be obtained by setting u = 0.5, and the skewness, kurtosis, quartiles, octaves, and deciles can also be found.

$$Me(X)_{\text{EChEW}} = \sqrt[\delta]{-log(1 - \left(log\left[1 - \left(\frac{1}{\tau}log\left(1 - \frac{1}{A}(1-0.5)^{1/\gamma}\right)\right)^{1/\theta}\right]\right)^{1/a}} \tag{25}$$

Order Statistic

This property is used to find the lowest and highest value from ordered statistics for a new EChEW distribution. It is one of the basic tasks in statistics because it is used in several fields, including quality control and reliability tests, to predict the failure of future elements. We take a number of random variables, let them be $X_1, X_2, \ldots, X_n$ where n is a set of identical and independent random variables taken from the distributive function of the new distribution EChEW (Sahoo, 2013) [8]:

$$g_{i:n}(X) = \sum_{j=0}^{n-i} \frac{n!}{(i-1)!(n-i)!}(-1)^j \binom{n-i}{j} [F(X)]^{j+i-1} f(X) \tag{26}$$

where $k = \frac{n!}{(i-1)!(n-i)!}$

Thus, the equation by substituting the distribution function in Eq. (5) and the probability density function in Eq. (6) for the new EChEW distribution is as follows:

$$g_{i:n}(x) = k\sum_{j=0}^{n-i}(-1)^j\binom{n-i}{j}[[A(1-e^{\{\tau[(1-e^{(1-e^{-x^\delta})^a})^\theta]\}})]^\gamma]^{j+i-1}\gamma A^\gamma \tau\theta a\delta(X)^{\delta-1}e^{-x^\delta}$$

$$\left(\left[1-e^{-x^\delta}\right]^a\right)^{\theta-1} e^{\left[1-e^{-x^\delta}\right]^a} e^{\tau\left(1-e^{\left(\left[1-e^{-x^\delta}\right]^a\right)^\theta}\right)} \left[1-e^{\tau\left(1-e^{\left(\left[1-e^{-x^\delta}\right]^a\right)^\theta}\right)}\right]^{\gamma-1} \left[1-e^{-x^\delta}\right]^{a-1} \tag{27}$$

When $i = 1$ in Eq. (27), we obtain the lowest value of the ordered statistic for the EChEW distribution.

$$g_{i:n}(X) = k\sum_{j=0}^{n-i}(-1)^j\binom{n-i}{j}\left[\left[A\left(1-e^{\left\{\tau\left[\left(1-e^{(1-e^{-X^\delta})^a}\right)^\theta\right]\right\}}\right)\right]^\gamma\right]^{j+i-1}$$

$$\gamma A^\gamma \tau\theta a\delta(X)^{\delta-1}e^{-X^\delta}$$

$$\left(\left[1-e^{-X^\delta}\right]^a\right)^{\theta-1} e^{\left[1-e^{-X^\delta}\right]^{a\theta}} e^{\tau\left(1-e^{\left(\left[1-e^{-X^\delta}\right]^a\right)^\theta}\right)} \left[1-e^{\tau\left(1-e^{\left(\left[1-e^{-X^\delta}\right]^a\right)^\theta}\right)}\right]^{\gamma-1} \left[1-e^{-X^\delta}\right]^{a-1} \tag{28}$$

Estimation

The maximum likelihood method is used to estimate the parameters of the new distribution because it is the best way to estimate the parameters if we have a possible sample of size n from the population. $X_1, X_2, \ldots, X_n$ which has the probability distribution EChEW, the maximum potential function is according to the following formula. (Oguntunde et al., 2018) [9]

$$L(x;\gamma,\tau,\theta,a,\delta)_{\text{EChEW}} = \prod_{i=1}^{n} f(x;\gamma,\tau,\theta,a,\delta)_{\text{EChEW}} \tag{29}$$

$$L(x;\gamma,\tau,\theta,a,\delta)_{\text{EChEW}}=\prod_{i=1}^{n}\left[\gamma A^{\gamma}\tau\theta g(x)G(x)^{\theta-1}e^{G(x)^{\theta}}e^{G(x)^{\theta}}e^{\tau(1-e^{G(x)^{\theta}}}[1-e^{\tau(1-e^{G(x)^{\theta}}}]^{\gamma-1}\right]$$

$$L(x;\gamma,\tau,\theta,a,\delta)_{\text{EChEW}}=$$

$$\prod_{i=1}^{n}\left(\left[\gamma A^{\gamma}\tau\theta a\delta(x)^{\delta-1}e^{-x^{\delta}}[1-e^{-x^{\delta}}]^{a-1}e^{\left[1-e^{-x^{\delta}}\right]^{a\theta}}e^{\tau\left(1-e^{\left(\left[1-e^{-x^{\delta}}\right]^{a}\right)^{\theta}}\right)}\right]\right.$$

$$\left.\left[1-e^{\tau\left(1-e^{\left(\left[1-e^{-x^{\delta}}\right]^{a}\right)^{\theta}}\right.}\right]^{\gamma-1}\left(\left[1-e^{-x^{\delta}}\right]^{a}\right)^{\theta-1}\right)$$

$$\ln L(x;\gamma,\tau,\theta,a,\delta)_{\text{EChEW}}=n\log\left(\gamma A^{\gamma}\tau\theta a\delta\right)+(\delta-1)\sum_{i=1}^{n}(\ln(x_i))-\sum_{i=1}^{n}x_i^{-\delta}$$
$$+(a-1)\sum_{i=1}^{n}\ln\left[1-e^{-x_i^{-\delta}}\right]+\sum_{i=1}^{n}\left[\left[1-e^{-x_i^{\delta}}\right]^{a\theta}\right]+\sum_{i=1}^{n}\left[\tau\left(1-e^{\left([1-e^{-x_i\delta}]^{a}\right)^{\theta}}\right)\right]$$
$$+(\gamma-1)\sum_{i=1}^{n}\left[1-e^{\tau(1-e}\left([1-e^{-x_i}]^{a}\right)^{\theta}\right]+(a\theta-a)\sum_{i=1}^{n}\left[1-e^{-x_i\delta}\right]\tag{30}$$

Now we extract the partial derivatives for each parameter, and thus they are as follows:

$$\frac{\partial\ln L(x;\gamma,\tau,\theta,a,\delta)_{\text{EChEW}}}{\partial\gamma}=\frac{n}{\gamma}\left(\gamma(A^{\gamma}lnA)+(A^{\gamma})+\sum_{i=1}^{n}\left[1-e^{\tau(1-e^{([1-e^{-x_i^{\delta}}]^{a})^{\theta}}}\right]\right.\tag{31}$$

$$\frac{\partial\ln L(x;\gamma,\tau,\theta,a,\delta)_{\text{EChEW}}}{\partial\tau}=\frac{n}{\tau}+\sum_{i=1}^{n}e^{\theta[1-e^{-x_i^{\delta}}]^{a}}-\sum_{i=1}^{n}\tau(1-e^{([1-e^{-x_i^{\delta}}]^{a})^{\theta}})e^{(1-e^{([1-e^{-x_i^{\delta}}]^{a})^{\theta}})}\tag{32}$$

$$\frac{\partial\ln L(x;\gamma,\tau,\theta,a,\delta)_{EChEW}}{\partial\alpha}=\frac{n}{\alpha\alpha}-\sum_{i-1}^{n}\ln(1-e^{-x_i^{\delta}})+\theta\sum_{i=1}^{n}\ln(1-e^{-x_i^{\delta}})([1-e^{-x_i^{\delta}}]^{a})^{\theta}$$
$$+\tau\theta\sum_{i=1}^{n}\ln(1-e^{-x_i^{\delta}})(e)^{\theta[1-e^{-x_i^{\delta}}]^{a}}[1-e^{-x_i^{\delta}}]^{a}+(\theta-1)\sum_{i=1}^{n}(1-e^{-x_i^{\delta}})\tag{33}$$

$$\frac{\partial\ln L(x;\gamma,\tau,\theta,a,\delta)_{\text{EChEW}}}{\partial\theta}=\alpha\sum_{i=1}^{n}\ln(1-e^{-x_i^{\delta}})([1-e^{-x_i^{\delta}}]^{a})^{\theta}+\tau\sum_{i=1}^{n}(1-e^{-x_i^{\delta}})^{\alpha}([1-e^{-x_i^{\delta}}]^{a})^{\theta}$$
$$+\tau\alpha\sum_{i=1}^{n}\ln(1-e^{-x_i^{\delta}})([1-e^{-x_i^{\delta}}]^{a})^{\theta}e^{\tau(1-e^{([1-e^{-x_i^{\delta}}]^{a})^{\theta}+([1-e^{-x_i^{\delta}}]^{a})^{\theta}}}+\sum_{i=1}^{n}\alpha(1-e^{-x_i^{\delta}})^{\theta}\tag{34}$$

$$\frac{\partial\ln L(x;\gamma,\tau,\theta,a,\delta)_{\text{EChEW}}}{\partial\delta}=\sum_{i=1}^{n}\ln(x_i)+\sum_{i=1}^{n}\frac{\ln(x_i)}{x_i^{\delta}}-\sum_{i=1}^{n}\frac{(\alpha-1)\ln(x_i)e^{-x_i^{\delta}}}{x_i^{\delta}(1-e^{-x_i^{\delta}})}-$$
$$\sum_{i=1}^{n}\frac{(\alpha\theta)\ln(1-e^{-x_i^{\delta}})^{\alpha\theta-1}e^{-x_i^{\delta}}}{x_i^{\delta}}+\sum_{i=1}^{n}(\alpha\tau\theta)x_i^{\delta}\ln(1-e^{-x_i^{\delta}})^{\alpha-1}e^{\theta(1-e^{-x_i^{\delta}})-x_i^{\delta}}+$$
$$\sum_{i=1}^{n}(\alpha\tau\theta)x_i^{\delta}\ln(x_i)(1-e^{-x_i^{\delta}})^{\alpha\theta-1}e^{\theta(1-e^{-x_i^{\delta}})-x_i^{\delta}}e^{\tau(1-e^{([1-e^{-x_i^{\delta}}]^{a})^{\theta}+([1-e^{-x_i^{\delta}}]^{a})^{\theta}-x_i^{\delta}}}\tag{35}$$

We note that the equations have been zeroed in order to obtain an estimate of the maximum potential:

$$\frac{\partial \ln L(x;\gamma,\tau,\theta,a,\delta)\text{EChEW}}{\partial \delta} = \frac{\partial \ln L(x;\gamma,\tau,\theta,a,\delta)\text{EChEW}}{\partial \theta} = \frac{\partial \ln L(x;\gamma,\tau,\theta,a,\delta)\text{EChEW}}{\partial \alpha} = \frac{\partial \ln L(x;\gamma,\tau,\theta,a,\delta)\text{EChEW}}{\partial \gamma} = \frac{\partial \ln L(x;\gamma,\tau,\theta,a,\delta)\text{EChEW}}{\partial \tau} = 0$$

And after it is equal to zero. We find the value of the random variable. Therefore, we need to use numerical methods because of their complexity, and numerical methods require the use of computer programs because they are difficult to accomplish manually. Therefore, the R program, which is one of the important and easy statistical programs, was used to estimate these parameters for a new distribution EChEW.

Application

The study sample represents the end-stress life of Kevlar373/epoxy synthetic fibers subjected to continuous pressure at 90% compression level until total failure and the data are as follows: (Merovci et al.,2017) [10], (Khan et al., 2020) [11].

0.0251, 0.0886, 0.0891, 0.2501, 0.3113, 0.3451, 0.4763, 0.5650, 0.5671, 0.6566, 0.6748, 0.675, 0.6753, 0.7696, 0.8375, 0.8391, 0.8425, 0.8645, 0.8851, 0.9113, 0.9120, 0.9836, 1.0483, 1.0596, 1.0773, 1.1733, 1.2570, 1.2766, 1.2985, 1.3211, 1.3503, 1.3551, 1.4595, 1.4880, 1.5728, 1.5733, 1.9316, 1.7083, 1.7263, 1.7460, 1.7630, 1.7746, 1.8275, 1.8375, 1.8503, 1.8808, 1.8878, 1.8881,, 2.3513, 2.34701.9558, 2.0048, 2.0408, 2.0903, 2.1093, 2.1330, 2.2100, 2.2460, 2.2878, 2.3203, 2.4951, 2.5260, 2.9911, 3.0256, 3.2678, 3.4045, 3.4846, 3.7433, 3.7455, 3.9143, 4.8073, 5.4005, 5.4435,

Some measures for this sample of data were extracted using the describe function provided by the R statistical program. The table shows descriptive statistics for the first data.

Table 1. Descriptive statistics for the real data sample space.

n	mean	Median	Min	Max	Skew	Kurtosis	Standard deviation (Sd)
76	1.959	1.736	0.025	9.09	1.940	4.947	.320

We notice from Table 1 that the skewness (skew) is positive, and this indicates that the data is skewed and extended more to the right. As for kurtosis, it is also positive, and this indicates that the data is pointy and close to a normal distribution.

From Table 2, it is clear that the new distribution, Exponentiated Chen Exponentiated Weibull distribution, is more superior, flexible, and appropriate than the rest of the other distributions, while the EGEW distribution came in last place with all criteria. The new distribution is an accurate representation of the data because it contains the lowest values for the statistical measures (A^*, W^*,AIC, CAIC, BIC, HQIC).

Representing the points close to the unit line, this is an indication that the data is close to matching the new distribution, Exponentiated Chen Exponentiated Weibull

Table 2. The Parameter values for the real dataset with EChEW distribution and competing distributions.

Distributions	Statistical Criteria					
	AIC	CAIC	BIC	HQIC	0000W*	0000A*
EChEW	250.789	251.64	262.443	255.44	0.067	0.39
GamEW	251.924	252.487	261.247	255.65	0.1105	0.655
BEW	254.04	254.898	265.694	258.698	0.1139	0.672
KuEW	254.13	254.987	265.783	258.787	0.1135	0.672
EGEW	254.131	254.988	265.785	258.788	0.1136	0.6731
WeEW	251.458	252.315	263.112	256.116	0.0776	0.468
TEEEW	252.234	253.091	263.887	256.891	0.085	0.5078
EW	250.327	250.60	257.319	253.121	0.1167	0.6912

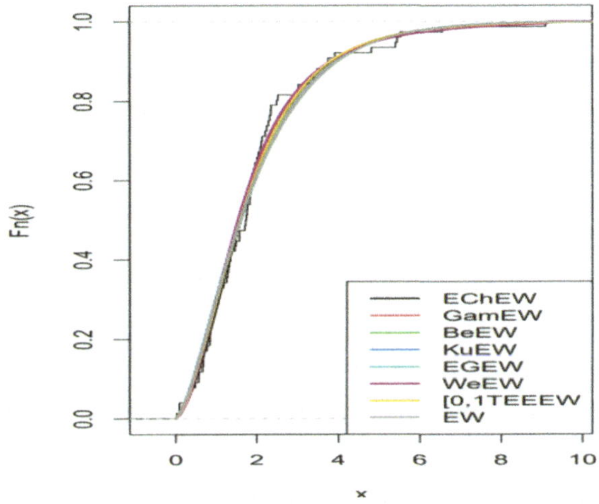

Fig. 5. An illustration of the histogram, distribution curves, and competing distributions.

distribution, which is a graphical method for comparing two probability distributions and expresses the confrontation of theoretical partitions with experimental partitions, and this matching is an indication of the superiority of the new distribution, the EChEW distribution.

We learn about the properties of the studied data and it turns out that there is positive skewness and flatness that varies according to the values of the given features and has a shape close to a normal distribution. The deviation value was not large. We also obtained the values of the first quartile and the third quartile. It is also possible to give an illustration of the curves of the density function of the distribution that were compared with the proposed distribution EChEW. Figure 5 shows the extent to which the distributions used in the research match the histogram of the studied data. It turns out that the distribution has a higher match and suitability to the data than other distributions, which indicates the effectiveness of this distribution in modeling data of this type in many fields.

Conclusion

Statistical distributions play a crucial role in inferences and applications to fit advanced real-world phenomena. In this study, a new distribution called EChEW was proposed and it was applied to real data taken from the site. Several other distributions were used in the research to determine the suitability of the proposed distribution with these distributions. Through the results shown in the experiment, it is clear that the new distribution is preferable to the distributions compared to it and is distributable, and its great flexibility in modeling data of this type. We recommend that researchers experiment with the distribution on other data and hope that the results will be positive or that they conduct an expansion that improves the accuracy of the model's match with real data.

Future Works

Although our current study has added a new expansion to some continuous distributions and studied some statistical properties related to them, there are more theoretical aspects that the researcher did not address and which are included among the limitations of the research work, including increasing or decreasing the number of parameters, which explains more types of natural phenomena. Which falls within the scope of these distributions. In addition, changing the default values to increase the diversity of the generated data and using other types of statistical criteria such as the mean square error and the least square error. Applying the proposed distributions within the statistical family to other types and groups of real data can help support the principle of flexibility of distributions and their suitability to diversity in areas of research and application.

References

1. Awodutire, P.: Statistical properties and applications of the exponentiated chen-G family of distributions: exponential distribution as a baseline distribution. Austrian J. Stat. **51**(2), 57–90 (2022)
2. Refaie, M.K., Yaqoob, A.A., Selim, M.A., Ali, E.I.: A Novel Version of the Exponentiated Weibull Distribution: Copulas, Mathematical Properties and Statistical Modeling. Pakistan J. Stat. Operat. Rese. 491–519 (2023)
3. Gupta, R.C., Gupta, P.L., Gupta, R.D.: Modeling failure time data by Lehman alternatives. Communications in Statistics-Theory and methods **27**(4), 887–904 (1998)
4. Fox, E., Russo, R., Dutton, K.: Attentional bias for threat: Evidence for delayed disengagement from emotional faces. Cogn. Emot. **16**(3), 355–379 (2002)
5. Johnson, N.L., Kotz, S.: A vector multivariate hazard rate. J. Multivar. Anal. **5**(1), 53–66 (1975)
6. Artin, E.: The gamma function. Courier Dover Publications (2015)
7. Hyndman, R.J., Fan, Y.: Sample quantiles in statistical packages. Am. Stat. **50**(4), 361–365 (1996)
8. Sahoo, P.: Probability and Mathematical Statistics. University of Louisville (2013)
9. Oguntunde, P.E., Adejumo, A.O., Khaleel, M.A., Okagbue, H.I., Odetunmibi, O.A.: The logistic inverse exponential distribution: Basic structural properties and application. World Congress on Engineering (2018)

10. Merovci, F., Alizadeh, M., Yousof, H.M., Hamedani, G.G.: The exponentiated transmuted-G family of distributions: theory and applications. Communications in Statistics-Theory and Methods **46**(21), 10800–10822 (2017)
11. Khan, M.S., King, R., Hudson, I.L.: Transmuted Burr type X distribution with covariates regression modeling to analyze reliability data. Am. J. Math. Manag. Sci. **39**(2), 99–121 (2020)

Study of Hemodynamics of Arterial System Using Atangana Baleanu Caputo (ABC) Fractional Order Two Element Wind-Kessel Model

A. Meenakshi[1](✉) and E. Renuga[2]

[1] Vel Tech Rangarajan Dr. Sagunthala R & D Institute of Science and Technology, Chennai, India
drameenakshi@veltech.edu.in
[2] Department of Mathematics, M.O.P. Vaishnav College for Women (Autonomous), Chennai 600 034, India
renugae.math@mopvaishnav.ac.in

Abstract. Mathematical modeling is vital for scientists to understand the dynamics of blood circulation in the cardiovascular system. For modeling complex systems, including phenomena in physics, engineering, biology, economics, and other fields, fractional calculus offers a more versatile framework. Through the use of fractional differentiation order, fractional-order approaches successfully offer substantial ways to describe flexibility by expanding the principles of differentiability and adding the non-local and memory elements. Fractional (non-integer order) calculus can offer a succinct model for explaining the dynamics of biological tissues. In this work, we examine how numerical simulations may be used to improve the circulation model by utilizing fractional-order operators. This article aims to provide insight into Bahloul's fractional order two-element Wind-Kessel(W-K) model for modeling the study of the hemodynamics of arterial systems and its smaller variation of the same is studied by Atangana Baleanu Caputo (ABC) fractional derivative and the results are simulated by using LADM (Laplace Adomian Decomposition Method.

Keywords: Aortic blood flow · two-element Wind-Kessel model · ABC Fractional-order derivative

1 Introduction

Cardiovascular disease detection is more difficult in the biomedical sciences since symptoms may take time to manifest and are sometimes discovered only after the disease has progressed. Because it is dangerous to experiment on the circulatory system, creating a model of the cardiovascular system that can be used to identify anomalies early on is essential [1, 2]. A mathematical model is a mathematical description of real phenomena. To forecast or offer insight, it describes developing a mathematical representation of a real-world situation [12, 18–20]. Mathematical modeling is critical in understanding

S. Pal and Á. Rocha (Eds.): ICMMCS 2025, LNNS 1398, pp. 216–224, 2025.
https://doi.org/10.1007/978-3-031-90998-6_21

blood circulation dynamics in the cardiovascular system and analyzing blood flow. There are several physiological and medicinal applications for which studying the behavior of the circulatory system is essential. In the late 1800s, German physiologist Otto Frank developed the W-K Model [6]. The wind-Kessel (W-K) model is a mathematical model used to investigate blood dynamics. Lumped parameter models, or W-K models, are frequently employed to describe the pulmonary arterial and vascular systems [7]. The real-world scenario was mainly developed using classical models. The viscoelasticity characteristics are disregarded in the classical arterial W-K model, which causes errors in the model and inevitably inaccurate predictions. A redesigned fractional order W-K model is offered [3] to reflect this property.

A generalization of traditional integer calculus is fractional calculus. These days, fractional computations are often employed to study the behaviors of numerous real-world phenomena [4, 5]. Fractional calculus is incorporated into the classical W-K model to create the arterial fractional order modified W-K model. According to several studies, CVS (cardiovascular study) wall material exhibits viscoelastic behavior instead of pure elastic behavior, comparable to most biological tissues. Arteries are generally considered as viscoelasticity reservoirs [8].

A more precise description of the change from viscosity to elasticity is provided by the fractional derivatives. Over time, many mathematicians have come up with various definitions, each with its own notation and approach. This paper uses the Atangana Baleanu Caputo (ABC) Fractional derivative which is one of its variations. Because it more precisely represents memory effects and long-range dependencies in dynamic systems than integer-order derivatives, the ABC fractional derivative is used. Notable prior work on the Atangana-Baleanu operator and its applications in several technical and applied scientific systems is included in references [22]. These studies offer helpful insights into how well the Atangana-Baleanu operator performs and how flexible it is in different situations. The aim of this work is to express the differential equations of the Wind-Kssel model using the ABC operator by the use of the fractional operator of order α concept. It is an essential tool.

The structure of this research article is as follows. The fundamental definitions and results of fractional derivatives, an introduction to the Wind-Kessel model for simulating arterial blood circulation, and an overview of Bahulol's fractional order arterial two-element Wind-Kessel model are covered [9, 10] in Sect. 2, which also serves as the foundation for our research. In Sect. 3, ABC Fractional derivative [22] is used to study a variation of Bahulol's fractional order two element artery Wind-Kessel model [12–15]. The ABC fractional order two-element Wind-Kessel model is numerically simulated using LADM [11, 16–21] in Sect. 4.

1.1 Objective

In order to simulate the two-element Wind-Kessel model using an ABC fractional order derivative, this article aims to investigate the behavior of fractional order differential equations.

2 Fundamental Definitions and Results

Numerous scientific and technical domains employ fractional calculus to simulate intricate phenomena, and these domains use the ABC fractional derivative. It is used to describe the behavior of materials that combine viscous and elastic characteristics. The non-local and memory-dependent characteristics of these materials are better captured by the fractional derivative. Many mathematicians have come across many definitions appropriate for the notion of a time-dependent, non-integer order integral or derivative. One version that is used widely is the Riemann-Liouville definition. The primary distinctions between fractional derivatives are their various kernels, depending upon their applications. Since the nonlocal dynamics could not be ascertained by derivative employing a singular kernel, the notion of non-singular derivatives is presented. It was incorporated in 2016 by Caputo and Fabrizo, who involved an exponential function in the non-singular derivative [10]. Later the Atangana-Baleanu-Caputo derivative, or ABC derivative, generalized the aforementioned derivative. It was found that the operator is non-local and does not have a distinct differential operator or unique kernel. Several real-world occurrences are being modeled using these findings [10]. Numerical solutions of fractional differential equations with nonsingular kernels, particularly those involving the Atangana-Baleanu-Caputo derivative, have been the subject of limited research to date. This is partially caused by the mathematical intricacy of these equations and the challenge of creating efficient numerical techniques. Nonetheless, the absence of exact numerical techniques and the demand for more precise and effective procedures have spurred academics to delve deeper into this field. A major addition to the field would be creating a new numerical frame fractional differential equations containing the Atangana-Baleanu Caputo derivative with a better degree of precision and convergence [23].

Definition 2.1. [4] The Riemann-Liouville Fractional integral operator of order $\alpha > 0$ of a function $f \in L^1(\Re^+)$ is defined by

$$D_t^{(-\alpha)} f(t) = \frac{1}{\Gamma(q)} \int_0^t (t-x)^{\alpha-1} dx$$

where Γ *(.) denotes the Euler Gamma function.*

Definition 2.2. [9] The Riemann-Liouville Fractional Derivative

$$D_t^{\alpha} y(t) = \frac{1}{\Gamma(n-\alpha)} \frac{d^n}{dt^n} \int_0^t (t-x)^{(n-\alpha-1)} y(x) dx.$$

Definition 2.3. [12] Tshe Caputo integral of the function y: $\Re^+ \rightarrow \Re$ is defined by

$${}^{C}D_t^{-\alpha} y(t) = \frac{1}{\Gamma(\alpha)} \int_0^t (t-x)^{\alpha-1} y(x) dx$$

where Γ (.) denotes the Gamma function and $0 < \alpha \leq 1$ shows the fractional order parameter.

Definition 2.4. [9, 11] The Caputo derivative with order α, $0 < \alpha \leq 1$ is defined by

$$ {}^{C}D_t^{\alpha} y(t) = \frac{1}{\Gamma(n-\alpha)} \int_0^t (t-x)^{n-\alpha-1} \frac{d^n}{dx^n} y(x)dx \quad \text{where } n-1 < \alpha < n, n \in N $$

Because of its unique characteristics, such as its nonlocal and nonsingular kernel, which is represented by the Mittag-Leffler function, the Atangana-Baleanu operator was selected. This operator best captures the complex processes present in the model under consideration [22]

Definition 2.5. [10] Atangana-Baleanu-Caputo (ABC) Fractional Derivative:
If $\Psi(t) \in H^1(0, \zeta)$ and $\alpha \in (0,1)$

$$ {}_0^{ABC}D_t^{\alpha}(y(t)) = \frac{N(\alpha)}{\Gamma(1-\alpha)} \int_0^t y(x) E_{\alpha}^{\frac{-\alpha}{1-\alpha}(t-x)^{\alpha}} dx $$

where $E_{\alpha}^{\frac{-\alpha}{1-\alpha}(t-x)^{\alpha}}$ is the Mittag-Leffler, the generalization of exponential function.

3 Wind-Kessel Model

3.1 Aortic Blood Flow

The aorta, which originates in the left ventricle of the heart and passes through the belly before dividing into smaller arteries, is the biggest artery in the human circulatory system. The closed-loop, pulsatile cardiac cycle is produced by the heart pumping blood in a manner akin to a pulse wave throughout the body. Ventricular diastole, the first cardiac phase, begins when the ventricles dilate, allowing newly oxygenated blood to fill the atria. Ventricular systole occurs when the ventricles contract following ventricular diastole and release blood into the body through the aorta. Systolic pressure is the maximum pressure ventricles can produce as they contract, forcing blood into the aorta [7].

3.2 Wind-Kessel (W-K) Model

For a hydraulic system, the W-K Model is comparable to Poiseuille's Law. It compares the movement of fluid via pipes to the flow of blood via arteries [6]. The smooth muscle fibers, elastin fibers, and collagen fibers that make up the arterial wall are the three primary features that determine the biomechanical properties of the arteries. It has long been known that artery tissues, like other soft collagenous tissue, behave viscoelastically rather than solely elastically [7]. The W-K models can be divided into two categories: viscoelastic and elastic models. Elastic W-K models consider the purely elastic mechanical characteristics of the vessel wall, which are modeled by an ideal capacitor. Viscoelastic Wind-Kessel (VWK) models, on the other hand, assume that blood vessel motions exhibit both elastic and viscous properties [4]. A more suitable description of the viscoelastic characteristics of arteries is provided by the fractional order Wind-Kessel model, which adds fractional derivatives and integrals. When trying to model phenomena that show memory effects or long-range interdependence, this method is quite pertinent [1].

3.3 Simple Wind-Kessel Model

The wind-Kessel model shown in Figs. 1 and 2. Was designed to depict the systemic artery system and the heart as a closed hydraulic circuit. The circuit in this comparison included a water pump that was attached to a chamber that was mostly filled with water but had an air pocket in it. Pumping the water causes the air in the chamber to compress, forcing the water out of it. The workings of the heart are comparable to this analogy. Models called Wind-Kessel models are frequently used to simulate the load that the heart experiences during a cardiac cycle. It describes the arterial compliance, peripheral resistance of the valves, and inertia of the blood flow that links blood pressure and blood flow in the aorta [3].

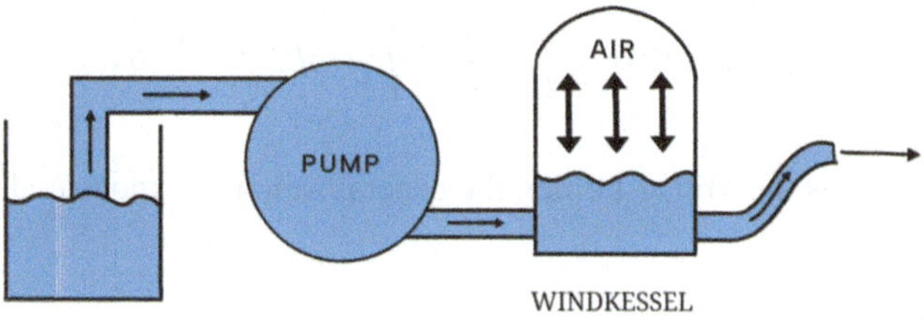

Fig. 1. The simple Wind-Kessel model and hydraulic analogy [3]

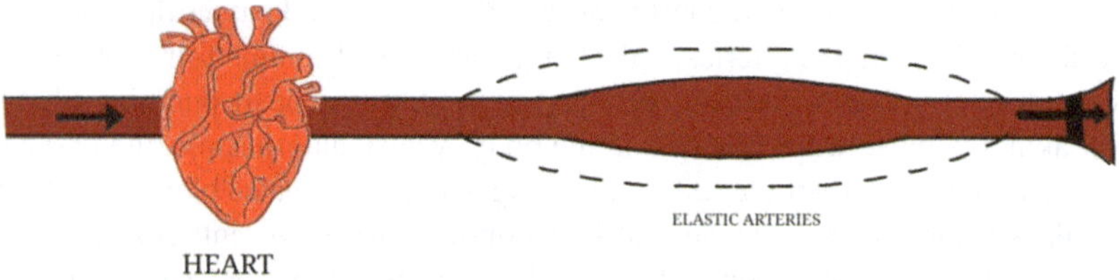

Fig. 2. The simple Wind-Kessel model and hydraulic analogy [3]

3.4 Two Element Arterial Wind-Kessel Model -WK2

In hemodynamics, this two-element model is frequently utilized to understand blood pressure and blood flow dynamics. It is important to remember that more complex models with extra components can accurately depict the intricate functioning of the cardiovascular system. "The most basic mathematical model that might potentially describe the overall behavior of human arterial systematic trees is the historical Wind-Kessel, which makes use of fundamental properties like total arterial compliance (C), aortic characteristic impedance (Zc), and total peripheral resistance (Rp). The primary goal of WK methods is to model the aortic input impedance as an afterload on the heart [3].

It is typical to use the electrical analogy while using WK, where resistance and compliance are represented, respectively, by a resistor and a capacitor." Additionally, there is a correlation between electrical voltage, flow rate, and blood pressure. As a result, the electrical impedance represented in the frequency domain as the voltage- to-current ratio is also equal to the aortic input impedance. The WK idea was initially presented with the remarkable two-element wind-Kessel (WK2), which essentially uses a capacitor coupled in parallel to a resistor to represent the entire artery network and account for peripheral resistance and overall arterial compliance, respectively [3].

3.5 Fractional-Order Arterial Wind-Kessel Model [1]

Various biological, physical, and dynamical processes may be effectively described using mathematical concepts FDEs [10]. "For the last 20 years, the Wind-Kessel representation technique has been inspired by the electrical circuit analogy, in which the electrical voltage symbolizes arterial blood pressure and the current represents blood flow via the arteries [10]. In [7], Mohamed A. Bahloul1 introduced a Fractional-order two-element Wind-Kessel (FWK2) model comparable to the traditional WK2. The resistor (R_p) and the frequency-dependent arterial compliance (FOC, c_α) in parallel reflect the arterial peripheral resistance and total complex arterial compliance, respectively, in the proposed model. By defining the fractional two-element Wind-Kessel based on the conserved mass and applying Kirchhoff's Current Law to the fractional circuit, the arterial blood flow pumped from the heart to the arterial vascular bed (q_{in}) may be expressed as follows:

$$q_{in}(t) = q_{stored}(t) + q_{out}(t) \tag{1}$$

[7] where q_{stored} is the blood stored in the artery, and q_{out} is the blood flow out of the arterial system, which can be expressed as follows:

$$q_{out} = \frac{1}{R_p} p_a(t) \tag{2}$$

where, R_p is the peripheral vascular resistance, p_a is the pressure in the artery. Typically using the conventional definition, it can be determined as the rate of flow by taking the first derivative of the volume equation for the time, However, we allow the differentiation order of the blood volume for time to be real ($\alpha \in [0\ 1]$) to account for the fractional properties of both RBC and the collagenous tissues forming the arterial bed. As a result, we apply the fractional-order derivative to this differential equation and using Kirchhoff's current law on the fractional circuit."

$$q_{stored}(t) = D_t^\alpha V(t) = \frac{d^\alpha V(t)}{dt^\alpha} \tag{3}$$

$$q_{stored}(t) = \frac{d^\alpha V(t)}{d^\alpha p_a(t)} \frac{d^\alpha p_a(t)}{dt^\alpha} \quad where \ C_\alpha = \frac{d^\alpha V(t)}{d^\alpha p_a(t)} \tag{4}$$

[7] where C_α is fractional order proportionality constant. Substituting (4) and (6) into (3) yield:

$$q_{in}(t) = C_\alpha \frac{d^\alpha p_a(t)}{dt^\alpha} + \frac{1}{R_p} p_a(t)$$

[7]

$$q_{in}(t) = C_\alpha \frac{d^\alpha p_a(t)}{dt^\alpha} + \frac{1}{R_p} p_a(t) \tag{5}$$

$$y(t) + \tau D_t^\alpha y(t) = u(t) \tag{6}$$

where $u(t) = R_p q_{in}(t)$, $y(t) = p_a(t)$, and $\tau = R_p C_\alpha$

Assuming null starting circumstances, (8) may be transformed using the Laplace transform to get the fractional aortic input impedance in the frequency domain.

4 Atangana Baleanu Caputo (ABC) Fractional Order Arterial Wind-Kessel (W-K) Model

In this article variation to Bahulol's fractional order two-element arterial W-K model is studied using ABC Fractional derivative. The ABC fractional derivative is applied because, in comparison to integer-order derivatives, it captures memory effects and long-range dependencies in dynamic systems more accurately. It is an essential tool in fractional calculus because of its adaptability and capacity to simulate complex dynamics. The ABC fractional derivative is a useful tool in biological systems research. It may be used to simulate complex biological processes more accurately, such as anomalous transport in biological tissues. Essentially, the Atangana-Baleanu-Caputo (ABC) fractional derivative is a tool that helps to better express and comprehend processes that display fractional order by expanding the possibilities of conventional calculus.

In the above Eq. (9) when we consider ABC Fractional Derivative the equation becomes.

$$y(t) + \tau\, {}_0^{ABC}D_t^\alpha(y(t)) = u(t) \tag{7}$$

where $u(t) = R_p q_{in}(t)$, $y(t) = p_a(t)$, and $\tau = R_p C_\alpha$ [1].

thus fractional aortic input impedance in the frequency domain can be obtained by using LADM [5, 15, 16].

The basic concept of LADM is to assume an infinite solution of the form type

$$y(t) = \sum_{n=0}^{\infty} y_n(t) \tag{8}$$

where the nonlinear term is decomposed into terms of Adomian polynomials if any [17], and replace y (t) in (9) and take Laplace transformation.

After obtaining the inverse Laplace transformation,

$$y(t) = y_0 + y_1 + y_2 + y_3 + \ldots.. \tag{9}$$

We only need the first few terms to approximate the actual solution; the numerical result is very accurate [17].

5 Conclusion

In conclusion, this research article intends to provide insight into the variations in the behavior of the wind-Kessel model for simulating arterial blood circulation. This approach involves constructing and analyzing a fractional-order differential equations in modeling Wind-Kessel model utilizing the Atangana-Baleanu-Caputo (ABC) fractional derivative and solving it using the Laplace Adomian Decomposition Method (LADM). The two-element Wind-Kessel model will not simulate better when compared to higher-element W-K models. However, due to the complexity of the Atangana Balenau fractional derivative, the two-element wind-kessel model is considered. Studying the higher element model is our future study. Moving forward, future work is suggested to involve an extensive examination of solving the derived equation using LADM. Additionally, MATLAB can be used to visualize and compare the solutions for various fractional orders α. This approach will provide insights into the influence of the fractional differential operator on the behavior of the arterial blood circulation model, thereby contributing to the understanding and refinement of fractional-order modeling techniques in biomedical engineering and related fields.

References

1. Bahloul, M.A., Benencase, M., Belkhatir, Z., Laleg Kirati, T.M.: Finite-time simultaneous estimation of aortic blood flow and differentiation order for fractional-order arterial Windkessel model calibration. IFAC-PapersOnLine **54**(15), 538–543 (2021). https://doi.org/10.1016/j.ifacol.2021.10.312
2. Resmi, V.L., Selvaganesan, N.: Study on fractional order arterial Windkessel model using optimization method. IETE J. Educ. **64**(2), 103–111 (2023). https://doi.org/10.1080/09747338.2023.2210093
3. Westerhof, N., Lankhaar, J.W., Westerhof, B. E.: The arterial Windkessel. Med. Biol Eng. Comput. **47**(2), 131–141 (2008, June 10). https://doi.org/10.1007/s11517-008-0359-2
4. Atangana, A.: Blind in a commutative world: simple illustrations with functions and chaotic attractors. Chaos Solitons Fractals **114**, 347–363 (2018). https://doi.org/10.1016/j.chaos.2018.07.022
5. Khuri, S.A.: A Laplace decomposition algorithm applied to a class of nonlinear differential equations. J. Appl. Math. **1**(4), 141–155 (2001). https://doi.org/10.1155/s1110757x01000183
6. Catanho, M., Sinha, M., Vijayan, V.: Model of aortic blood flow using the Windkessel effect. In: BENG 221
7. Mohamed A.B., Taous-Meriem, L.-K.: Fractional order models of arterial Windkessel as an alternative in the analysis of the left ventricular afterload (2019)
8. Bahloul, M.A., Laleg-Kirati, T.M.: Three-element fractional-order viscoelastic arterial Windkessel model. In: 2018 40th Annual International Conference of the IEEE Engineering in Medicine and Biology Society (EMBC) (2018). https://doi.org/10.1109/embc.2018.8513473
9. Podlubny, I.: Fractional Differential Equations. Elsevier (1998, October 27). http://books.google.ie/books?id=K5FdXohLto0C&printsec=frontcover&dq=.Fractional+differential+equations.Podlubny+I&hl=&cd=1&source=gbs_api
10. Khan, Z.U., Ali, A., Khan, F.M., Shah, K.: Existence theory and numerical simulation of ebola model corresponding to nonlocal fractional order derivative. Arab J. Basic Appl. Sci. **27**(1), 375–388 (2020). https://doi.org/10.1080/25765299.2020.1826668

11. Guo, P.: The adomian decomposition method for a type of fractional differential equations. J. Appl. Math. Phys. **07**(10), 2459–2466 (2019). https://doi.org/10.4236/jamp.2019.710166
12. Magin, R.L.: Fractional calculus in bioengineering, Part 1. Crit. Rev. Biomed. Eng. **32**(1), 1–104 (2004). https://doi.org/10.1615/CritRevBiomedEng.v32.10
13. Escalante-Martínez, J.E., Gómez-Aguilar, J.F., Calderón-Ramón, C., Aguilar-Meléndez, A., Padilla-Longoria, P.: A mathematical model of circadian rhythms synchronization using fractional differential equations system of coupled van der Pol oscillators. Int. J. Biomath. **11**(01), 1850014 (2018). https://doi.org/10.1142/S1793524518500146
14. Ullah, S., Altaf Khan, M., Farooq, M.: A fractional model for the dynamics of TB virus, Chaos. Solitons Fractals 116, 63-71. ISSN 0960-0779 (2018). https://doi.org/10.1016/j.chaos.2018.09.001
15. Farman, M., Saleem, M.U., Ahmad, A., Ahmad, M.: Analysis and numerical solution of SEIR epidemic model of measles with non-integer time fractional derivatives by using Laplace Adomian Decomposition Method. Ain Shams Eng. J. **9**(4), 3391–3397 (2018). https://doi.org/10.1016/j.asej.2017.11.010
16. Guo: The adomian decomposition method for a type of fractional differential equations. J. Appl. Math. Phys. (2019). https://doi.org/10.4236/jamp.2019.710166
17. Annamalai, M., Renuga, E., Čep, R., Karthik, K.: Analysis of caputo fractional-order co-infection COVID-19 and influenza SEIR epidemiology by laplace adomian decomposition method. Mathematics **12**(12), 1876. (2024). https://doi.org/10.3390/math12121876
18. Peng, S.-L., Jhanjhi, N.Z., Pal, S., Amsaad, F. (eds.).: Proceedings of 3rd International Conference on Mathematical Modeling and Computational Science. Advances in Intelligent Systems and Computing. Springer Nature Singapore (2023). https://doi.org/10.1007/978-981-99-3611-3
19. Peng, S.-L., Lin, C.-K., Pal, S. (eds.): Proceedings of 2nd International Conference on Mathematical Modeling and Computational Science. Advances in Intelligent Systems and Computing. Springer Nature Singapore (2022). https://doi.org/10.1007/978-981-19-0182-9
20. Peng, S.-L., Hao, R.-X., Pal, S. (eds.): Proceedings of First International Conference on Mathematical Modeling and Computational Science. Advances in Intelligent Systems and Computing. Springer Singapore (2021). https://doi.org/10.1007/978-981-33-4389-4
21. Azzh Saad, A., Yasmin, H., Ghani, F., Shah, R., Nonlaopon, K.: Comparative analysis of advection–dispersion equations with atangana–baleanu fractional derivative. Symmetry **15**(4), 819 (2023). https://doi.org/10.3390/sym15040819
22. Yadav, P., Jahan, S., Shah, K., Peter, O.J., Abdeljawad, T.: Fractional-order modelling and analysis of diabetes mellitus: utilizing the Atangana-Baleanu Caputo (ABC) operator. Alexandria Eng. J. Alexandria Eng. J. **81**, 200–209 (2023). https://doi.org/10.1016/j.aej.2023.09.006
23. Wali, M., Arshad, S., Eldin, S.M., Siddique, I.: Numerical approximation of Atangana-Baleanu Caputo derivative for space-time fractional diffusion equations. AIMS Math. **8**(7), 15129–15147 (2023). https://doi.org/10.3934/math.2023772

An Ensemble Approaches to Improve Diabetic Retinopathy (DR) Detection and Severity Classification from Retinal Fundus Images

K. Kayathri[1,2](✉) and K. Kavitha[1]

[1] Department of Computer Science, Mother Teresa Women's University, Kodaikanal, Tamil Nadu, India
gayathri.vijayaanand@gmail.com, Kavitha.urc@gmail.com
[2] Department of Computer Science, Agurchand Manmull Jain College, Meenambakkam, Chennai, Tamil Nadu, India

Abstract. To find a treatment that works and avoids severe vision loss, Diabetic Retinopathy (DR) must be effectively detected and graded. However, DR screening programmers face difficulties due to the accessibility of human graders and the inefficiency of manual fundus image analysis. The research presented here introduces a comprehensive approach that combines both Machine Learning (ML) and Deep Learning (DL) approaches for enhancing the advanced detection of diabetic retinopathy (DR) and screening diabetics' eyes for damage. Convolutional Neural Networks (CNNs) with Grey Wolf Optimization (GWO) and classifiers like Support Vector Machines (SVM) and XGBoost can be combined in an innovative methodology called the CGSX ensemble. Our proposed work has the best overall performance, 0.95 F1 score achieving 0.95 precision, 0.94 recall, and 0.94 accuracy.

Keywords: Machine Learning · Deep Learning · Diabetic Retinopathy · Convolutional Neural Networks · Support Vector Machines · XGBoost

1 Introduction

World Institute of Ophthalmology and the WHO suggest that people with diabetes undergo eye tests no less than once yearly to avoid inconveniences from diabetic retinopathy. Recognizing diabetics with retinopathy is one step toward preventing visual loss. In addition to early medical intervention, an important healthy lifestyle can help prevent diabetic retinopathy (DR). The several stages of diabetic retinopathy (DR) are depicted in Fig. 1.

A series of preprocessing processes, including CLAHE and ADF, are performed after the initial step of the method, which is the capture of retinal images. The current study intends to develop a combination method for the automated diagnosis of DR, which uses DL and ML approaches to extract features and categorize diabetic retinopathy. Our mission is to build an advanced and efficient system that reliably identifies various stages

S. Pal and Á. Rocha (Eds.): ICMMCS 2025, LNNS 1398, pp. 225–234, 2025.
https://doi.org/10.1007/978-3-031-90998-6_22

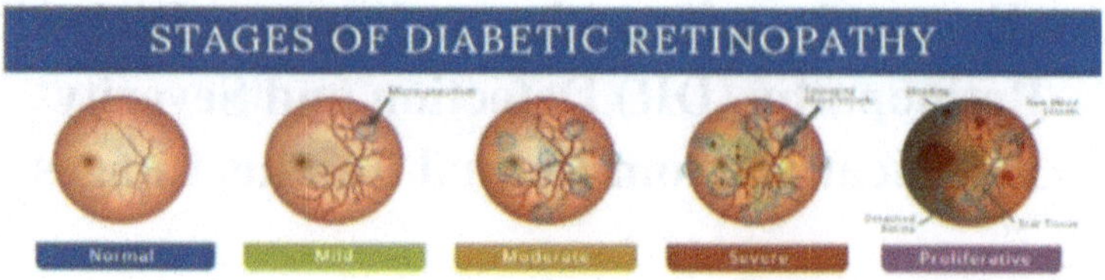

Fig. 1. Stages of Diabetics Retinopathy (DR)

of diabetic retinopathy from retinal images. This comprehensive method has the potential to fundamentally change the analysis and treatment of DR, eventually improving the lives of lots of individuals affected by this visualization impairing condition. The Lists of the Work's Novel Contributions are:

1) Image quality was improved using CLAHE and ADF algorithms. 2) A combination of the Fuzzy C-Means (FCM) and Firefly Algorithm (FA) for segmentation to precisely identify and classify anomalies of the retina.3) A combined machine learning and DL methodology (CGSX) to increase the classification of (DR) diabetic retinopathy severity status level. 4) Earlier stages (DR Stages 0–4) are more precisely identified.

2 Literature Review

Table 1 shows the Literature Review.

Table 1. Describe Literature Review

Authors & Year	Dataset	Methodology
Wang et al. [1], 2021	Private dataset	Lesion-Net architecture on InceptionV3 integrates lesion
Qummaret al. [2], 2019	Kaggle EyePACS	Ensemble DenseNet121, ResNet50, DenseNet169, Xception, InceptionV3for feature improvement
Cortes et al. [4], 2020	EyePACS, Messidor-2	InceptionV3 encoder and Gaussian hybrid model
Sugeno et al. [5], 2021	APTOS, DIARETDB11	EfficientNet - B3
Chenrui Zhang et al. [6], 2022	APTOS	Source-Free Transfer Learning (SFTL);
K. Kayathri ed al. [7], 2024	Kaggle EyePACS	Dynamic SwishNet-181
Boix et al. [8], 2021	APTOS	CNNs

3 Methodology

To detect diabetic retinopathy, retinal pictures must be preprocessed using CLAHE and ADF. Then, deep learning models such as Dynamic SwishNet-181, Efficient-NET, RESNET, and VGG16 are used for classification. To provide a comprehensive assessment of preprocessing and several deep learning models for improved Diabetic Retinopathy detection, the performance evaluation evaluates using important variables like F1-Score, Accuracy, Precision, and Recall.

3.1 Dataset Description

Table 2A and 2B shows the APTOS 2019 dataset class and image Descriptions.

Table 2A. APTOS 2019 Dataset

S. No.	Samples Images	Severity Levels
1	1805	Class - [Normal DR] - 0
2	370	Class - [Mild DR] - 1
3	999	Class - [Moderate DR] - 2
4	193	Class - [Severe DR] - 3
5	295	Class - [Proliferative DR] - 4
Overall Images 3662		

Table 2B. APTOS 2019 Dataset Overall Work Descriptions

APTOS 2019 Dataset	Dataset Descriptions
Actual Clinical Examinations from an Indian Eye Clinic	Origin
50° View Angle, 4288 × 2848 Pixels, JPG File Type	Image Characteristics
516	Quantity of Images
None, Moderate, Severe, Proliferate, and No_DR	DR severity Classes
Accurately Labeled DR Lesions and Normal Structures	Annotations
DR levels were provided	Disease Severity
resized for deep learning models to 224 x 224 pixels	Image Preprocessing

3.2 Pre-processing: (CLAHE and ADF Algorithms)

The image processing techniques CLAHE and ADF are utilized for noise reduction and contrast enhancement. ADF uses a controlled diffusion technique based on picture gradients, while CLAHE locally equalizes histograms in limited image regions. Both methods increase image quality for feature extraction, medical image processing, and precise analysis and aging [4]. Figure 2a demonstrates the input DR image, Fig. 2b expresses the noise filtering, and Fig. 2c displays the contrast enhancement output image. When CLAHE and ADF work together, retinal photos have better contrast, less noise, and the ability to retain important characteristics, which improves the diagnosis of diabetic retinopathy (DR) [7].

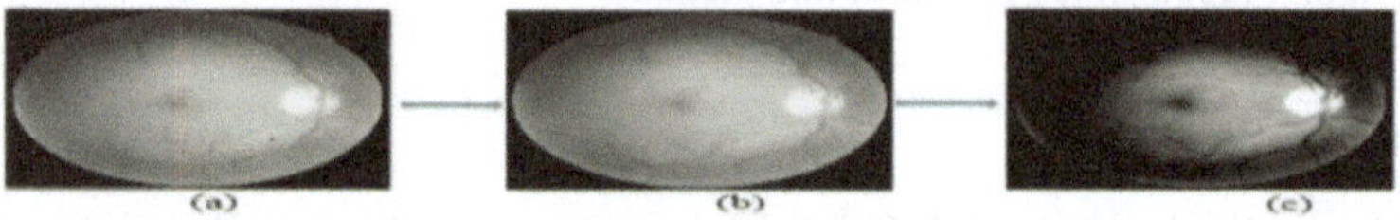

Fig. 2. a) Input b) After Noise Image Filtering c) After Image Contrast Enhancement.

3.3 Image Segmentation

Fuzzy C-Means (FCM) and the Firefly Algorithm (FA) were combined in diabetic retinopathy lesion segmentation to identify and characterize retinal abnormalities. FA adjusts parameters such as intensity thresholds for improved segmentation, whereas the FCM method classifies pixels and differentiates between abnormal and normal retinal structures. This hybrid technique improved the accuracy of lesion segmentation, a crucial step general in the initial identification and monitoring of diabetic retinopathy [11]. Figure 3a and 3b show the original and segmented images.

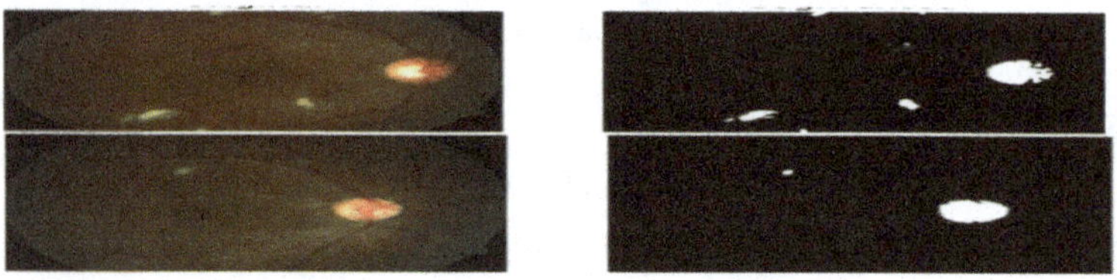

Fig. 3. a. Original Fundus DR Images. b. Segmented Fundus DR Images

3.4 Future Extraction Proposed Work

An essential first step in autonomously analyzing medical photos for illness identification is feature extraction. Factors gathered from diabetic retinopathy patients may be associated with lesions, optic nerve, and blood vessels. By detecting these characteristics, researchers can automatically train machine learning systems to detect DR popular retinal images [3]. Figure 4 illustrates feature extraction using GLCM, GLRM, and GLDM.

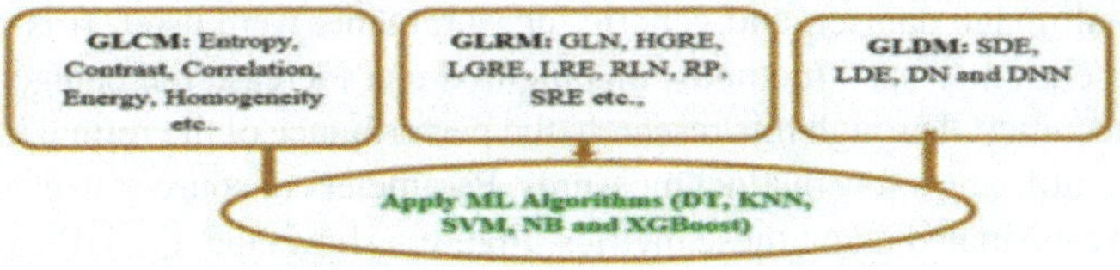

Fig. 4. Future Extraction DR

3.5 CGSX (CNN, GWO, SVM, and XGBoost): Our Proposed Work

Machine learning (ML) techniques (SVM, RF, KNN, DT, XGBoost) and Deep Learning methodology (CGSX, D-SwishNet-181, CNN, EfficientNet, SwishNet-181, -Proposed work) are used to classify images of diabetic retinopathy fundus. Figure 5 shows that CNN architecture consists of two entirely connected layers that practice max-pooling, Adam optimizer, definite cross-entropy loss, and ReLU activations with five convolutional layers. Using 20 wolves throughout 100 iterations, the GWO is used for feature selection. RBF kernel with C=1.0, and gamma='scale' is used by the SVM. XGBoost are the settings for learning rate=0.3, 100 estimators, and maximum depth=6. Using a weighted voting mechanism, the CGSX Ensemble integrates CNN for feature extraction, GWO for feature selection, and SVM and XGBoost for classification. Grid search was used to adjust the hyperparameters, and the CNN was trained using XGBoost 1.3.3, 0.001is learning rate, a 30-batch size, 30 epochs, Scikit-learn 0.24.1, Python 3, Keras 2.4.3, and TensorFlow 2.4.1were used in the training, and use GWO (Grey Wolf Optimizer) for future extraction.

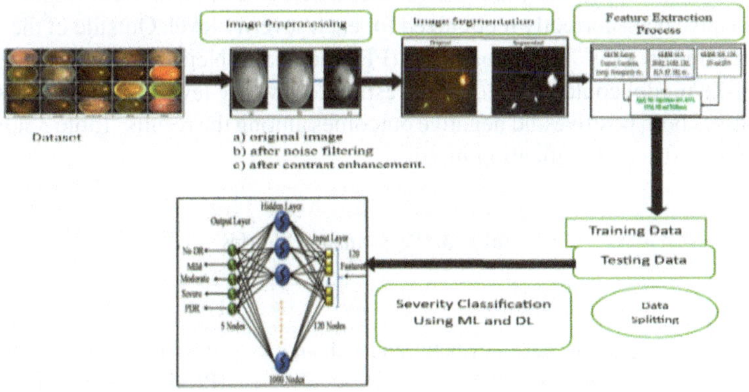

Fig. 5. Proposed Architecture Work

4 Result and Discussion

The findings of this review propose a new way to deal with the classification of diabetic retinopathy. This was achieved using a unique CNN architecture to related gathering classifiers (SVM/XGBoost) and feature selection (GWO). The strategy was assessed using a retinal image dataset, and generic measurements were used. It is possible that compared to currently used methods, this method can increase the categorization efficiency and accuracy. Through this research, the performance of the retinal image dataset was surveyed utilizing a few distinct measures. Parameters of source, size, and class distribution were evaluated using these metrics. Figure 6 shows that the DR extracted data. Ten (10) percent of the images were utilized for testing, 80% were retinal images for training, and 10 percent were chosen randomly as a validation set to assess the efficacy and preserve the optimal weight compositions.

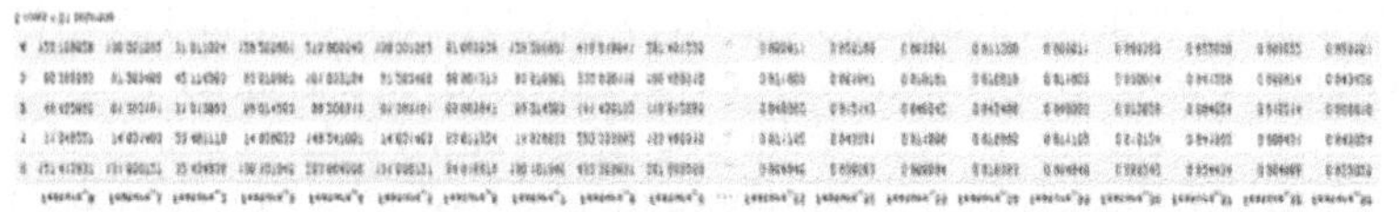

Fig. 6. Extracted Data using GWO

4.1 Performance Evaluation

This confusion matrix displays the accuracy of the model's predictions for various DR severity stages diagonally five different classes: 0, 1, 2, 3, and 4. The confusion matrix is shown below. The courses expected to be offered are displayed in each column, whereas the classes taught are displayed in each row. The model demonstrates that it is exact in recognizing different DR severity grades by introducing "True Negatives" (T - N) and True Positives (TP) in the diagonal. These True Positives and True Negatives reflect cases that have been correctly recognized for each severity level. Outside of the diagonal, instances referred to as "False Positives" (FP) and "False Negatives" (F N) are instances in which the model could not effectively estimate severity levels. This misclassification encompasses both positive and negative outcomes among the results. Table 3 summarizes the commonly used classification metrics.

Table 3. Performance Metrics

Metric	Description	Formulas
F1-Score	The harmony between recall and precision	(2 * ((Precision) * (Recall))) / (Precision + Recall)
Precision	The proportion of instances that were expected to be positive	(TP) / ((T P) + F P)
Sensitivity (Recall)	Correct identification of the proportion of actual positive cases	(TP) / (TP + F N)
Accuracy	The overall proportion of accurate predictions	((TP) + (T N)) / (T P + T N + F P + F N)

4.2 ML Classification Techniques Results for APTOS Dataset

Table 4 and Figure 7 show that the accuracy, recall, precision, and F1 score of Machine Learning models the performance analysis of ML classification algorithms. Methods such as SVM, RF, KNN, DT, and XGBoost fall into this category. Many metrics are carried out as part of this specific research project. The greatest accuracy (0.71) is the SVM model, even though the maximum accuracy (0.66 is the RF model and the F1 score reached (0.66). Both scores are the highest possible. The recall numbers of SVM and RF are not different; it's 0.71.

Table 4. Results of the ML Classification Algorithms

ML Models	Precision	Re - Call	Accuracy	F1 - Scores
DT	0. 61	0. 62	0. 62	0. 61
KNN	0. 60	0. 64	0. 64	0. 61
XGBoost	0. 65	0. 69	0. 69	0. 65
RF	0. 66	0. 71	0. 70	0. 66
SVM	0. 57	0. 71	0. 71	0. 63

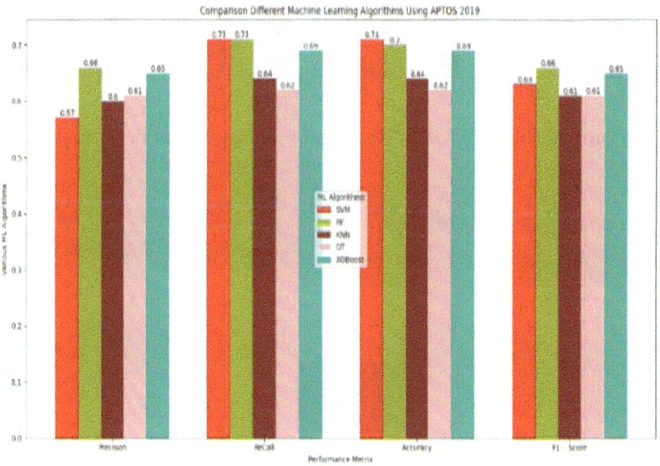

Fig. 7. ML – Classifications

4.3 Results for DL Classification Techniques Using Dataset APTOS

The evaluation of Deep learning (DL) classification in comparison is displayed in Table 5. The F1 score, accuracy, recall, and precision of Deep Learning Models in the performance analysis of DL classification algorithms (CGSX: CNN, GWO, SVM, and XGBoost) are shown in Figure 8 and the confusion matrix (CM) used for the proposed work CGSX is represented in Figure 9.

Table 5. Comparative evaluation of models for DL classification

DL and CGSX Models	F1 Score	Recall	Precision	Accuracy
CGSX (Our Model Work)	0.95	0.94	0.95	0.94
CNN	0.87	0.81	0.94	0.81
SwishNet-181	0.91	0.90	0.92	0.91
EfficientNet	0.92	0.85	0.90	0.88
D-SwishNet-181	0.93	0.92	0.94	0.92

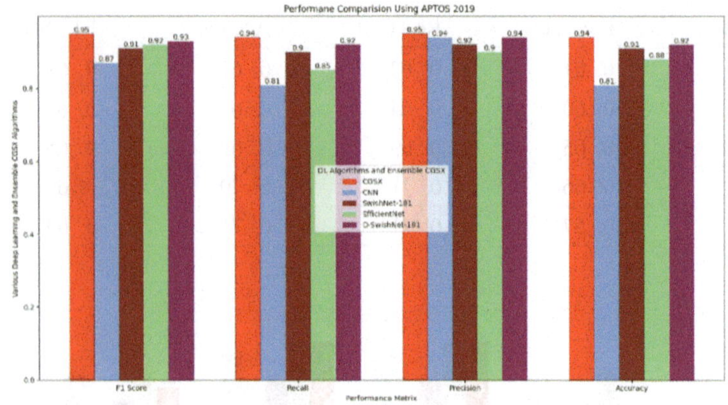

Fig. 8. Various DL and CGSX Methodology Comparison Performance Metrix

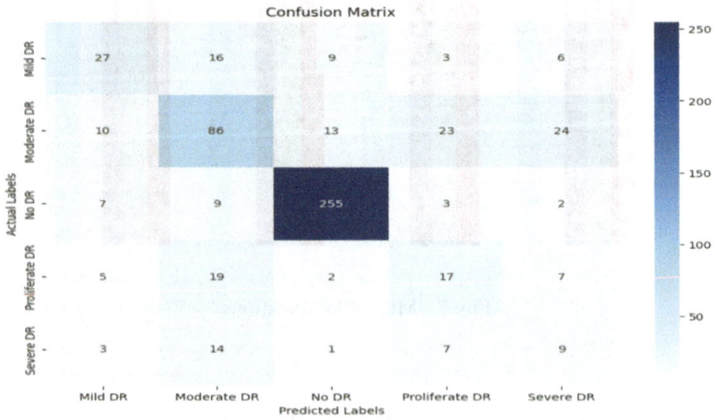

Fig. 9. CGSX (CNN, GWO, SVM, and XGBoost) - Confusion Matrix for DR classifications classes

5 State of the Art Comparison

The SCL model [10] performed experiments in the dataset such as APTOS 2019 with percent, a precision of 0.73, a recall of 0.71, 0.84 accuracy, and an F1-score (0.70) for multiple-stage classification in DR. On the same dataset, the "Error Correction Output Code SVM" (ECOC-SVM) Ensemble [9] experiments yielded precision(0.80), recall(0.81), accuracy, and F1 scores of 0.80, and 0.82. Figure 10 shows that our proposed CGSX model's recall, F1 scores, accuracy, and precision are 0.94, 0.95, 0.94, and 0.95 more than other state-of-the-art techniques.

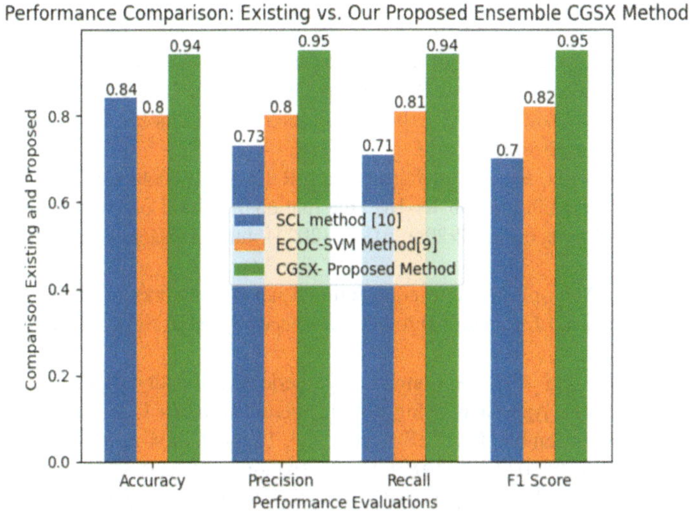

Fig. 10. Performance Comparison: Existing vs. Our Proposed CGSX

The best results from the above scenario in our work are CGSX with the highest predicted values for all metrics.

6 Conclusion with Future Work

In conclusion, this research presents an extensive way to deal with classified diabetic retinopathy (DR) that coordinates DL and ML strategies. According to the study findings, deep learning models usually perform better than traditional machine learning techniques. This structure can assist ophthalmologists in promptly identifying and assessing DR, lessening diagnostic responsibilities and human mistakes, and boosting DR screening effectiveness in clinical environments. Overall, the CGSX (CNN, GWO, SVM, and XGBoost) ensemble has made it possible to develop DR diagnosis, and treatment, and support improved clinical judgment and patient care. Further future work will apply other DR and real-time datasets.

References

1. Wang, S., et al.: Lesion-net: A lesion-aware deep learning architecture for diabetic retinopathy severity grading. IEEE Trans. Med. Imaging **40**(12), 3317–3327 (2021)
2. Qummar, A., et al.: An ensemble of deep learning models for classification of diabetic retinopathy. Sensors (Switzerland) **19**(19), 4480 (2019)
3. Shaukat, N., Amin, J., Sharif, M.I., Sharif, M.I., Kadry, S., Sevcik, L.: Classification and Segmentation of Diabetic Retinopathy: A Systemic Review. Appl. Sci. **13**, 3108 (2023). https://doi.org/10.3390/app13053108
4. Cortes, E., et al.: Addressing uncertainty in diabetic retinopathy prediction using a hybrid deep learning approach. In: 2020 IEEE 33rd International Conference on Artificial Neural Networks and Neural Systems (ICANN), pp. 1–8 (2020)
5. Sugeno, M., et al.: Efficientnet-b3 based deep learning model for diabetic retinopathy detection. Biocybernetics and Biomedical Engineering **4**(1), 127–137 (2021)
6. Chenrui Zhang, Y., et al.: Diabetic Retinopathy Grading by a Source-Free Transfer Learning Approach, Volume 73, March 2022, 103423. Biomed. Signal Process. Control (2022). https://doi.org/10.1016/j.bspc.2021.103423
7. Kayathri, K., Kavitha, K.: Advancing Diabetic Retinopathy Detection and Severity Classification using Dynamic SwishNet-181. International Journal of Intelligent Systems and Applications in Engineering **12**(20s), 61–77 (2024). https://ijisae.org/index.php/IJISAE/article/view/5119
8. Boix, E., et al.: Meta-plasticity in convolutional neural networks for diabetic retinopathy classification. In: 2021 International Joint Conference on Neural Networks (IJCNN), pp. 1–8. IEEE (2021)
9. Wong, W.K., Juwono, F.H., Apriono, C.: Diabetic Retinopathy Detection and Grading: A Transfer Learning Approach Using Simultaneous Parameter Optimization and Feature-Weighted ECOC Ensemble. In: IEEE Access, vol. 11, pp. 83004–83016 (2023). https://doi.org/10.1109/ACCESS.2023.3301618. https://ieeexplore.ieee.org/document/10207039
10. Islam, R., et al.: Applying supervised contrastive learning for the detection of diabetic retinopathy and its severity levels from fundus images. Computers in Biology and Medicine, 105602 (2022). https://doi.org/10.1016/j.compbiomed.2022.105602
11. Keerthiveena, B., Veerakumar, T., Esakkirajan, S., Subudhi, B.N.: Computer-aided diagnosis for Diabetic Retinopathy based on Firefly algorithm. 11th International Conference on Advanced Computing (ICoAC), pp. 310–315. Chennai, India (2019). https://ieeexplore.ieee.org/abstract/document/9087279

Hand Detection and Finger Counting Using Python

P. Anitha(✉), S. Vignesh, A. Vigneshwaran, R. Vasanthakumar, E. Yogesh, and S. Tamilselvan

KSR College of Engineering, Namakkal, Tamil Nadu, India
psp03ster@gmail.com

Abstract. This project is to develop a robust hand detection and finger counting system using Python, leveraging the Media Pipe library and computer vision. The primary goal is to enable touchless interfaces in public spaces, particularly in elevators, where users can intuitively select floors by raising specific fingers. This technology aims to enhance hygiene and user experience by eliminating the need for physical contact with buttons. The project seeks to provide an educational exploration of computer vision while showcasing a practical application with real-world implications for creating more intuitive and hygienic human-machine interactions in various environments.

Keywords: HSV Color Space · CNN · Pattern Matching

1 Introduction

In a technologically advancing world, the integration of computer vision has revolutionized human-computer interaction, opening doors to innovative applications. This Python-based project leverages the capabilities of the Media Pipe library to delve into the realm of hand detection and finger counting. By harnessing the power of realtime webcam input, the program not only showcases the intricacies of hand tracking and landmark identification but also introduces a touchless interface for potential deployment in spaces like elevators. The unique proposition lies in the seamless fusion of computer vision and practical utility, envisioning a future where users can navigate elevator controls effortlessly by raising specific fingers, contributing to enhanced hygiene in public settings. Beyond its educational significance in computer vision, this project serves as a testament to the transformative potential of technology in redefining conventional human-machine interactions for a more intuitive and hygienic user experience.

2 Existing System

In existing system, the user must press the lift buttons on the lift, which means user enters the lift and press lift button which floor they want to go. If any persons affected by the covid it is problem for who are pressing the button. For avoiding this inconvenience, this proposed application was developed.

S. Pal and Á. Rocha (Eds.): ICMMCS 2025, LNNS 1398, pp. 235–244, 2025.
https://doi.org/10.1007/978-3-031-90998-6_23

2.1 Limitation of the Existing System

- Users must operate manually.
- It is problem for every user because any one affect covid infection means it will spread.

3 Proposed System

The drawbacks, which are faced during existing system, can be eradicated by using the proposed Finger counting and hand detection using OpenCV. The main objective of the proposed system is to provide a user-friendly interface for society people and user don't want to press the button, and an effectively implements Finger Counting and Hand Detection using open cv. In our application web camera will integrate user can show the finger Sign Language. This application automatically captures sign language and compare with train dataset and effectively move the GUI lift image.

3.1 Advantages of the Proposed System

- It is a user-friendly method.
- User can show the finger Sign Language.
- Automatically capture sign language.

4 Modules Description

The implementation has following modules,

1. Camera Configuration and Capture Process
2. HSV Color Space and Background Elimination
3. Segmentation
4. Sign Identification using CNN
5. Pattern Matching

4.1 Camera Configuration and Capture Process

This application is fully applied with the camera. Once the person wants to use the lift, this system is going to configure the camera in this application. The camera will continuously be running. This is fully control on user. If the person enters the lift the user will show the sign to the camera. The camera will capture the image. This module is used to capture the image.

4.2 HSV Color Space and Background Elimination

Since the images obtained are in RGB color space, it becomes more difficult to segment the hand gesture based on the skin color only. It is a model which splits the color of an image into 3 separate parts namely: Hue, Saturation and value. HSV is a powerful tool to improve stability of the images by setting apart brightness from the chromaticity. The Hue element is unaffected by any kind of illumination, shadows and shadings and can thus be considered for background removal.

4.3 Segmentation

Division After that, the original image is converted to grayscale. As much as this process will result in the loss of color in the region of the skin gesture, it will also enhance the robustness of our system to changes in lighting or illumination. Non-black pixels in the transformed image are binarized while the others remain unchanged, therefore black.

4.4 Sign Identification Using CNN

In this module helps the camera to identify the sign image. If the user shows the one finger means the camera will capture the image and identify. To predict hand gestures and extract characteristics from the frames, a CNN model is employed. It is a multilayered feed forward neural network mostly used in image recognition. Each convolution layer in the CNN design comprises of a pooling layer, an activation function, and an optional batch normalization. It also has a set of fully connected layers. One of the photos gets smaller as it travels around the network.

4.5 Pattern Matching

The method of determining if a particular data sequence occurs within the provided training data is known as pattern analysis. It is supervised learning and identifies the suitable match from different group of data and enhances the number of events to improve the performance and efficiency of all events (Fig. 1).

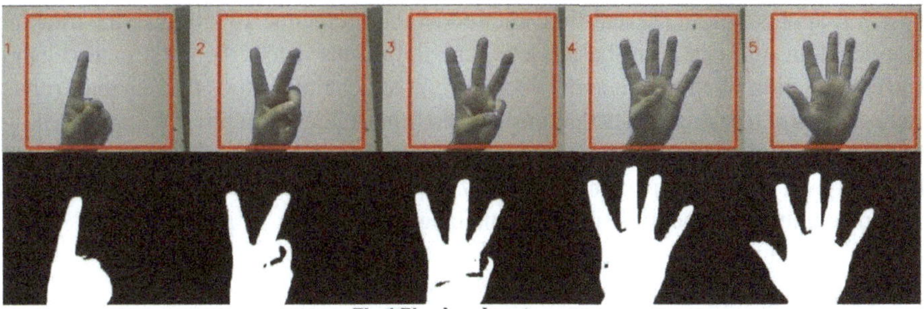

Fig 4 Five hand gestures

Table 4: Accuracy of each gesture with plain background and non-plain background

Gesture	Accuracy with plain background (in %)	Accuracy with non-plain background (in %)
1	95	45
2	94	48
3	96	46
4	91	40
5	92	41

Fig. 1. Hand Gestures

5 System Design

The process of defining a system's architecture, modules, interfaces, and data in order to meet predetermined requirements is known as system design. One way to apply systems theory to product creation is through systems design.

5.1 Input Design

User-oriented inputs are transformed into computer-based representations through input design. Errors in data processing are primarily caused by inaccurate input data. Error data entered by the data operator can be controlled by the input design. The Making data entry as simple, logical, and error-free as feasible is the aim of input design. The proposed system is completely menu driven. It is a powerful tool for interactive design. It helps the user comprehend the range of alternatives available and prevents them from making an invalid selection. All entry screens are interactive in nature. It has been designed considering all the constraints of the end-user.

Some other features included are:

- The form title clearly states the purpose of the form.
- Adequate space is given for data entry.

5.2 Hand Tracking Algorithm

Although hand tracking algorithms have been widely used in virtual reality and HCI system, it is still a challenging problem in vision-based research area. The majority of hand tracking algorithms require specialized equipment to provide results that are suitable since VR applications have real-time and robust requirements. To initialize the tracking zone, contour and curvature-based detection algorithms detect the outstretched hand. The combination of color cue and velocity-weighted characteristics then enables robust multi-cue hand tracking. Tests demonstrate that the suggested multi-cue hand tracking method produces consistent real-time outcomes even in busy background situations. The approach fulfils the speed and accuracy requirements of frontal- view vision-based human computer interactions like machine language (Fig. 2).

5.3 Output Design

The most significant and direct information source for management and customers is the output. The intelligent output design will improve the system's user interface. Which also facilitates decision-making. The results are permanently copied into hard copy using the outputs for future reference. The system's output is frequently used as the benchmark for assessing how well the system is doing.

The following criteria formed the basis of the output design.

- Effectiveness in selecting the different outputs that the system user should print.
- separating the outputs that are meant to be printed and those that are meant to be shown.
- The manner in which the results will be displayed.

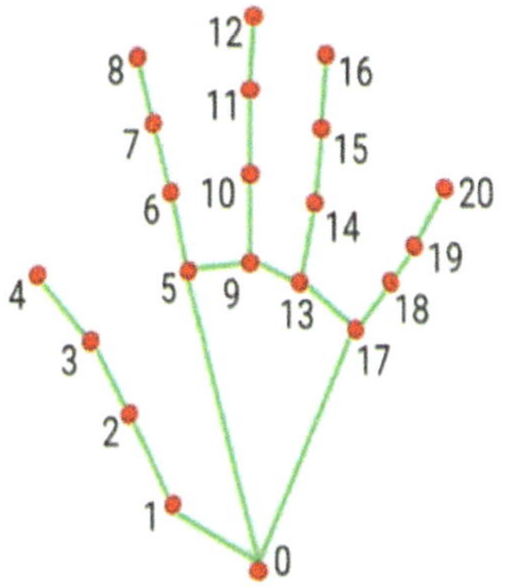

0. WRIST
1. THUMB_CMC
2. THUMB_MCP
3. THUMB_IP
4. THUMB_TIP
5. INDEX_FINGER_MCP
6. INDEX_FINGER_PIP
7. INDEX_FINGER_DIP
8. INDEX_FINGER_TIP
9. MIDDLE_FINGER_MCP
10. MIDDLE_FINGER_PIP
11. MIDDLE_FINGER_DIP
12. MIDDLE_FINGER_TIP
13. RING_FINGER_MCP
14. RING_FINGER_PIP
15. RING_FINGER_DIP
16. RING_FINGER_TIP
17. PINKY_MCP
18. PINKY_PIP
19. PINKY_DIP
20. PINKY_TIP

Fig. 2. Structure of Hand Algorithm

5.4 Code Design

Code is an ordered collection of symbols designed to provide unique identification of an attribute. Codes can be used for various purposes. They can specify an object's physical, or performance characteristics and they can be used to give operational instructions. They also can show inter relationships and may sometimes use to achieve secrecy or confidentiality. Codes are created to maximize both machine and human efficiency. Codes possess uniqueness, expandability, conciseness, uniform nets, simplicity, versatility, sort ability, meaningfulness and operability. Sufficient effort and time are spent in the preliminary study of the problem to design an efficient code. Activate serve scripting is object oriented. The source code is designed so that it can do transaction efficiently. It is the code that dose all the updating, modifications, etc. for all object used in the project there exist an associated source code, which explains the work of that object. It also describes the flow of the project. Source code is enhanced by structured coding techniques by good internal comments and features provided by the language. The code design in this project is made modular. The modular behavior enables easy debugging and testing. Inserting comment statement wherever enhances the coding. This is done during the documentation process coding is done in such a way that errors can be trapped easily. Also, modifications can easily be appended due to the code's modular behavior.

6 Dataflow Diagram

LEVEL-0:
Figure 3.

Level-1:
Figure 4.

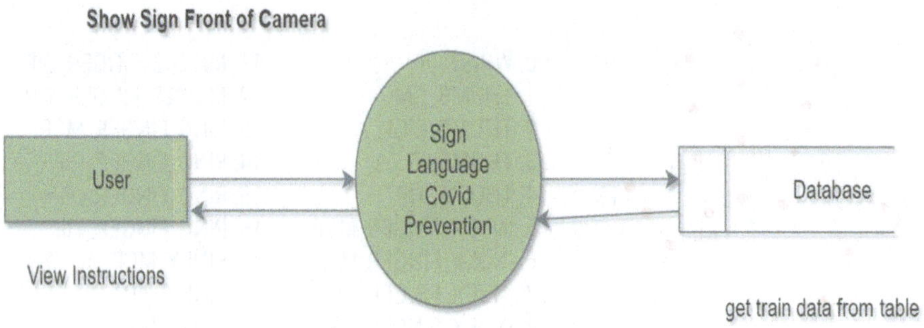

Fig. 3. Dataflow Diagram (Level-0)

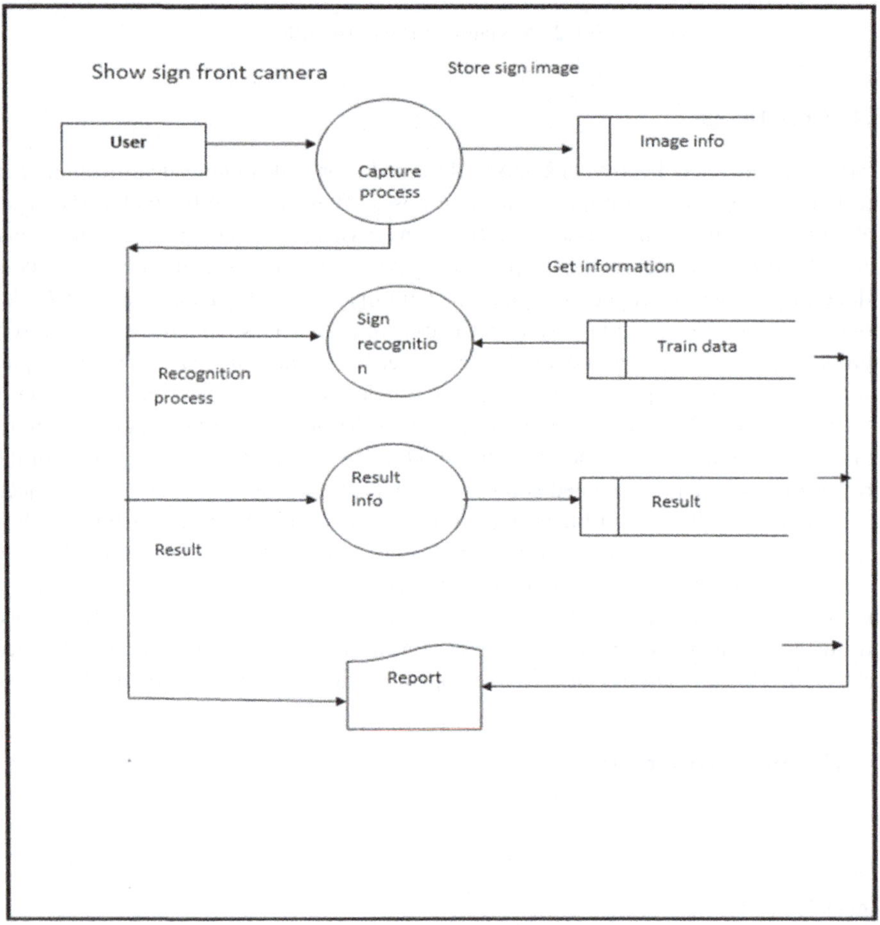

Fig. 4. Dataflow Diagram (Level-1)

6.1 Systemflow Diagram

Figure 5.

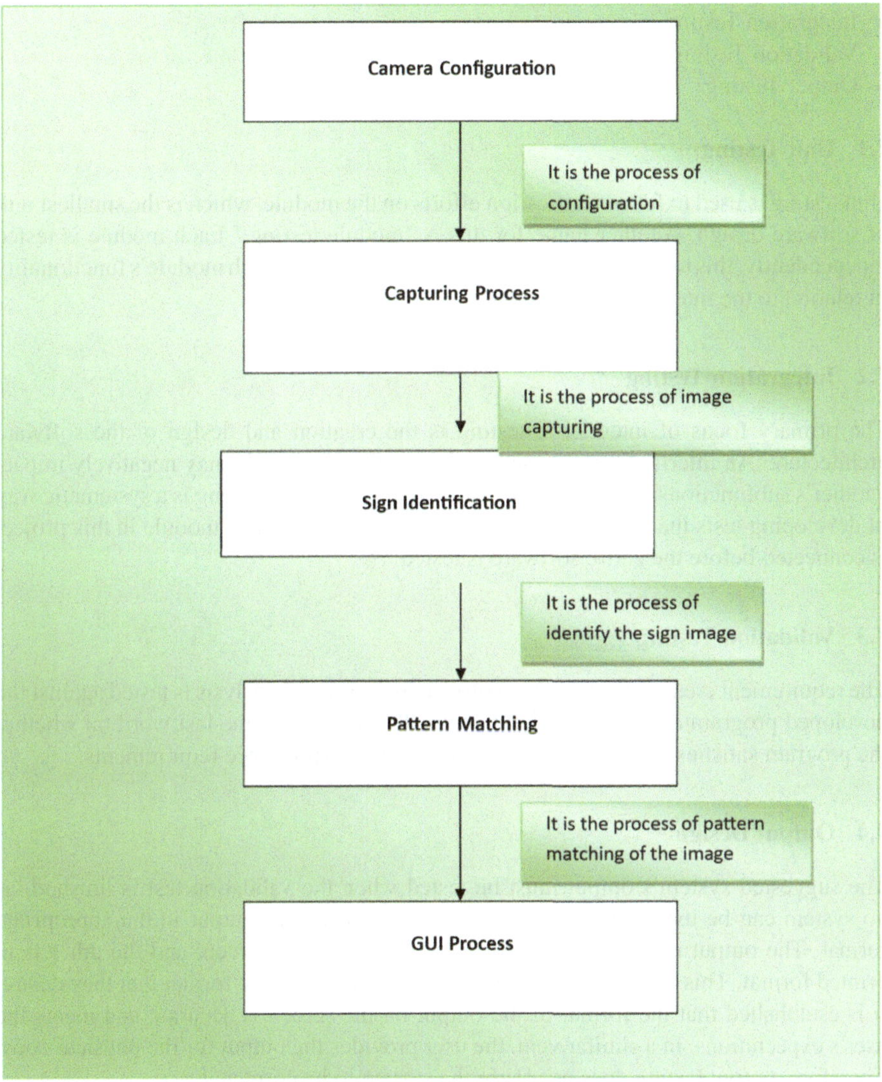

Fig. 5. System Flow Diagram

7 System Testing Methodologies

The different types of testing are:

- Unit Testing
- Integration Testing
- Validation Testing
- Output Testing

7.1 Unit Testing

Unit testing is used to focus verification efforts on the module, which is the smallest unit of software design. Another name for this is "module testing." Each module is tested independently; this testing is done at the programming stage. Each module's functionality in relation to the intended output is determined in this step.

7.2 Integration Testing

The primary focus of integration testing is the creation and design of the software architecture. An interface may cause data loss, and one module may negatively impact another's subfunctions and display on. Therefore, integration testing is a systematic way of developing tests that find issues connected to interfaces. Each module in this project is connected before the entire software is tested.

7.3 Validation Testing

The requirement created as part of the software requirement analysis is tested against the developed program during validation testing. This test offers the last word on whether the program satisfies all behavioral, functional, and performance requirements.

7.4 Output Design

The suggested system's output must be tested when the validation test is finished, as no system can be useful if it cannot generate the necessary output in the appropriate format. The output is regarded in two ways: one is on the screen, and the other is in printed format. This is tested by questioning the users about the format that they desire. It is established that the format of the output on the screen is accurate and meets the user's expectations; in a similar vein, the user provides the output for the physical copy. Therefore, output testing does not cause the system to be corrected.

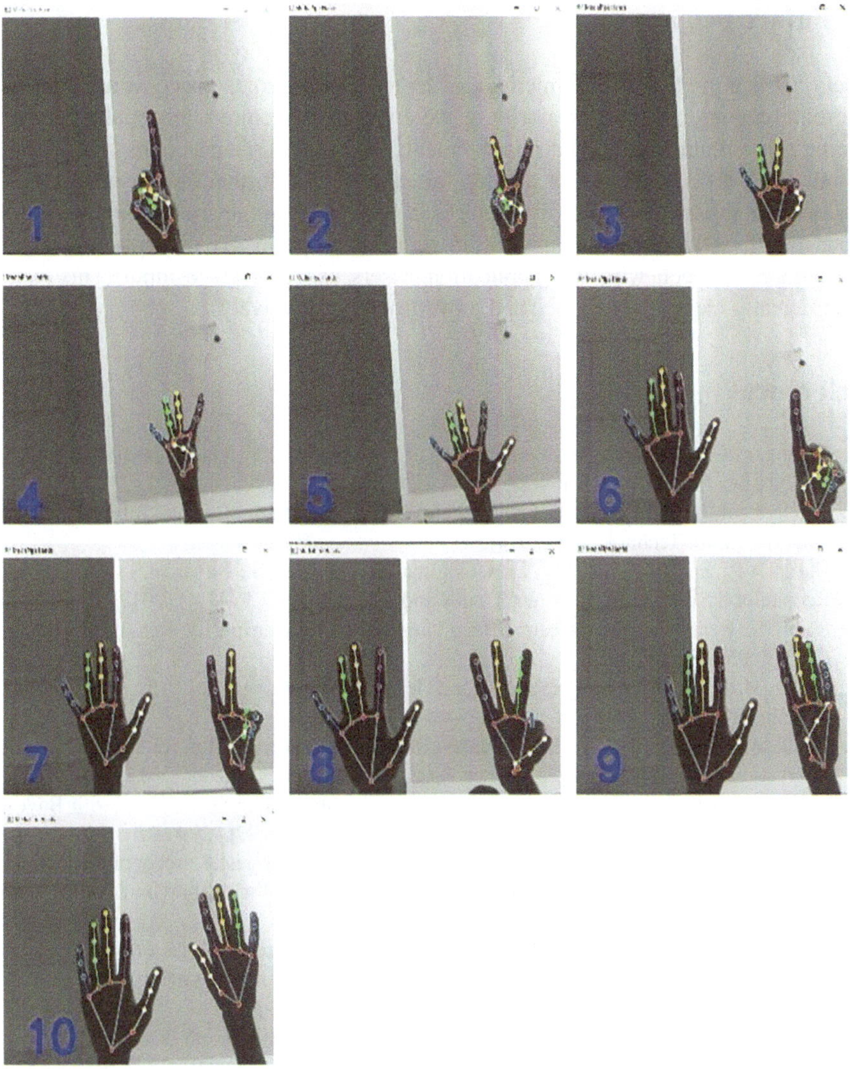

8 Conclusion

In conclusion, this project successfully demonstrates the integration of computer vision, utilizing the Media Pipe library for hand detection and finger counting. The touchless interface designed for elevators showcases the transformative potential of technology in addressing real- world challenges. By offering a practical solution for intuitive floor selection without physical contact, the project emphasizes the significance of hygienic and user-friendly interfaces in public spaces. This endeavor not only contributes to the advancement of computer vision applications but also underscores the importance of innovation in enhancing everyday interactions for a safer and more convenient future.

9 Future Scope

Every application has its own merits and demerits. The project has covered almost all the requirements. Since the majority of the coding is structured or modular in nature, adding more requirements and improvements is simple. Enhancements can be added by introducing new modules or altering the current ones further enhancements can be made to the application with Voice interface, Sometime Big apartment and malls have more than five floors now identified the sign is more than five difficult processes. It can integrate speech with this application. Users can give voice input. This system automatically identifies the voice and mover to floor effectively.

References

1. Badgujar, S.D., Talukdar, G., Gondhalekar, O., Kulkarni, S.Y.: Hand Gesture Recognition System. Int. J. Sci. Res. Publ. **4**(2) (2014)
2. Fidelis Paul, J., Seth, D., Paul, C., Dastidar, J.G.: Hand Gesture Recognition Library. Int. J. Sci. Appl. Info. Technol. **3**(2) (2014)
3. Haria, A.P., Sri Subramanian, A., Asokkumar, N., Poddar, S., Nayak, J.S.: Hand Gesture Recognition System. Int. J. Comp. Trends and Technol. (IJCTT) **47**(4) (2017)
4. Singh, S.: Real Time Hand Gesture Recognition Using Fingertips. Int. Res. J. Eng. Technol. (IRJET) **05**(06) (2018)
5. Ahuja, M.K., Singh, A.: A Survey of Hand Gesture Recognition. Int. J. Adv. Res. Comp. Sci. Manage. Stud. **3**(5) (2015)
6. Rautaray, S.S., Agarwal, A.: Vision Based Hand Gesture Recognition for Human Computer Interaction: A Survey. Springer Science & Business Media Dordrecht (2012)
7. Ashfaq, T., Khurshid, K.: Classification of hand gestures using Gabor filter with Bayesian and naïve bayes classifier. Int. J. Adv. Comp. Sci. Appl. **7**(3), 276–279 (2016)
8. Chaikhumpha, T., Chomphuwiset, P.: Real—time two hand gesture recognition with condensation and hidden Markov models. In: 2018 international workshop on advanced image technology (IWAIT), pp. 14. Chiang Mai (2018)
9. Hong, J., Kim, E.S., Lee, H.-J.: Rotation-invariant hand posture classification with a convexity defect histogram. In: IEEE international symposium on circuits and systems (ISCAS) 20–23 May 2012, pp. 774–777 (2012)
10. Liu, K., Chen, C., Jafari, R., Kehtarnavaz, N.: Fusion of inertial and depth sensor data for robust hand gesture recognition. Sens. J. IEEE **14**(6), 1898–1903 (2014)
11. Peng, S.-L., Jhanjhi, N.Z., Pal, S., Amsaad, F., (eds.): Proceedings of 3rd International Conference on Mathematical Modeling and Computational Science. Advances in Intelligent Systems and Computing. Springer Nature Singapore (2023)
12. Peng, S.-L., Lin, C.-K., Pal, S., (eds.): Proceedings of 2nd International Conference on Mathematical Modeling and Computational Science. Advances in Intelligent Systems and Computing. Springer Nature Singapore (2022)
13. Peng, S.-L., Hao, R.-X., Pal, S., (eds.): Proceedings of First International Conference on Mathematical Modeling and Computational Science. Advances in Intelligent Systems and Computing. Springer Singapore (2021)

Advanced Portable Radar System Using ESP32: A Hardware Perspective

Bhukya Arun Kumar[1](✉), Ananya Tripathi[2], Ban Salman Shukur[3], Indu Bala[1], Maad M. Mijwil[4], and Mostafa Abotaleb[5]

[1] School of Electronics and Electrical Engineering, Lovely Professional University, Phagwara, Punjab, India
arunbhukya@gmail.com, i.rana80@gmail.com
[2] School of Computer Application, Lovely Professional University, Phagwara, Punjab, India
[3] Computer Science Department, Baghdad College of Economic Sciences University, Baghdad, Iraq
Dr_bansalman@baghdadcollege.edu.iq
[4] Computer Techniques Engineering Department, Baghdad College of Economic Sciences University, Baghdad, Iraq
mr.maad.alnaimiy@baghdadcollege.edu.iq
[5] Department of System Programming, South Ural State University, Chelyabinsk, Russia
abotalebmostafa@bk.ru

Abstract. This article proposes a portable radar system combining multiple sensors to enhance monitoring capabilities in various applications. Servo motors enable directional control, while ultrasonic sensors enable precise distance measurements. Also, environmental awareness can be enhanced by the combination of temperature sensors (DHT22), light sensors (LDR), gas sensors (MQ-2), and a camera module, which will be helpful as we can see the images associated with the data. GPS is used to find the location, which maps to the actual coordinates of the sensor. This article concluded that these devices could be used to design an advanced radar system and transmit data in real-time.

Keywords: ESP32 4G LTE Module · Gas Sensor (MQ-2) · GPS (Global Positioning System) · Light Sensor (LDR) · Servo Motor · Temperature Sensor · Ultrasonic Sensor · User-friendly Design

1 Introduction

The Internet of Things (IoTs) and cloud computing are two very different technologies that are already a part of our daily lives [1]. According to IoTs, sensor data is transferred to the cloud for remote access and storage. IoTs are invaluable for real-time monitoring and data insights for decision-making because of their widespread availability across several locations and the vast amounts of data and information they hold [2–6]. IoT applications include monitoring air pollution levels in urban areas, wildlife movements in natural habitats, and enhancing security and surveillance in commercial and residential areas.

S. Pal and Á. Rocha (Eds.): ICMMCS 2025, LNNS 1398, pp. 245–256, 2025.
https://doi.org/10.1007/978-3-031-90998-6_24

Data is collected from sensors that sense the changes in the real-time environment, with the help of microcontrollers processing the data and transmitting the data with communication protocols to cloud storage. The data is presented in a useful manner in any application so that users can monitor remotely. In such a way, IoT has many applications in industries such as agriculture, healthcare, and many more [7–10]. Cloud computing, which offers on-demand storage and compute capabilities over the Internet, is contributing to the growing trend of implementing the Internet of Things (IoT) in industries such as healthcare, smart cities, Industry 4.0, and more. [11].

This article thoroughly discusses a groundbreaking invention: a portable radar with several varied sensors expertly crafted for environmental monitoring. Compared to traditional radar systems used for monitoring and surveillance, the technology presented here is deemed groundbreaking, especially in light of the growing need for adaptive sensing systems. At the heart of this distinctly novel technology is the combination of portable radar equipment and contemporary sensors, increasing the application area's versatility for environmental monitoring, security, and surveillance. Parts of our system include light-temperature sensors, servo motors, and ultrasonic sensors. The unique value proposition of the system stretches beyond just sensor variety to include real-time data transmission made possible by integrating the ESP32 4G LTE module [12–14]. This makes it possible for information to be transmitted fast over a long distance, facilitating easy retrieval of important data—one great thing in crisis management situations. In today's ever-changing environment, there is always a need for adaptive, portable, and user-oriented monitoring tools, without which survival would not be easy [15].

An extensive technical examination of this system with regard to its architecture, sensor integration method and how they communicate using data transfer techniques is provided. The case study and field test results demonstrate how efficient and versatile the system can be applied in many different scenarios. The multi-sensor portable radar is priceless in terms of security, surveillance, and environmental monitoring due to 4G long-term evolution (LTE) connectivity [16–19]. It can collect multi-modal data, transfer them immediately, and adapt them to address various problems [20–22]. Since it is insightful, the study also desires innovations that will lead to more improvements and applications.

The literature review profiles the latest IoT technologies as well as sensors in use across different sectors. For instance, several researchers have suggested cheap smart security systems, IoT-based gas leak detection methods, and temperature monitoring. The main objective is to build an advanced radar system with many sensors to enhance its accuracy and understanding, especially towards disaster relief efforts. As a whole, these publications show how sensors can be deployed for realistic, environmentally friendly solutions as well as intelligence surveillance through IoE integration. The proposed advanced radar system aims to enhance environmental conservation and disaster response monitoring capabilities by integrating technology and IoT.

The suggested portable radar device manages critical rules of existing monitoring gear by integrating various sensors, including LDR, MQ2, DHT11, servo vehicles, and ultrasonic sensors, to capture comprehensive environmental facts. Employing ESP32 4G LTE SIM modules guarantees instant information transmission to far-off servers, overcoming latency troubles in conventional systems and permitting real-time tracking. The system's adaptive nature allows for customizable sensor configurations and adaptable tracking techniques, improving flexibility and addressing the inflexibility of conventional monitoring tools. This progressive approach gives a flexible, efficient, and adaptive monitoring answer capable of addressing diverse application requirements and improving general monitoring effectiveness.

The main contribution of this article is the design and development of an advanced portable radar system to enhance surveillance capabilities in various applications. To accomplish this work, a group of sensors are integrated, and the Internet of Things and cloud computing technologies are utilized in developing this system and transmitting data in real-time. This work includes a combination of LDR, MQ2, DHT11 devices, servo motors and ultrasonic sensors. In addition, this work provides mechanisms to address monitoring system limitations and enhance capabilities to ensure accurate and highly efficient data collection.

2 Methodology

The functionality of the Advance Portable Radar System, equipped with multiple sensors and 4G LTE connectivity, can be dissected into several fundamental components and processes:

Data Collection from Multiple Sensors:

- Ultrasonic Sensors: In this type of sensor, the actual distances to the targets are measured and the sensor parameters are adjusted as they generate ultrasonic waves in order to determine how long it takes to return after colliding with something. These sensors are practical for mapping the surrounding areas and identifying barriers.
- Servo Motors: It is used to provide precise direction control, focusing on collecting information by moving sensors or cameras towards the desired direction. To obtain angle and distance measurements, the servo motors' angle positions must be set, and the movement verified using a special calibration tool.
- Gas sensors: These devices are used to determine gas leakage or pollution levels, as a response is recorded in the event of dangerous gas levels. The MQ2 sensor is installed in areas exposed to contamination, and a calibration curve is created to give accurate measurements.
- Light sensor: This device measures the intensity of ambient light, and this data can be used in various applications, such as monitoring natural lighting in agriculture. The LDR is set up in different conditions, measuring resistance at each level and obtaining accurate light intensity readings.
- Temperature Sensor: It is critical to get readings from temperature sensors that allow us to know what climate we are living in, HVAC control, and many other things. Compare sensor readings with a calibrated reference instrument and apply correction factors to adjust readings accordingly. Deploy DHT11 sensors in diverse environments to capture variations in temperature and humidity levels

- GPS Receivers: Location data is collected by GPS receivers, which help the system accurately position itself, thereby giving the data being gathered geospatial context.
- Camera module: The camera module provides a way to see what's going on. It can capture real-time images and possibly videos of the monitored environment.

3 Data Transmission

This paper closely aligns the perspective of the advanced portable radar system with the Internet of Things (IoT) paradigm, emphasizing the integration and connectivity of diverse sensors and components. The portable radar system achieves seamless interoperability, real-time data transmission, and remote monitoring capabilities by leveraging IoT technologies. This interconnected approach enhances system adaptability, scalability, and efficiency, paving the way for intelligent monitoring solutions that address various applications and environmental challenges.

Integrating cloud computing and IoT permits the system to collect, examine, and share data effectively, facilitating knowledgeable decision-making and improving typical monitoring effectiveness in diverse domains. After all the sensor records have been gathered and processed, they are transferred to a server via the ESP32 4G LTE SIM A7672/A7670 module. The ESP32 module acts as a communication gateway to create a constant and quick connection with the server using the 4G LTE network. Thanks to this actual-time data transmission, the server receives the latest information from the sector, allowing quick evaluation and selection. The ESP32 module employs MQTT and HTTP/HTTPS protocols for data transmission to the server, ensuring green and stable communication. Data encryption uses TLS/SSL protocols to guard sensitive information for transmission duration. Error-dealing with mechanisms, including retransmission and information integrity checks, ensures reliable data switch and system overall performance. The module logs error messages and reports to directors of conversation troubles for a timely resolution, improving the overall reliability and integrity of the superior portable radar system. Figures 1, 2, 3 and 4 show the connection method and settings. The methodology applied in this work can be traced through these figures.

3.1 Remote Data Analysis and Action

The raw data is received, processed, and analyzed in real-time on the server side, or it is kept for examination at a later time. Several approaches and data processing strategies can be used to get important insights from the gathered data. The system may send notifications or initiate real-time activities depending on the application. For example:

- In security and surveillance applications, it can identify suspicious activities or intrusions and send alerts to relevant authorities.
- In environmental monitoring, it can detect pollution levels exceeding thresholds and initiate corrective actions.
- In wildlife tracking, it can provide researchers with precise location data for animal movement analysis.
- The results of this analysis can also be made accessible remotely, allowing users to monitor the situation and make informed decisions in real-time.

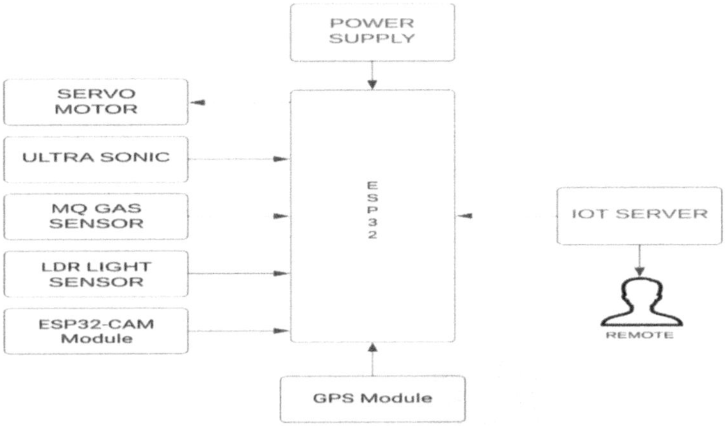

Fig. 1. Block diagram of the proposed system.

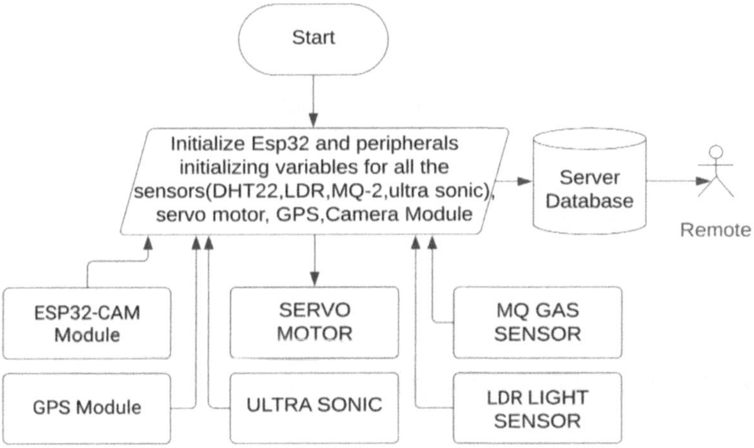

Fig. 2. Data Flow diagram of system proposed.

The purpose was to evaluate sensors' accuracy, reliability, and responsiveness through careful analysis of the collected data and the efficiency of real-time data transmission and remote monitoring capabilities. Simulation tools and modeling techniques were used to simulate system behavior, evaluate system implementation under various scenarios, and validate the system design and operational capabilities.

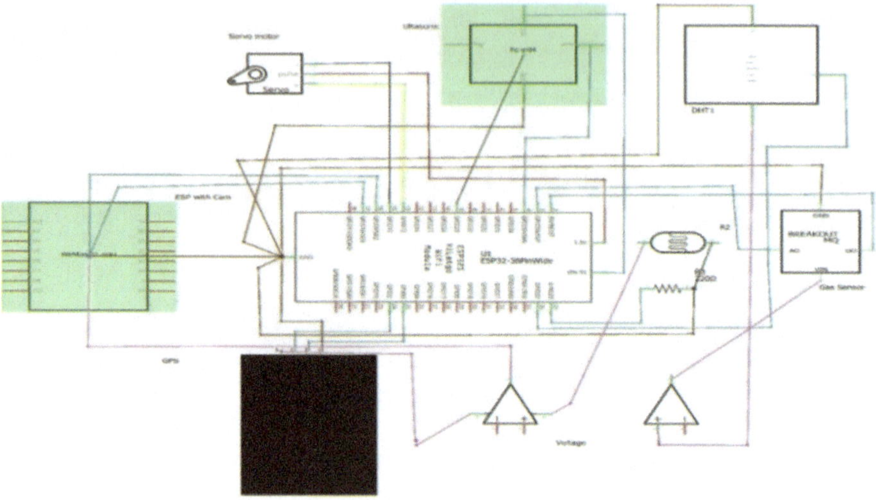

Fig. 3. Schematic Diagram of proposed system.

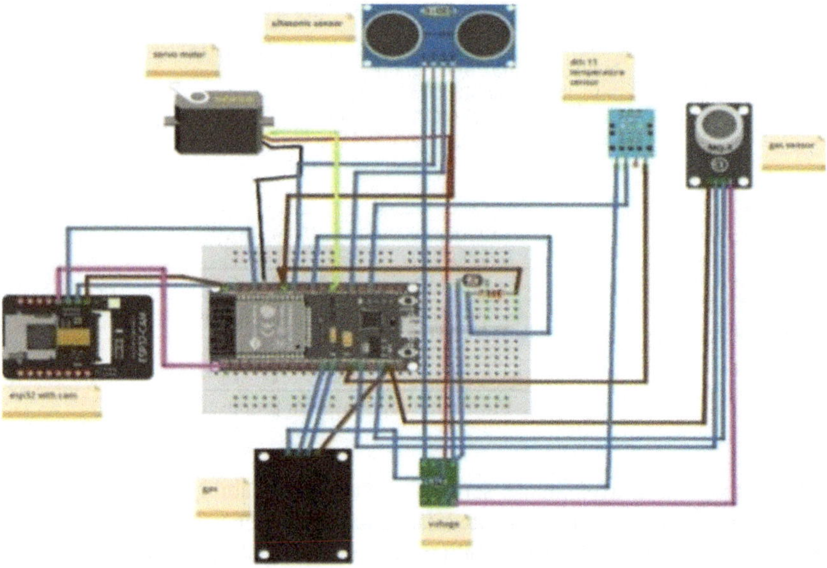

Fig. 4. Circuit Diagram of proposed system.

4 Results and Discussions

After implementing the hardware ESP32 4G LTE SIM A7672/A7670, we can now send the data over the internet; this data can be analyzed and used further. Currently, we are getting distance from the ultrasonic, direction from the servo motor, humidity and temperature from the DHT22 sensor, light from the LDR sensor, MQ-2 gas sensor, and visuals of the surroundings from the camera module. With this data, we can find objects in a particular direction, temperature, humidity, and gases in the surroundings with visuals. Figures 5, 6, 7, 8, 9 and 10 show the results obtained. Table 1 shows the difference between the Advanced Portable Radar System using ESP32 and existing systems.

Here, we are analyzing the data in different formats:

- Light intensity (LDR), Gas Sensor (MQ-2 Sensor), Distance of Object (Ultra Sonic), GPS Coordinates (GPS Module), Direction of Object (Servo Motor), Temperature and humidity (DHT22), Distance Control from the Blynk app
- Object detection Frequency (Ultra Sonic), Camera Module

There is plenty of room for development, and by adding more sensors, more data may be gathered. The Advance Portable Radar System is operational and produces the following output results:

Fig. 5. Light intensity received from LDR sensor; Gas/Smoke received from MQ2 Sensor.

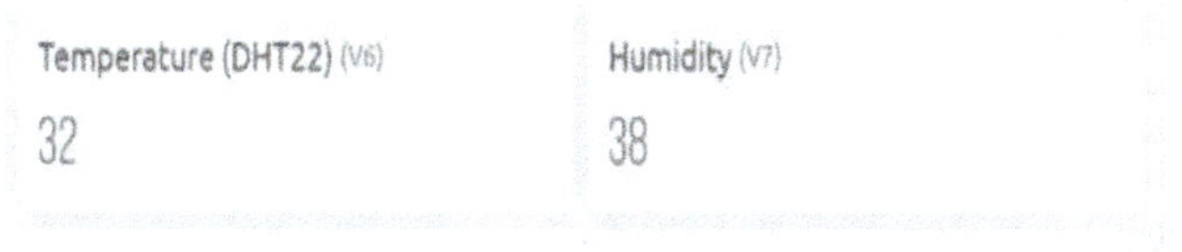

Fig. 6. Temperature and Humidity received from DHT22 sensor.

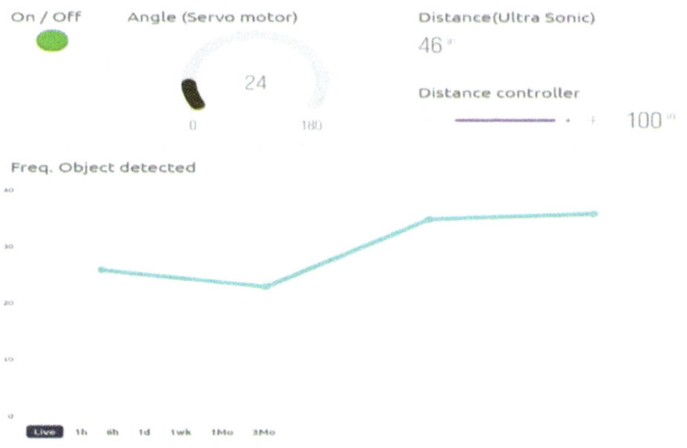

Fig. 7. Direction and Distance received from servo motor and ultra-sonic, Distance can be controlled from Blynk app.

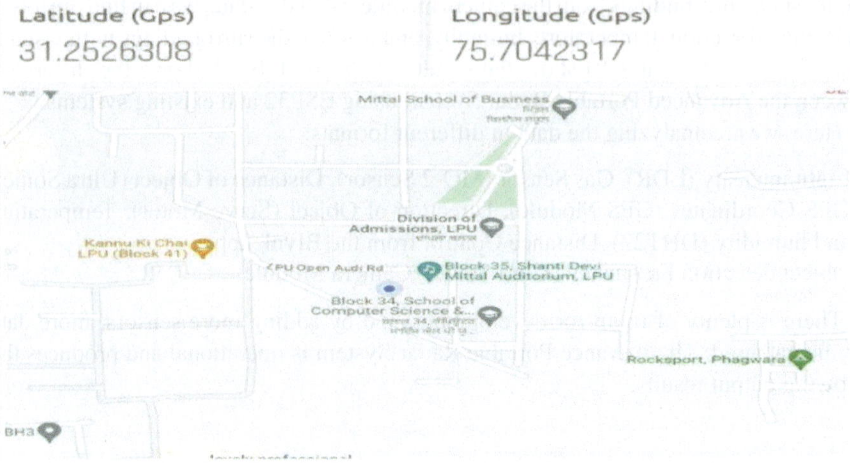

Fig. 8. GPS Coordinates received from GPS Module displayed using Map.

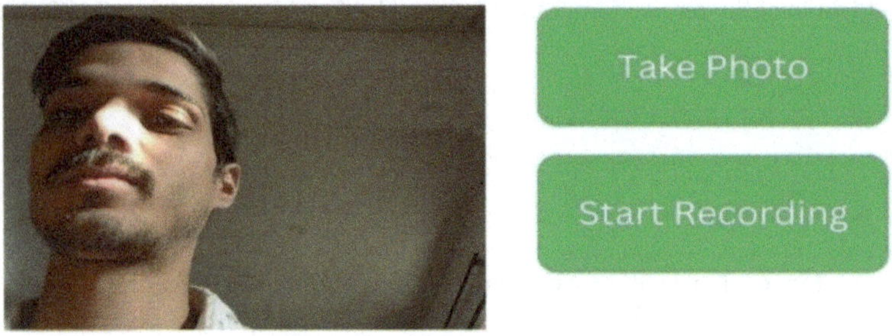

Fig. 9. Camera Module interfacing.

The system performance section for the advanced portable radar includes rigorous sensor calibration tests and benchmarking against industry-standard instruments to validate accuracy. Extended testing over various conditions and implementation of advanced error-handling mechanisms ensures high reliability. Real-time monitoring tests and statistics transmission speed assessments show the device's fast responsiveness. The device is famous for its high accuracy, reliability, and responsiveness, effectively meeting the tracking necessities for diverse applications.

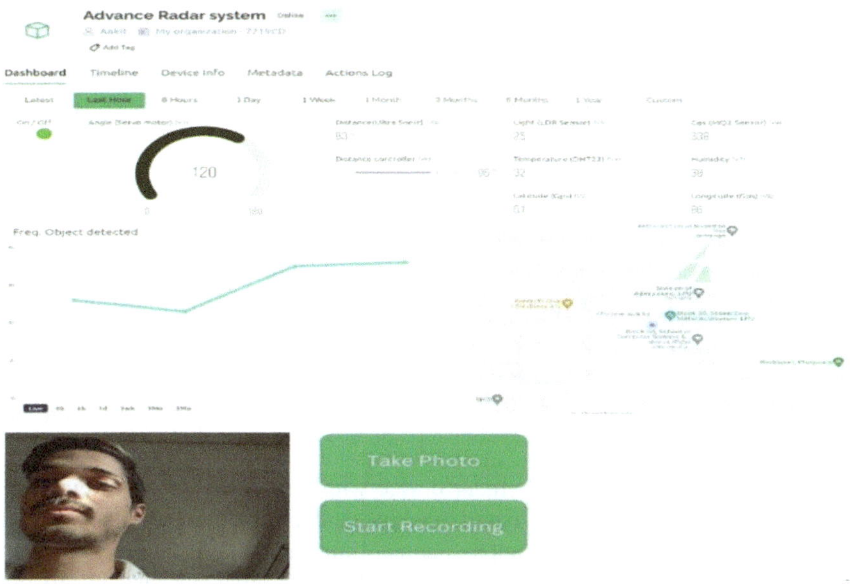

Fig. 10. Web Dashboard on Blynk App.

Table 1. The difference between the Advanced Portable Radar System using ESP32 and existing systems.

Features/Benefits	Advanced Portable Radar System using ESP32	Existing Systems	Comparison/Advantage
Sensor Integration	Enhanced sensor integration and variety	Limited sensor integration	Comprehensive data capture with improved accuracy and reliability
Communication and Data Transmission	Efficient communication protocols and real-time transmission	Traditional communication methods	Seamless handling of large data volumes and multiple connections
Performance and Reliability	Superior performance and reliability	Standard performance and reliability	High accuracy, rapid response, and consistent operation across environments

(continued)

Table 1. (*continued*)

Features/Benefits	Advanced Portable Radar System using ESP32	Existing Systems	Comparison/Advantage
Innovative Features and Technologies	Incorporation of innovative features and technologies	Conventional features and technologies	Enhanced functionalities, user experience, and new capabilities
Versatility and Adaptability	Highly versatile and adaptable	Limited versatility &and adaptability	Customizable solutions tailored to specific needs and diverse applications

5 Conclusion

Integrating a complicated portable radar device with a server indicates a remarkable development in sensor-primarily based tracking and information evaluation. These fusion yields various benefits, encompassing real-time data transmission, remote monitoring, and data storage for future reference, which are essential attributes in situations requiring heightened situational understanding. To enhance capabilities, our task integrates sensors, including LDR, ultrasonic, servo motor, DHT22, and a digicam module. Future studies need to put extra emphasis on improving data processing capabilities for more incredible unique records to enhance our system similarly. Combining gadget mastering and sophisticated algorithms can improve goal detection, lower false alarms, and boost usual reliability. Likewise, more excellent thorough expertise of the surroundings is promised with the aid of combining complementary technology like lidar and sonar.

Merging those technologies results in better device performance and a complete perspective for decision-making. Ultimately, the aggregate of a cutting-edge transportable radar system and a server affords significant potential for various packages, and persevered study will likely result in an increasing number of complex and advanced systems to deliver decision-makers more precise data for critical decision-making.

References

1. Botta, A., Donato, W., Persico, V., Pescapé, A.: Integration of Cloud computing and Internet of Things: A survey. Futur. Gener. Comput. Syst. **56**, 684–700 (2016). https://doi.org/10.1016/j.future.2015.09.021
2. Karthik, S.A., Hemalatha, R., Aruna, R., Deivakani, M., Reddy, R.V.K., Boopathi, S.: Study on Healthcare Security System-Integrated Internet of Things (IoT). In Perspectives and Considerations on the Evolution of Smart Systems **1**, 21 (2023). https://doi.org/10.4018/978-1-6684-7684-0.ch013
3. Zhao, W., Wang, M., Pham, V.T.: Unmanned Aerial Vehicle and Geospatial Analysis in Smart Irrigation and Crop Monitoring on IoT Platform. Mob. Inf. Syst. **2023**(4213645), 1–12 (2023). https://doi.org/10.1155/2023/4213645
4. Soori, M., Arezoo, B., Dastres, R.: Internet of things for smart factories in industry 4.0, a review. Internet of Things and Cyber-Physical Systems **3**, 192–204 (2023). https://doi.org/10.1016/j.iotcps.2023.04.006

5. Kamyab, H., et al.: The latest innovative avenues for the utilization of artificial Intelligence and big data analytics in water resource management. Results in Engineering **20**, 101566 (2023). https://doi.org/10.1016/j.rineng.2023.101566
6. Sallam, K., Mohamed, M., Mohamed, A.W.: Internet of Things (IoT) in Supply Chain Management: Challenges, Opportunities, and Best Practices. Sustainable Machine Intelligence Journal **2**(3),1–32 ((2023). https://doi.org/10.61185/SMIJ.2023.22103
7. Kumar, J., Ramesh, P.R.: Low Cost Energy Efficient Smart Security System with Information Stamping for IoT Networks. In: Proceedings of International Conference on Internet of Things: Smart Innovation and Usages (IoT-SIU), pp. 1–6 (2018). https://doi.org/10.1109/IoT-SIU.2018.8519875
8. Ahmad, Y.A., et al.: On the Evaluation of DHT22 Temperature Sensor for IoT Application. In: Proceedings of International Conference on Computer and Communication Engineering (ICCCE), pp. 1–6 (2021). https://doi.org/10.1109/ICCCE50029.2021.9467147
9. Al-Mistarehi, A.H., et al.: Artificial Intelligence Solutions for Health 4.0: Overcoming Challenges and Surveying Applications. Mesopotamian Journal of Artificial Intelligence in Healthcare, 15–20 (2023). https://doi.org/10.58496/MJAIH/2023/003
10. Rajak, P., Ganguly, A., Adhikary, S., Bhattacharya, S.: Internet of Things and smart sensors in agriculture: Scopes and challenges. J. Agricult. Food Res. **14**, 100776 (2023). https://doi.org/10.1016/j.jafr.2023.100776
11. Jeyaraj, R., Balasubramaniam, A., Kumara, M.A.A., Guizani, N., Paul, A.: Resource Management in Cloud and Cloud-influenced Technologies for Internet of Things Applications. ACM Comput. Surv. **55**(12), 1–37 (2023). https://doi.org/10.1145/3571729
12. Sorif, S.M., Saha D., Dutta, P.: Smart Street Light Management System with Automatic Brightness Adjustment Using Bolt IoT Platform. In: Proceedings of International Conference on IOT, Electronics and Mechatronics Conference (IEMTRONICS), pp. 1–6 (2021). https://doi.org/10.1109/IEMTRONICS52119.2021.9422668
13. Fayos-Jordan, R., Felici-Castell, S., Segura-Garcia, J., Alcaraz-Calero, J.M., Cervello-Duato, A.: Exploiting Multihoming Capabilities in 5G-Enabled IoT Nodes. IEEE Access **11**, 134940–134950 (2023). https://doi.org/10.1109/ACCESS.2023.3338180
14. Lousado, J.P., Pires, I.M., Zdravevski, E., Antunes, S.: Monitoring the health and residence conditions of elderly people, using LoRa and the things network. Electronics **10**(14), 1729 (2021). https://doi.org/10.3390/electronics10141729
15. Kodali, R.K., Greeshma, R.N.V., Nimmanapalli, K.P., Borra, Y.K.Y.: IOT Based Industrial Plant Safety Gas Leakage Detection System. In: Proceedings of International Conference on Computing Communication and Automation (ICCCA), pp. 1–6 (2019). https://doi.org/10.1109/CCAA.2018.8777463
16. Shrestha, S., Anne, V.P.K., Chaitanya, R.: IoT Based Smart Gas Management System. In: Proceedings of International Conference on Trends in Electronics and Informatics (ICOEI), pp. 1–6 (2019). https://doi.org/10.1109/ICOEI.2019.8862639
17. Hyun, E., Jin, Y., Ju, Y., Lee, J.: Development of short-range ground surveillance radar for moving target detection. In: Proceedings of Asia-Pacific Conference on Synthetic Aperture Radar, pp. 1–6 (2015). https://doi.org/10.1109/APSAR.2015.7306300
18. Jha, A., Saha, D.: Mobile broadband for inclusive connectivity: what deters the high-capacity deployment of 4G-LTE innovation in India? Inf. Syst. Front. **24**, 1305–1329 (2021). https://doi.org/10.1007/s10796-021-10128-6
19. Prakash, M., Abdrabou, A., Zhuang, W.: Stochastic delay guarantees for devices with dual connectivity. IEEE Internet Things J. **11**(2), 2126–2138 (2024). https://doi.org/10.1109/JIOT.2023.3292296
20. Himeur, Y., Sayed, A.N., Alsalemi, A., Bensaali, F., Amira, A.: Edge AI for internet of energy: challenges and perspectives. Internet of Things **25**, 101035 (2024). https://doi.org/10.1016/j.iot.2023.101035

21. Kasera, R.K., Gour, S., Acharjee, T.: A comprehensive survey on IoT and AI based applications in different pre-harvest, during-harvest and post-harvest activities of smart agriculture. Comput. Electron. Agric. **216**, 108522 (2024). https://doi.org/10.1016/j.compag.2023.108522
22. Merenda, M., Porcaro, C., Iero, D.: Edge machine learning for AI-Enabled IoT devices: a review. Sensors **20**(9), 2533 (2020). https://doi.org/10.3390/s20092533

Hiding Data Using Efficient Combination of RSA Cryptography, and Compression Steganography Techniques

P. Anitha(✉), S. Surya, S. Swetha, N. Suganeshwaran, and V. Suriya

KSR College of Engineering, Tiruchengode, Tamil Nadu, India
psp03ster@gmail.com

Abstract. The study of content-based image retrieval (CBIR) algorithms has been extensive due to the exponential rise in digital photo storage. Processing and storage resources are frequently heavily depleted in CBIR services. Thus, it makes sense to take use of the cloud server's enormous capacity to manage CBIR services. However, privacy protection becomes a huge concern since there is a lack of confidence in the cloud server. In this paper, we provide an innovative outsourced CBIR technique based on the bag-of-encrypted-words (BOEW) concept. The picture is encrypted using three different techniques: color value replacement, block permutation, and intra-block pixel permutation. The local histograms are then calculated by the cloud server using the encrypted photo blocks. It can be useful to combine RSA cryptography with DWT, Huffman Coding, and steganography.

Keywords: Public Key Encryption · Data Hiding · Compression Steganography · RSA Cryptography

1 Introduction

As a result, the quantity of digital photos rises dramatically. It is appealing to outsource CBIR services to the cloud server because of these demands. The image owner can quickly and easily obtain the needed photos from the cloud server in this manner, negating the need to store the image database locally. The primary worry for the owner of the photograph is its privacy, even with the many advantages of CBIR outsourcing. It is important to appropriately protect both the query image and the image database. A proposed BOEW model for outsourcing CBIR is presented. Our suggestion is to encrypt photos in blocks and ensure that the encrypted blocks can be immediately extracted to extract secure and valuable local features. The encrypted visual words are produced by applying the k-means clustering technique.

2 Existing System

In existing system, the image-based data is becoming more significant, such as object recognition, disease detection, and face identification. As a result, image-based data typically requires more storage than text-based data. The approach's premise that users

S. Pal and Á. Rocha (Eds.): ICMMCS 2025, LNNS 1398, pp. 257–267, 2025.
https://doi.org/10.1007/978-3-031-90998-6_25

can be completely trusted in the actual world makes it impractical, even if it has the capacity to encrypt and compare ciphertexts efficiently. It also may have a key leakage issue

2.1 Limitation of the Existing System

- Decreased safety.
- Problem with key leakage

3 Proposed System

The image owner uses TrapGen to create a trapdoor and uploads it to the cloud server in order to obtain comparable images. Following the image owner's query request, the cloud server searches for θ similar images, which it then returns to the image owner as search results. The owner of the image uses ImgDec to decrypt the obtained photos after obtaining the search results.

3.1 Advantages of the Proposed System

- combines steganography with RSA encryption to improve security.
- Ensures that only authorized access is granted by hiding and encrypting data
- Adaptable to new cryptographic and steganographic techniques.

4 Modules Description

The implementation has following modules,

1. Public Key Encryption
2. Data Hiding
3. Compression Steganography
4. RSA Cryptography
5. Image Retrival

4.1 Public Key Encryption

The cryptographic approach known as asymmetric encryption, often called public key encryption, uses a pair of keys for safe communication: a public key and a private key. The public key is used during the encryption process, while the private key is used during the decryption process. This method ensures that the information can only be decrypted and accessed by the designated recipient by using the private key.

4.2 Data Hiding

The process of hiding private or sensitive information such that it is not immediately visible within other files or media is known as data hiding. Security, privacy, and copyright protection are the usual reasons for using it. Data concealment is achieved by using a variety of techniques and algorithms, from straightforward cryptographic methods to intricate ways.

4.3 Compression Steganography

This technique makes use of steganography, which conceals secret data within non-secret data to avoid discovery, and data compression, which minimizes file size by eliminating redundancies. Compression steganography makes sure that the changes are less obvious and the concealed information is secure by embedding the hidden data within the compressed portions of the media, such as the least significant bits of quantized DCT coefficients in JPEG images.

4.4 RSA Cryptography

RSA cryptography, which bears the names of its creators, Ron Rivest, Adi Shamir, and Leonard Adleman, is a fundamental component of contemporary secure communication. It uses a pair of keys—a public key for encryption and a private key for decryption—to provide a secure method for encrypting and decrypting data, ensuring both confidentiality and authenticity. The security of RSA relies on the computational difficulty of factoring large composite numbers, specifically the product of two large prime numbers.

4.5 Image Retrival

Here, users can look up photographs by using the names of the images that are related to them. Users can obtain photographs using this search by providing the pixel points that relate to the images. Thus, in the Image generation module, admin can accompany (n) number of photos as shown in Fig. 1.

Fig. 1. RSA Cryptography and compression steganography

5 System Design

The process of establishing a system's architecture, modules, interfaces, and data in order to meet predetermined criteria is known as systems design. One way to apply systems theory to product creation is through systems design.

5.1 Input Design

The input design serves as a link between the user and the information system. To convert transaction data into a format that can be processed, it involves developing standards and data preparation techniques. Users can accomplish this by directly entering data into the system, or computers can read data from written or printed documents.The input design prioritizes minimizing errors, minimizing the amount of input needed, preventing delays, reducing unnecessary processes, and maintaining process simplicity. The input is made in a method that maintains privacy while offering security and usability.

5.2 RSA Algorithm

A cornerstone of contemporary cryptography, the RSA (Rivest-Shamir-Adleman) algorithm Fig. 2 enables safe communication and digital signatures in the digital sphere. Its asymmetric encryption method ensures data exchange's validity and confidentiality by using a public key for encryption and a private key for decryption. In order to generate the public and private key pairs, huge prime integers must be chosen and corresponding values must be computed. Using the recipient's public key, encryption converts plaintext into ciphertext; utilizing the recipient's private key, decryption does the opposite. RSA has several uses, from protecting communications and transactions over the internet to using digital signatures to confirm the authenticity of digital documents. Because of its strong security, adaptability, and broad use, it is an essential instrument in the field of cybersecurity.

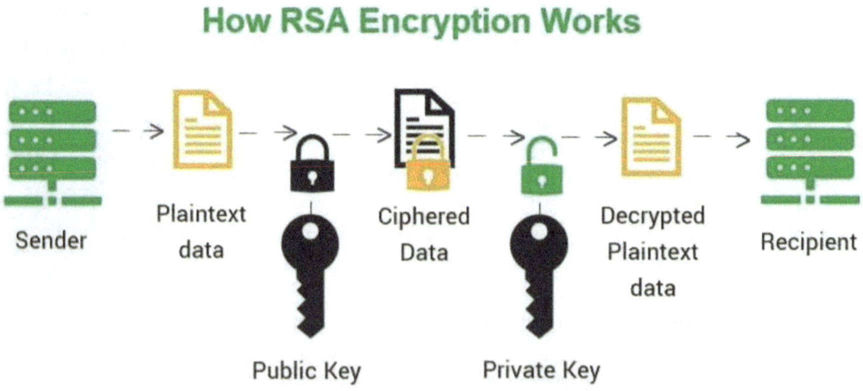

Fig. 2. RSA Encryption technique

5.3 Output Design

An output that meets the needs of the end user and effectively conveys the information is considered high quality. The way that processing results are communicated to users and other systems is through any system's outputs. The output design determines how data is moved for instant usage and the hard copy output. It is the most important and accessible source of information for the user. Effective and thoughtful output design strengthens the connection between the system and facilitates user decision-making. The process of designing computer output should be methodical and well-planned; appropriate output should be created while making sure that every output component is made in a way that makes the system simple and efficient for users to utilize. When using a computer for analysis and design. The output design was based on the following factors.

- Provide details about previous endeavors, present circumstances, or future outlooks.
- Indicate significant occasions, chances, issues, or cautions.
- Verify an activity.

5.4 Code Design

A codebase's design for the RSA algorithm implementation must take a number of factors into account in order to guarantee effectiveness, security, and user friendliness. Modular code structures with separate functions or classes for the encryption, decryption, and key generation procedures are ideal. To improve readability and maintainability, naming conventions should be simple and unambiguous First, the public and private keys should be computed, prime numbers should be generated, and key storage and retrieval should be managed by the key generation module. To ensure that strong and secure key pairs are generated, these functions should follow the RSA key creation algorithm The essential RSA cryptographic functions should then be encapsulated in the encryption and decryption modules. The encryption function ought to generate ciphertext after receiving a public key and a plaintext message as input comparably, the ciphertext and private key should be accepted as input by the decryption function, which should then output the original plaintext message. Furthermore, it is imperative to incorporate error handling mechanisms across the entire codebase to address potential exceptions, such as incorrect inputs or cryptographic malfunctions. Dataflow diagram has been described in figures [3–6].

6 Dataflow Diagram

LEVEL-0:
See Fig. 3.

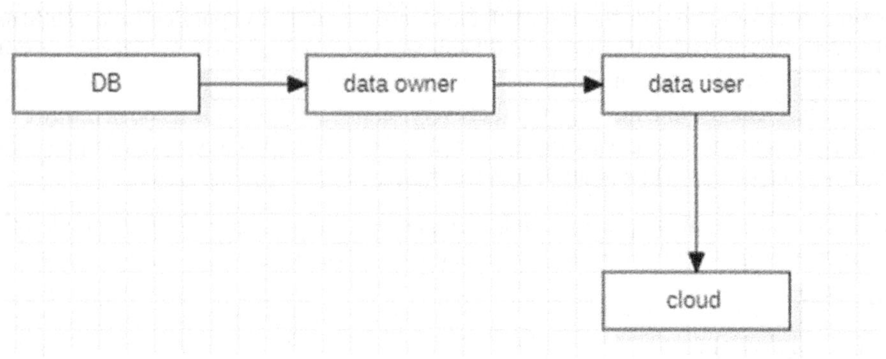

Fig. 3. Level -0 dataflow Diagram

LEVEL-1:
See Fig. 4.

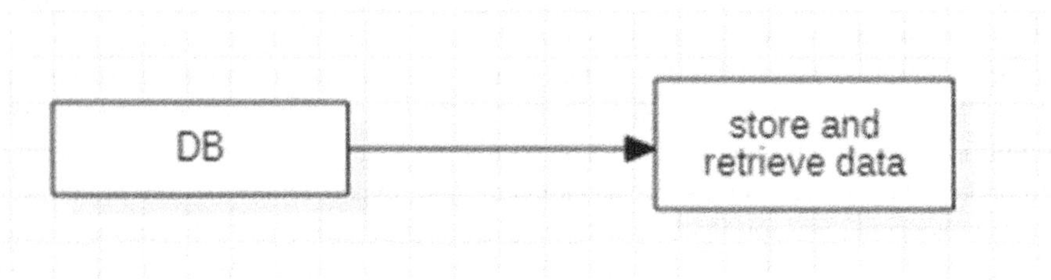

Fig. 4. Level -1 dataflow Diagram

LEVEL-2:
See Fig. 5.

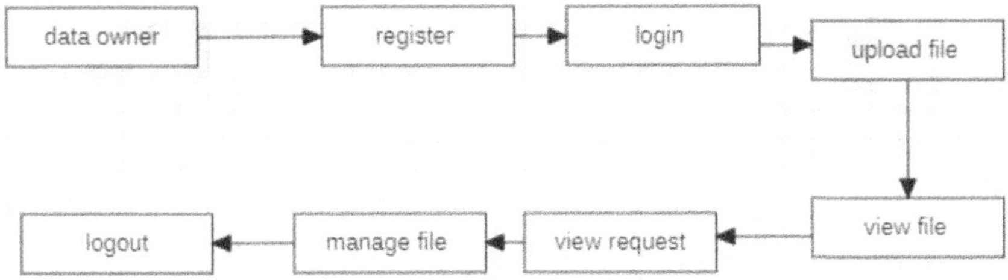

Fig. 5. Level -2 dataflow Diagram

LEVEL-3:
See Fig. 6.

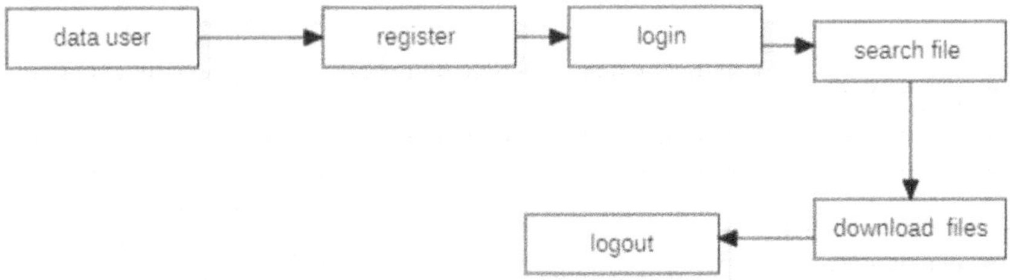

Fig. 6. Level -3 dataflow Diagram

6.1 Systemflow Diagram

The figure [7] depicts a cloud-based data access control system involving a Data Owner, Data User, and Cloud Storage. The Data Owner uploads files to the cloud, views user requests, and manages file access. The Data User requests files from the Data Owner, searches for and downloads files from the cloud, but requires authorization and a key to access them. The Cloud Storage manages data ownership, user authorization, and file access. The process flow involves the Data User sending a request, the Data Owner reviewing and authorizing access, and the Cloud providing the requested file after verification. The system ensures secure data sharing through a controlled authentication mechanism (Fig. 7).

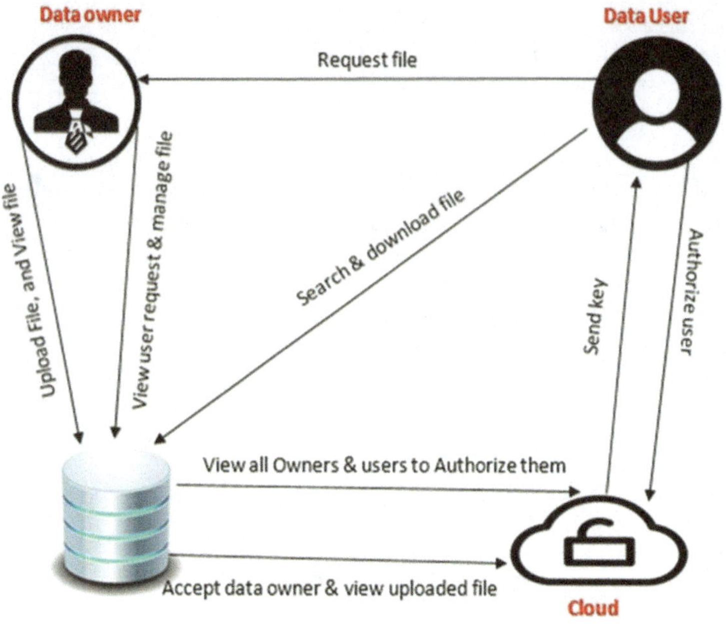

Fig. 7. System flow Diagram

7 System Testing Methodologies

The different types of testing are:

- Unit Testing
- Integration Testing
- Validation Testing
- Output Testing

7.1 Unit Testing

Unit testing focuses verification efforts on the smallest unit of software design, the module. This is also known as "Module Testing" The modules are tested separately this testing is carried out during programming stage itself. In this step each module is found to be working satisfaction as regard to the expected output from the module.

7.2 Integration Testing

The planning and building of the software architecture is the main emphasis of integration testing. An interface may cause data loss, and one module may negatively impact another's subfunctions and display on. As a result, integration testing is a methodical approach to creating tests that identify interface-related problems. Each module in this project is tested individually before the complete software is put to the test.

7.3 Validation Testing

System testing confirms that the integrated software system as a whole satisfies all criteria. It challenges an arrangement to ensure consistent results. One example of a system test is the configuration-oriented system integration test. Based on process flows and descriptions, system testing prioritizes pre-driven process links and integration points.

7.4 Output Design

When the keys are generated, the public key and the private key should be output by the system as tuples or objects, respectively, with the public exponent and modulus and the private exponent and modulus, respectively. For these keys to work with other systems, they need be provided in standardized formats like PEM or DER. The result of encrypting a message is the ciphertext, which is an array or list of numbers that corresponds to the encrypted data. The result for decryption should be the plaintext message in its original

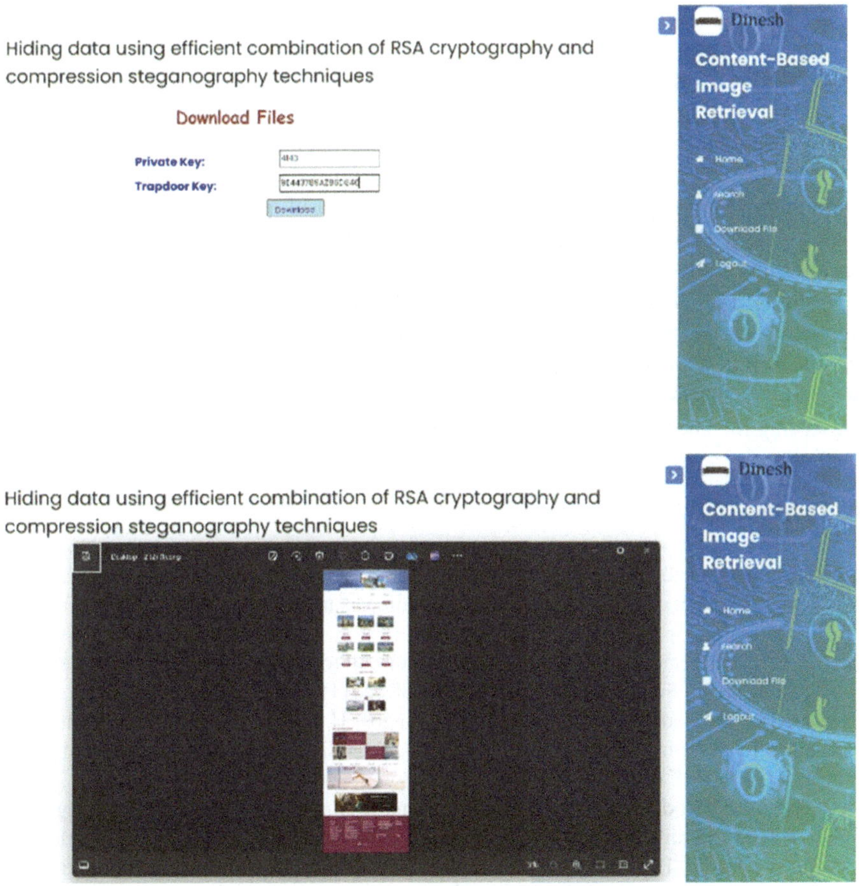

Fig. 8. Output Design

format. These outputs' [Figure 8] unambiguous labeling and formatting make them simple to use and comprehend, which promotes safe and effective data management and communication throughout the entire process.

8 Conclusion

This study suggests a novel CBIR method that protects privacy. High retrieval accuracy is the aim of the novel bag-of- encrypted-words (BOEW) paradigm. As case studies, we employ intra-block pixel permutation, block permutation, and color value replacement to protect the picture content. Local histograms are the result of computing local attributes. The k-means method is used to generate encrypted visual words. The histogram of the visual words is produced to represent the image. The Manhattan distance between feature vectors may be used to calculate the similarity between two photographs on the cloud server side. Other than the process of searching. In our method, the cloud server can also be used to generate the index. There is room for improvement in the suggested plan. First of all, designing it might be worthwhile in the future.

9 Future Scope

The security, performance, and usefulness of the RSA algorithm implementation could all be greatly enhanced in the future. To protect against the possible risks posed by quantum computing, one crucial area is the incorporation of quantum-resistant algorithms, such as hash- or lattice-based cryptography. Computation time can be decreased by streamlining the key generation process using sophisticated prime number generation algorithms, particularly for higher key sizes. Enhanced key management procedures will improve security overall. These procedures include automatic key rotation, secure key storage solutions, and key expiration policies. Furthermore, using bigger key sizes—4096 bits or more—can boost defenses against factorization and brute-force assaults, preserving the encryption system's resilience against changing security risks.

References

1. Lewin, J.M., et al.: Comparison of full-field digital mammography with screen-film mammography for cancer detection: results of 4,945 paired examinations. Radiology **218**(3), 873–880 (2001)
2. Lu, C.S.: Homomorphic encryption-based secure sift for privacy preserving feature extraction. Proc. SPIE Int. Soc. Optical Eng. **7880**(2), 788 (2011)
3. Ferreira, B., Rodrigues, J., Leitao, J., Domingos, H.: "Privacy-˜ preserving content-based image retrieval in the cloud. In: IEEE 34th Symposium on Reliable Distributed Systems, pp. 11–20. IEEE (2015)
4. Ferreira, B., Rodrigues, J., Leitao, J., Domingos H.: Practical privacy-preserving content-based retrieval in cloud image repositories. IEEE Trans. Cloud Comput. **99**, 1 (2017)
5. Rui, Y., Huang, T.S., Ortega, M., Mehrotra, S.: Relevance feedback: a power tool for interactive content-based image retrieval. IEEE Trans. Circuits Syst. Video Technol. **8**(5), 644–655 (1998)

6. Liu, Y., Zhang, D., Lu, G., Ma, W.-Y.: A survey of content-based image retrieval with high-level semantics. Pattern Recogn. **40**(1), 262–282 (2007)
7. Akgul, C.B., Rubin, D.L., Napel, S., Beaulieu, C.F., Greenspan, H., Acar, B.: Content-based image retrieval in radiology: current status and future directions. J. Digi. Imaging **24**(2), 208–222 (2011)
8. Zhang, X., Liu, W., Dundar, M., Badve, S., Zhang, S.: Towards large-scale histopathological image analysis: Hashing-based image retrieval. IEEE Trans. Med. Imaging **34**(2), 496–506 (2015)
9. Curtmola, R., Garay, J., Kamara, S., Ostrovsky, R.: Searchable symmetric encryption: Improved definitions and efficient constructions. J. Comput. Secur. **19**(5), 79–88 (2011)
10. Li, J., et al.: Proceedings IEEE INFOCOM. IEEE **2010**, 1–5 (2010)
11. Cao, N., Wang, C., Li, M., Ren, K., Lou, W.: Privacy-preserving multi-keyword ranked search over encrypted cloud data. IEEE Trans. Parallel Distrib. Syst. **25**(1), 222–233 (2013)
12. Peng, S.-L., Jhanjhi, N. Z., Pal, S., Amsaad, F. (eds.): Proceedings of 3rd International Conference on Mathematical Modeling and Computational Science. Advances in Intelligent Systems and Computing. Springer Nature Singapore (2023). https://doi.org/10.1007/978-981-99-3611-3
13. Peng, S.-L., Lin, C.-K., Pal, S. (eds.): Proceedings of 2nd International Conference on Mathematical Modeling and Computational Science. Advances in Intelligent Systems and Computing. Springer Nature Singapore (2022). https://doi.org/10.1007/978-981-19-0182-9
14. Peng, S.-L., Hao, R.-X., Pal, S. (eds.): Proceedings of First International Conference on Mathematical Modeling and Computational Science. Advances in Intelligent Systems and Computing. Springer, Singapore (2021). https://doi.org/10.1007/978-981-33-4389-4

Simple LPG Gas Leakage Detection System with Alarm and Lighting Indication for Domestic Household Application and Its Mathematical Modelling

R. Gowri Shankar Rao[1](✉), N. Kalaivani Rajam[2], and S. Gayathri[1]

[1] Department of Physics, Vel Tech Rangarajan Dr. Sagunthala R & D Institute of Science and Technology, Chennai, India
gshankarrao@veltech.edu.in

[2] Department of Mathematics, Vel Tech Rangarajan Dr. Sagunthala R & D Institute of Science and Technology, Chennai, India

Abstract. A survey has been conducted during village visit as part of Unnat Bharat Abhiyan extension activity. During the visit and communication with the local village people it was observed that many house hold had LPG gas cylinders being used for their cooking needs. In some of the families there were more elderly people who take care of the cooking. It was found that these elderly people had very little knowledge about gas leakage and detection. Since undetected gas can lead to fire hazards, explosion and even major accidents due to the leakage of the methane gas. Hence we have planned for development of a simple LPG Gas Leakage Detection System with Alarm and Lighting Indication for Domestic House hold Application. The principle behind LPG detection is based on chemiresistors typically a metal oxide semiconductor whose electrical resistance changes when exposed to garget gas molecules whenever the threshold exceeds a predefined value, system triggers an buzzer alarm and emergency light is flashed. This would provide both audible and visual indication for the handicapped people particularly who are blind and deaf. A successful demonstration was carried out practically and tested. A mathematical model is also illustrated. With features like smartphone notifications, visual alerts, and audible sirens, the main contribution of the MQ-5 based LPG detection system is to assist those with disabilities by making sure they receive early warnings in an accessible way.

Keywords: Methane gas sensor · Buzzer · LED lighting · Microcontroller · ppm

1 Introduction

Natural sources, agricultural practices, waste management, and the energy industry are the main sources of methane dangers in India. Landfills, cattle, rice paddies, and coal mining are some of the main causes. Methane is released by the huge wetlands of India and the thawing permafrost in the Himalayas. The problem is made worse by insufficient

S. Pal and Á. Rocha (Eds.): ICMMCS 2025, LNNS 1398, pp. 268–277, 2025.
https://doi.org/10.1007/978-3-031-90998-6_26

sewage treatment and poor waste management. Regulatory obstacles and erratic enforcement make mitigation initiatives even more difficult. Methane is a powerful greenhouse gas that contributes to the development of ground-level ozone, which has a considerable influence on climate change and human health. Methane capture devices, stronger legislation and regulations, better waste management, and improved agricultural practices are all necessary to reduce methane emissions.

The goal of creating an affordable methane gas detection system [1–3] for rural areas is to prevent explosions, improve waste management, mitigate the effects of climate change, safeguard public health by lowering respiratory risks, and enable communities to adopt sustainable practices while maintaining regulatory compliance and financial gains.

A reasonably priced methane gas detection device for rural regions reduces the risk of explosions, boosts respiratory health by lowering explosive hazards, and slows down global warming [4–6]. It encourages resource conservation, effective waste management, and sustainable farming methods while providing communities with the means to comply with environmental laws and reap financial rewards.

A rural methane gas detection system's components include reasonably priced sensors [7–9], data analytics for tracking emissions, solar-powered units for energy efficiency, wireless communication networks for real-time alerts, and integration with agricultural and waste management practices in the area to provide thorough monitoring and mitigation. The incorporation of machine learning methods [10] into gas sensing technologies, emphasizing the ways in which these developments boost the precision of detection, lower false alarms, and enhance sensor performance.

Li et al. [11] presented a very sensitive methane detection technique based on QEPAS, wavelet filtering, and a high-power diode laser.

The development of a polymer-coated fiber optic methane sensor that is enabled by metal-organic frameworks (MOFs). This invention offers a viable remedy for home applications by greatly enhancing the sensitivity and selectivity of methane detection [12]. In their analysis of current innovations in methane gas detection sensor [13] talked about improvements in sensor materials, technologies, and smart system integration. The evaluation highlights how crucial it is to create dependable, affordable sensors for home use. [14] paper describes a smart home gas detector and alarm system that integrates with home automation networks. This system enables remote monitoring and control, providing real-time alerts and enhancing user convenience and safety.

1.1 Objective

Our proposed system's goal is to demonstrate a design that can autonomously identify, signal, which could alert handicapped people in rural villages also provide an alert in the form of buzzing in addition with LED light indication. The system is also designed so that the over all cost is very economical and affordable.

2 System Design

Gases such as butane, propane, and methane can be detected with the MQ-5 methane sensor. Tin dioxide (SnO_2), the sensing element used in it, has a decreased conductivity in clean air. The conductivity of the sensor rises in direct proportion to the gas concentration when exposed to methane. It uses a heating coil to keep the temperature between 300 °C and 500 °C while it is in operation. An analogue voltage signal matching to the gas concentration is output by the sensor. The output of the sensing systems gives an voltage of the range of 1 to 5 v based on the gas concentration. The output read by the system will be higher based on the higher gas concentration and it has very good response to the gas leakage environment.

2.1 Components Used

The following components are included in a basic circuit design for a methane gas detection system based on MQ-5 sensing system.

Methane gas concentration is detected with the MQ-5 Methane Gas Sensor.
Microcontroller: Analyses the signal from the sensor.
Resistors are used in pull-up/pull-down combinations and to set the proper voltage levels.
Capacitors: For stabilizing and filtering noise.
Buzzer: Produces a sound alert.
LED: Shows the current gas concentration status (optional).
Power Supply: Usually 3.3 V or 5 V, this supply gives the components the necessary power.

2.2 Novelty and Circuit Design

The innovative aspect of the proposed model is the incorporation of an affordable, user-friendly LPG detection system with alarm and illumination functions, especially made for home household safety. This system might make use of sophisticated mathematical modelling to arrange sensors optimally and enhance response times and detection accuracy. In order to improve overall safety and convenience, the design may be adopted with smart home technology, which offers remote monitoring and automatic shutoff capabilities. The system could be a major improvement over current methods by emphasizing accessibility and simplicity, making effective gas leak detection more widely available and simpler for typical homeowners to install and operate.

Indeed, the integration of both audio alerts and visual led light blinking mechanism can assist those with limitations related to hearing and vision. The audio signal will help individuals with visual impairments, while the visual lighting alert will give a clear warning to those with hearing impairments.

The essential parts that make up the circuit design for methane sensing with a MQ-5 sensor are the MQ-5 sensor, a microcontroller (such an Arduino), a power source, resistors, a buzzer, and optionally an LED. When ever there is a physical leakage in the methane gas which is detected by the sensor there by resulting in the proportionate analog voltage which can be read to an simple multimeter.

The MQ-5 LPG sensor has quick response times, stability, and excellent sensitivity to LPG, natural gas, and town gas. It is perfect for home gas leak detection systems since it has a straightforward drive circuit, minimal operating costs, and an easy interface with microcontrollers.

Because of its great sensitivity to hydrogen, methane, and LPG, the MQ-5 LPG sensor ensures accurate detection of gas leaks. It can detect a variety of gases, indicating that its selectivity is moderate. However, for some applications, filtering may be necessary. The sensor is energy-efficient for continuous usage because it runs at a low power consumption of usually 150 mA. With its great sensitivity, the MQ-5 sensor can identify quantities of hydrogen, methane, and LPG ranging from 200 to 10,000 parts per million.

The MQ-5 sensor limitation to be addressed is that it typically needs to be calibrated every six to twelve months in order to provide accurate results. The sensitivity and reaction time of the sensor can be impacted by temperature and humidity.

2.3 Flow Chart

See Fig. 1.

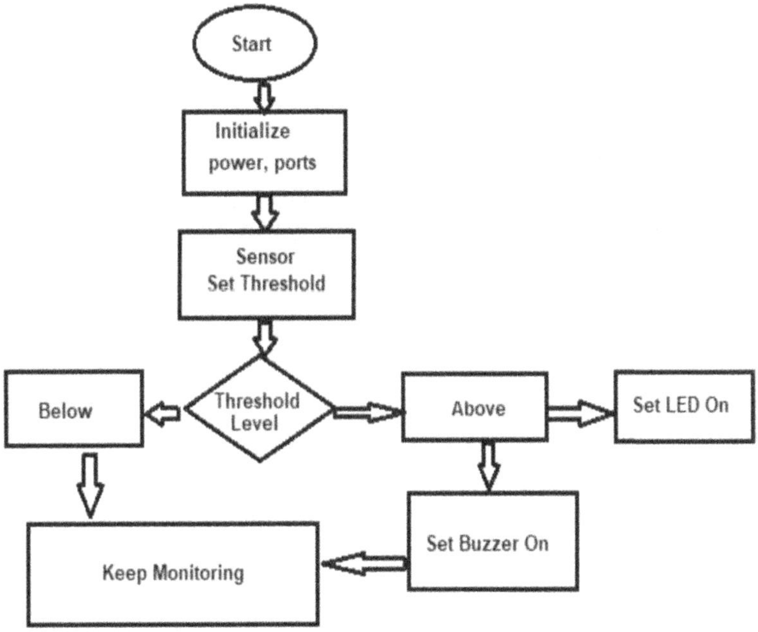

Fig. 1. Flow chart of proposed gas sensing system

3 Mathematical Model of the Sensing and Alarm System

The model will include the following key elements:

1. **Sensor Response**: The relationship between gas concentration and sensor output.
2. **Threshold Detection**: The logic for activating the buzzer.
3. **Output Signal**: The signal to the buzzer based on the sensor reading.

3.1 Sensor Response

The MQ-5 sensor provides an analog voltage output that is related to the concentration of methane (CH_4) in the air. The sensor's datasheet typically provides a sensitivity curve that can be approximated by a logarithmic function. The general form of the sensor output (V_{out}) in terms of methane concentration (C_{CH_4}) is:

$$V_{out} = A \cdot \log\left(B \cdot C_{CH_4} + 1\right)$$

Here, A and B are constants derived from the sensor's characteristics and calibration data.

3.2 Threshold Detection

For a defined threshold voltage V_{th} corresponding to a critical methane concentration C_{th}, the buzzer should be activated when V_{out} exceeds V_{th}.

$$V_{th} = A \cdot \log(B \cdot C_{th} + 1)$$

3.3 Buzzer Activation

The buzzer is activated based on the comparison between V_{out} and V_{th}. Define the output signal S_{buzzer} as a binary variable where:

$$S_{buzzer} = 1;\ if\ V_{out} > V_{th}$$

$$\boldsymbol{S_{buzzer} = 0;\ if\ V_{out} \leq V_{th}}$$

3.4 Combining the Equations

Combining the sensor response and the threshold detection logic, we get the following mathematical model for the methane sensing system:

$$V_{out} = A \cdot \log\left(B \cdot C_{CH_4} + 1\right)$$

$$V_{th} = A \cdot \log(B \cdot C_{th} + 1)$$

$$S_{buzzer} = 1;\ if\ C_{CH_4} > C_{th}$$

$$S_{buzzer} = 0;\ if\ C_{CH_4} \leq C_{th}$$

3.5 Final Model

The final mathematical model for the MQ-5 methane gas sensing system with buzzer indication is:

$$V_{out} = A \cdot \log\left(B \cdot C_{CH_4} + 1\right)$$

$$S_{buzzer} = 1;\ if\ C_{CH_4} > C_{th}$$

$$S_{buzzer} = 0;\ if\ C_{CH_4} \leq C_{th}$$

The above equations describe the analog output that defines the concentration of methane gas. This model also represents the condition for the annunciator system to be high or low based on the sensor output.

4 In-house Prototype Testing

A MQ-5 sensor is used in the gas sensing system deployment procedure to measure methane levels. The sensor outputs an analog signal that is commensurate to the gas concentration. This signal is read by a microcontroller, like an Arduino, which then decodes the data. When the microcontroller detects a methane concentration above a preset threshold, users have the option of an LED or buzzer alert. The system has been calibrated to ensure accuracy. Crucial components are the MQ-5 sensor, microcontroller, power supply, resistors, buzzer, and LED. This setup provides a reliable and effective method for locating gas leaks in applications pertaining to safety.

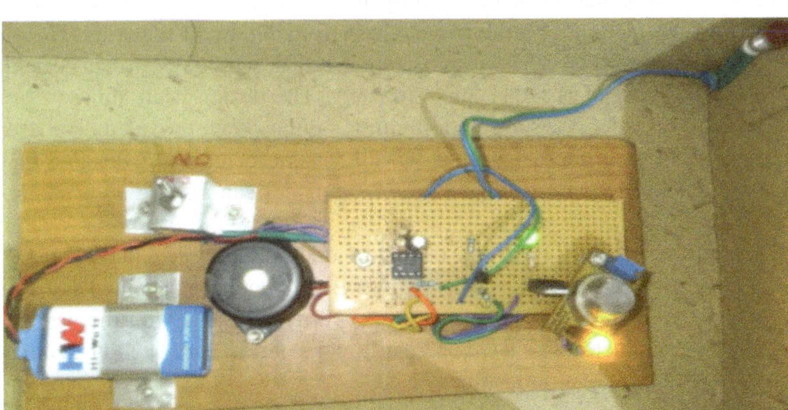

Fig. 2. Testing carried out by manually switch on/off the LPG stove

The Fig. 2 Shows the simplest prototype model which finds the best suit for the Indian kitchen. The resonse time and quick response of the sensor is recorded by keeping the detector close to the applicance. The testing is carried by opening and closing the gas

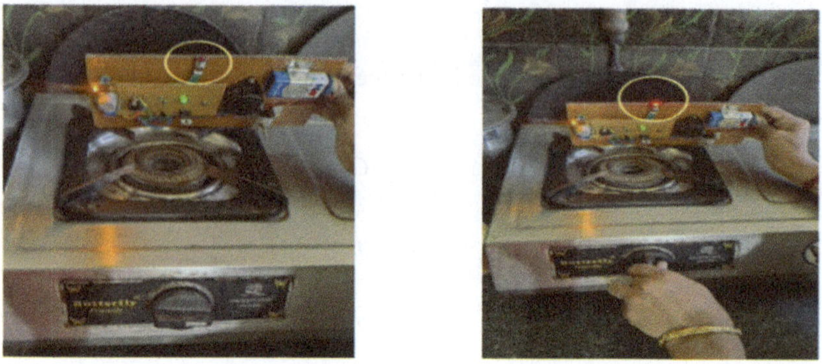

Fig. 3. Testing carried out by manually switch on/off the LPG stove

stove knob several times and the sensor response was observed repeatedly. It was noted that the ouput voltage of the gas sensor was varying in a proportionate manner when the leakage was less the output was in the order of few volts. When the opening of the knob was fully open the ouput of the sensor system has increased tremendously. The use of simple Arduino controller makes it more feasible that any kind of user with basic knowledge in microcontroller could use it more effectively. The use is also given an option with LED alert as some of the users has the disability of hearing. This was overcome with the flashing of LED for every two seconds. The challenging part was tuning the system to a particular threshold and provide a safety application.

4.1 Economic Analysis of the Proposed System

As its known the population of the country is too high, any system developed and deployed has to be met with optimum cost and affordable to all societies. This model includes a nominal sensor which is very cost effective. The whole system can be deployed and tested with a cost of Rs.1041 approximately. The design itself show to be very straight forward open loop system with multiple options been provided. The user can choose the system with microcontroller or without microcontroller. Base on the customers requirement the features of the prototype model can also be tailor made with optimum design.

The below Tables 1 and 2 shows the detailed economic analysis of the design system. As it is very evident the cost of each and every component is specified individually making a transparent system. The user can also verify the same with the vendor. The only monitoring parameter which the user has to take utmost care would be calibration of the sensor for every six months time period.

4.2 Long Term Stability and Performance

The sensor used for LPG detection has a proven track of long term stability and also features very good performance. There are several other factors on the design such as the user access, proper maintenance and also the environmental conditions. On the whole this LPG sensor has very good sensitivity due to which it could be more reliable and

Table 1. Cost Analysis in Indian rupee with microcontroller

Sno	Product name	Quantity	Cost in INR
1	Gas sensor MQ-5	1	86
2	Aruduino mcu	1	575
3	Buzzer alarm	1	40
4	Led	5	10
5	Resistors	5	10
6	Capacitors	5	20
7	Power adapter	1	150
8	PCB board	1	50
9	Casing	1	100
Total			1041.00

Table 2. Cost Analysis in Indian rupee without microcontroller

Sno	Product name	Quantity	Cost in INR
1	Gas sensor MQ-5	1	86
2	Buzzer alarm	1	40
3	Led	1	10
4	Resistors	5	10
5	Capacitors	5	20
6	Power adapter	5	150
7	PCB board	1	50
8	Casing	1	100
Total			466.00

provides good accuracy in the field testing. The other factor which could affect the system ability is the high temperature exposure, unclean environment, humidity which can slightly lower the performance. If the user maintains with proper cleaning and takes care of proper calibration then there is no question of poor performance. The user can has long term operation and gas detection without worn-out issues.

4.3 Integration with Existing Infrastructure and Design for Disabled Individuals

An LPG detector needs to be connected to the home's electrical system in order to work with existing power supply. Either standard AC power outlets or a DC power supply with the proper converter can be used for this. Battery backups may be incorporated to ensure system operation during power interruptions. In remote areas where electricity supply may be unpredictable, solar energy or other renewable energy sources can provide a reliable power source. Since the system eliminates mishaps related to LPG in homes (Fig. 3) and maintains constant monitoring even in the event of erratic power situations, its applicability in rural locations is notable.

LPG detection technology must be integrated with unique design considerations for people with impairments in order to improve accessibility and safety. Features like visual alerts like flashing lights, loud alarms with various frequencies, and vibration notifications for the hearing impaired can be added. Voice activation, intuitive user interfaces, and remote monitoring via smartphone apps are possible additional features of the system. These adaptations provide an inclusive safety solution that may meet a range of needs in both urban and rural settings by ensuring that individuals with disabilities can respond to gas leak alerts effectively.

5 Results and Discussion

The proposed gas sensing system effectively detects methane concentrations using the MQ-5 sensor. The microcontroller can accurately read and analyze the analogue signal that the system continuously emits when it is in operation, indicating the gas concentration levels. The buzzer and LED continuously turned on when methane levels above the preset threshold during testing, demonstrating the system's responsiveness and dependability. Calibration ensured accurate detection, which decreased false alarms.

6 Conclusion

The above project has been specifically designed for LPG gas detection for the village where people are careless in operating due to their field works. This project demo provides a successful and practical innovative solution where the gas sensor senses the leak or any kind of gas leakage in the kitchen premises. The annunciator system along with the light indication would definitely benefit the people who are physically challenged there by aligning the safety goals.

References

1. Shi, J., Jiang, Y., Duan, Z., Li, J., Yuan, Z., Tai, H.: Designing an optical gas chamber with stepped structure for non-dispersive infrared methane gas sensor. Sens. Actuators A **367**, 115052 (2024)
2. Wojtas, J.: Towards the highest sensitivity and selectivity of the Earth's and Space's laser-based methane sensors. Measurement **225**, 114064 (2024)
3. Chakraborty, S., Aithal, S.: Smart LPG leakage monitoring and control system using gas sensor (MQ-X), AWS IoT, and ESP Module. Int. J. Appl. Eng. Manag. Lett. 101–109 (2024)
4. Olivieri, M., et al.: Characterization of H2S QEPAS detection in methane-based gas leaks dispersed into environment. Photo Acoustics **29**, 100438 (2023)
5. Peng, S.-L., Jhanjhi, N.Z., Pal, S., Amsaad, F. (eds.): Proceedings of 3rd International Conference on Mathematical Modeling and Computational Science. Advances in Intelligent Systems and Computing. Springer Nature Singapore (2023). https://doi.org/10.1007/978-981-99-3611-3

6. Nureni, Y., Adigun, O., Oloyede, A., Akinade, O.: Gas leakage detector and monitoring system. Int. J. Eng. Manuf. **12**, 56 (2022)
7. Peng, S.-L., Lin, C.-K., Pal, S. (eds.): Proceedings of 2nd International Conference on Mathematical Modeling and Computational Science. Advances in Intelligent Systems and Computing. Springer Nature Singapore (2022). https://doi.org/10.1007/978-981-19-0182-9
8. Okonkwo, P., Obi, A., Solomon, I.: Design and construction of cooking gas (LPG) leakage detector. In: Conference: Proceedings of the 2020 Sustainable Engineering and Industrial (2021)
9. Johnson, M.R., Tyner, D.R., Szekeres, A.J.: Blinded evaluation of airborne methane source detection using Bridger photonics LiDAR. Remote Sens. Environ. **259**, 112418 (2021)
10. Umar, A., Baballe, M.A.: Automatic gas leakage monitoring system using MQ-5 sensor. Rev. Comput. Eng. Res. **2**, 1–7 (2021)
11. Peng, S.-L., Hao, R.-X., Pal, S. (eds.): Proceedings of First International Conference on Mathematical Modeling and Computational Science. Advances in Intelligent Systems and Computing. Springer Singapore (2021). https://doi.org/10.1007/978-981-33-4389-4
12. Jain, A., Sharma, P., Malhotra, B.D.: Energy-efficient gas sensors: a review. ACS Sens. **5**(8), 2448–2460 (2020)
13. Ding, M., Fu, X., Li, Y.: Application of machine learning in intelligent gas sensing. Sens. Actuators B Chem. **303**, 127159 (2020)
14. Li, Y., Wang, R., Tittel, F.K., Ma, Y.: Sensitive methane detection based on quartz-enhanced photo acoustic spectroscopy with a high-power diode laser and wavelet filtering. Opt. Lasers Eng. **132**, 106155 (2020)
15. Cao, R., et al.: Metal-organic framework functionalized polymer coating for fiber optical methane sensors. Sens. Actuators B Chem. **324**, 128627 (2020)
16. Aldhafeeri, T., Tran, M.-K., Vrolyk, R., Pope, M., Fowler, M.: A review of methane gas detection sensors: recent developments and future perspectives. Inventions **5**(3), 28 (2020)
17. Sabillaa, S.I., Sarnoa, R., Siswantoro, J.: Estimating gas concentration using artificial neural network for electronic nose. 4th Information Systems International Conference Indonesia (2017). https://doi.org/10.1016/j.procs.2017.12.145
18. Li, X., Zhang, H., Zhang, Y.: Smart home gas detector and alarm system. Int. J. Smart Home **10**(1), 65–74 (2016)
19. The European Space Agency. https://www.esa.int/Applications/Observing_the_Earth
20. GSM Based LPG Detection. https://ukdiss.com

On Blow-Up Solution of the Nonlinear Parabolic Partial Differential Equation

Farah Thaer Al-Khayyat(✉) and Amal Jasim Mohammed

Department of Mathematics, College of Education for Pure Sciences, University of Mosul, Mosul, Iraq

farah.22esp31@student.uomosul.edu.iq, a.j.moha7@uomosul.edu.iq

Abstract. This research aims to solve the nonlinear parabolic differential equation approximately which has not been previously solved analytically using the explicit forward time-centered space and the implicit generalized finite difference method which may give an actual and reliable approach for approximating the solution. In this paper, the approximate solution will be the first test of this type of equation, where, the results show that different cases depend on the convergent condition for each method. This approximate solution can be utilized to understand and analyze the properties of the equation and study the abilities to reach the optimal solution. Moreover, it can also be employed to examine the behavior of equations and to recognize their stability properties. The two methods show their accuracy through the numerical results that allow the solution for different cases: unsteady, singular, and blow-up solutions. Two examples are examined to present these types of solutions and study their behavior with illustrations of the methods supported by figures generated by MATLAB, along with the possibility the comparison of the results. Finally, the approximate solutions which are obtained from the implemented methods indicate that the technique used to determine the equation is accurate, leading to more accurate and efficient equation determination.

Keywords: partial differential equation · explicit method · implicit method · forward time-centered space method · generalized finite difference method

1 Introduction

Most physical phenomena, whether fluid dynamics, electricity, magnetism, mechanics, optics, or heat flow, can usually be described by partial differential equations (PDEs). This makes most mathematical physicists work on a system of partial differential equations which are actual phenomena in our life. Although simplifications can be made that reduce the PDEs to the system of ordinary differential equations, a complete description of these systems falls within the general scope of partial differential equations [1]. In dynamic, applications that we do not realize, such as Wen Yu proposed in his research controlling the frequency using different equations to present different notes in the song [2]. Furthermore, in Biology, the diffusion equation, whose applications include the spread of bacteria and viruses, such as the Coronavirus, as researched [3]. For that reason, the PDEs, are important for qualifications and applications in mathematics and

S. Pal and Á. Rocha (Eds.): ICMMCS 2025, LNNS 1398, pp. 278–289, 2025.
https://doi.org/10.1007/978-3-031-90998-6_27

physics related to mathematics, engineering, and electrical and mechanical sciences. The mathematical method can not be determined and generalized for solving differential equations, but some methods are generalized to a particular group of differential equations. Also, some approximation methods may be used as semi-analytical methods for solving effectively and easily ordinary and partial differential equations, and among such methods is the well-known variational iteration method, [4–7]. Sometimes it is impossible to find analytical solutions for some partial differential equations especially when they are nonlinear or solved concerning time. Then, such equations can be solved numerically. The finite difference method (FDM) is one of the numerical methods that can be applied to irregular grids. These numerical methods are divided into two general classes: explicit methods (EM) and implicit methods (IM). Lizska and Orkisz proposed the generalized of the FDM or what can be called the Beta Formulation (BF) [8, 9]. Authors in [10] used the generalized finite difference (GFD) to the explicit solutions of PDEs (parabolic and hyperbolic equations). Other researchers have published problems that become infinite in a given time, called Blow-ups (see [11, 12]), including solutions to the Cauchy problem and PDEs. After the concept Blow-up, one more was introduced, which has more general meaning than Blow-up, which includes finding some sufficient conditions to extinguish solutions to the initial boundary problem that is address in our research which is called Quenching [13].

This paper will introduce the nonlinear parabolic initial boundary problem (NPIBP):

$$\mathcal{U}_t(r,t) = n\mathcal{U}_{rr}(r,t) + \frac{1}{(1-\mathcal{U}(r,t))^{\rho}}, \quad t > 0, r \in (0,\varphi), \tag{1}$$

we will study and solve Eq. (1) numerically for $\rho = 1$ and $\rho = \frac{1}{2}$ that have not previously been solved analytically by using two dependable numerical methods which are: the explicit forward time-centered space (FTCS) and the implicit GFD methods. In addition, n is real and represents the coefficient of diffusion, also, α, ρ and φ are real. The initial and the boundary conditions (IC and BC) are defined as:

$$\mathcal{U}(0,t) = \mathcal{U}(r,\varphi) = 0 \quad \mathcal{U}(0,r) = \alpha. \tag{2}$$

Kawarada [13] showed analytical discussion and a note about the Eq. (1) when $\rho = 1$: $u(\varphi/2, t)$ reaches one in a certain time when $\varphi > 2\sqrt{2}$, and if this condition is satisfied, then $u_t(\varphi/2, t)$ becomes unbounded in a certain time. But this condition was not enough, for that, some results were obtained by several interested authors (see [14–18]). In this article [19], authors complete the second part of Kawarada's proof, while in [15], authors illustrated theorems and lemmas about Eq. (1) which include the function of the form $\frac{1}{(1-\mathcal{U}(r,t))^{\rho}}$ for $\rho = 1$ and $\rho = \frac{1}{2}$.

Following is the structure of how the paper is organized: Section 2, present the explicit FTCS and implicit GFD methods including the algorithm for each method to calculate the values of the unknown functions. In Section 3, two different examples of nonlinear PDEs will be displayed using the FTCS and the GFD methods supported by figures for different cases. Section 4 will discuss the results of the different cases.

2 Mathematical Model

In this paragraph, the explicit FTCS and implicit GFD methods are illustrated on the NPIBP Eq. (1)

$$\mathcal{U}_t(r,t) = n\mathcal{U}_{rr}(r,t) + \frac{1}{(1-\mathcal{U}(r,t))^{\rho}},$$

where, $\mathcal{U}(r,t)$ is a function with independent variables r and time t. The second part of Eq. (1), $\frac{1}{(1-\mathcal{U}(r,t))^{\rho}}$, represents the nonlinear part. Our aim is to use the explicit and implicit methods to find the numerical solution of the PDE. The method FTCS has different forms than the GFD method. The next section presents the algorithm and the scheme for each method.

2.1 The FTCS Method

In this method, the forward difference on time derivative while central on continuous space derivative is applied [20, 21] (Fig. 1).

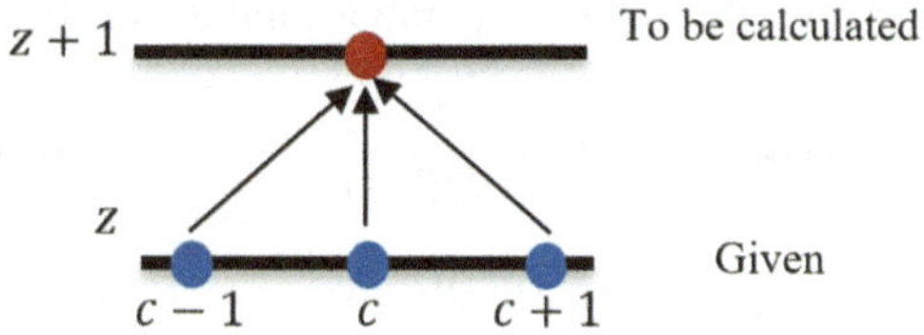

Fig. 1. Strategy scheme for the explicit scheme.

using the finite difference scheme

$$\mathcal{U}_t(r,t) = \mathcal{U}_c^{z+1} - \mathcal{U}_c^z, \mathcal{U}_{rr}(r,t) = \mathcal{U}_{c+1}^z - 2\mathcal{U}_c^z + \mathcal{U}_{c-1}^z \tag{3}$$

on Eq. (1) with simplification, to obtain

$$\mathcal{U}_c^{z+1} - \mathcal{U}_c^z = \frac{n\mathrm{N}_2}{(\mathrm{N}_1)^2}(\mathcal{U}_{c+1}^z - 2\mathcal{U}_c^z + \mathcal{U}_{c-1}^z) + \frac{N_2}{(1-\mathcal{U}_c^z)^{\rho}}, \tag{4}$$

and r and t are defined as $s_1 \le r \le e_1$ and$s_2 \le t \le e_2$, respectively, where s_1, s_2, e_1 and e_2 are constants. Defined $N_1 = \frac{e_1 - s_1}{n_1}$ and$N_2 = \frac{e_2 - s_2}{n_2}$, where n_1 and n_2 are integers. Where $d = \frac{nN_2}{(N_1)^2}$ is a convergence constant, then Eq. (4) becomes,

$$\mathcal{U}_c^{z+1} = d\mathcal{U}_{c+1}^z + (1-2d)\mathcal{U}_c^z + d\mathcal{U}_{c-1}^z + \frac{N_2}{(1-\mathcal{U}_c^z)^{\rho}}. \tag{5}$$

Substitute c from 1 to $n_1 - 1$ in Eq. (4) to get a new system of equations:

$$\begin{aligned}
\mathcal{U}_1^{z+1} &= d\mathcal{U}_2^z + (1-2d)\mathcal{U}_1^z + d\mathcal{U}_0^z + \frac{N_2}{(1-\mathcal{U}_1^z)^\rho}, \\
\mathcal{U}_2^{z+1} &= d\mathcal{U}_3^z + (1-2d)\mathcal{U}_2^z + d\mathcal{U}_1^z + \frac{N_2}{(1-\mathcal{U}_2^z)^\rho}, \\
&\vdots \\
\mathcal{U}_{n_1-1}^1 &= d\mathcal{U}_{n_1}^0 + (1-2d)\mathcal{U}_{n_1-1}^0 + d\mathcal{U}_{n_1-2}^0 + \frac{N_2}{\left(1-\mathcal{U}_{n_1-1}^z\right)^\rho},
\end{aligned} \tag{6}$$

the values of z from 0 to $n_2 - 1$ should be substituted in Eq. (6), to find other unknown functions ($\mathcal{U}_n^z$).

2.2 GFD Method

A general form of the finite difference approximation of the diffusion equation can be easily applied on the PDEs using the following scheme: [20, 21] (Fig. 2):

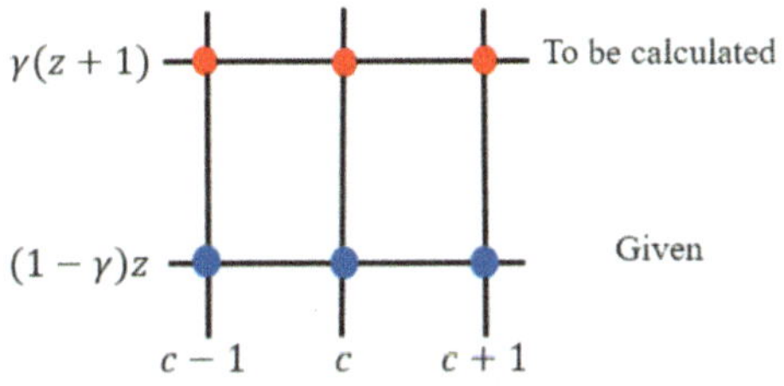

Fig. 2. Strategy scheme for the implicit scheme.

when $\gamma = 0$, the GFD method returned to be the implicit FTCS method, and for $\frac{1}{2} \le \gamma \le 1$, the GFD method is unconditionally steady otherwise it is unsteady, we will show that in the numerical examples section.

Applying the implicit GFD method on Eq. (1), yield:

$$\frac{\mathcal{U}_c^{z+1} - \mathcal{U}_c^z}{\mathrm{N}_2} = n\frac{\gamma(\mathcal{U}_{c+1}^{z+1} - 2\mathcal{U}_c^{z+1} + \mathcal{U}_{c-1}^{z+1}) + (1-\gamma)(\mathcal{U}_{c+1}^z - 2\mathcal{U}_c^z + \mathcal{U}_{c-1}^z)}{(\mathrm{N}_1)^2} + \frac{1}{(1-\mathcal{U}_c^z)^\rho}, \tag{7}$$

where, N_2 is time steps and N_1 is the space steps. Where $d = \frac{nN_2}{(N_1)^2}$ is a convergence constant, after some simplification, Eq. (7) becomes,

$$-d\gamma\mathcal{U}_{c+1}^{z+1} + (1+2d\gamma)\mathcal{U}_c^{z+1} - d\gamma\mathcal{U}_{c-1}^{z+1} = d(1-\gamma)\mathcal{U}_{c+1}^z + (1-2d(1-\gamma))\mathcal{U}_c^z + \frac{N_2}{(1-\mathcal{U}_c^z)^\rho} \tag{8}$$

Eq. (8) can be written in a general form as:

$$A\mathcal{U}^{z+1} = \mathcal{W}^{z}, \tag{9}$$

where, $A, \mathcal{U}$ and W are matrices and defined as:

$$A = \begin{bmatrix} 1+2d\gamma & -d\gamma & 0 & 0 & 0 & 0 \\ -d\gamma & 1+2d\gamma & -d\gamma & 0 & \cdots 0 & 0 \\ 0 & -d\gamma & 1+2d\gamma & -d\gamma & 0 & 0 \\ 0 & 0 & -d\gamma & 1+2d\gamma & 0 & 0 \\ \vdots & \vdots & \vdots & \vdots & \ddots \vdots & \vdots \\ 0 & 0 & 0 & 0 & -d\gamma & 0 \\ 0 & 0 & 0 & 0 & \cdots 1+2d\gamma & -d\gamma \\ 0 & 0 & 0 & 0 & -d\gamma & 1+2d\gamma \end{bmatrix}$$

$$\mathcal{U}^{z+1} = \begin{bmatrix} \mathcal{U}_1^{z+1} \\ \mathcal{U}_2^{z+1} \\ \mathcal{U}_3^{z+1} \\ \mathcal{U}_4^{z+1} \\ \vdots \\ \mathcal{U}_{n_1-3}^{z+1} \\ \mathcal{U}_{n_1-2}^{z+1} \\ \mathcal{U}_{n_1-1}^{z+1} \end{bmatrix},$$

and

$$\mathcal{W}^{z} = \begin{bmatrix} d\gamma\mathcal{U}_0^{z+1} + d(1-\gamma)\mathcal{U}_0^{z} + (1-2d(1-\gamma))\mathcal{U}_1^{z} + d(1-\gamma)\mathcal{U}_2^{z} + \frac{N_2}{(1-\mathcal{U}_1^{z})^{\rho}} \\ d(1-\gamma)\mathcal{U}_1^{z} + (1-2d(1-\gamma))\mathcal{U}_2^{z} + d(1-\gamma)\mathcal{U}_3^{z} + \frac{N_2}{(1-\mathcal{U}_2^{z})^{\rho}} \\ d(1-\gamma)\mathcal{U}_2^{z} + (1-2d(1-\gamma))\mathcal{U}_3^{z} + d(1-\gamma)\mathcal{U}_4^{z} + \frac{N_2}{(1-\mathcal{U}_3^{z})^{\rho}} \\ d(1-\gamma)\mathcal{U}_3^{z} + (1-2d(1-\gamma))\mathcal{U}_4^{z} + d(1-\gamma)\mathcal{U}_5^{z} + \frac{N_2}{(1-\mathcal{U}_4^{z})^{\rho}} \\ d(1-\gamma)\mathcal{U}_4^{z} + (1-2d(1-\gamma))\mathcal{U}_5^{z} + d(1-\gamma)\mathcal{U}_6^{z} + \frac{N_2}{\left(1-\mathcal{U}_5^{z}\right)^{\rho}} \\ d(1-\gamma)\mathcal{U}_5^{z} + (1-2d(1-\gamma))\mathcal{U}_6^{z} + d(1-\gamma)\mathcal{U}_7^{z} + \frac{N_2}{(1-\mathcal{U}_6^{z})^{\rho}} \\ \vdots \\ d(1-\gamma)\mathcal{U}_{n_1-2}^{z} + (1-2d(1-\gamma))\mathcal{U}_{n_1-1}^{z} + d\gamma\mathcal{U}_{n_1}^{z+1} + d(1-\gamma)\mathcal{U}_{n_1}^{z} + \frac{N_2}{\left(1-\mathcal{U}_{n_1-1}^{z}\right)^{\rho}} \end{bmatrix}.$$

2.3 Algorithm

In this paragraph, the steps that need to be taken for each method to achieve a numerical solution to a given problem.

2.3.1 Proposed Algorithm of the FTCS Method

S1: Input $n, n_1, n_2, \varphi, \alpha, a_1, a_2, b_1, b_2, \rho$.
S2: Estimate $L = b_1 - a_1$ and $T = b_2 - a_2$.
S3: Evaluate $N_1 = \frac{L}{n_1}$ and $N_2 = \frac{T}{n_2}$.
S4: Evaluate $d = \frac{nN_2}{(N_1)^2}$.
S5: for $c = 1 : n_1 + 1$, compute:

$$r(c) = 0 + (c-1)N_1.$$

S6: for $z = 1 : n_2 + 1$, compute:

$$t(z) = 0 + (z-1)N_2.$$

S7: set $\mathcal{U} = 0$.
S8: Compute: I.C. $\mathcal{U}_0^0, \mathcal{U}_1^0, \mathcal{U}_2^0, \ldots, \mathcal{U}_c^0$.
Compute: R.B.C. $\mathcal{U}_c^1, \mathcal{U}_c^2, \mathcal{U}_c^3, \ldots, \mathcal{U}_c^z$.
Compute: L.B.C. $\mathcal{U}_0^1, \mathcal{U}_0^2, \mathcal{U}_0^3, \ldots, \mathcal{U}_0^z$.
S9: for $c = 2 : n_1$ and $z = 1 : n_2$, compute:

$$\mathcal{U}_c^{z+1} - \mathcal{U}_c^z = \frac{n\mathrm{N}_2}{(\mathrm{N}_1)^2}(\mathcal{U}_{c+1}^z - 2\mathcal{U}_c^z + \mathcal{U}_{c-1}^z) + \frac{N_2}{(1-\mathcal{U}_c^z)^\rho}.$$

S10: Print the numerical solution $\mathcal{U}(r, t)$.

2.3.2 Proposed Algorithm of the FTCS Method

S1: Input $n, n_1, n_2, \varphi, \alpha, a_1, a_2, b_1, b_2, \rho, \gamma$.
S2: Estimate $L = b_1 - a_1$ and $T = b_2 - a_2$.
S3: Evaluate $N_1 = \frac{L}{n_1}$ and $N_2 = \frac{T}{n_2}$.
S4: Evaluate $d = \frac{nN_2}{(N_1)^2}$.
S5: for $c = 1 : n_1 + 1$, compute:

$$r(c) = 0 + (c-1)N_1.$$

S6: for $z = 1 : n_2 + 1$, compute:

$$t(z) = 0 + (z-1)N_2.$$

S7: set $\mathcal{U} = 0$.
S8: Compute: I.C. $\mathcal{U}_0^0, \mathcal{U}_1^0, \mathcal{U}_2^0, \ldots, \mathcal{U}_c^0$.
Compute: R.B.C. $\mathcal{U}_c^1, \mathcal{U}_c^2, \mathcal{U}_c^3, \ldots, \mathcal{U}_c^z$.
Compute: L.B.C. $\mathcal{U}_0^1, \mathcal{U}_0^2, \mathcal{U}_0^3, \ldots, \mathcal{U}_0^z$.

S9: set $A = 0$ and $B = 0$.
S10: for $c = 2 : n_1 - 2$ evaluate: $A(1, 1) = 1 + 2d\gamma$, $A(1, 2) = -d\gamma$, $A(n_1 - 1, n_1 - 2) = -d\gamma$ and $A(n_1 - 1, n_1 - 1) = 1 + 2d\gamma$.
S11: for $z = 1 : n_2$ evaluate:

$$B(1) = d\gamma\mathcal{U}_0^{z+1} + d(1-\gamma)\mathcal{U}_0^z + (1 - 2d(1-\gamma))\mathcal{U}_1^z + d(1-\gamma)\mathcal{U}_2^z + \frac{N_2}{(1-\mathcal{U}_1^z)^\rho}.$$

S12: for $c = 2 : n_1 - 2$ and $z = 1 : n_2$ evaluate:

$$B(c) = d(1-\gamma)\mathcal{U}_c^z + (1 - 2d(1-\gamma))\mathcal{U}_{c+1}^z + d(1-\gamma)\mathcal{U}_{c+2}^z + \frac{N_2}{(1-\mathcal{U}_{c+1}^z)^\rho}.$$

and

$$B(n_1 - 1) = d(1-\gamma)\mathcal{U}_{n_1-2}^z + (1 - 2d(1-\gamma))\mathcal{U}_{n_1-1}^z + d\gamma\mathcal{U}_{n_1}^{z+1} + d(1-\gamma)\mathcal{U}_{n_1}^z + \frac{N_2}{\left(1-\mathcal{U}_{n_1-1}^z\right)^\rho}.$$

S13: Print the numerical solution $\mathcal{U}(r, t)$.

3 Numerical Results

In this section, two different examples are presented using the FTCS and GFD methods.

Example 1: Consider the following problem [19].

$$\mathcal{U}_t(r, t) = \mathcal{U}_{rr}(r, t) + \frac{1}{1 - \mathcal{U}(r, t)}, \tag{10}$$

with left boundary condition $\mathcal{U}(-\varphi, t) = 0$, right boundary condition $\mathcal{U}(\varphi, t) = 0$ and initial boundary condition $\mathcal{U}(r, 0) = \alpha$. Where $-\varphi < r < \varphi, 0 < t < 1, n_1 = 15, n_2 = 3$. Here, n_1 and n_2 are numbers of time intervals by FTCS and GFD methods, respectively. By comparing Eq. (1) with Eq. (10), $n = 1$and$\rho = 1$. Let $\varphi = 4\sqrt{2} > 2\sqrt{2}$ and $N_1 = 0.7542$, $N_2 = 0.3333$, and $d = 0.5859$.

Example 2: Consider the following problem [15].

$$\mathcal{U}_t = \mathcal{U}_{rr} + \frac{1}{(1 - \mathcal{U})^{\frac{1}{2}}}, \tag{11}$$

with left boundary condition $\mathcal{U}(-\varphi, t) = 0$, right boundary condition $\mathcal{U}(\varphi, t) = 0$ and initial boundary condition $\mathcal{U}(r, 0) = \alpha$. Solving Eq. (11) when $-\varphi < r < \varphi, 0 < t < 2$, $n_1 = 9, n_2 = 3$ and comparing Eq. (1) with Eq. (11), $n = 1, \rho = \frac{1}{2}$. Let $\varphi = 4\sqrt{2}$. Then, $N_1 = 0.7542$, $N_2 = 0.4$ and $d = 0.7031$.

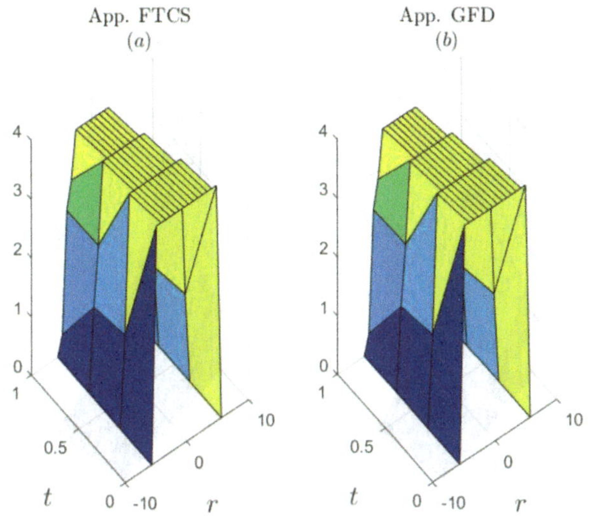

Fig. 3. (a) represents the numerical solution by the FTCS method and (b) represents the numerical solution by the GFD method where $\boldsymbol{\alpha} = 4$, $\boldsymbol{\gamma} = 0$.

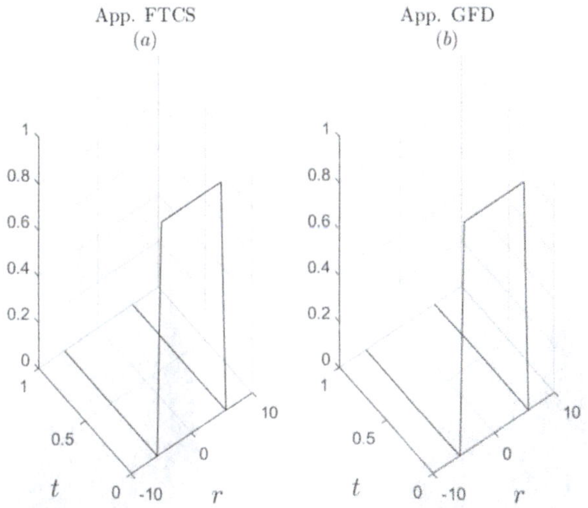

Fig. 4. (a) represents the numerical solution by the FTCS method and (b) represents the numerical solution by the GFD method where $\alpha = 1$, $\gamma = 0.3$.

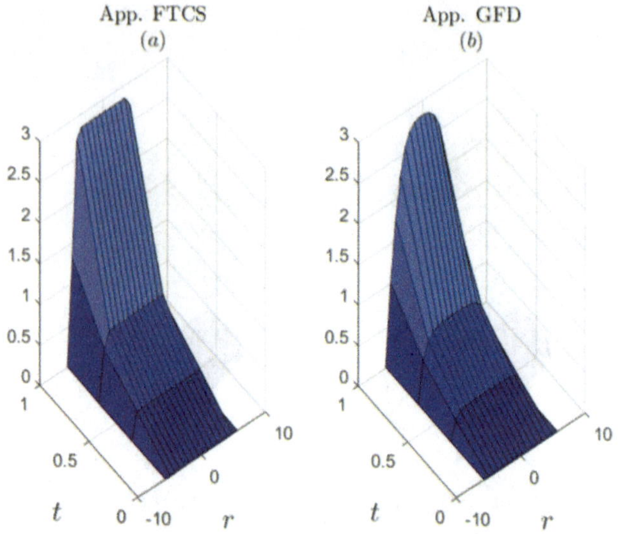

Fig. 5. (a) represents the numerical solution by the FTCS method and (b) represents the numerical solution by the GFD method where $\alpha = 0$, $\gamma = 1$.

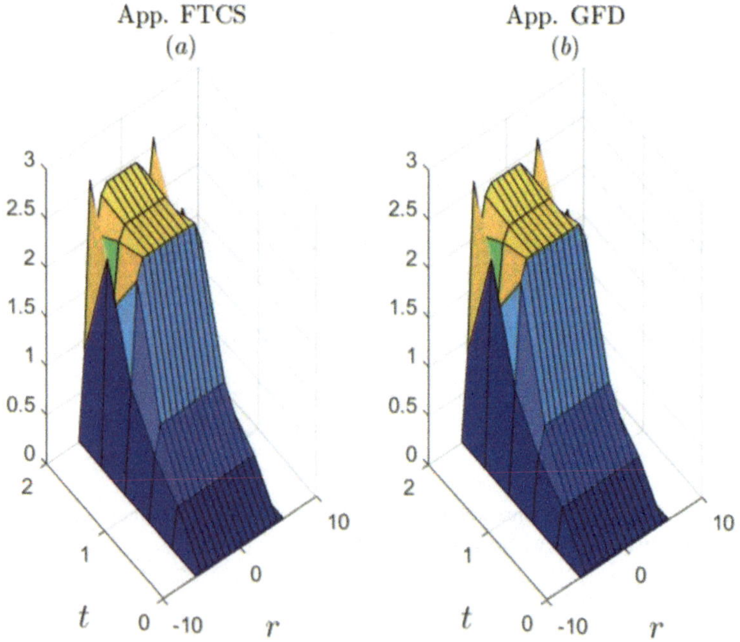

Fig. 6. (a) represents the numerical solution by the FTCS method and (b) represents the numerical solution by the GFD method where $\alpha = 0$, $\gamma = 0$.

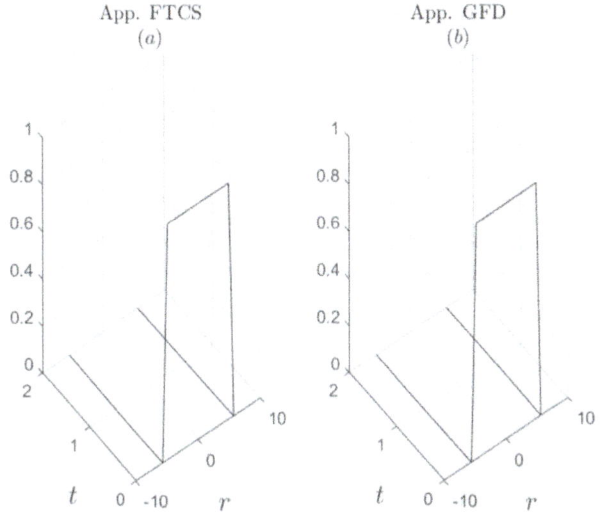

Fig. 7. (a) represents the numerical solution by the FTCS method and (b) represents the numerical solution by the GFD method where $\alpha = 1, \gamma = 0.3$.

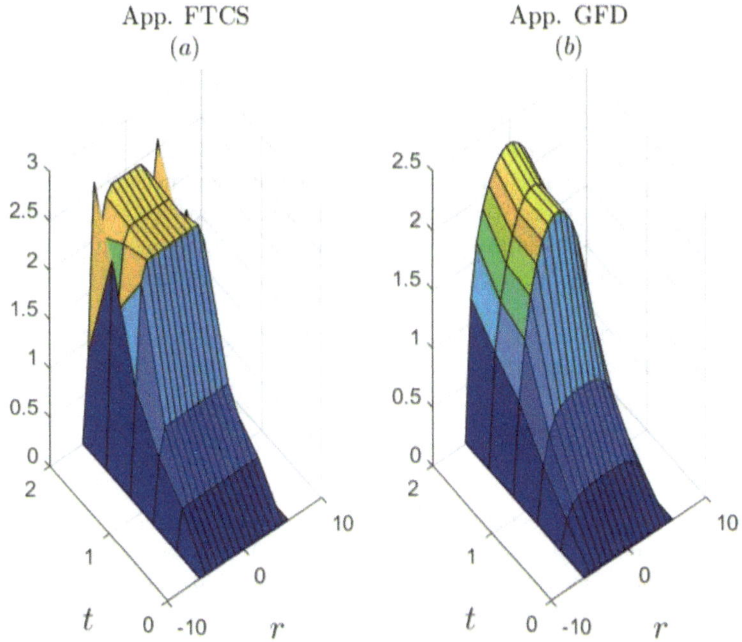

Fig. 8. (a) represents the numerical solution by the FTCS method and (b) represents the numerical solution by the GFD method where $\alpha = 0, \gamma = 2$.

4 Discussion of the Results

In Example 1, when $n = 1$, $\rho = 1$, Eq. (1) becomes Eq. (10). Figures 3 and 4 show that Eq. (10) has different types of solutions: unsteadied, singular, and blow-up. Figure 3 shows that, when $\gamma = 0$, the two methods have similar results and unsteady solutions. Figure 4, presents the case when $\alpha = 1$, the solution in both methods has singularity no matter if we change the value of γ. Figure 5 shows that the blow-up quenching type solution appeared. Also, Fig. 5 shows that the approximate solution approaching a number $\mu \in (2.5 < r < 3)$ in a finite time and it is important to note that μ does not change regardless of how we change the value of γ.

While in Example 2, $n = 1$, $\rho = \frac{1}{2}$, Eq. (1) becomes Eq. (11). Figures 6 and 7 show that, Eq. (11) have unsteadied, singular, and blow-up solutions. In Fig. 6 when $\gamma = 0$, the two methods have similar results and an unsteady solution. Figure 7 shows that, when $\alpha = 1$, the solution in both methods has singularity no matter if we change the gamma value. While, Fig. 8 shows that, the blow-up quenching type solution appeared. Figure 8 shows that, the approximate solution approaching a number $\mu \in (2 < r < 2.5)$ in a finite time and it is important to note that μ does not change regardless of how we change the value of γ.

5 Conclusion

Using dependable numerical methods the explicit FTCS and the implicit GFD methods to solve two examples Eq. (1) introduces and discusses the numerical solution using two credible methods: the explicit FTCS and the implicit GFD methods. In this paper, two examples are displayed based on the value of $\rho = 1$ and $\rho = \frac{1}{2}$, Eqs. (10) and (11) respectively. In both examples, three different types of solutions have been obtained: unsteady, singular, and blow-up quench solutions. In conclusion, the implicit GFD method is better than the FTCS method in the case $\alpha = 0$, and $\gamma > 1$, where we have a blow-up quenching type solution.

Acknowledgments. The Authors sincerely thank the "University of Mosul College of Education for Pure Science" for supporting Master's students.

References

1. Farlow, S.J.: Partial differential equations for scientists and engineers. Courier Corporation (1996). https://doi.org/10.1142/p018
2. Yu, W.: The composition system of pop music movement based on finite element differential equations. Appl. Math. Nonlinear Sci. **8**(1), 1221–1230 (2023). https://doi.org/10.2478/amns.2022.2.0109
3. Logeswari, K., Ravichandran, C., Nisar, K.S.: Mathematical model for spreading of COVID-19 virus with the Mittag–Leffler kernel. Numer. Methods Partial Differ. Equ. **40**(1), e22652 (2024). https://doi.org/10.1002/num.22652
4. Hussain, A.K., Fadhel, F.S., Rusli, N., Yahya, Z.R.: Iteration variational method for solving two-dimensional partial integro-differential equations. J. Phys. Conf. Ser. 12091 (2020). https://doi.org/10.1088/1742-6596/1591/1/012091

5. Hussain, A.K., Rusli, N., Fadhel, F.S., Yahya, Z.R.: Solution of one-dimensional fractional order partial integro-differential equations using variational iteration method. In: AIP Conference Proceedings, AIP Publishing (2016). https://doi.org/10.1063/1.4965216
6. Abdulsahib, A.A., Fadhel, F.S., Abid, S.H.: Modified approach for solving random ordinary differential equations. J. Theor. Appl. Inf. Technol. **97**(13), 3574–3584 (2019). https://www.researchgate.net/publication/348677511_MODIFIED_APPROACH_FOR_SOLVING_RANDOM_ORDINARY_DIFFERENTIAL_EQUATIO
7. Abdulsahib, A.A., Fadhel, F.S., Eidi, J.H.: Approximate solution of linear fuzzy random ordinary differential equations using laplace variational iteration method. Iraqi J. Sci. (2024). https://doi.org/10.24996/ijs.2024.65.2.18
8. Liszka, T.: An interpolation method for an irregular net of nodes. Int. J. Num. Meth. Eng. **20**(9), 1599–1612 (1984). https://doi.org/10.1002/nme.1620200905
9. Liszka, T., Orkisz, J.: The finite difference method at arbitrary irregular grids and its application in applied mechanics. Comput. Struct. **11**(1–2), 83–95 (1980). https://doi.org/10.1016/0045-7949(80)90149-2
10. Benito, J.J., Ureña, F., Gavete, L.: Solving parabolic and hyperbolic equations by the generalized finite difference method. J. Comput. Appl. Math. **209**(2), 208–233 (2007). https://doi.org/10.1016/j.cam.2006.10.090
11. Keller, J.B.: On solutions of nonlinear wave equations. Comm. Pure Appl. Math. **10**(4), 523–530 (1957). https://doi.org/10.1002/cpa.3160100404
12. Kawarada, H.: On solutions of nonlinear wave equations. J. Phys. Soc. Japan **31**(1), 280–282 (1971). https://doi.org/10.1143/JPSJ.31.280
13. Kawarada, H.: On solutions of initial-boundary problem for $u_t = u_{xx} + \frac{1}{1-u}$. Publ. Res. Inst. Math. Sci. **10**(3), 729–736 (1974). https://doi.org/10.2977/prims/1195191889
14. Acker, A., Walter, W.: On the global existence of solutions of parabolic differential equations with a singular nonlinear term. Nonlinear Anal.: Theory, Meth. Appl/ **2**(4), 499–504 (1978). https://doi.org/10.1016/0362-546X(78)90057-3
15. Levine, H.A., Montgomery, J.T.: The quenching of solutions of some nonlinear parabolic equations. SIAM J. Math. Anal. **11**(5), 842–847 (1980)
16. Levine, H.A.: The quenching of solutions of linear parabolic and hyperbolic equations with nonlinear boundary conditions. SIAM J. Math. Anal. **14**(6), 1139–1153 (1983). https://doi.org/10.1137/0514088
17. Levine, H.A.: The phenomenon of quenching: A survey. In: Trends in The Theory and Practice of Non-Linear Analysis, Proceedings of the VIth International Conference on Trends in the Theory and Practice of Non-Linear Analysis, pp. 275–286. Elsevier (1985). https://doi.org/10.1016/S0304-0208(08)72720-8
18. Levine, H.A.: Quenching, nonquenching, and beyond quenching for solution of some parabolic equations. Annali di Matematica Pura ed Applicata **155**(1), 243–260 (1989). https://doi.org/10.1007/BF01765943
19. Chan, C.Y., Kwong, M.K.: Quenching phenomena for singular nonlinear parabolic equations. Nonlinear Anal. Theory, Methods Appl. **12**(12), 1377–1383 (1988). https://doi.org/10.1016/0362-546X(88)90085-5
20. Lapidus, L., Pinder, G.F.: Numerical solution of partial differential equations in science and engineering. Math. Comput. **41**(164), 771 (1983). https://doi.org/10.2307/2007712
21. Causon, D.M., Mingham, C.G.: Introductory finite difference methods for PDEs. Bookboon (2010). https://www.researchgate.net/publication/311764945_Introductory_finite_volume_methods_for_PDEs

Evolving Educational Paradigms: Blended Learning as a Tool for Digital Literacy in Philology Teacher Education

Thi Tuyet Anh Nguyen[1,2], Hai Anh Le[1], and Quang Cuong Ton[1](✉)

[1] VNU University of Education, Vietnam National University, Hanoi, Vietnam
{tuyetanhnt,lehaianh,cuongtq}@vnu.edu.vn

[2] VNU Institute for Education Quality Assurance, Vietnam National University, Hanoi, Vietnam

Abstract. This study examines the impact of blended learning on developing digital literacy among pre-service Philology teachers at the University of Education, Vietnam National University, Hanoi. A case study involving 47 students and focus group interviews with 12 participants revealed that blended learning significantly enhanced digital competencies across seven key areas and 23 components. The research highlights the importance of cultivating digital literacy in pre-service Philology teachers to prepare them for future teaching careers. Blended learning emerged as an effective method, seamlessly integrating technology into the curriculum and equipping students with essential digital skills. This study contributes to the broader conversation on the role of blended learning in teacher education, ultimately helping future educators confidently navigate the digital world.

Keywords: Blended Learning · Digital Literacy · Pre-service Philology Teacher Education · Educational Paradigms

1 Introduction

Blended learning is becoming an inevitable trend in modern education due to its superior advantages in developing essential competencies for contemporary students, including digital literacy [1]. In Vietnamese education, organizing blended learning for pre-service Philology teachers focusing on digital literacy development aligns with the modern educational trend amidst the digital transformation era. It also meets the requirements for innovating teaching methods outlined in the new 2018 General Education Curriculum (GEC) for Linguistics and Literature promulgated by the Ministry of Education and Training of Vietnam [2]. The 2018 GEC for Linguistics and Literature focuses on developing students' linguistic, literary, and digital literacy skills. The blended learning model helps pre-service Philology teachers practice active teaching methods using ICT to create engaging lessons, enhance teaching effectiveness, and prepare them to meet the new curriculum's requirements, including innovative methods and assessments.

Research on digital literacy in teacher education has gained attention recently. McGarr and McDonagh [3] assessed 208 pre-service teachers in Ireland. They found

S. Pal and Á. Rocha (Eds.): ICMMCS 2025, LNNS 1398, pp. 290–304, 2025.
https://doi.org/10.1007/978-3-031-90998-6_28

that, while active on social media, they needed more critical digital skills for teaching and had limited knowledge of digital ethics. The study highlighted the need for teacher education programs to enhance both practical and ethical digital competencies, as pre-service teachers often view technology as a tool for efficiency rather than a pedagogical resource. Akayoglu et al. [4] emphasized the crucial role of instructors and mentors in fostering digital literacy among pre-service and novice teachers. When instructors and mentors use digital tools in the classroom, they serve as role models, encouraging pre-service teachers to adopt them in their future practice. Gutierrez-Angel et al. [5] advocated optimizing university curricula to enhance digital literacy, emphasizing content creation, critical evaluation, and ethical technology use in preparing future educators for the digital age.

Recent studies have focused on developing digital literacy through blended learning [6, 7]. Tang and Chaw [8–10] emphasized that while students often use digital tools for social purposes, effective academic use in blended learning requires advanced skills such as information management, critical thinking, and responsible online behavior, making digital literacy crucial for success. Claire McGuinness [11] emphasizes that e-tutorials within blended learning provide flexible access to materials, promote active learning, and enhance digital competencies like information literacy, content creation, communication, and problem-solving, fostering critical thinking and hands-on practice.

In Vietnam, existing research has adapted global digital literacy frameworks to develop models suitable for students at various educational levels, including primary, secondary, and higher education [12–15]. However, while the development of digital literacy in teacher education has gained attention, more research needs to be done that is explicitly focused on pre-service Philology teachers [16]. Most pre-service Philology teachers recognize technology's importance and use various tools daily, but they need a clearer understanding of digital literacy's specific requirements for their field. Despite lecturers' efforts to integrate technology, the practical application remains challenging due to the absence of a focused training paradigm for digital literacy development. In the context of rapidly evolving educational technology, there is a need for innovative training models for pre-service Philology teachers at universities to enhance the quality of Linguistics and Literature teaching in the digital age. Given the specific requirements of this group, it is crucial to identify the relevant digital competencies and component skills essential for their training. This study aims to explore how blended learning can enhance the development of digital literacy among pre-service Philology teachers. Specifically, this research aims to:

- Identify critical aspects of digital literacy for pre-service Philology teachers to help lecturers in training programs recognize the necessary competencies for this group.
- Evaluate the use of blended learning to develop digital literacy in pre-service Philology teachers, assessing its effectiveness and offering practical insights for improving teaching in the digital education era.

To achieve these objectives, the study seeks to answer the following questions:

- What are the specific manifestations of digital literacy for pre-service Philology teachers?

- How can the blended learning model be applied to develop digital literacy for pre-service Philology teachers effectively?

2 Theoretical Frameworks

2.1 Digital Literacy Framework

Law and colleagues [17] propose the following definition for digital literacy: "*Digital literacy is the ability to access, manage, understand, integrate, communicate, evaluate and create information safely and appropriately through digital technologies for employment, decent jobs and entrepreneurship. It includes competences that are variously referred to as computer literacy, ICT literacy, information literacy and media literacy*". Digital literacy is crucial for success in learning, research, and future career development [18]. The importance of digital literacy is evidenced by the many national and regional efforts to develop and implement digital literacy frameworks and strategic plans to bolster citizens' digital literacy.

The Digital Competence Framework for Citizens (DigComp), developed by the European Commission, aims to help citizens improve their digital literacy. The latest version, DigComp 2.2, includes five key areas with 21 specific digital literacy: 1) Information and Data Literacy; 2) Communication and Collaboration; 3) Digital Content Creation; 4) Safety; 5) Problem Solving. DigComp 2.2 defines several proficiency levels to assess and guide the development of digital literacy, including: Foundation (Basic understanding and limited practical application of digital competencies); Intermediate (Increased understanding and ability to apply digital competencies in routine tasks); Advanced (Strong understanding and effective application of digital competences in a variety of contexts); Highly Specialized (Expertise in digital competences, enabling innovation and problem-solving in complex and specialized contexts) [19]. These levels help evaluate and promote continuous improvement in digital literacy, ensuring individuals are equipped to handle various technological demands effectively. DigComp 2.2 provides a comprehensive framework for developing and assessing essential digital literacy for citizens in the digital age, supporting educators and organizations in designing training programs to enhance digital literacy.

UNESCO's Institute of Statistics Digital Literacy Global Framework builds on the five digital competence areas in DigComp 2.0 - Information and data literacy, Communication and collaboration, Digital content creation, Safety, and Problem-solving - while adding two additional areas: Devices and software operations, and Career-related competences. The devices and software operations competence focuses on identifying and using hardware tools and technologies and managing data, information, and digital content for software tools. Career-related competencies encompass the digital literacy skills required for specific occupations, including operating specialized digital technologies and analyzing and evaluating specialized data and information for particular fields [17].

2.2 Blended Learning

There are various concepts of blended learning, and understanding the term "blended learning" remains a topic of debate. One of the most popular definitions of "blended learning" comes from Graham, who describes blended learning as learning systems that "combine face-to-face instruction with computer-mediated instruction" [20]. Blended learning uses a combination of theories, methods, and technologies to optimize learning in a specific context [21, 22]. Researchers generally agree on the ratio of face-to-face to online components, with blended learning typically involving 30% to 79% of online content [23, 24].

2.3 Backward Design Principle

The learning design adheres to the principles of backward design, as outlined in Grant Wiggins and Jay McTighe's book "Understanding by Design" [25]. This approach begins by identifying the learning goals, which include the knowledge and skills students should acquire by the course's end. By prioritizing the learning process over teaching methods, backward design fosters a student-centered approach. It involves three main steps: determining the desired outcomes (Learning goals or objectives), collecting evidence of learning (Assessments), and designing the Learning activities and Instructional materials.

3 Research Design

To answer the two research questions posed above, the study implemented three steps as follows (Fig. 1):

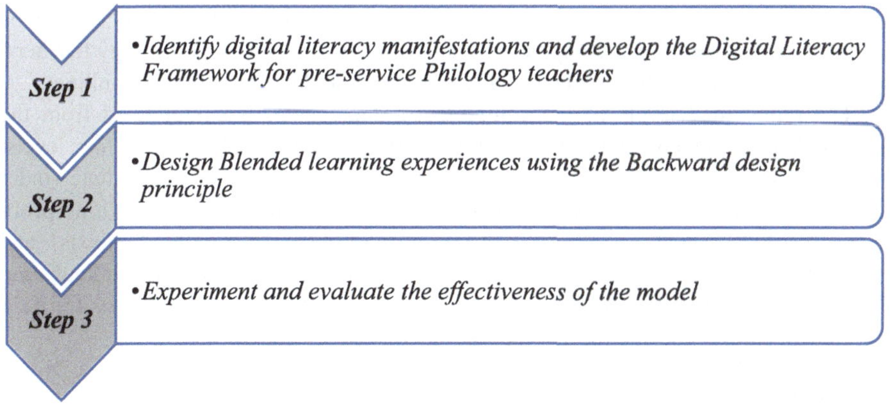

Fig. 1. Research Design

The first step involves determining the specific manifestations of digital literacy for pre-service Philology teachers and developing a tailored Digital Literacy Framework. This framework will outline the essential digital literacy and knowledge that these students need to acquire.

Once these goals are established, the next step is to determine the assessment methods to measure students' achievement of these goals. Only after these stages are completed do we consider the instructional methods and activities. This intentional approach ensures that the course is designed to meet the defined learning goals effectively. The flipped learning model is selected to achieve the digital literacy development objectives for pre-service Philology teachers. This model combines traditional face-to-face instruction with online learning activities supported by technology, leveraging both strengths to provide a comprehensive learning experience.

The final step involves implementing the blended learning model and evaluating its effectiveness in achieving the desired educational outcomes. This includes assessing how well the model helps students develop their digital literacy and making any necessary adjustments to improve its effectiveness. The evaluation will provide insights into the strengths and areas for improvement in the teaching approach, contributing to the ongoing enhancement of digital literacy in Philology teacher education.

4 Results and Discussion

4.1 Digital Literacy for Pre-service Philology Teachers

In this study, we use the Digital Literacy Global Framework (DLGF) introduced by UNESCO (2018) as the primary basis for identifying competence areas and component competences. The reason for choosing UNESCO's framework is that it proposes a set of competencies crucial for university students, particularly in teacher education: career-related competencies. However, the requirements and manifestations of digital literacy for pre-service Philology teachers differ from other fields due to the specific nature of the discipline. Some component competencies in the UNESCO digital literacy framework identified as irrelevant or less related to pre-service Philology teachers will not be included in the customized competency framework for pre-service Philology teachers, specifically Managing digital identity, Programming, and Computational thinking.

The proficiency levels of the component competencies are referenced from the level scale recommended by the DigComp 2.2 framework, which includes four levels: Basic (performing basic activities with guidance where needed), Intermediate (independently meeting personal needs and solving clearly defined and non-routine problems), Advanced (meeting the needs of oneself and others in complex contexts), and Highly Specialized (the highest and most specialized level). We also refer to DigComp 2.2's method of providing descriptions of specific competence manifestations for each component competence, tailored to pre-service Philology teachers.

Based on these perspectives and approaches, the research team proposes a Digital Literacy Framework for pre-service Philology teachers with seven competence areas and 23 component competencies with specific manifestations as follows (Table 1):

4.2 Blended Learning Strategy for Pre-service Philology Teachers to Develop Digital Literacy

Identify Learning Objectives and Learning Outcomes

The described competencies above are the digital literacy requirements for pre-service Philology teachers and can be viewed as learning objectives for their digital literacy skills.

Table 1. Digital Literacy Framework for Pre-service Philology Teachers

Competence area	Competences	Specific manifestations of digital competence for pre-service Philology teachers
C0. Devices and software operations	*C0.1 Physical operations of digital devices*	Students are able to identify and use the functions and features of hardware tools and technologies such as projectors, laptops, and smart interactive boards; they can use a projector to display learning materials, a laptop to prepare lessons, and a tablet to search for online resources, making the lessons more engaging and easier to understand
	C0.2 Identifying software to operate digital devices	Students can efficiently navigate a learning management system (LMS) to upload course materials, track learning progress, and manage assignments; adept at accessing digital libraries to find relevant scholarly articles/ e-books and integrating these resources into the lesson plans
C1. Information and data literacy	*C1.1 Browsing, searching and filtering data, information, and digital content*	Students can clearly define the information required for a lesson on an author or literary genre, specifying the need for critical analyses, historical context, and contemporary interpretations; effectively use academic databases, online journals, and digital libraries to locate relevant articles, e-books, and multimedia resources about the author's works; refine search strategies based on the needs of their curriculum and the availability of new digital resources
	C1.2 Evaluating data, information, and digital content	Students can identify reputable sources, distinguish between scholarly and unreliable information, and evaluate that content to ensure its accuracy and educational value
	C1.3 Managing data, information and digital content	Students can systematically manage digital libraries or databases of literary resources, ensuring easy access and structured content handling for teaching and learning purposes

(*continued*)

Table 1. (*continued*)

Competence area	Competences	Specific manifestations of digital competence for pre-service Philology teachers
C2. Communication and collaboration	*C2.1 Interacting through digital technologies*	Students can engage in an online forum to discuss literature, share opinions, critique works, and exchange teaching experiences. They may also create and manage a website, blog, or podcast to publish analyses, evaluate works, and provide study materials, using digital technologies to enhance interaction and communication in teaching and learning Linguistics and Literature
	C2.2 Sharing through digital technologies	Students can share information and collaborate through digital technologies. They use platforms like Google Drive, social media, and forums to distribute materials, research, and teaching experiences. They create digital content such as videos, blogs, and presentations to enhance learning and contribute to the academic community. Additionally, they collaborate on group projects using digital tools, promoting active engagement and knowledge sharing
	C2.3 Engaging in citizenship through digital technologies	Students can use social media and online platforms to engage in literary forums, such as Facebook groups dedicated to authors or communities interested in Linguistics and Literature. They can share literary analyses, participate in discussions on current literary topics, and contribute opinions in cultural activities online to promote cultural and educational development within the online community
	C2.4 Collaborating through digital technologies	Students might use digital platforms (Padlet, Google Drive, Microsoft Sway…, etc.) to collectively develop teaching materials such as lesson plans, presentations, or multimedia resources related to Linguistics and Literature. They can also engage in online discussions or virtual meetings to brainstorm ideas, share insights, and collectively construct knowledge about literary themes, authors, or teaching methodologies

(*continued*)

Table 1. (*continued*)

Competence area	Competences	Specific manifestations of digital competence for pre-service Philology teachers
	C2.5 Netiquette	Students understand and can adapt their communication strategies to specific audiences. They must also be mindful of cultural and generational diversity in digital settings. When participating in online forums to discuss Linguistics and Literature, students should use suitable language and communication styles for the audience to prevent cultural or generational misunderstandings, thereby fostering an effective online literary exchange environment that respects diversity
C3. Digital content creation	*C3.1 Developing digital content*	Students can create digital stories, vlogs, or podcasts analyzing literature by integrating text, images, and sound; write and publish digital essays or blog posts using multimedia elements like videos and infographics; create interactive e-books or digital portfolios with hyperlinks and annotations; and participate in online forums or social media groups, sharing and discussing literary topics with multimedia support
	C3.2 Integrating and re-elaborating digital content	Students can create a digital library on a specific literary topic, incorporating various sources, annotations, and multimedia elements; collaborate on digital projects like annotated bibliographies or thematic websites to present comprehensive views on literary subjects; and adapt literary works into new formats such as digital storyboards, podcasts, or video series, offering different interpretations and perspectives
	C3.3 Copyright and licenses	Students can comprehend the application of copyright and licenses to data, information, and digital content
C4. Safety	*C4.1 Protecting devices*	Students are able to safeguard devices and digital content, understand risks and threats in digital environments, be aware of safety and security measures, and prioritize reliability and privacy concerns

(*continued*)

Table 1. (*continued*)

Competence area	Competences	Specific manifestations of digital competence for pre-service Philology teachers
	C4.2 Protecting personal data and privacy	Students understand how to responsibly use and share personally identifiable information while ensuring protection against potential harm to themselves and others. Additionally, they must understand that digital services utilize privacy policies to outline how personal data is handled
	C4.3 Protecting health and well-being	Students can protect themselves and others from potential risks in digital environments, such as cyberbullying, and understand how digital technologies can promote social well-being and inclusion
	C4.4 Protecting the environment	Students can choose the most appropriate solutions to protect the environment from the impact of digital technology and its use
C5. Problem-solving	*C5.1 Solving technical problems*	Students can evaluate technical issues when operating devices and using digital environments and resolve them with the most appropriate solutions
	C5.2 Identifying needs and technological responses	Students can evaluate needs, select the most appropriate digital tools and technological solutions to address those needs and decide on the best ways to refine and customize the digital environment for personal requirements
	C5.3 Creatively using digital technologies	Students proficiently use digital tools to create knowledge and innovate educational processes, demonstrating creativity and critical thinking in content development and interactive learning. They engage individually and collaboratively in digital environments to solve conceptual problems, ensuring effective problem-solving and application of digital technologies in education

(*continued*)

Table 1. (*continued*)

Competence area	Competences	Specific manifestations of digital competence for pre-service Philology teachers
	C5.4 Identifying digital competence gaps	Students can decide on the most appropriate ways to improve or update their digital competence needs, assess the digital competence development of others, select the most suitable opportunities for personal development, and stay updated with new advancements
	C5.4 Identifying digital competence gaps	Students can decide on the most appropriate ways to improve or update one's digital competence needs, assess the digital competence development of others, select the most suitable opportunities for personal development, and stay updated with new advancements
C6. Career-related competences	*C6.1 Operating specialized digital technologies for a particular field*	Students can proficiently use specialized digital technologies relevant to language and Literature education, such as digital text analysis tools, online linguistic databases, and e-learning platforms, integrate these tools into practice to enhance learning experiences, conduct literary analysis, and create engaging lessons
	C6.2 Interpreting data, information and digital content for a particular field	Students can understand, analyze, and evaluate specialized data, information, and digital content related to Linguistics and Literature. This includes interpreting digital texts and using analytical tools to assess the content's quality and relevance, navigating digital environments effectively to access, curate, and synthesize information from diverse sources, and ensuring the ability to support teaching and research with reliable digital resources

Applying the Backward Design principle, once these learning objectives are identified, course designers can establish the learning modalities and create assessment activities to measure the achievement of these digital literacy outcomes. Subsequently, they can design suitable course content, materials, and teaching methods (Fig. 2).

Establish Learning Modality (Flipped Learning)

In a typical blended course, half of the classes are conducted online asynchronously, allowing students to work at their own pace, while the remaining sessions are held

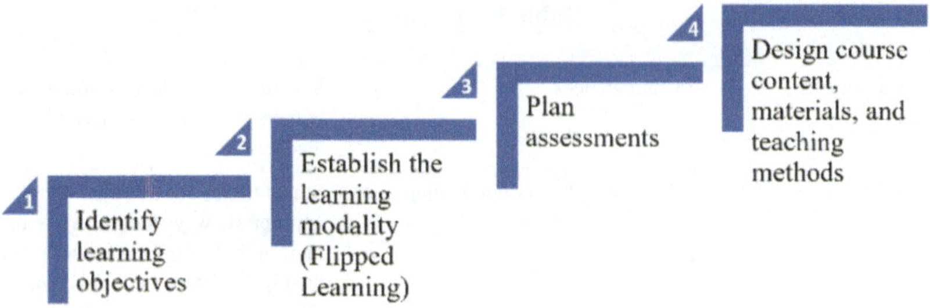

Fig. 2. Process of course design

in person. To develop the digital literacy of pre-service Philology teachers, the study implemented the flipped classroom model, a specific form of blended learning. The teaching process includes pre-class preparation, where students engage with course content through readings, videos, or other materials on the LMS. In-class active learning sessions focus on interactive activities such as project-based learning, co-authoring, debates, demonstrations, small group discussions, role-play-based learning, scenario-based learning, simulations, student presentations, and problem-based learning, aiming to deepen understanding through application, analysis, and synthesis of online concepts. Finally, post-class consolidation involves students reflecting on their learning and completing follow-up assignments.

Plan Assessments

The study implements fundamental principles for practical assessment in a blended learning environment:

- *Align assessments with course learning goals:* Ensure assessments are directly tied to the course learning objectives, especially those focusing on digital literacy skills.
- *Use formative assessment:* Formative assessments foster an engaging, student-focused learning environment. Techniques used in a typical blended classroom of Linguistics and Literature include: Think-pair-share (Students contemplate a question, discuss with a partner, and share with the class); One-minute papers (Brief written reflections at the end of a lesson); 3-2-1 summaries (Students note three things they learned, two interesting points, and one question); Four corners (Students move to different corners of the room based on their responses); Fishbowl discussions (A small group discusses while others observe); Sticky note feedback (Students provide feedback or questions on sticky notes for the instructor to address)…
- *Incorporate various assessment types*: Use a mix of quizzes, projects, discussions, debates, and exams to evaluate student learning from different perspectives.
- *Leverage technology for assessment*: Utilize online tools like LMS, Padlet, Google Forms, and Quizlet for quizzes, discussions, and projects to monitor progress and provide feedback.

Design Learning Activities and Instructional Materials

These strategies are applied to help organize course content, foster engagement, and build community in blended Philology courses to develop digital literacy:

- *Organize course content on LMS:* Structure content into units or chapters on the LMS, maintain consistency, and include varied materials and activities, clearly indicating face-to-face and online tasks.
- *Create visually engaging online content:* Use high-quality images, infographics, videos, and screen capture tutorials to cater to different learning styles and enhance interactivity.
- *Create social presence:* Establish a sense of belonging by self-introducing and encouraging student introductions, fostering interaction through group projects and forums, and using diverse multimedia channels for communication and feedback.
- *Encourage technology use in learning tasks:* Design challenging tasks that allow students to choose their presentation formats, promoting the use of technology.
- *Provide clear learning guides:* Offer accessible learning guides with technological support to help students complete tasks.
- *Use a variety of active learning methods:* Incorporate group work, pair work, project-based learning, experiential learning, and field trips to actively engage students in both online and offline settings.

4.3 Experiment Results

To evaluate the effectiveness of the proposed blended learning model for developing digital literacy skills among pre-service Literature teachers, we conducted a case study with 47 fourth-year students enrolled in the course of the Bachelor of Linguistics and Literature Teacher Education Program at the University of Education, Vietnam National University, Hanoi, during the 2023–2024 academic year. This course is designed to be taught using a flipped classroom model, utilizing the online learning management system of Vietnam National University (VNU LMS). The research methods included classroom observation, a feedback survey of the 47 students (via Google Forms), and in-depth interviews with 12 students to assess their improvement in digital literacy skills. Quantitative and qualitative data were then collected, compiled, and analyzed. The results indicated that the majority of students self-assessed as having achieved very good or fairly good improvement in their digital literacy after completing the course (over 80%). Only two component competencies, *C3.3. Copyright and licenses*, and *C4.2. Protecting personal data and privacy*, were rated as having no improvement or partial improvement by 41% and 32% of the students, respectively (Fig. 3).

The in-depth interviews with 12 students revealed their satisfaction with the learning experience in the blended learning environment. One student expressed, *"I had designed videos before, but never for a learning task. Our group project in this course taught me how to create instructional materials in video format, and it was an exciting and useful experience"*. Another student shared, *"The LMS system makes it easy for me to follow the course content and review the learning material when needed. My group also frequently worked online and had no difficulties in collaboration. The most impressive part was our group's podcast project on Vietnamese folk songs, which received high praise, and all members put in great effort to complete the learning task"*.

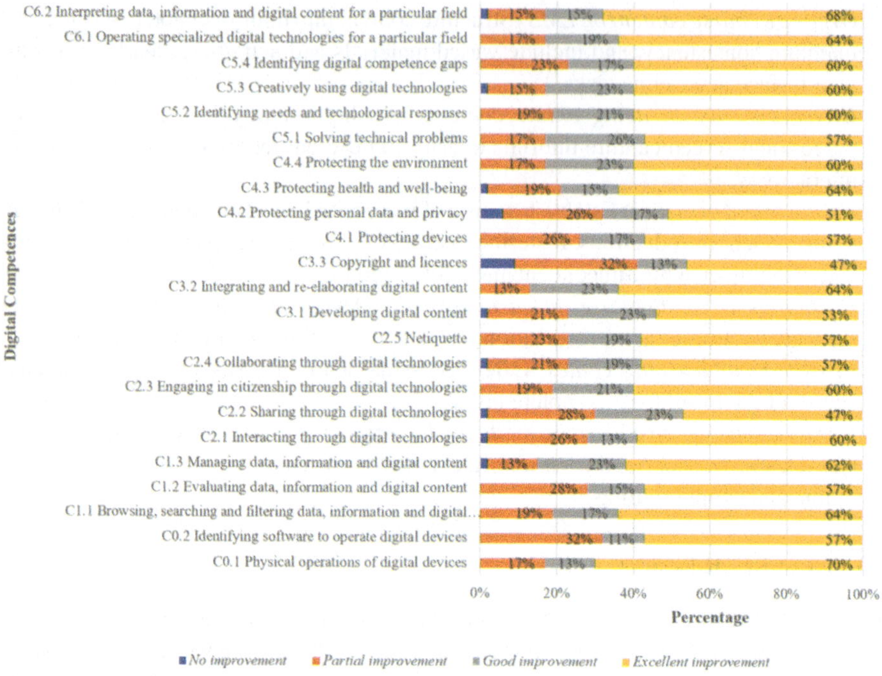

Fig. 3. Teachers digital competences and their improvement levels

5 Conclusion and Recommendations

Blended learning is a contemporary teaching approach that aligns seamlessly with today's educational landscape, where technology plays a crucial role. The research results further clarify that integrating technology with appropriate pedagogical methods in the curriculum helps develop digital literacy in learners [2, 9, 10, 16]. In Vietnam, incorporating a blended learning model into teacher training, particularly for future educators, enhances teaching innovation at higher education institutions and supports the development of human resources in line with current guidelines for Linguistics and Literature education [2, 16]. Digital literacy is a vital skill for modern educators, and blended learning - integrating in-person and online methods with technological support - is an effective model for fostering digital literacy in Philology teacher education.

This study explores the digital literacy required for pre-service Philology teachers, identifying seven competency areas and 23 specific competencies. It provides detailed descriptions and examples of these competencies and suggests strategies for designing blended learning courses focused on developing digital literacy. Experimental findings indicate improved digital literacy among pre-service Philology teachers and positive experiences with the blended learning model. Future research on digital literacy for pre-service Philology teachers should explore the long-term impact of blended learning on skill retention and practical application in teaching. Comparative studies across teacher education disciplines could highlight distinct digital competencies, enabling more tailored training. Additionally, examining the digital literacy needs of in-service Philology teachers would provide insights for ongoing professional development, enhancing the integration of digital literacy in Philology education and preparing educators for the digital age.

Acknowledgments. Nguyen Thi Tuyet Anh was funded by the Master, PhD Scholarship Programme of Vingroup Innovation Foundation (VINIF), code [VINIF.2023.TS.005].

References

1. Pham, T.Q., Dang, T.M., Van Nguyen, H., Nguyen, H.T.: Training elementary teachers in Vietnam by blended learning model. In: Peng, S.L., Jhanjhi, N.Z., Pal, S., Amsaad, F. (eds.) Proceedings of 3rd International Conference on Mathematical Modeling and Computational Science. ICMMCS 2023. Advances in Intelligent Systems and Computing, vol. 1450. Springer, Singapore (2023). https://doi.org/10.1007/978-981-99-3611-3_18
2. Training, V.M.o.E.a.: 2018 General Education Curriculum (GEC) for Linguistics and Literature (2018). Available from: https://chantroisangtao.vn/he-tai-nguyen/chuong-trinh-gdpt-2018-mon-ngu-van/
3. McGarr, O., McDonagh, A.: Exploring the digital competence of pre-service teachers on entry onto an initial teacher education programme in Ireland. Irish Educ. Stud. **40**(1), 115–128 (2021)
4. Akayoglu, S., et al.: Digital literacy practices of Turkish pre-service EFL teachers. Australas. J. Educ. Technol. **36**(1), 85–97 (2020)
5. Gutierrez-Angel, N., et al.: Digital literacy in the university setting: a literature review of empirical studies between 2010 and 2021. Front. Psychol. **13**, 896800 (2022)
6. Patmanthara, S., Hidayat, W.N.: Improving vocational high school students digital literacy skill through blended learning model. J. Phys. Conf. Ser. **1028**, 012076 (2018)
7. Makarova, E.A., Makarova, E.L.: Blending pedagogy and digital technology to transform educational environment. Int. J. Cogn. Res. Sci. Eng. Educ. (IJCRSEE) **6**(2), 57–66 (2018)
8. Chaw, L.Y., Tang, C.M.: Exploring the relationship between digital competence proficiency and student learning performance. Eur. J. Educ. **59**(1), e12593 (2024)
9. Tang, C.M., Chaw, L.Y.: Digital literacy and effective learning in a blended learning environment. Academic Conferences and Publishing International (2015)
10. Tang, C.M., Chaw, L.Y.: Digital literacy: A prerequisite for effective learning in a blended learning environment? Electron. J. E-learn. **14**(1), 54–65 (2016)
11. McGuinness, C., Fulton, C.: Digital literacy in higher education: a case study of student engagement with e-tutorials using blended learning. J. Inf. Technol. Educ. Innov. Pract. **18**, 001–028 (2019)
12. Tấn, ĐN., Marquet, P.: Năng lực công nghệ số đáp ứng nhu cầu xã hội: Các mô hình quốc tế và hướng tiếp cận tại Việt Nam. Tạp chí Khoa học Xã hội TP. HCM **244**(12), 23–39 (2018)
13. Lê Anh Vinh, B.D.Q., Lân, Đ.Đ., Lai, Đ.T., Trí, T.N.: Xây dựng khung năng lực số cho học sinh phổ thông Việt Nam. Tạp chí Khoa học giáo dục Việt Nam. Tập 17, S1, tr.1–11 (2021)
14. Hùng, T.ĐV., Hòa, T.T.Đ: Năng lực số dành cho giảng dạy và học tập trực tuyến. Tạp chí Thông tin và Tư liệu **2**, 3–11 (2022)
15. Vinh, T.D. Design and application of virtual reality technology in digital pedagogy. In: Proceedings of 2nd International Conference on Mathematical Modeling and Computational Science: ICMMCS 2021. Springer (2022)
16. Nga, N.T.T.: Phát triển năng lực số cho học sinh trung học qua môn Ngữ văn. 2022. Tạp chí Khoa học giáo dục Việt Nam, Số 11, tr.6–12 (2022). https://doi.org/10.15625/2615-8957/122 11102

17. Law, N., Woo, D.J., De la Torre, J., Wong, K.: A global framework of reference on digital literacy skills for indicator 4.4. 2 (2018)
18. Killen, C.: Collaboration and coaching: powerful strategies for developing digital capabilities. Digital Literacy Unpacked, pp. 29–44 (2018). https://doi.org/10.29085/9781783301997.005
19. Vuorikari, R., Kluzer, S., Punie, Y.D.: 2.2: The Digital Competence Framework for Citizens—With New Examples of Knowledge. Skills and Attitudes (2022). https://doi.org/10.2760/115376
20. Bonk, C.J., Graham, C.R.: The Handbook of Blended Learning: Global Perspectives, Local Designs. Wiley + ORM (2012)
21. Cronje, J.: Towards a new definition of blended learning. Electron. J. e-Learn. **18**(2), 114–121 (2020). https://doi.org/10.34190/ejel.20.18.2.001
22. McCarthy, S., Palmer, E.: Defining an effective approach to blended learning in higher education: a systematic review. Aust. J. Educ. Technol. **39**(2), 98–114 (2023). https://doi.org/10.14742/ajet.8489
23. Hew, K.F., Cheung, W.S.: Using Blended Learning: Evidence-Based Practices, vol. 20. Springer (2014)
24. Owston, R., York, D.N.: The nagging question when designing blended courses: Does the proportion of time devoted to online activities matter? Internet Higher Educ. **36**, 22–32 (2018). https://doi.org/10.1016/j.iheduc.2017.09.001
25. Wiggins, G.: Understanding by Design. Association for Supervision and Curriculum Development (2005)

Advancement in Communication Efficiency Within Federated Learning

Praveer Dubey(✉), Mansi Gupta, and Mohit Kumar

Information Technology, Dr B R Ambedkar National Institute of Technology, Jalandhar 144011, Punjab, India
{praveerd.it.23,mansig.it.21,kumarmohit}@nitj.ac.in

Abstract. In the burgeoning field of Federated Learning (FL), communication efficiency is pivotal for the practical deployment of collaborative machine learning models. The paper reviews the latest advancements aimed at reducing communication overhead in FL. We explore cutting-edge model compression techniques, including pruning and quantization, which minimize the size of model updates. We also assess the effectiveness of sparse updates, which focus on transmitting only significant changes and quantized updates, which decrease the precision of transmitted data to conserve bandwidth. Additionally, we investigate asynchronous communication methods that enable clients to update the central model at different times, reducing idle periods and enhancing overall system efficiency. Our analysis is based on a comparative evaluation framework that measures the impact of these advancements on FL system performance. Through a case study, we illustrate the practical benefits of these communication-efficient approaches. The results highlight significant improvements in bandwidth utilization and model convergence times, emphasizing the importance of communication optimization in FL. This paper contributes to the discussion on improving FL scalability and identifies opportunities for future research in this vital area of distributed machine learning.

Keywords: Federated Learning · Communication Efficiency · Internet of Things · Artificial Intelligence · Client Selection

1 Introduction

Federated learning (FL) is a distributed machine learning method that has garnered significant attention recently for its ability to tackle growing concerns about data privacy and security. In the age of big data and the proliferation of connected devices, traditional centralized machine learning models often require the collection of large amounts of sensitive user data on a central server, which raises substantial privacy concerns. FL addresses this issue by allowing multiple parties to collaboratively train a shared model without exchanging raw data. Instead, the training data stays on local devices and only model updates are sent

S. Pal and Á. Rocha (Eds.): ICMMCS 2025, LNNS 1398, pp. 305–323, 2025.
https://doi.org/10.1007/978-3-031-90998-6_29

to a central server for aggregation. This approach preserves user data privacy while also offering several additional benefits.

FL can effectively handle the heterogeneity of data across different devices and users, which is a common challenge in real world applications. This allows the system to adapt to the diverse data distributions present in the federated environment. By distributing the training process across a large number of devices, federated learning can scale to handle massive amounts of data without the need for a centralized data repository. This distributed approach enables the system to leverage the collective computational resources of the participating devices. Federated learning can leverage the computational power of edge devices, reducing the load on central servers and improving the overall efficiency of the system. The decentralized approach helps to optimize resource utilization and minimize the strain on the central infrastructure. The localized training on user devices in federated learning can lead to more personalized and tailored models, better suited to the specific needs and preferences of individual users. The personalization aspect is important for creating solutions that meet the distinct requirements of each user or client [1]. In today's world of ubiquitous connectivity, the rise of edge computing and increasing concerns about data privacy, federated learning has become an essential tool for developing robust and privacy preserving machine learning models. Its applications span healthcare, finance, smart cities and IoT, where protecting sensitive user data is critical.

As Federated Learning enhances data privacy and security by allowing models to be trained in a decentralized manner. However, this decentralized approach often requires frequent communication between clients and the central server, leading to significant communication overhead. As a result, improving communication efficiency has become a crucial area of research to ensure scalability and cost-effectiveness in FL. Several strategies have been developed to address this challenge. Model compression techniques, such as quantization, sparsification and knowledge distillation are commonly used to reduce the amount of data that needs to be transmitted. For instance, one approach integrates compression methods like TopK and quantization during local training to minimize communication overhead in diverse settings. Another method uses client voting and model aggregation to streamline the communication process [2].

Lossy compression frameworks also play a role in improving efficiency. By employing techniques like data partitioning, lossy model parameters and efficient serialization, these frameworks reduce the volume of data exchanged. Additionally, methods that only transmit the most significant gradient updates help lower communication costs by decreasing uploaded the data volume. Federated distillation techniques focus on reducing the need to transfer full model parameters by using a public dataset to transfer knowledge from a larger global model to smaller local models. Optimizing the number of communication rounds is another strategy, where introducing momentum updates can help stabilize and accelerate convergence, thereby reducing the frequency of required communications.

Client selection is another approach that enhances efficiency by dynamically choosing clients based on their computational power or network conditions.

Moreover, hierarchical FL structures, where models are aggregated at intermediary edge servers before being sent to the central server, help reduce the communication load on the core network, further improving overall communication efficiency. Recent advancements in federated learning have focused on enhancing communication efficiency, recognizing it as crucial for the success and widespread adoption of this approach. Key techniques explored include:

Compression Techniques: Strategies like structured updates, gradient compression and knowledge distillation reduce the amount of data transmitted during the federated learning process.

Quantization: Methods such as ternary compression and low bit quantization significantly reduce the size of model updates by representing parameters with fewer bits, leading to substantial communication savings.

Sparsification: Techniques that use sparsification reduce communication costs in heterogeneous federated learning networks by transmitting only the most important model updates.

Layer wise Asynchronous Updates: Allowing asynchronous updates at different model layers and applying temporally weighted aggregation can reduce communication overhead while maintaining model performance.

By addressing the challenge of communication efficiency, these advancements have made federated learning a more practical and scalable solution, paving the way for its broader adoption in real world applications where data privacy and security are paramount.

The structure of the paper includes the following sections: The Introduction provides an overview of Federated Learning (FL), highlighting the importance of communication efficiency for practical FL deployment, including faster training, better model performance, and real world feasibility. The Advancements in Communication Efficient FL section explores various strategies to address communication inefficiency, such as model compression techniques, sparse and quantized updates, and asynchronous communication. The Comparative Evaluation Framework section analyzes the impact of these advancements on FL system performance, focusing on bandwidth utilization and model convergence times. The Real World Applicability section presents a case study illustrating the benefits of communication efficient strategies in improving FL scalability. Finally, the Future Research Directions section identifies potential avenues for further research in enhancing communication efficiency within distributed machine learning frameworks.

1.1 Objective of the Survey

The aim of this paper is to explore and assess communication-efficient techniques in Federated Learning that minimize communication overhead while maintaining model performance. It focuses on evaluating model compression methods, sparse and quantized updates and asynchronous communication strategies through a comparative evaluation framework. Furthermore, the paper seeks to illustrate the practical impact of these techniques on bandwidth utilization and model convergence times using a real-world case study, thereby enhancing the scalability and effectiveness of FL systems.

1.2 Motivation and Contributions

The motivation behind this paper lies in addressing the critical challenge of communication efficiency in Federated Learning (FL). As the volume of data generated by connected devices continues to grow exponentially, traditional centralized machine learning approaches face significant hurdles in terms of data privacy and scalability. FL offers a decentralized alternative by allowing model training to occur locally on devices, thereby preserving user privacy. However, the efficient exchange of model updates between client devices and a central server is pivotal for FL's practical deployment. Effective communication optimization not only enhances training speed and model performance but also ensures the feasibility of FL in real world applications, particularly in sectors such as healthcare, finance and IoT, where data privacy is paramount.

The review contributes significantly to distributed machine learning by proposing a comprehensive framework to enhance communication efficiency in Federated Learning. It explores techniques such as model compression, sparse and quantized updates and asynchronous communication methods. Additionally, it presents a comparative evaluation framework that analyzes the impact of these strategies on FL system performance, using metrics like bandwidth utilization and model convergence times to demonstrate their effectiveness. Through a detailed case study, the paper illustrates how these strategies can be practically implemented, highlighting their benefits in improving FL scalability and operational efficiency across various application scenarios. Furthermore, the paper outlines future research directions to further optimize communication efficiency in FL, addressing emerging challenges and adapting strategies to evolving technological landscapes and application domains. Overall, it aims to advance the field by fostering scalable and privacy preserving machine learning solutions for next generation IoT and other data intensive applications.

2 Communication Efficiency in FL

Recent research emphasizes the critical importance of communication efficiency for deploying Federated Learning (FL) practically. Efficient communication significantly impacts training speed, model performance and real world feasibility. Efficient communication protocols are essential for minimizing the latency and overhead associated with transmitting model updates between client devices and central servers. This reduction in communication overhead accelerates the overall training process, enabling models to converge more quickly and adapt to new data patterns, thereby enhancing their predictive accuracy. Frequent transmission of large model updates can significantly slow down the training process. By reducing the size and frequency of data transfers, communication efficient FL algorithms can accelerate training, especially in large scale deployments with numerous devices. Communication efficiency can indirectly improve model performance. Lower communication overhead allows for more frequent updates from devices, leading to a more refined and accurate global model. Devices like smartphones and Internet of Things (IoT) sensors often have limited bandwidth and battery life. Communication efficient FL techniques are essential for real world applications where these devices are prevalent. They ensure the training process is feasible without draining device resources.

In the context of FL, efficient communication ensures that decentralized training can proceed smoothly without excessive delays, which is particularly crucial in applications where timely updates are critical. For example, in healthcare, where FL is used to train predictive models on patient data from multiple hospitals, minimizing communication delays ensures that the models can be updated in near real time, leading to more accurate and relevant predictions. Similarly, in financial services, efficient communication enables banks to update fraud detection models promptly across distributed branches, improving the models' effectiveness in identifying fraudulent activities. Recent research has focused on various techniques to enhance communication efficiency in FL. One such technique is model compression, which involves reducing the size of the model updates that need to be transmitted. This can be achieved through methods like weight pruning, which removes redundant parameters or quantization, which reduces the precision of the model weights, thereby decreasing the amount of data that needs to be sent. These compression techniques help to lower the bandwidth requirements and speed up the communication process.

Another approach is the use of sparse updates, where only the most significant changes in the model parameters are transmitted, rather than the entire model. This selective transmission reduces the volume of data exchanged and can lead to substantial communication savings. Additionally, asynchronous communication methods are being explored, allowing client devices to update the central server independently and at different times. This reduces the need for synchronization across devices, thereby minimizing delays and improving the overall efficiency of the FL system. Adaptive communication protocols are also gaining attention, as they can dynamically adjust to varying network conditions. These protocols optimize data transfer rates based on the available bandwidth and latency, ensuring efficient communication even in environments with fluctuating network performance. By adapting to real time conditions, these protocols help maintain the efficiency and reliability of the FL process. Despite these advancements, challenges remain in optimizing communication efficiency for FL. Scalability is a significant concern, as FL systems need to handle an increasing number of devices and participants. Ensuring efficient communication in such large scale deployments requires robust and scalable frameworks that can manage diverse data sources and network configurations without compromising performance. Additionally, maintaining the security and robustness of communication channels is critical. Researchers are exploring advanced encryption techniques and error handling mechanisms to protect data integrity and mitigate potential vulnerabilities in FL deployments.

Therefore, communication efficiency is crucial for realizing the full potential of Federated Learning in various domains, from healthcare and finance to IoT and beyond. Current research trends emphasize developing robust, scalable and privacy preserving communication protocols that enhance FL's speed, performance, and applicability in real world scenarios. By addressing the challenges of scalability, security and dynamic network conditions, FL continues to evolve as a transformative approach in distributed machine learning, paving the way for next generation data driven solutions that prioritize privacy and efficiency.

3 Advancements in Communication Efficient FL

The recent advancements have explored various strategies to address the communication inefficiency challenge in federated learning, which is a critical factor for the practical deployment of this distributed machine learning approach. The key advancements in communication efficient federated learning include:

- **Model Compression Techniques:** Researchers have developed several model compression techniques to reduce the size of the data transmitted during the federated learning process.
  Structured updates: Techniques like structured updates can reduce the amount of data transmitted by exploiting the inherent structure and sparsity of the model parameters.
  Gradient compression: Methods that compress the gradients or model updates before transmission, such as gradient quantization and sparsification, can significantly reduce the communication overhead.
  Knowledge distillation: Approaches that leverage knowledge distillation can train smaller "student" models from larger "teacher" models, reducing the size of the transmitted updates while preserving model performance.
- **Sparse and Quantized Updates:** Sparse updates techniques that transmit only the most relevant or important model updates, rather than the full set of parameters, can help to reduce the communication costs in federated learning. Moreover, Quantizing the model parameters to use fewer bits for representation can greatly reduce the size of the transmitted data, without significantly impacting the model's performance.
- **Asynchronous Communication:** Asynchronous communication methods allow client devices to update the server independently, without strict synchronization requirements. This approach can help minimize communication delays and reduce the overall communication overhead, as clients do not need to wait for all other devices to complete their updates before transmitting their own.

Therefore, Advancements in Communication Efficient federated learning have been pivotal in overcoming communication bottlenecks, serving as crucial solutions to reduce communication overhead and enable the practical deployment of

FL in real world applications. By significantly reducing the amount of data transmitted and minimizing communication delays, these techniques have enhanced the overall efficiency, scalability and feasibility of FL systems. These improvements are critical for ensuring that federated learning can be effectively implemented in diverse and dynamic environments, maintaining high performance while preserving data privacy.

3.1 Model Compression Techniques

To enhance communication efficiency in Federated Learning, various model compression techniques have been developed. These methods aim to reduce the volume of data exchanged between client devices and the central server, addressing one of the major bottlenecks in FL systems. Three notable techniques are structured updates, gradient compression, and knowledge distillation.

1. Structured Updates

Structured updates exploit the inherent structure and sparsity of model parameters to reduce the amount of data transmitted during FL.

Exploiting Sparsity: Many machine learning models, especially deep neural networks, have parameters that are not uniformly important. Some weights contribute more significantly to the model's performance, while others are redundant or near zero. By identifying and transmitting only the significant weights, structured updates reduce the communication overhead. Techniques such as pruning, where unimportant weights are removed, leverage this sparsity to create more compact models.

Parameter Sharing: Another approach is to identify and transmit only the most critical parts of the model, such as specific layers or components that have the most impact on performance. This can be particularly useful in convolutional neural networks (CNNs), where certain layers may contain more redundant information than others.

Low Rank Approximation: This method involves decomposing weight matrices into products of lower rank matrices. For instance, a large weight matrix can be approximated by multiplying two smaller matrices, which significantly reduces the number of parameters that need to be transmitted.

Therefore, Structured updates ensure that only the most relevant and non redundant information is communicated, thereby optimizing the use of available bandwidth and reducing transmission time.

Structured updates emerge as a powerful approach to address this challenge. These techniques exploit the inherent sparsity present in many machine learning models, particularly deep neural networks. Sparsity refers to the observation that many model weights (A) have minimal impact on the model's performance and can be near zero in value. Mathematically, the sparsity ratio (SR) of a model can be quantified as:

$$SR = (Number of zero weights)/(Total number of weights) \quad (1)$$

Structured updates leverage this sparsity by focusing on transmitting only the most critical information - the non zero weights. This significantly reduces communication overhead compared to transmitting the entire model update.

Consider a model with d total weights. During a standard FL update, the communication cost would be $O(d)$. However, with structured updates, where only k non zero weights (w_i) are transmitted, the communication cost reduces to O(k). This substantial reduction in data transmission paves the way for faster training speeds and improved resource efficiency on participating devices [3]. The key lies in identifying these insignificant weights for efficient communication. Pruning techniques like magnitude pruning and gradient pruning achieve this by analyzing the weight values or their gradients during training [4,5]. Magnitude pruning sets weights with magnitudes below a certain threshold (τ) to zero, while gradient pruning focuses on weights whose gradients fall below another threshold (γ). These techniques effectively remove redundant information from the model updates, leading to communication efficiency gains.

However, finding the optimal level of sparsity (k) remains crucial. Excessive pruning can lead to a significant drop in model accuracy. Future research directions involve exploring adaptive pruning methods that can adjust the sparsity level based on the training progress [6]. This would ensure a balance between communication efficiency and model performance in FL deployments. Structured updates offer a promising approach for Communication Efficient FL. By leveraging the inherent sparsity of models and employing techniques like pruning and quantization, researchers can pave the way for faster, more scalable, and resource friendly FL applications in various real world scenarios [7].

2. Gradient compression

During FL training, devices perform local updates on their data and transmit these updates (often in the form of gradients) to the central server for aggregation. However, transmitting the complete gradient vector for every model parameter can be highly inefficient, especially for large models with many parameters (d). The communication cost associated with sending the full gradient vector is proportional to d. Gradient compression addresses this issue by reducing the size of the transmitted data. This can be achieved through various techniques, each with its own mathematical underpinnings and trade offs:

(a) Gradient Quantization: This approach reduces the number of bits used to represent each gradient value. Imagine a gradient vector with values between a minimum and maximum. By using fewer bits (e.g., 8 bits), we map the original values to a smaller set of discrete values. This reduces communication cost but can introduce information loss due to discretization, potentially affecting accuracy. Consider a d dimensional gradient vector g with values ranging from g_{min} to g_{max}. Let b be the number of bits used for quantization (e.g., b = 8 for 256 discrete values). The quantization process can be formulated as follows [8]:

$$g_i(quantized) = round((g_i \cdot g_{min})/(g_{max} \cdot g_{min}) * (2^b.1)) \quad (2)$$

where, g_i is the i_{th} element of the gradient vector and $g_i(quantized)$ is the corresponding quantized value. Quantization techniques in Federated Learning reduce communication costs by transmitting fewer bits per value. However, they may lead to information loss through discretization, which can affect both convergence rates and model accuracy.

(b) Gradient Sparsification: This technique focuses on transmitting only the most significant gradients, setting the rest to zero. Many gradients have minimal impact on the model update, so this approach leverages this inherent sparsity. However, careful selection of the sparsity level (number of transmitted gradients) is crucial to maintain convergence and accuracy. Sparsification can be implemented using various techniques like magnitude based pruning or gradient magnitude thresholding. One approach involves defining a threshold (τ) and setting gradients with absolute values below the threshold to zero [9]:

$$g_i(sparse) = \begin{cases} 0, & \text{if } |\Delta g_i| < \tau \\ g_i, & \text{otherwise} \end{cases} \tag{3}$$

Gradient sparsification offers a substantial decrease in communication costs by sending only essential gradients. However, maintaining convergence and accuracy requires meticulous adjustment of the sparsity level, impacting overall performance [10].

(c) Hybrid Compression: While gradient quantization and sparsification offer significant communication cost reductions on their own, researchers are constantly pushing the boundaries. Enter hybrid approaches, which combine these techniques for even greater compression. Imagine taking quantized gradients, which are already smaller due to reduced bit usage, and then further sparsifying them to transmit only the most critical values. Squeezing the data inevitably leads to some information loss. This loss can potentially hinder the training process, impacting the final accuracy of the model [11].

Finding the balance between communication efficiency and acceptable accuracy is crucial. Techniques like adaptive compression algorithms come to the rescue. These algorithms act like a safety net on the tightrope, dynamically adjusting the compression level based on the training progress. This ensures you achieve good results without sacrificing too much communication efficiency.

3. Knowledge distillation:

Federated Learning offers a promising approach for collaborative machine learning on distributed devices. However, ensuring efficient knowledge transfer between participating devices while preserving privacy can be challenging. Knowledge Distillation (KD) emerges as a powerful technique to address this challenge. By leveraging the knowledge from a pre trained "teacher" model, KD empowers less powerful "student" models on participating devices to achieve comparable or even better performance [12].

Traditional Knowledge Distillation involves training a complex, high capacity "teacher" model on a large dataset. The knowledge encapsulated in this teacher model is then distilled into a smaller, less resource intensive "student" model. This distillation process involves two main components [13]:

(a) Soft Targets: In standard classification tasks, models predict class probabilities for a given input. During distillation, the teacher model not only provides the predicted class label (hard target) but also a softer, more informative distribution over all possible classes (soft target). This soft target captures the teacher's confidence in each class prediction.
(b) Distillation Loss: A loss function is defined to encourage the student model to mimic the teacher's behavior. This loss function typically combines two components:

$$L_{distill} = \lambda * KL(\sigma(z_s), \sigma(z_t)) \tag{4}$$

where, $L_{distill}$ is the knowledge distillation loss, λ is a hyperparameter controlling the relative importance of distillation compared to the cross entropy loss, KL is the Kullback Leibler divergence, a measure of similarity between two probability distributions, $\sigma(z)$ represents the softmax function, which converts model outputs (z) into class probabilities, z_s and z_t are the logits (pre softmax outputs) of the student and teacher models, respectively.

Knowledge distillation ensures privacy by utilizing a pre trained central model's knowledge without exposing raw data, addressing confidentiality concerns. Additionally, it accommodates device heterogeneity by empowering less capable devices with insights from a robust central model, thereby enhancing their performance. Several methods have been devised to implement Knowledge Distillation (KD) in Federated Learning (FL) scenarios. One approach, FedKD, entails training a compact student model on individual devices while simultaneously distilling insights from a larger teacher model hosted on the central server. This process involves transmitting updated student models to enable collaborative learning through aggregation. Another strategy involves Client side Distillation, where knowledge distillation occurs locally on each device. The teacher model can originate from a pre trained central model or a locally trained model from previous FL rounds, optimizing learning efficiency and preserving data privacy [14].

Therefore, by leveraging knowledge distillation techniques, Federated Learning can achieve improved performance while ensuring privacy and efficient knowledge transfer across participating devices. Understanding the mathematical aspects and exploring the available references can guide researchers and developers in implementing effective knowledge distillation solutions for their specific FL applications.

3.2 Sparse and Quantized Updates

Sparse and Quantized Updates emerge as powerful allies in the fight against communication inefficiency in FL. These techniques leverage the inherent properties of model parameters to significantly reduce the amount of data transmitted during training.

Sparsification:
Sparsification techniques exploit the inherent sparsity present in many machine learning models, particularly deep neural networks. Many weights (w) in these models have minimal impact on the model's performance and can be near zero in value. Mathematically, the sparsity ratio (SR) of a model quantifies this phenomenon:

$$SR = (Numberofzeroweights)/(Totalnumberofweights) \tag{5}$$

Sparsification focuses on transmitting only the most critical information - the non zero weights. This significantly reduces communication overhead compared to transmitting the entire model update.

Imagine a model with d total weights. During a standard FL update, the communication cost would be proportional to d. However, with sparsification, where only$k$ non zero weights (w_i) are transmitted, the communication cost reduces to k. This substantial reduction improves training efficiency for resource constrained devices and enables large scale FL deployments with minimal communication bottlenecks.

Popular sparsification techniques include:

Magnitude Pruning: Weights with magnitudes below a certain threshold (τ) are set to zero, effectively eliminating redundant information. This approach can be formulated as

Gradient Pruning: Weights whose gradients fall below a threshold (γ) are pruned, focusing on weights with minimal contribution to learning in the current iteration. Here, the update is modified as [3]:

$$w_i = w_i.\eta * \Delta w_i \tag{6}$$

$$w_i = \begin{cases} 0, & \text{if } |\Delta w_i| < \gamma \\ w_i, & \text{otherwise} \end{cases} \tag{7}$$

where Δw_i is the gradient of the loss function with respect to w_i and η is the learning rate.

Quantization(Encoding with Fewer Bits):
Quantization offers another powerful tool for reducing communication overhead in FL. This technique focuses on reducing the number of bits used to represent model weights. Traditional model weights are often stored using 32 bit floating point numbers, which can be quite large. Quantization aims to represent these weights using a lower number of bits, such as 8 bits or even fewer. By reducing the number of bits used per weight (b), quantization significantly reduces the size of each weight update. This translates to a decrease in communication cost compared to transmitting full precision weights. However, finding the optimal bit width for quantization is crucial. While lower bit width quantization leads to smaller updates (communication cost proportional to $b * k$), it can also introduce slight inaccuracies in the model weights. Here is a trade off between communication efficiency and maintaining model accuracy [15].

Some Popular quantization techniques includes **Post Training Quantization**, which pre trained model weights after training with full precision. While,**Quantization Aware Training (QAT)** is trained with lower precision weights from the beginning, allowing the model to adapt to the reduced representation.

3.3 Asynchronous Communication

Federated Learning (FL) offers a promising approach for collaborative machine learning on distributed devices. However, the constant need for synchronization between devices and the central server during model updates can create a communication bottleneck. This is especially relevant in situations involving numerous participating devices or unstable network connections. Asynchronous communication emerges as a powerful alternative for efficient and scalable FL deployments [16] (Table 1).

Synchronous vs. Asynchronous Communication
In synchronous communication, all devices participating in the FL training process wait for each other to complete their local updates before the central server aggregates the results. This approach guarantees that all devices contribute to the global model update in each round [17]. However, it can be inefficient in scenarios with:

- Heterogeneous Devices: Devices with varying processing power may complete their updates at different times, leading to idle waiting periods for faster devices.
- Unreliable Network Connections: Devices with intermittent connectivity may experience delays in uploading updates, further hindering the progress of the slower devices waiting for synchronization.

Asynchronous communication addresses these limitations by allowing devices to upload their local model updates to the central server independently, without waiting for others to finish. This approach offers several advantages [18]:

- Improved Scalability: Asynchronous FL can handle large scale deployments with a high number of devices by eliminating idle waiting periods.
- Enhanced Flexibility: Devices with slow processing power or unreliable connections can still participate in the training process without significantly impacting the overall progress.
- Reduced Communication Overhead: By eliminating unnecessary synchronization messages, asynchronous communication can potentially reduce communication costs.

However, asynchronous communication also poses challenges such as devices uploading updates based on outdated local models due to significant time intervals between updates, potentially hindering convergence. Additionally, delays caused by processing limitations or network issues can result in devices becoming stragglers, further impacting convergence.

Table 1. Communication-Efficient Techniques in Federated Learning: A Comparative Analysis

Technique	Description	Benefits	Challenges	Recent Research Focus
Sparse and Quantized Updates	Sparse: Transmit only a subset of significant gradients/weights (e.g., magnitude based pruning) Quantized: Reduce precision of transmitted values using techniques like gradient quantization (fewer bits)	Reduced communication cost	Potential accuracy loss due to information loss	Adaptive methods for selecting sparsity level and quantization bit width. Exploiting structured sparsity patterns for improved efficiency
Asynchronous Communication	Clients update the server independently, eliminating synchronization delays	Improved scalability for large deployments. Reduced idle waiting time for faster devices	Stale gradients can impact convergence. Straggler problem (slow devices) disrupts training	Techniques for handling stale gradients and stragglers (e.g., deadline based approaches, model averaging with timestamps). Asynchronous model averaging for better convergence guarantees
Knowledge Distillation (KD)	Leverage knowledge from a pre trained "teacher" model to improve student model performance on participating devices	Improved model accuracy on devices with limited resources	Requires pre trained teacher model	Techniques for privacy preserving distillation in FL (e.g., Secure Aggregation, Differential Privacy). Federated Knowledge Distillation with Selective Sharing
Model Pruning	Permanently remove unimportant weights/connections from the model	Significant model size reduction	Accuracy loss if pruning is aggressive	Finding optimal pruning strategy for minimal accuracy loss. Lottery Ticket Hypothesis: Identifying a sparse subnetwork that can be trained to high accuracy
Code Compression	Compress the model code itself using techniques like sparsification or quantization	Reduced model storage and deployment size	Limited applicability depending on model architecture	Exploring code compression techniques specifically for FL models (e.g., exploiting sparsity patterns induced by other compression methods). Neural Architecture Search (NAS) for designing Communication Efficient models
Low Rank Methods	Utilize techniques like matrix factorization to decompose model updates into lower dimensional representations	Reduced communication overhead by transmitting compressed updates	Can be computationally expensive for large models	Efficient low rank decomposition algorithms for federated setting. Combining low rank methods with other compression techniques for better efficiency
Federated Feature Engineering	Focuses on compressing and learning informative features on devices before transmitting them to the central server	Reduces communication cost by sending processed features instead of raw data	Requires careful design of feature engineering techniques to preserve relevant information	Techniques for collaborative feature selection and learning in a federated setting

Several strategies have been developed to mitigate the challenges posed by asynchronous communication in Federated Learning. These include periodic averaging, where the central server aggregates updates at regular intervals within specific time windows. Delay compensation techniques, such as timestamping updates or predicting stale gradients, help to address timing discrepancies between local updates and aggregation. Additionally, robust aggregation algorithms are designed to handle outdated or unreliable updates from devices that experience delays, ensuring more resilient and effective model aggregation in FL systems.

4 Comparative Evaluation Framework

The advancements in communication-efficient techniques for Federated Learning are pivotal for improving the overall performance and scalability of FL systems. This section provides a comprehensive evaluation framework to assess the impact of these advancements (as per the Table 2). The evaluation focuses on key metrics such as bandwidth utilization and model convergence times, which are essential for understanding the effectiveness of communication optimization strategies. The evaluation framework for communication efficiency in Federated Learning systematically examines various strategies to optimize communication overhead. It focuses on reducing data size through model compression, sparse updates and quantization, while also assessing the impact on communication overhead by tracking changes in the number of communication rounds and network load. Additionally, the framework evaluates model convergence times to understand how these techniques affect the speed and efficiency of achieving model accuracy. It balances the trade-offs between communication efficiency and model performance, examines scalability in large-scale FL environments, and tests the practical applicability of techniques in real-world scenarios. The framework provides insights and recommendations on the most effective methods based on specific conditions and requirements.

Bandwidth Utilization
Bandwidth utilization is a critical factor in FL, especially when dealing with large-scale deployments involving numerous devices. Efficient use of bandwidth ensures that the communication overhead is minimized, making the FL process more practical and feasible in real-world applications.

- Model Compression Techniques: These techniques significantly reduce the size of the data that needs to be transmitted during the training process. By compressing model updates, the amount of information exchanged between client devices and the central server is minimized. For instance, methods like weight pruning and quantization reduce the number of parameters that need to be sent, leading to lower bandwidth consumption. Evaluating the impact of these techniques involves measuring the reduction in data size and comparing it with the baseline scenario where no compression is applied.

- Sparse and Quantized Updates: Sparse updates involve transmitting only the most significant gradients, while quantization reduces the precision of the transmitted updates. Both approaches aim to cut down the amount of data exchanged. The evaluation framework should assess the extent to which these methods decrease bandwidth usage without compromising the model's performance. Metrics such as the percentage reduction in data size and the corresponding impact on bandwidth utilization are crucial for this assessment.
- Asynchronous Communication: Allowing clients to update the server independently reduces synchronization delays and potentially lowers the volume of simultaneous data transmissions. By evaluating the bandwidth utilization in asynchronous communication scenarios, researchers can determine the effectiveness of this approach in reducing network congestion and improving communication efficiency.

Model Convergence Times

Model convergence time is another vital metric for evaluating the performance of FL systems. It measures the time taken for the model to reach a specified level of accuracy or loss. Faster convergence times indicate more efficient training processes and better utilization of resources.

- Impact of Compression Techniques: While compression techniques reduce data size and bandwidth usage, they may also affect the convergence rate of the model. For example, aggressive weight pruning or quantization might lead to information loss, which can slow down convergence. The evaluation framework should compare the convergence times of models trained with and without compression to determine the trade-offs involved.
- Effectiveness of Sparse and Quantized Updates: Sparse updates transmit only the most critical gradients, which can impact the model's ability to converge quickly. Similarly, quantization might introduce errors due to reduced precision. The framework should assess how these techniques influence the convergence time by comparing the number of communication rounds required to achieve a target accuracy with and without these optimizations.
- Asynchronous Communication and Convergence: Asynchronous updates can lead to stale gradients, where updates are based on outdated local models. This issue can potentially hinder convergence. The evaluation should include scenarios with different levels of asynchrony to understand its impact on convergence times. Metrics such as the average number of communication rounds and the total training time required to reach the desired model performance are essential for this analysis.

Comparative Analysis

To provide a comprehensive evaluation, the framework should compare the different communication optimization techniques across various dimensions as follows:

Table 2. Comprehensive Evaluation of Communication-Efficient Techniques in Federated Learning

Aspect	Metric	Description	Assessment Criteria
Bandwidth Utilization	Data Size Reduction	Evaluates how much the data size is reduced using different techniques (e.g., model compression, sparse updates, quantization).	- Measure percentage reduction in data size compared to baseline (no compression). - Assess impact on overall bandwidth consumption. - Compare effectiveness of each technique in reducing data size.
Bandwidth Utilization	Communication Overhead	Analyzes the communication overhead and efficiency gains achieved through optimization techniques.	- Evaluate decrease in communication rounds needed. - Examine changes in network load and congestion. - Compare bandwidth savings across different scenarios (e.g., synchronous vs. asynchronous communication).
Model Convergence Times	Convergence Rate	Measures the time and number of communication rounds required for the model to reach a specified accuracy or loss threshold.	- Compare convergence times of models trained with and without compression techniques. - Analyze the impact of sparse and quantized updates on convergence speed. - Assess the effect of asynchronous communication on convergence rate, considering factors like stale gradients and stragglers.
Comparative Analysis	Efficiency vs. Accuracy	Balances the trade-offs between communication efficiency and model accuracy.	- Compare model performance (accuracy) before and after applying each optimization technique. - Identify the trade-offs in terms of potential accuracy loss due to data compression or quantization. - Highlight scenarios where certain techniques provide the best balance of efficiency and accuracy.
Comparative Analysis	Scalability	Evaluates how well the techniques perform in large-scale FL deployments with numerous participating devices.	- Assess changes in bandwidth usage and convergence times as the number of clients increases. - Analyze the scalability of each technique in handling larger datasets and more participants. - Determine the robustness and effectiveness of techniques in different scale scenarios.
Comparative Analysis	Real-World Applicability	Considers the practical application of these techniques in real-world FL scenarios, taking into account network reliability and device heterogeneity.	- Evaluate performance in case studies or simulations that reflect realistic conditions (e.g., varying network reliability, device computational power). - Analyze the adaptability of techniques to different environmental factors. - Highlight practical challenges and solutions for deploying these techniques in real-world settings.
Trade-Offs and Recommendations	Effectiveness and Suitability	Provides insights and recommendations based on the trade-offs and effectiveness of each technique in different FL scenarios.	- Discuss the specific conditions under which each technique is most beneficial. - Offer practical recommendations for choosing the appropriate method based on network conditions, device capabilities, and desired model performance. - Summarize best practices for implementing communication-efficient FL strategies in various application domains.

(a) Efficiency vs. Accuracy: Balancing communication efficiency with model accuracy is a critical aspect. Techniques that aggressively reduce data size might lead to a loss in accuracy. The evaluation should present a comparative analysis showing how each technique impacts both bandwidth utilization and model performance.

(b) Scalability: The scalability of these techniques in large-scale FL deployments is another important consideration. The framework should assess how well each method performs as the number of participating devices increases. This involves evaluating the impact on bandwidth usage and convergence times in scenarios with varying numbers of clients.

(c) Real-World Applicability: The practical applicability of these techniques in real-world FL scenarios should be a key focus. This includes considering factors such as network reliability, device heterogeneity, and computational constraints. The evaluation should include case studies or simulations that reflect realistic conditions to provide insights into the effectiveness of these techniques in actual deployments.

(d) Trade-Offs and Recommendations: Finally, the framework should highlight the trade-offs associated with each communication optimization technique. This includes discussing the conditions under which a particular method is most effective and providing recommendations for practitioners. For instance, weight pruning might be more suitable for scenarios with limited bandwidth but sufficient computational power on client devices, while asynchronous communication might be ideal for environments with unstable network connections.

By establishing a detailed comparative evaluation framework, this paper aims to provide a comprehensive understanding of the impact of various communication-efficient techniques on Federated Learning systems. By assessing metrics such as bandwidth utilization and model convergence times, researchers can demonstrate the importance of optimizing communication in FL. This framework not only highlights the benefits and challenges associated with each technique but also offers practical insights and recommendations for deploying FL in real-world applications. The ultimate goal is to enhance the efficiency, scalability, and feasibility of FL systems, paving the way for their widespread adoption in diverse domains.

5 Future Research Directions

Communication efficiency is crucial in Federated Learning (FL) as it ensures faster training speeds, better model performance and overall system feasibility. Efficient communication minimizes data transmission and delays, which is essential for scaling FL in real-world applications with numerous participating devices. Therefore, to advance communication efficiency in Federated Learning, researchers should prioritize the development of sophisticated model compression techniques that minimize data size while preserving model accuracy. Adaptive compression algorithms that dynamically adjust based on network conditions and model requirements are promising [19]. Additionally, integrating multiple compression methods, such as combining quantization with pruning or sparse updates, could lead to more significant efficiency gains.

Addressing asynchronous communication challenges is another critical area for future research. Techniques to mitigate stale gradients and the straggler problem are essential [20]. Predictive modeling could be utilized to adjust for delayed updates, while hybrid communication models that blend synchronous and asynchronous methods could enhance the robustness and efficiency of FL systems. These approaches are vital for improving the overall performance and scalability of FL in diverse environments. Robust aggregation algorithms that withstand outdated or unreliable updates from straggler devices are needed. These algorithms must maintain model performance despite heterogeneous device capabilities and varying network conditions [21]. Investigating adaptive aggregation techniques through machine learning can provide more resilient solutions, ensuring the effectiveness and reliability of FL systems in practical, real-world applications.

6 Conclusion

Communication overhead remains a significant challenge for collaborative machine learning despite the promising approach of Federated Learning. Utilizing methods that improve communication can significantly improve the scalability, privacy and usability of federated learning systems. This study explored the impact of communication efficiency on FL performance, highlighting the importance of a comprehensive evaluation system that takes into account bandwidth usage, speed of convergence, model accuracy and privacy measures. We analyzed recent progress in communication-efficient techniques including gradient sparsification, quantization, asynchronous communication, knowledge distillation and low-rank methods. These approaches have potential in overcoming communication obstacles and maximizing the use of FL for practical scenarios.

Ongoing enhancements in communication efficiency are anticipated as research on Federated Learning advances. Future research could concentrate on creating adaptive methods that modify communication approaches depending on network conditions and device capabilities. Exploring how to improve FL for different networks with different bandwidth and latency is a promising area of research. Moreover, it is essential to incorporate strong security measures along with communication efficient methods to guarantee data integrity and user privacy. Addressing these obstacles and promoting additional research can greatly increase the practicality of FL in areas such as personalized healthcare and smart city management, potentially revolutionizing collaborative learning models.

References

1. Vora, J., Bouacida, N., Krishnan, A., Mohapatra, P.: FedDM: enhancing communication efficiency and handling data heterogeneity in federated diffusion models. arXiv preprint arXiv:2407.14730 (2024)
2. Almanifi, O.R.A., Chow, C.-O., Tham, M.-L., Chuah, J.H., Kanesan, J.: Communication and computation efficiency in federated learning: a survey. Internet Things **22**, 100742 (2023). https://doi.org/10.1016/j.iot.2023.100742

3. Yin, C., Alqahtani, A., Noel, M.S.: The next frontier in pancreatic cancer: targeting the tumor immune milieu and molecular pathways. Cancers **14**(11) (2022). https://doi.org/10.3390/cancers14112619
4. Xia, Q., Ye, W., Tao, Z., Wu, J., Li, Q.: A survey of federated learning for edge computing: research problems and solutions. High-Confidence Comput. **1**(1), 100008 (2021). https://doi.org/10.1016/j.hcc.2021.100008
5. Caron, M., Morcos, A., Bojanowski, P., Mairal, J., Joulin, A.: Pruning Convolutional Neural Networks with Self-Supervision (2020). https://arxiv.org/abs/2001.03554
6. Frankle, J., Carbin, M.: The Lottery Ticket Hypothesis: Finding Sparse, Trainable Neural Networks (2019). https://arxiv.org/abs/1803.03635
7. Shahid, O., Pouriyeh, S., Parizi, R.M., Sheng, Q.Z., Srivastava, G., Zhao, L.: Communication Efficiency in Federated Learning: Achievements and Challenges (2021). https://arxiv.org/abs/2107.10996
8. Liu, H., He, F., Cao, G.: Communication-Efficient Federated Learning for Heterogeneous Edge Devices Based on Adaptive Gradient Quantization (2022). https://arxiv.org/abs/2212.08272
9. Han, P., Wang, S., Leung, K.K.: Adaptive Gradient Sparsification for Efficient Federated Learning: An Online Learning Approach (2020). https://arxiv.org/abs/2001.04756
10. Wangni, J., Wang, J., Liu, J., Zhang, T.: Gradient Sparsification for Communication-Efficient Distributed Optimization (2017). https://arxiv.org/abs/1710.09854
11. Konečný, J., McMahan, H.B., Yu, F.X., Richtárik, P., Suresh, A.T., Bacon, D.: Federated Learning: Strategies for Improving Communication Efficiency (2017). https://arxiv.org/abs/1610.05492
12. Gou, J., Yu, B., Maybank, S.J., Tao, D.: Knowledge distillation: a survey. Int. J. Comput. Vis. **129**(6), 1789–1819 (2021). https://doi.org/10.1007/s11263-021-01453-z
13. Mora, A., Tenison, I., Bellavista, P., Rish, I.: Knowledge Distillation for Federated Learning: A Practical Guide (2022). https://arxiv.org/abs/2211.04742
14. Wu, C., Wu, F., Lyu, L., Huang, Y., Xie, X.: Communication-efficient federated learning via knowledge distillation. Nat. Commun. **13**(1), 2032 (2022)
15. Zhao, Z., et al.: Towards Efficient Communications in Federated Learning: A Contemporary Survey (2022). https://arxiv.org/abs/2208.01200
16. Marfoq, O., Neglia, G., Kameni, L., Vidal, R.: Federated Learning for Data Streams (2023). https://arxiv.org/abs/2301.01542
17. Wilhelmi, F., Giupponi, L., Dini, P.: Analysis and Evaluation of Synchronous and Asynchronous FLchain (2022). https://arxiv.org/abs/2112.07938
18. Yang, D., Ji, Y., Kou, Z., Zhong, X., Zhang, S.: Asynchronous Federated Learning with Incentive Mechanism Based on Contract Theory (2023). https://arxiv.org/abs/2310.06448
19. Zhang, X., Zhu, X., Wang, J., Yan, H., Chen, H., Bao, W.: Federated learning with adaptive communication compression under dynamic bandwidth and unreliable networks. Inf. Sci. **540**, 242–262 (2020). https://doi.org/10.1016/j.ins.2020.05.137
20. Xu, C., Qu, Y., Xiang, Y., Gao, L.: Asynchronous federated learning on heterogeneous devices: a survey. Comput. Sci. Rev. **50**, 100595 (2023). https://doi.org/10.1016/j.cosrev.2023.100595
21. Pillutla, K., Kakade, S.M., Harchaoui, Z.: Robust aggregation for federated learning. IEEE Trans. Signal Process. **70**, 1142–1154 (2022). https://doi.org/10.1109/TSP.2022.3153135

Study of Two-Dimensional Unsteady Natural Convection of a Fluid in a Square Cavity with a Solid Body in the Presences of Thermal Radiation

Rasha N. Fadel and Alaa A. Hammodat(✉)

Department of Mathematics, College Education of Pure Science, Mosul University, Mosul, Iraq
rasha.22esp36@student.uomosul.edu.iq,
alaahammodat@uomosul.edu.iq

Abstract. This study aims to accurately analyze the radiative effects on incompressible fluid flow and unsteady magnetohydrodynamic natural convection inside a square cavity containing a solid body in the center. Both the upper and bottom barriers of the hollow preserve heat. Energy, mass, and motion equations are all governed by the model. The non-dimensional equations of continuity, momentum, and energy are found by applying an appropriate transformation. A MATLAB algorithm which simulates the mathematical problem and solves the governing equations is shown. It is constructed using the alternative direction explicit method. Additionally, examining the effects of varying radiation parameter, Rayleigh number, Hartmann amount, Darcy number, and Prandtl magnitude.

Keywords: Unsteady state · Thermal radiation · Natural convection · Alternative direction explicit method · Rayleigh number

1 Introduction

The demand for researchers exploring novel studies, testing, and possibly potential uses has risen significantly due to the immense impact technological advancement has had on how we live today. Modeling real experiments as differential equations with suitable parameters is an example of methodology. These problems with differential equations can be calculated more quickly and sensibly with the use of mathematical modeling tools. A number of convections hear transport tests in cavities with various sizes and inclinations could be conducted. The bodies are considered sources of warmth, and it is clear that the cavity's heat is influenced by its position.

The topic of stretched terrain has generated a great deal of excitement for the subject and a sizable amount of scholarly work in 2015 [6]. The impact of slide and convective boundary conditions on heat transfer emanating from a Casson nanofluid traveling with a stretched sheet in magnetohydrodynamic (MHD) stagnation point flow is looked into by Ibrahim and Makinde in 2015 [8]. The consequences of thermal radiation and chemical changes on boundary layer slip flow and melting heat transfer of nanofluid caused by a

S. Pal and Á. Rocha (Eds.): ICMMCS 2025, LNNS 1398, pp. 324–334, 2025.
https://doi.org/10.1007/978-3-031-90998-6_30

dynamic stretching sheet have been explored by Krishnamurthy et al. in 2016 [7]. In 2017, according to Ghosh and Mukhopadhyay [1], nanofluid has been used in numerous energy systems, including continuous stretching of plastic films, artificial cellulose, cooling nuclear systems, drawing copper wires, hot rolling, natural convection in enclosures, and continuous stretching of plastic films.

In the situation of stretched sheets, Manjunatha et al. in 2019 [4] detected an important growth in the conductivity of heat when adding metallic nanoparticles to the foundation fluids.Due to studies done by Bahiraei et al. in 2020 [2], nanofluids' greater thermal conductivity in comparison to that of typical liquids causes them to have fewer thermal resistances, which in consequence results in more notable heat transfer rates. By reconciling both the pros and cons of separate suspensions, hybrid nanofluids have the goal of further improving heat transfer and pressure drop properties. This is due to their improved aspect ratio, expanded thermal network, and synergistic effect of nanomaterials. Solid dispersion of composite nanomaterials into base liquids is known as composite nanofluids in 2021 [3].The heat effectiveness of a horizontal tube was analytically evaluated in 2021 by Akbar et al. [5] by taking into consideration hybrid nanofluids (alumina and titania) floating within water. It was discovered that when hybrid nanofluid concentration and rate of flow rise, so does the enhancement of heat communication.

In research that was published, Hammodat A. et al. in 2021 [9] examined the fluid flow in a bridge under the consequences of electromagnetism (EMF) and then used the line strategy to solve the complex partial differential equations that detailed the issue. The effects of heat radiation on a Cu-water nanofluid technology's convection flow behind a differentially sustained container were investigated, and the findings are reported in 2021 [10]. A sealed square with a similar depth, lx, and length, ly ($lx = ly = 0.2$) is located in the center of the housing2. Mebarek-Oudina F. and Chabani I. give a discussion on a variety of heat transfer technologies involving porous media, domains of magnetism, and nanoscale fluids, focusing primarily on the effect of the solar thermal field in2022 [11].

A numerical study of the thermal and temporal associated mass in a flexible, identical square layer is proposed by Algwauish Gh. et al. in 2023, using a combination of convection flow [12]. The goal is to investigate the thermal radiation effect on the fluid flow of unsteady natural convection in a differentially heated cavity with solid block inserts under a magnetic flow field influence. It is found that the impact of various parameters is significant in this case.

1.1 The Objectives of the Study

The objective of this study is to explore how thermal radiation affects the natural convection of an incompressible fluid within a square cavity that contains a solid body at its center. The research focuses on understanding the complex interactions between heat transfer, fluid flow, and magnetic fields by analyzing key parameters like the radiation factor, Rayleigh no., Hartmann no., Darcy no., and Prandtl number. Using a numerical model, the study aims to provide detailed insights into how these factors influence the behavior of the fluid and heat distribution in the system.

2 Mathematical Modelling of the Problem

A 2-D square cavity with width and height ⌈, as well as an equilateral isolated body with side length centered in the cavity, are schematically depicted in Fig. 1. T_{hot} and T_{cold}, the left and right walls of the outer hollow, have constant temperatures despite variations. It is thought that the enclosure's top and bottom walls are totally isolated. There is liquid inside the cavity. It can be found in the § and † directions, respectively, to define the velocities, ⊓ and ⊑. Two factors contribute to the downward trend: gravity and the magnetic field. It is assumed that the fluid's thermophysical properties are constant and that density changes in the buoyancy duration are excluded using the Boussinesq approximation. Moreover, the present investigation assumes that glutinous dissipation is disregarded. The following continuity, momentum, and energy equations explain the fluids in the enclosure and their heat transfer during two-dimensional unsteady natural convection:

The formula for continuity:

$$u_x + v_y = 0 \tag{1}$$

Formula of momentum:

$$u_t + uu_x + uu_y = -\frac{p_x}{\rho} + \upsilon\nabla^2 u - \frac{\upsilon}{k}u \tag{2}$$

$$v_t + uv_x + vv_y = -\frac{p_y}{\rho} + \upsilon\nabla^2 v - \frac{v_o^2}{\rho}v - g\beta(T - T_c) \tag{3}$$

Consequently, the general momentum equation looks like this:

$$\begin{aligned} &(v_x - u_y)_t + \left((uv_x + vv_y)_x - (uu_x + vu_y)_y\right) \\ &= \upsilon\nabla^2(v_x - u_y) + \frac{v\Theta_o^2}{\rho}v_x - g\beta T_x + \frac{\upsilon}{k}u_y \end{aligned} \tag{4}$$

Formula for energy:

$$T_t + uT_x + vT_y = \alpha\nabla^2 T - \frac{1}{\rho C_p}(q_{rx} + q_{ry}) \tag{5}$$

In both the § and † directions, radiation-induced heat flow is given by: $\amalg_\nabla = \frac{-4\wr^*}{3k^*}T_§^4 and \amalg_\nabla = \frac{-4\wr^*}{3k^*}T_†^4$, so that in Eq. (5):

$$T_t + uT_x + vT_y = \left(\alpha + \frac{16o^* T_c^3}{3k^*\rho C_p}\right)\nabla^2 T \tag{6}$$

$\rho, \upsilon, k, \mathcal{P}, \mathcal{B}_o, \wr, \}, \beta, T, \alpha, C_{\surd}, k^*, \wr^*$ The following factors need to be assessed: density, kinematic viscosity, media permeability, pressure, constant magnetic field, electrical conductivity, acceleration, volumetric coefficient of thermal expansion, temperature, thermal diffusivity, specific heat at constant pressure, mean absorption coefficient, and Stefan-Boltzmann constant. The pertinent boundary circumstances at are listed below:

$$\left.\begin{array}{l} \text{every solid boundary} \quad u = v = 0 \\ at\, x = 0, d, 0 \leq y \leq d : T = T_h, T_c \\ at\, y = 0, d, 0 \leq x \leq d : \quad T_y = 0 \end{array}\right\} \tag{7}$$

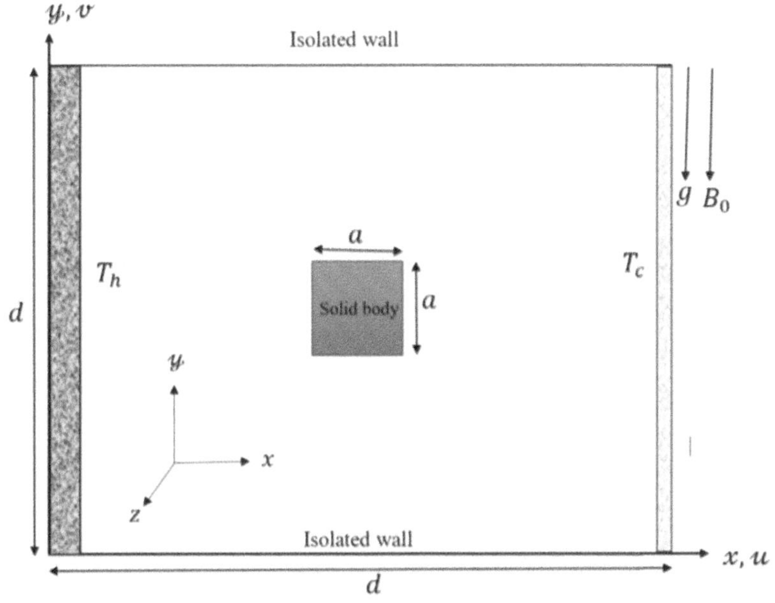

Fig. 1. A Diagrammatic View of the Enclosure

3 Formulating Mathematically

Dimensionality needs to be added to the governing equations (1), (4), and (6) because the finite difference method will be applied to solve them. For this aim, let us now propose dimensionless quantities [13].

$$\left.\begin{array}{l} u = \frac{\alpha}{d}\mathcal{U}, \ v = \frac{\alpha}{d}\mathcal{V}, \ Rd = \frac{3k^*k+4o^*T_c^3}{3k\mu C_p} \\ x = d\mathcal{X}, \ y = d\mathcal{Y}, \quad Ra = \frac{g\beta\Delta T d^3}{\alpha^2} \\ w = \frac{\alpha}{d^2}\Omega, \ Pr = \frac{\upsilon}{\alpha}, \ M = \mathcal{B}_o d\left(\frac{o}{\mu}\right)^{1/2} \\ Da = \frac{k}{d^2}, \ \psi = \alpha\Psi, \ \theta = \frac{T-T_c}{T_h-T_c}, \tilde{t} = \frac{\alpha}{d^2}t \end{array}\right\} \tag{8}$$

where $w, \Omega, , \mu, , \psi, \Psi, \theta, Rd, Ra, Pr, M .Da$ include viscosity, temperature, temperature without dimensions, stream function, vorticity, and dimensionless vorticity. Radiant heat, The Darcy, Hartmann, Rayleigh, and Prandtl numbers. The governing equations' dimensionless representations are as follows:

$$\mathcal{U}_{\mathcal{X}} + \mathcal{V}_{\mathcal{Y}} = 0. \tag{9}$$

$$\begin{aligned}(\mathcal{V}_{\mathcal{X}} - \mathcal{U}_{\mathcal{Y}})_{\bar{t}} + \left((\mathcal{U}\mathcal{V}_{\mathcal{X}} + \mathcal{V}\mathcal{V}_{\mathcal{Y}})_{\mathcal{X}} - (\mathcal{U}\mathcal{U}_{\mathcal{X}} + \mathcal{V}\mathcal{U}_{\mathcal{Y}})_{\mathcal{Y}}\right) = PrDa^{-1}\mathcal{U}_{\mathcal{Y}} + \\ +Pr\nabla^2(\mathcal{V}_{\mathcal{X}} - \mathcal{U}_{\mathcal{Y}}) + M^2 Pr\mathcal{V}_{\mathcal{X}} - Ra\theta_{\mathcal{X}}\end{aligned} \tag{10}$$

$$\theta_{\bar{t}} + \mathcal{U}\theta_{\mathcal{X}} + \mathcal{V}\theta_{\mathcal{Y}} = Rd\nabla^2\theta \tag{11}$$

This is how the equations' boundary conditions look:

$$\left.\begin{array}{ll} \text{on the lef wall}: & \mathcal{U} = \mathcal{V} = 0, \theta = 1 \\ \text{on the right wall}: & \mathcal{U} = \mathcal{V} = 0, \theta = 0 \\ \text{on the insulated}: & \mathcal{U} = \mathcal{V} = 0, \theta_{\mathcal{Y}} = 0 \end{array}\right\} \tag{12}$$

According to [14], the stream and vorticity functions are calculated.

$$\mathcal{U} = \Psi_{\mathcal{Y}}\ \mathcal{V} = -\Psi_{\mathcal{X}}\ \Psi_{\mathcal{X}\mathcal{X}} + \Psi_{\mathcal{Y}\mathcal{Y}} = -\Omega\} \tag{13}$$

In order to find the governing equations and boundary conditions (12), dimensionless parameters are the subject of equations (8) and (13).

$$\Psi_{\mathcal{X}\mathcal{X}} + \Psi_{\mathcal{Y}\mathcal{Y}} = -\Omega \tag{14}$$

$$\begin{aligned}\Omega_{\bar{t}} - Pr\nabla^2\Omega = -(\Psi_{\mathcal{Y}}\Omega_{\mathcal{X}} - \Psi_{\mathcal{X}}\Omega_{\mathcal{Y}}) - M^2 Pr(\Psi_{\mathcal{X}\mathcal{X}}) \\ -Ra(\theta_{\mathcal{X}}) + \ldots + PrDa^{-1}(\Psi_{\mathcal{Y}\mathcal{Y}})\end{aligned} \tag{15}$$

$$\theta_{\bar{t}} - \nabla^2\theta = -\frac{1}{Rd}(\Psi_{\mathcal{Y}}\theta_{\mathcal{X}} - \Psi_{\mathcal{X}}\theta_{\mathcal{Y}}) \tag{16}$$

Under the conditions of constraint:

$$\left.\begin{array}{ll} \text{on the solid wall} & \Psi = \quad 0 \\ \text{on}\ \mathcal{X} = 0, 1, 0 \le \mathcal{Y} \le 1 : \theta = 1 & \Omega = -\Psi_{\mathcal{X}\mathcal{X}} \\ \text{on}\ \mathcal{Y} = 0, 1, 0 \le \mathcal{X} \le 1 : \theta = 0 & \Omega = -\Psi_{\mathcal{X}\mathcal{X}} \\ \text{on}\ \mathcal{Y} = 0, 1, 0 \le \mathcal{X} \le 1 : \theta_{\mathcal{Y}} = 0 & \Omega = -\Psi_{\mathcal{X}\mathcal{X}} \end{array}\right\} \tag{17}$$

4 Solution Approach

Using an explicit finite difference approach and the associated boundary conditions, a set of dimensionless governing equations for nonlinear partial differentials has been numerically obtained. The flow is divided into sections by a grid or mesh of lines parallel to the $\mathcal{X}$ – and $\mathcal{Y}$-axes, with the $\mathcal{X}$- axis indicating the plate's upward direction and the $\mathcal{Y}$-axes normal to the plate. If we take $\mathcal{Y}_{max}$ (=25) and take it to $\mathcal{Y} \to \infty$, it indicates that $\mathcal{Y}$ varies from 0 to 25. We then measure the height of the plate $\mathcal{X}_{max}$ (=100), i.e., $\mathcal{X}$ varied from 0 to 100. With the minor time period, let m = 250 and n = 250 be the grid spacing in the $\mathcal{X}$ and $\mathcal{Y}$ directions, respectively, and as follows: $\Delta\mathcal{X} = 0.4(0 \le \mathcal{X} \le 100)$ and $\Delta\mathcal{Y} = 0.4(0 \le \mathcal{Y} \le 25)$. The partial Eqs. (14)– (16) with boundary conditions (17) can be solved explicitly using the finite difference method, as shown in [15, 16].

$$\frac{\Psi_{i+1,j} - 2\Psi_{i,j} + \Psi_{i-1,j}}{(\Delta\mathcal{X})^2} + \frac{\Psi_{i,j+1} - 2\Psi_{i,j} + \Psi_{i,j-1}}{(\Delta\mathcal{Y})^2} = -(\Omega)_{i,j} \tag{18}$$

$$\left(\frac{\Omega_{i,j}-\Omega_{i-1,j}}{\Delta\tilde{t}}\right)-Pr\left(\frac{\Omega_{i+1,j}-2\Omega_{i,j}+\Omega_{i-1,j}}{(\Delta\mathcal{X})^2}+\frac{\Omega_{i,j+1}-2\Omega_{i,j}+\Omega_{i,j-1}}{(\Delta\mathcal{Y})^2}\right)=$$

$$\ldots=-\left(\frac{\Psi_{i,j+1}-\Psi_{i,j}}{\Delta\mathcal{Y}}\frac{\Omega_{i,j}-\Omega_{i-1,j}}{\Delta\mathcal{X}}-\frac{\Psi_{i,j}-\Psi_{i-1,j}}{\Delta\mathcal{X}}\frac{\Omega_{i,j+1}-\Omega_{i,j}}{\Delta\mathcal{Y}}\right)-\ldots$$

$$\ldots-M^2Pr\left(\frac{\Psi_{i+1,j}-2\Psi_{i,j}+\Psi_{i-1,j}}{(\Delta\mathcal{X})^2}\right)+Ra\left(\frac{\theta_{i,j}-\theta_{i-1,j}}{\Delta\mathcal{X}}\right)+PrDa^{-1}\left(\frac{\Psi_{i,j+1}-2\Psi_{i,j}+\Psi_{i,j-1}}{(\Delta\mathcal{Y})^2}\right) \quad (19)$$

$$\left(\frac{\theta_{i,j}-\theta_{i-1,j}}{\Delta\tilde{t}}\right)-\left(\frac{\theta_{i+1,j}-2\theta_{i,j}+\theta_{i-1,j}}{(\Delta\mathcal{X})^2}+\frac{\theta_{i,j+1}-2\theta_{i,j}+\theta_{i,j-1}}{(\Delta\mathcal{Y})^2}\right)=$$

$$\frac{1}{Rd}\left(\frac{\Psi_{i,j+1}-\Psi_{i,j}}{\Delta\mathcal{Y}}\frac{\theta_{i,j}-\theta_{i-1,j}}{\Delta\mathcal{X}}-\frac{\Psi_{i,j}-\Psi_{i-1,j}}{\Delta\mathcal{X}}\frac{\theta_{i,j+1}-\theta_{i,j}}{\Delta\mathcal{Y}}\right) \quad (20)$$

5 Results and Discussions

Within a square cavity containing a solid body at its center, the radiative effects of unsteady magnetohydrodynamic (MHD) natural convection and incompressible fluid flow are precisely examined in this paper. The thermal safety of the chamber is preserved by the top and bottom dividers. MATLAB code using ADE method is developed to display the problem and resolve the equations that govern the issue. The estimated result, together with the temperature and velocity forms, is displayed further down, along with a chemical rationale.

5.1 The Darcy Number Impact (*Da*)

Figures 2, 3, 4 and 5 show how changing the Darcy parameters changes the temperature and velocity profiles of the flow field while holding all other variables constant. Figures 2 and 3 shows that as Darcy values rise, so does the velocity of the flow field. The graph in Figures 4 and 5 shows how the rate of heat transfer slows as the Darcy number increases.

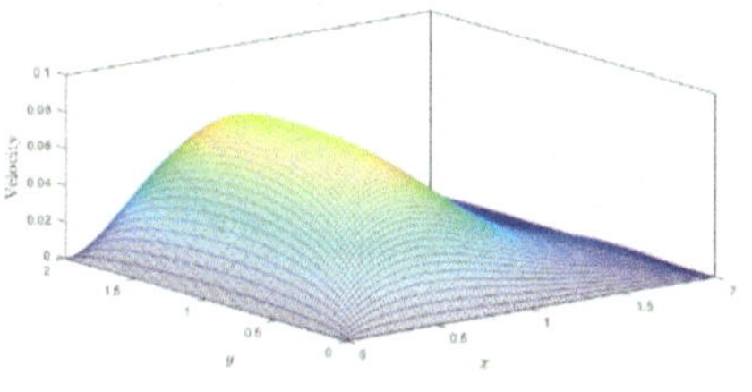

Fig. 2. Velocity profile when Darcy number Da = 10.

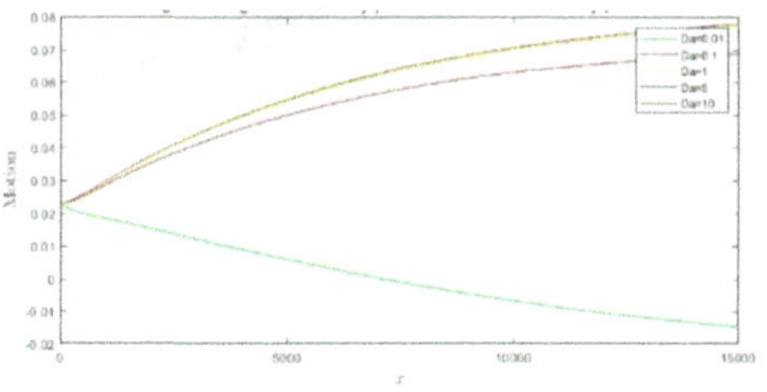

Fig. 3. Changes in the Darcy parameter affect the velocity profile.

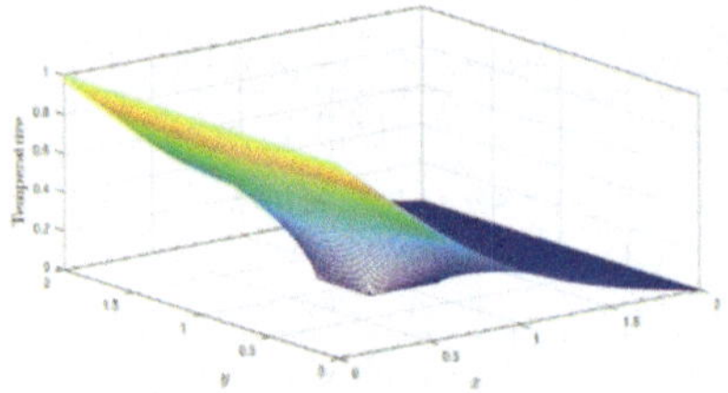

Fig. 4. Distribution of temperature when Darcy number Da = 10.

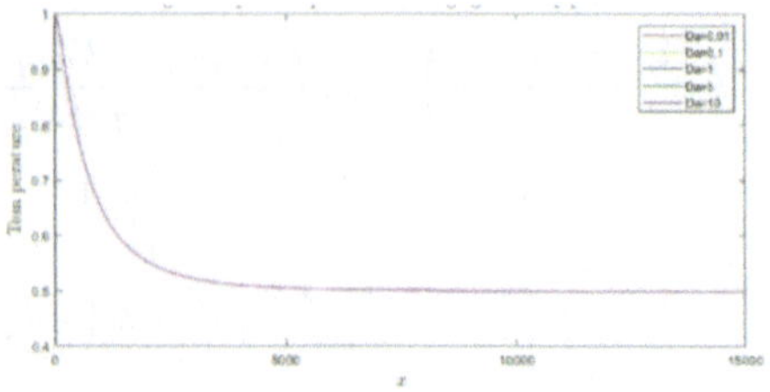

Fig. 5. Temperature profile after changing the Darcy parameter.

5.2 Effect of Rayleigh Parameter *Ra*

Figures 8 and 9, like Figures 6 and 7, show how the Rayleigh coefficient values affect the temperature and velocity shapes while all other parameters remain fixed. The above graphs show that as the Rayleigh number increases, the temperature and velocity profiles improve.

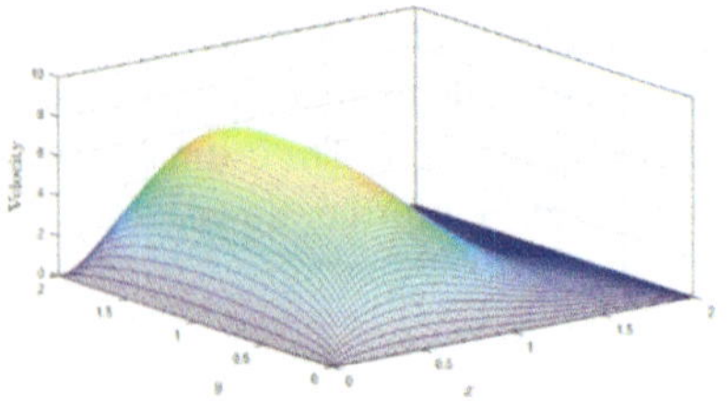

Fig. 6. Velocity distribution when Rayleigh parameter Ra = 10.

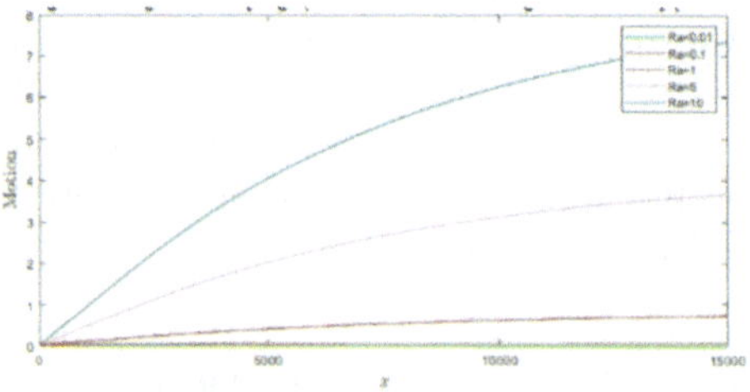

Fig. 7. A change in the Rayleigh parameter causes a change in the velocity profile.

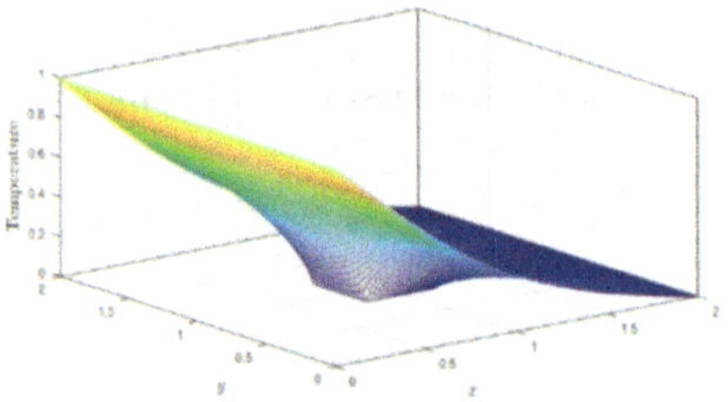

Fig. 8. Temperature profile when Rayleigh parameter Ra = 10.

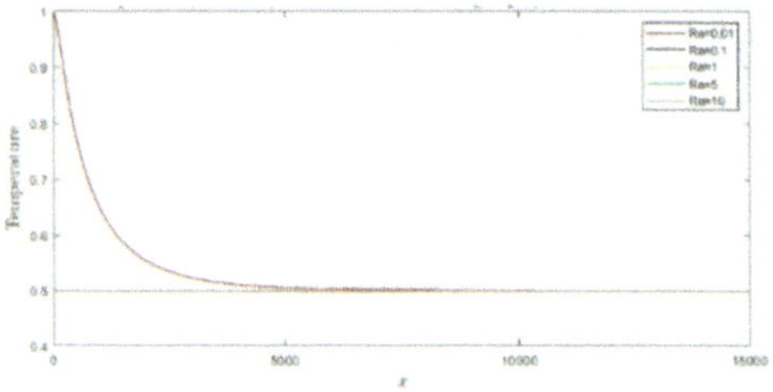

Fig. 9. Temperature profile as a result of Rayleigh parameter variation.

5.3 Effect of Prandtl parameter *Pr*

As shown in Figs. 10, 11, 12 and 13, both the velocity and temperature profiles of the flow field decreases as the Prandtl parameter *Pr* value increases.

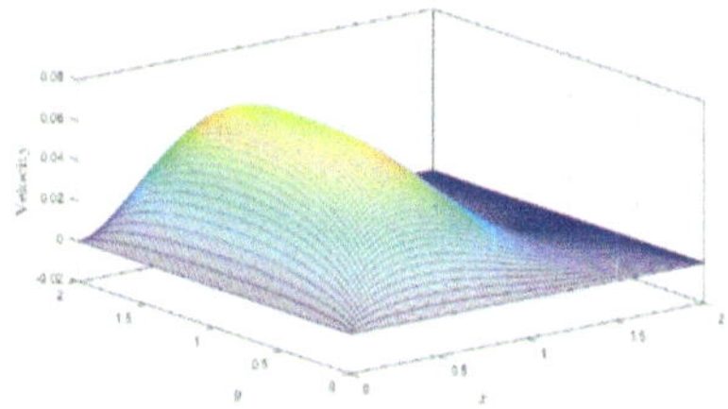

Fig. 10. Velocity distribution when Prandtle number Pr = 1.

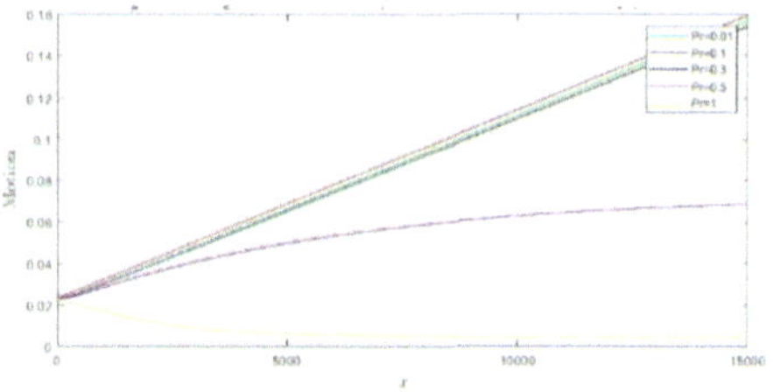

Fig. 11. Changes in the Prandtle parameter affect the velocity profile.

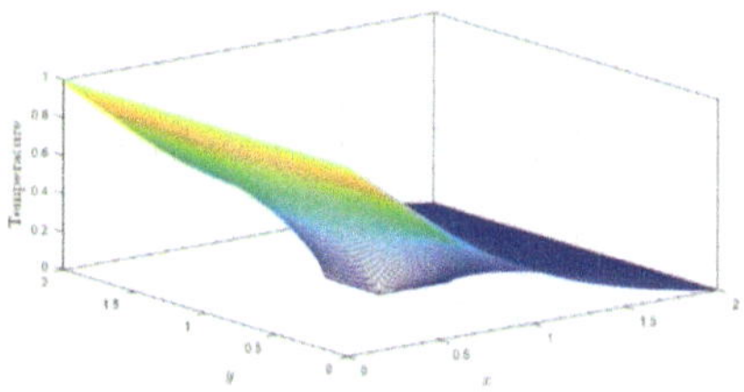

Fig. 12. Temperature profile when Prandtle number Pr = 1.

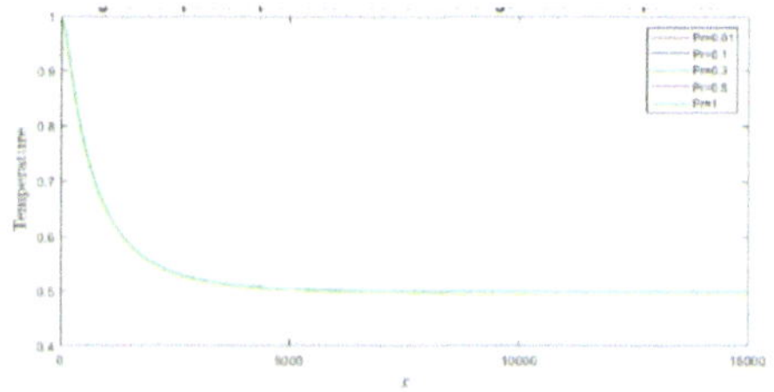

Fig. 13. Temperature profile as a result of a change in the Prandtle parameter.

5.4 *MM*'s Magnetic Field's Impact

All other parameters remain constant, and the velocity and temperature profiles of the flow field are shown as a function of the magnetic parameter M in Figs. 14, 15, 16 and 17. As shown in Figs. 14 and 15 the velocity of the flow field decreases as the magnetic field values M increase. Figures 16 and 17 depicts how temperature profiles rise as magnetic field M values fall.

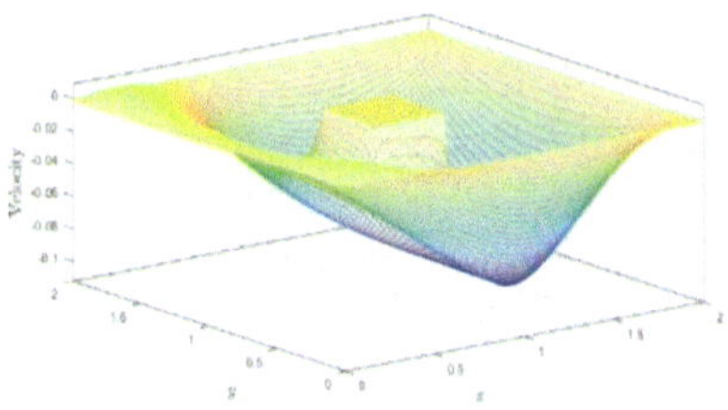

Fig. 14. Velocity profile when Hartmann number M = 10.

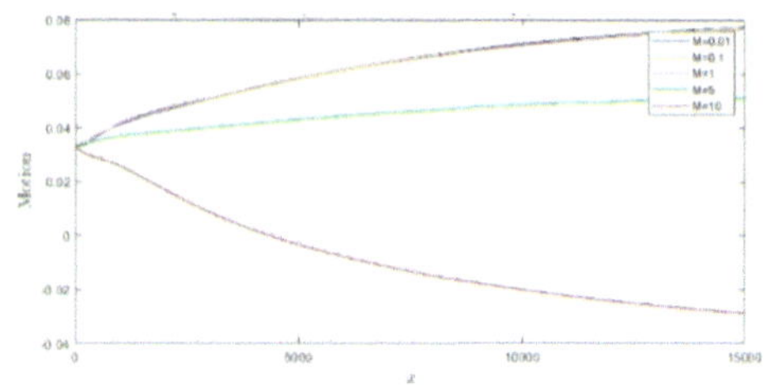

Fig. 15. Hartmann parameter related velocity profile variations.

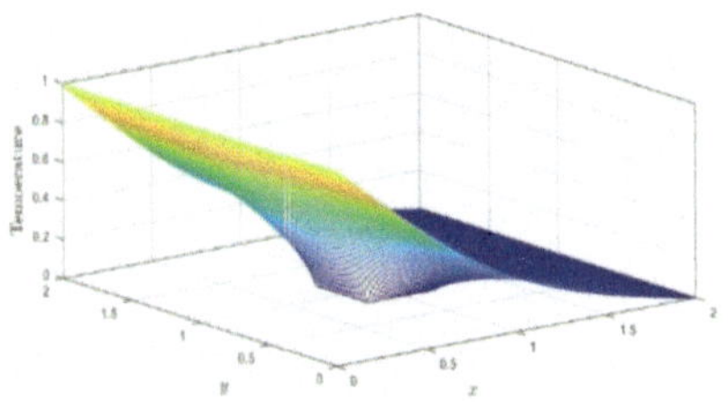

Fig. 16. Temperature profile when Hartmann number M = 10.

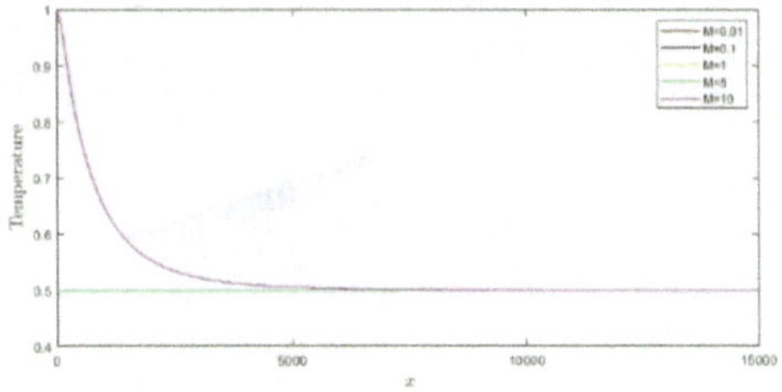

Fig. 17. Temperature profile due to change of Hartmann parameter.

5.5 Effect of Radiation Parameter *Rd*

As the radiation parameter *Rd* increases, the temperature and velocity profiles rise, as shown in Figs. 18, 19, 20 and 21.

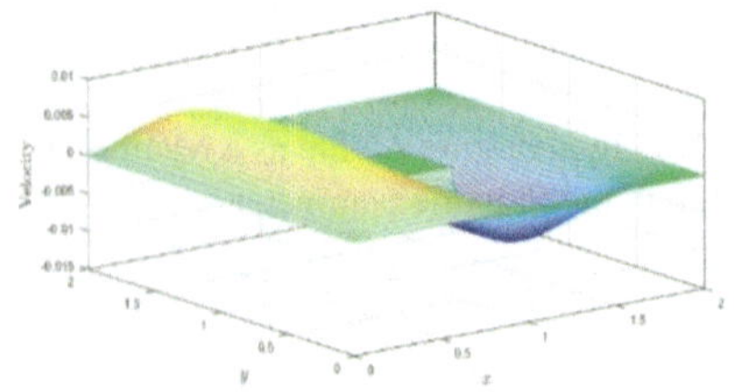

Fig. 18. Velocity distribution when Radiation parameter Ra = 10.

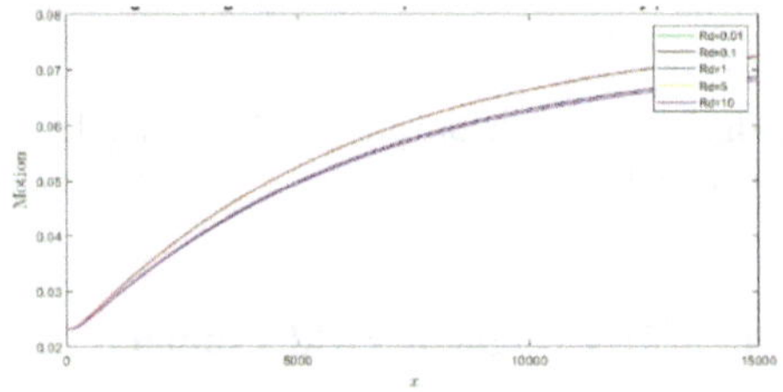

Fig. 19. Changes in the Radiation parameter affect the velocity profile.

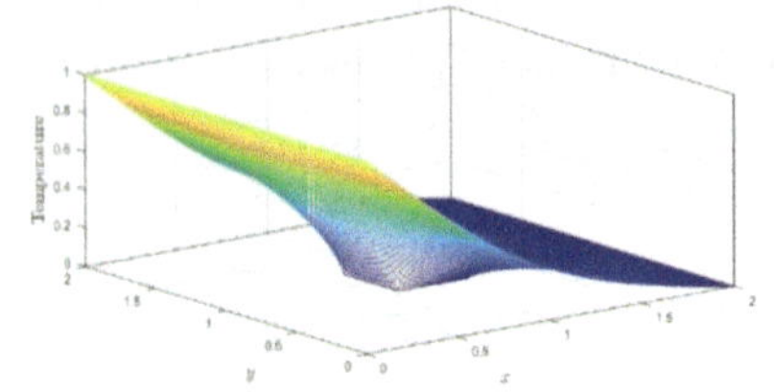

Fig. 20. Temperature profile when Radiation parameter Ra = 10.

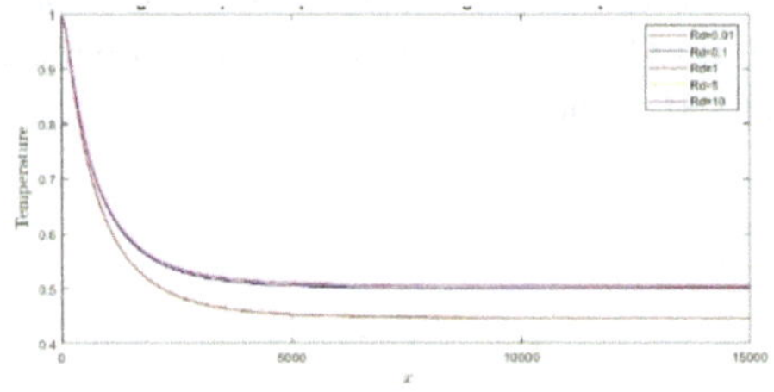

Fig. 21. Temperature profile due to change of Radiation parameter.

6 Conclusions

The research purpose is to focus on the thermal radiation's effect on the unsteady natural convection of fluid in a differentially heated cavity with solid block inserts under the influence of a magnetic field.

1. When the Darcy parameter *Da* is increased, so does the fluid velocity. While the temperature field decreases when the Darcy parameter *Da* is increased.
2. As the Rayleigh number *Ra* increases, so do the temperature and velocity fields.

3. As the Prandtl parameter value Pr increases, the velocity of the flow field decreases. While the temperature profiles rise, the Pr values fall.
4. As the magnetic field value M increases, the velocity profile of the flow field decreases. But the temperature profiles rise as magnetic field M values increase.
5. When the radiation parameter Rd is increases, both the temperature and velocity profiles rise.

Acknowledgements. The author expresses sincere gratitude to the "College of Education for Pure Sciences at the University of Mosul, Iraq" for their invaluable support and for giving him the chance to conduct this research.

References

1. Khader, M.M., Megahed, A.M.: Approximate Solutions for the Flow and Heat Transfer due to a Stretching Sheet Embedded in a Porous Medium with Variable Thickness, Variable Thermal Conductivity and Thermal Radiation using Laguerre Collocation Method. Appl. Appl. Math. An Int. J. **10**(2), 817–834 (2015)
2. Ibrahim, W., Makinde, O.D.: Magnetohydrodynamic stagnation point flow and heat transfer of casson nanofluid past a stretching sheet with slip and convective boundary condition. J. Aeros. Eng. **29**(2) (2015). https://doi.org/10.1061/(ASCE)AS.1943-5525.0000529
3. Krishnamurthy, M.R., Gireesha, B.J., Prasannakumara, B.C., Gorla, R.S.R.: Thermal radiation and chemical reaction effects on boundary layer slip flow and melting heat transfer of nanofluid induced by a nonlinear stretching sheet. Nonlinear Engineering (2016). https://doi.org/10.1515/nleng-2016-0013
4. Ghosh, S., Mukhopadhyay, S.: MHD slip flow and heat transfer of Casson nanofluid over an exponentially stretching permeable sheet. Int. J. Automot. Mechan. Eng. **14**(4), 4785–4804 (2017). https://doi.org/10.15282/ijame.14.4.2017.14.0375
5. Manjunatha, S., Ammani Kuttan, B., Jayanthi, S., Chamkha, A., Gireesha, B.J.: Heat transfer enhancement in the boundary layer flow of hybrid nanofluids due to variable viscosity and natural convection. Heliyon **5**(4), e01469 (2019). https://doi.org/10.1016/j.heliyon.2019.e01469
6. Bahiraei, M., Monavari, A., Moayedi, H.: Second law assessment of nanofluid flow in a channel fitted with conical ribs for utilization in solar thermal applications: effect of nanoparticle shape. Int. J. Heat Mass Transf. **151**, 119387 (2020). https://doi.org/10.1016/j.ijheatmasstransfer.2020.119387
7. Tlili, I., Nabwey, H.A., Girinath Reddy, M., Sandeep, N., Pasupula, M.: Effect of resistive heating on incessantly poignant thin needle in magnetohydrodynamic Sakiadis hybrid nanofluid. Ain Shams Eng. J. **12**(1), 1025–1032 (2021). https://doi.org/10.1016/j.asej.2020.09.009
8. Akbar, H.M., Mohammed, A.S., Ezzat, S.B.: Hybrid nanofluid to improve heat transfer and pressure drop through horizontal tube. Materials Today: Proceedings **42**, 1885–1888 (2021). https://doi.org/10.1016/j.matpr.2020.12.227
9. Hammodat, A., et al.: The effects of electrical conductivity on fluid flow between two parallel plates in a porous medium. Iraqi J. Sci. **113**, 4953–4963 (2021)
10. Sayyedah, A., et al.: Effect of thermal radiation on natural convection of a Nanofluid in a square cavity with a solid body. Therm. Sci. **25**(3A), 1949–1961 (2021)
11. Mebarek-Oudina, F., Chabani, I.: Review on nano-fluids applications and heat transfer enhancement techniques in different enclosures. Journal of Nanofluids. **11**, 155–168 (2022)

12. Algwauish, G., et al.: A Numerical Simulation of Convection and Conduction Heat Transfer for a Fluid in a Porous Medium. European Journal OF Pure and Applied Mathematics **16**(2), 833–846 (2023)
13. Alsabery, A.I., et al.: Natural Convection Flow of a Nanofluid in an Inclined Square Enclosure Partially Filled with a Porous Medium Scientific Reports **7**, 2357 (2017). https://doi.org/10.1038/s41598-017-02241-x
14. Rajput, R.K.: A Text Book of Fluid Mechanics and Hydraulic Machines in SI Units, S, Chand and Company Ltd, An Iso 9001:2008 Company (2008)
15. Lyengar, S.R.K., Jain, R.k.: Numerical Methods Copyright © 2009, New Age International (P) Ltd., Publishers Published by New Age International (P) Ltd., Publishers (2009)
16. Abdulsahib, A.A., Fadhel, F.S., Eidi, J.H.: Approximate solution of linear fuzzy random ordinary differential equations using laplace variational iteration method. Iraqi Journal of Science, 804–817 (2024)

Exam Grades Prediction Mechanism (EGpM) Using Machine Learning Algorithms

Basheer Riskhan[1(✉)], Noor Abdikhalak Hajj Noor[1], Hafiz Jibril[1], Hussein Waliyju[1], and Rahul Kumar[2,3]

[1] School of Computing and Informatics, Albukhary International University, Alor Setar, Kedah, Malaysia
b.riskhan@aiu.edu.my, {Noor.hajj,hafiz.jibril, waliyu.hussein}@student.aiu.edu.my

[2] University Centre of Research and Development, Chandigarh University, Mohali 140413, India

[3] Division of Research and Development, Lovely Professional University, Phagwara 144411, India

Abstract. Academic success is a goal for everyone in the academic community. However, this success can be hindered by various factors. Determining these factors is therefore crucial so that they can be dealt with beforehand and consequently come up with steps and solutions to deal with them. Artificial Intelligence has recently played a significant role in the academic field. Through what is now referred to as Educational Mining and Learning Analysis, big data from around the world are being put through machine learning algorithms to provide insights on achieving academic success. This study analyzes how a student's environmental factors affect their scores. We use a dataset that includes 30000+ data with various independent variables, including gender, parent's marital status, and study hours, i.e., to determine students' scores in three courses: Reading, Writing, and Math. The research uses Logistic Regression, Artificial Neural Networks (ANN), Naïve Bayes, and Decision Tree algorithms as machine learning models named EGpM. The study showed that of the four models, Logistic Regression and Naïve Bayes were the most accurate, with 73.89% and 74.38%, respectively. ANN wasn't far off with 73.63% accuracy, while Decision Tree was the least accurate with 64.12%.

Keywords: Educational Data Mining · Environmental Factors · Learning Analysis · Confusion Matrix · Logistic Regression · Artificial Neural Networks (ANN) · Naïve Bayes · Data Mining and Decision Tree

1 Introduction

Academic achievement has a significant impact on a student's motivation, persistence, and sense of self in higher education [1]. The academic performance which is poor or have high failure rates usually result in unacceptable levels of attrition and reduced graduate throughput while increased cost of education [1]. This means that it is essential for teachers to understand the student's way of life and find a strategy that would help

S. Pal and Á. Rocha (Eds.): ICMMCS 2025, LNNS 1398, pp. 335–346, 2025.
https://doi.org/10.1007/978-3-031-90998-6_31

bring out the best in the student. According to attribution theory, students must have a sense of control over how their academic assignments turn out [2]. This way, they will be more motivated to complete a task successfully [2]. In addition, they can work with teachers to get all the help they need to work on their weaknesses. Students tend to understand the reason for a specific outcome when they have this control. For example, suppose a student doesn't study well and subsequently fails their exams. In that case, they can attribute that factor to their poor performance and become motivated to study even more to pass their examinations. Tackling this would help students better their communities. Underprivileged students with no access to studying materials could get aid from government policies such as mobile platforms to facilitate knowledge sharing and collaborations [3] or also use new technologies such as virtual machines, which have increased performance for online learning [4] to better cater for students who are in remote areas. Other measures, such as speech-to-text language recognition systems, as discussed by [5], can also be used by tutors to help students affected by language barriers.

Academic failure can be attributed to factors that are both internal and external. Internal factors may be caused by the students, e.g., not studying well or skipping classes. In contrast, external factors may be caused by the immediate surroundings of the students, e.g., problems at home or difficult examinations. If educators become alert enough to link a student's failure in an examination to either an internal or external factor, they might be able to reduce the examination failure rate. Outcomes in education are crucial in indicating an individual's progress, thus influencing their future career opportunities and economic growth. Recently, there has been a need to use machine learning techniques to analyze educational data. This is what is referred to as Educational Data Mining (EDM). This is because these techniques have the potential to unearth patterns and insights that traditional methods may be blind to.

The process of extracting or "mining" knowledge from massive volumes of data is referred to as data mining, or knowledge discovery in databases [6]. EDM focuses on creating techniques for examining data from educational systems in order to find patterns in huge datasets that would be extremely challenging or impossible to examine otherwise because of the enormous amount of data they contain [7]. As of right now, computer science rather than education forms the basis of most data mining activity [8]. Due to the convenience of acquiring student log data and carrying out automatic data analysis, data mining techniques have been predominantly used in e-learning and web-based research within the academic area [8]. Yes, there is. To comprehend students' interactions with technology and the challenges involved in studying how kids with a variety of needs and cognitive traits operate with technology in these contexts, research on the application of EDM in actual classrooms is necessary [8].

The primary thesis of this study is that by using machine learning techniques, we can identify and quantify the most critical environmental factors affecting academic performance, thus providing insights for educators.

Having the right research questions is also crucial in ascertaining what this study tackles, and accordingly, this work answers the following questions:

- What are the main environmental factors affecting student exam scores?

- How can machine learning techniques analyze the relationship between a student's environment and exam scores?
- How can machine learning techniques analyze the relationship between a student's environment and exam scores?
- How do different environmental factors, or in this case variables, interact to impact a student's academic performance?

This study aims to identify the most common environmental factors affecting exam scores. By doing so, we can see how different factors affect student scores. The study also seeks to compare the performance of various machine learning algorithms in predicting exam scores. This could help see which algorithm provides the most accurate prediction. Finally, insights will be provided by the study based on the expected results to help those in the education sector make informed decisions.

1.1 Significance of Research

To develop an effective strategic mechanism (EGpM) in education, which is essential and helpful for the student's early update and warning to understand about the fourth-appearing exam and their grades, it will also be beneficial to students to be early aware of their grades and also alarm them to more concentrate towards their studies. One main factor that has impacted the students most is socio-economic status. Recent research has unearthed a correlation between academic success and SES [9]. The study used logistic regression to predict student achievement on socio-economic factors such as parental income and parental education status. This research helped pre-identify the students who are on the borderline and can be at risk.

1.2 Scope

Machine Learning techniques can identify patterns in datasets from education, potentially delivering informed decisions that enhance student performance. This research shows how different algorithms effectively analyze educational data and give policymakers and educators data-driven decisions that would help in proactive solutions. For policymakers, these decisions provide information that would help create informed policies that support students, specifically those from humble backgrounds, through various initiatives that would eventually help create a healthy student studying environment.

The main environmental factors to be used in the decision-making will be parental education, parental marital status, student's study plan, gender, and socioeconomic status. The research excludes factors such as motivation, cognitive abilities, and learning styles. The goal is to deliver insights that policymakers and educators could put into action.

1.3 Limitations

Using machine learning techniques provides a comprehensive approach to providing insights into educational data, but this approach also has several limitations. Firstly, the model uses historical data that was set, and considering how factors change, the study fails to give insights into current factors, hence introducing bias. Secondly, the study focuses solely on environmental factors and purposely excludes other vital factors, such as cognitive abilities and previous exam scores, which could be crucial in analyzing the scores. Additionally, the data provided may suffer from inaccurate or incomplete details. Lastly, despite their predictive power, machine learning models lack interpretability, and hence, it becomes very difficult to draw concise actions from the results provided.

2 Literature Review

Artificial intelligence techniques have recently been given more attention due to their ability to analyze the ever-growing amount of data available worldwide. Research shows increased evidence that artificial intelligence could benefit many fields, especially education [10]. According to [11], data mining can be used in the field of education to focus on identifying, extracting, and evaluating characteristics related to students' learning processes in order to better understand the learning process. [12] This book talks about data mining software that allows users to assess, summarize, and classify relationships discovered when mining data from different domains.

A study by [13] on student performance gathered 600 students from different colleges. The study used classification on category, language, and background qualification to predict students' scores [13]. Another study by [14] on students of sample size 300 (225 males, 75 females) was used to analyze student performance based on student attendance in class.

[15] divided the 400 upper school students of Aligarh Muslim University in India equally between the sexes for a performance study. The aim of the study was to determine the predictive usefulness of several personality, cognitive, and demographic indicators for success in the science stream [15]. While [17] utilized the decision tree model to forecast the final grade of students who studied the programming language course, [16] used student data to examine their learning behavior. By going through the data from 108-s (about 2 min)-year BSIT students, the research utilized the J48 algorithm to identify the most influential factors affecting student performance [18]. Previous research relied on variables obtained during the course or semester, limiting the ability to intervene early with struggling students, especially those facing economic challenges [19]. Educational technologies and digital tools have been increasingly used to enhance learning experiences, offering innovative approaches such as ontology-based systems [20, 21] and gamification [22, 23].

3 Methodology

The prediction process EGpM for this study entailed four significant processes: Data Collection, Data Pre-Processing and Labelling, Machine Learning Algorithm, and Prediction Results (Fig. 1).

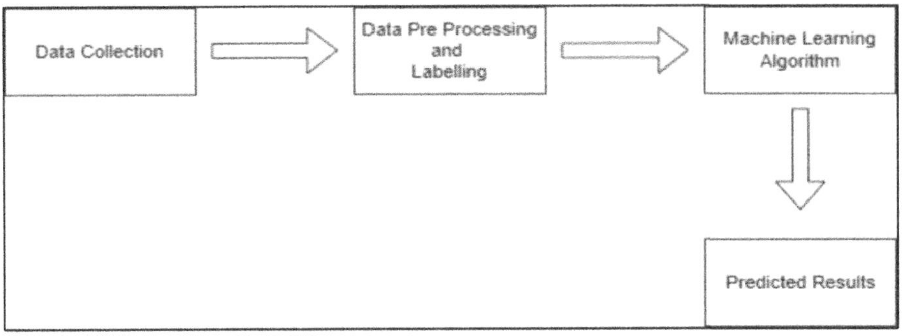

Fig. 1. EGpM Prediction Process Flow

3.1 Data Collection

During this process, we gathered data that we believed had sufficient variables to enable us to have the maximum number of factors that could help predetermine the results as needed. The data acquired had 30641 entries, and the variables include Gender, Ethnic Group, Parent's Education, Lunch type, Test Preparation, Parents' Marital Status, Involvement in Sports, Number of Siblings, Transport Means, Weakly Study Hours, and if the student is a first child or not. Whether some variables are relevant will be based on the variables chosen for the prediction process.

3.2 Data Preprocessing and Labelling

This process involves cleaning data by removing null values, comparing some data with the score variables, labeling it, and splitting it into the training and testing sets. All this is to ensure data consistency so that the results we get are as accurate as possible during the prediction process.

3.3 Machine Learning Algorithm Modelling

In this research, we chose four machine learning algorithms: Artificial Neural Network algorithm, Naive Bayes algorithm, Logistics Regression algorithm, and Decision Tree Algorithm. In the process, we introduced a confusion matrix to evaluate the performance of each algorithm. The components included:

- True Positives (T_P): The no. of correctly predicted as the positive class.
- True Negatives (T_n): The no. of correctly predicted as the negative class.
- False Positive (F_P): The no. of incorrectly predicted as the positive class
- False Negative (F_n): The no. of incorrectly predicted as the negative class.

The confusion matrix computed the accuracy, precision, recall, and F1 score.

3.3.1 Accuracy

This measures the proportion of correctly classified instances out of the total cases (Fig. 2).

$$\text{Accuracy} = \frac{TP + TN}{TP + TN + FP + FN}$$

Fig. 2. EGpM Accuracy Formula

3.3.2 Precision

Out of all instances predicted as positive, this measures the predicted proportion correctly (Fig. 3).

$$\text{Precision} = \frac{TP}{TP + FP}$$

Fig. 3. EGpM Precision Formula

3.3.3 Recall

This calculates the percentage of all positive events that were accurately predicted as positive (Fig. 4).

$$\text{Recall} = \frac{TP}{TP + FN}$$

Fig. 4. EGpM Recall Formula

3.3.4 F1 Score

This measures the means of precision and recall (Fig. 5).

$$F1\ \text{Score} = 2 \times \frac{\text{Precision} \times \text{Recall}}{\text{Precision} + \text{Recall}}$$

Fig. 5. EGpM Score Formula

4 Results

This study analyzed the impact of various environmental factors on student scores using several machine learning algorithms: Logistic Regression, Naive Bayes, Decision Tree, and ANN. The dataset included features such as gender, ethnic group, parental education, lunch type, test preparation, parental marital status, practice of sports, whether the student is the first child, number of siblings, means of transport to school, and weekly study hours. The target variable was the average score of students in Math, Reading, and Writing, which was binarized into pass/fail categories based on a threshold of 60.

Table 1. Comparison of the Three Models Based on Prediction Results

Algorithm	Accuracy	True Positives (TP)	True Negatives (TN)	False Positives (FP)	False Negatives (FN)	Comments
Logistic Regression	73.89%	2543	301	818	187	Balanced performance, high false positives
Naïve Bayes	74,38%	2513	350	769	217	Slightly higher accuracy, balanced position, and recall
Decision Tree	64.12%	1976	492	627	754	Lower accuracy means higher precision for the fail category
Artificial Neural Network (ANN)	73.63%	2480	354	765	250	Balanced model, strong generalization capabilities

4.1 Model Performance

Confusion matrices were created to offer a greater understanding of the performance with regard to true positives, true negatives, false positives, and false negatives. Accuracy was the major parameter used to evaluate each model. In Table 1, the final outcome is displayed.

4.1.1 Logistic Regression

Figure 6 shows the accuracy of 73.89% when using logistic regression.

Logistic Regression Accuracy: 0.7388932190179267
Logistic Regression Classification Report:

	precision	recall	f1-score	support
0	0.62	0.27	0.37	1119
1	0.76	0.93	0.84	2730
accuracy			0.74	3849
macro avg	0.69	0.60	0.60	3849
weighted avg	0.72	0.74	0.70	3849

Fig. 6. Results for Logistic Regression

Confusion Matrix:

- True Positives (T_p): 2543
- True Negatives (T_n): 301
- False Positives (F_p): 818
- False Negatives (F_n): 187

Logistic Regression indicated balanced performance with accuracy. The confusion matrix showed that the model effectively predicted both pass and fail categories but placed several failing students in the passing category.

4.1.2 Naïve Bayes Algorithm

This model performed similarly to Logistic Regression but showed slightly higher accuracy. It also placed many failing students in the passing category but had a balanced precision and recall for both classes.

Naive Bayes Accuracy: 0.7438295661210704
Naive Bayes Classification Report:

	precision	recall	f1-score	support
0	0.62	0.31	0.42	1119
1	0.77	0.92	0.84	2730
accuracy			0.74	3849
macro avg	0.69	0.62	0.63	3849
weighted avg	0.72	0.74	0.71	3849

Fig. 7. Results for Naïve Bayes

Figure 7 shows the accuracy of 74.38% when using the Naive Bayes Algorithm.

Confusion Matrix:

- True Positives (T_p): 2513
- True Negatives (T_n): 350
- False Positives (F_p): 769
- False Negatives (F_n): 217

4.1.3 Decision Tree

Decision Tree Accuracy: 0.6412055079241361
Decision Tree Classification Report:

	precision	recall	f1-score	support
0	0.39	0.44	0.42	1119
1	0.76	0.72	0.74	2730
accuracy			0.64	3849
macro avg	0.58	0.58	0.58	3849
weighted avg	0.65	0.64	0.65	3849

Fig. 8. Results for Decision Tree

Figure 8 shows the accuracy of 64.12% when using the Decision Tree algorithm.
Confusion Matrix:

- True Positives (T_p): 1976
- True Negatives (T_n): 492
- False Positives (F_p): 627
- False Negatives (T_n): 754

The Decision Tree model showed lower accuracy than Logistic Regression and Naive Bayes. It showed higher precision for the fail category but had a significant number of false negatives, indicating a tendency to misclassify passing students as failing.

4.1.4 Artificial Neural Network

Figure 9 shows the accuracy of 76.63% when using the Artificial Neural network.
Confusion Matrix:

- True Positives (T_p): 2480
- True Negatives (T_n): 354
- False Positives (F_p): 765
- False Negatives (F_n): 250

The ANN model showed accuracy comparable to logistic regression and naive Bayes. It demonstrated a good balance between precision and recall, with fewer false positives and false negatives, indicating a well-balanced model with strong generalization capabilities.

```
Epoch 1/10
1540/1540 ━━━━━━━━━━━━━━━━━━━━ 2s 821us/step - accuracy: 0.72
04 - loss: 0.5568
Epoch 2/10
1540/1540 ━━━━━━━━━━━━━━━━━━━━ 1s 803us/step - accuracy: 0.73
95 - loss: 0.5297
Epoch 3/10
1540/1540 ━━━━━━━━━━━━━━━━━━━━ 1s 852us/step - accuracy: 0.74
39 - loss: 0.5289
Epoch 4/10
1540/1540 ━━━━━━━━━━━━━━━━━━━━ 1s 949us/step - accuracy: 0.74
76 - loss: 0.5195
Epoch 5/10
1540/1540 ━━━━━━━━━━━━━━━━━━━━ 1s 796us/step - accuracy: 0.74
94 - loss: 0.5190
Epoch 6/10
1540/1540 ━━━━━━━━━━━━━━━━━━━━ 1s 844us/step - accuracy: 0.74
Epoch 7/10
1540/1540 ━━━━━━━━━━━━━━━━━━━━ 1s 817us/step - accuracy: 0.75
09 - loss: 0.5145
Epoch 8/10
1540/1540 ━━━━━━━━━━━━━━━━━━━━ 1s 786us/step - accuracy: 0.75
27 - loss: 0.5130
Epoch 9/10
1540/1540 ━━━━━━━━━━━━━━━━━━━━ 1s 768us/step - accuracy: 0.75
11 - loss: 0.5111
Epoch 10/10
1540/1540 ━━━━━━━━━━━━━━━━━━━━ 1s 787us/step - accuracy: 0.74
64 - loss: 0.5218
121/121 ━━━━━━━━━━━━━━━━━━━━ 0s 675us/step - accuracy: 0.7302
- loss: 0.5425
ANN Accuracy: 0.7362951636314392
```

Fig. 9. Results for Artificial Neural Network

5 Discussion

The results show that environmental factors serve a significant role in predicting student performance. Among the models used, Logistic Regression and Naive Bayes showed higher accuracy, suggesting that these models effectively capture the relationship between the independent and dependent variables. The Decision Tree model showed lower accuracy among the machine learning models used.

Despite being complex, The ANN did not substantially outperform the simpler models (Logistic Regression and Naive Bayes) in terms of accuracy. However, its performance was well-balanced with lower false positive and negative rates, which denotes strong generalization capabilities. All models tended to misclassify failing students as passing based on confusion matrix analysis based solely on environmental factors.

6 Conclusion and Future Works

This study significantly analyzed the proposed EGpM influence of environmental factors on student exam scores using advanced machine learning techniques, including logistic regression, artificial neural network (ANN), Naïve Bayes, and decision tree algorithms). Logistic regression and Naive Bayes models demonstrate high accuracy rates, highlighting their effectiveness in capturing the relationship between environmental factors and student performance. The study highlighted the critical impact of socioeconomic status, parental involvement, and school environment on academic success, providing valuable insights for educators and policymakers to develop targeted interventions. While admitting limitations and the need for advanced research, this study underscores the importance of exploiting machine learning to inform evidence-based educational strategies and policies, ultimately enhancing educational outcomes and supporting student success effectively.

References

1. Jayanthi, S.V., Balakrishnan, S., Ching, A.L.S., Latiff, N.A.A., Nasirudeen, A.M.A.: Factors contributing to academic performance of students in a tertiary institution in Singapore. Am. J. Educ. Res. **2**(9), 752–758 (2014)
2. Lim, H.-Y.: Effects of attributions and task values on foreign language use anxiety. J. Educ. Hum. Dev. **1**(2), 1–20 (2007)
3. Almassri, S., Hussain, K., Riskhan, B. : Study of mobile E-communities effectiveness for society and social networking development (2023). https://doi.org/10.13140/RG.2.2.32900.30081
4. Basheer Riskhan, R.M.: Virtual Machine Performance Approaches in the Online Education System. In: Proceedings of the International MultiConference of Engineers and Computer Scientists (2016)
5. Abd Jalil, H.B., Phyo, P.S., Sheikh, M.A.U., Basheer, R.: Speech-to-Text Conventional Myanmar (Burmese) Language Recognition System (2024)
6. Baradwaj, B.K., Pal, S.: Mining educational data to analyze students' performance. arXiv preprint arXiv:1201.3417 (2012)
7. Romero, C., Ventura, S.: Data mining in education. Wiley Interdiscip Rev Data Min Knowl Discov **3**(1), 12–27 (2013)
8. Angeli, C., Howard, S.K., Ma, J., Yang, J., Kirschner, P.A.: Data mining in educational technology classroom research: Can it make a contribution? Comput. Educ. **113**, 226–242 (2017)
9. Sirin, S.R.: Socioeconomic status and academic achievement: a meta-analytic review of research. Rev. Educ. Res. **75**(3), 417–453 (2005)
10. Hattie, J.: Visible learning: A synthesis of over 800 meta-analyses relating to achievement. Routledge (2008)
11. El-Halees, A.: Mining students data to analyze e-Learning behavior: A Case Study. Citeseer (2009)
12. Hsu, J.: Critical and future trends in data mining: a review of key data mining technologies/applications. Data mining: Opportunities and challenges, pp. 437–452 (2003)
13. Pandey, U.K., Pal, S.: Data Mining: A prediction of performer or underperformer using classification. arXiv preprint arXiv:1104.4163 (2011)
14. Naqvi, S.: Factors affecting students' performance a case of private colleges. Bangladesh e-Journal of Sociology 3 (2006)
15. Khan, Z.N.: Scholastic achievement of higher secondary students in science stream. Online Submission **1**(2), 84–87 (2005)
16. Ben-Zadok, G., Hershkovitz, A., Mintz, E., Nachmias, R.: Examining online learning processes based on log files analysis: a case study. In: 5th International Conference on Multimedia and ICT in Education (m-ICTE'09) (2009)
17. Al-Radaideh, Q.A., Al-Shawakfa, E.M., Al-Najjar, M.I.: Mining student data using decision trees. In: International Arab Conference on Information Technology (ACIT'2006). Yarmouk University, Jordan (2006)
18. Apolinar-Gotardo, M.: Using decision tree algorithm to predict student performance. Indian J. Sci. Technol. **12**(5) (2019)

19. Huynh-Cam, T.-T., Chen, L.-S., Le, H.: Using decision trees and random forest algorithms to predict and determine factors contributing to first-year university students' learning performance. Algorithms **14**(11), 318 (2021)
20. Majid, M., et al.: Ontology-Based System for Educational Program Counseling. Intelligent Automation & Soft Computing **30**(1) (2021)
21. Herani, G.M., Rajar, A.W., Zaman, N., Alam, A.: Knowledge Transformation and Economic Development: The Role of Digital Technology-An Analysis (2007)
22. Alamri, M.Z., Jhanjhi, N.Z., Humayun, M.: Digital curriculum importance for new era education. In: Employing recent technologies for improved digital governance, pp. 1–18. IGI Global (2020)
23. Alsubaie, A., Alaithan, M., Boubaid, M., Zaman, N.: Making learning fun: Educational concepts & logics through game. In: 2018 20th International Conference on Advanced Communication Technology (ICACT), pp. 454–459. IEEE (2018)

Mathematical Modeling Approximation Methods Accuracy

Aws Asaad Hamdi[1(✉)], Samer R. Yaseen[2], and Ghassan E. Arif[2]

[1] General Directorate of Education of Salahaldin-Tikrit University, Tikrit, Iraq
aws7331@gmail.com

[2] College of Education for Pure Sciences, Tikrit University, Tikrit, Iraq
{amer2017,ghasanarif}@tu.edu.iq

Abstract. Numerical approximation methods and their applications play an important role in many scientific and engineering fields. It enables us to solve mathematical equations that cannot be solved analytically, understand complex systems, and predict them. After the great development in mathematical modeling and programming, this role developed and grew. Since there are many vital applications, whether in engineering, physics, chemistry or economics, approximation methods combined with mathematical modeling have enabled us to solve complex problems that cannot be solved using only one method. Better understand and analyze complex systems, and solve complex problems. It also enables us to provide more accurate forecasts. The lenticular approximation method is based on dividing the problem into a series of small steps and then solving each step using simple analytical methods. Lenticular approximation methods can be divided into several types, including linear and nonlinear methods. In this study, this study aims to introduce the reader to numerical approximation methods in mathematical modeling. This study will discuss the types of numerical approximation methods and their applications in mathematical modeling. We studied a mathematical model and chose the appropriate method of approximation, then calculated the error rate and approximation error and verified the approximation. The results indicated that using numerical approximation methods and mathematical modeling together is a powerful tool that can be used to solve many complex problems and contributed to reducing the error rate by a very large percentage and reducing the standard deviation of the mathematical model.

Keywords: Numerical approximation · mathematical modeling · error rate · standard deviation

1 Introduction

In real-world applications, many equations are too complex to be solved analytically. Numerical approximation methods address this by providing practical solutions for problems in fields as diverse as climatology, where they help model weather patterns, and biomedicine, where they assist in simulating biological processes.

S. Pal and Á. Rocha (Eds.): ICMMCS 2025, LNNS 1398, pp. 347–357, 2025.
https://doi.org/10.1007/978-3-031-90998-6_32

Mathematical modeling is the manner of representing a real or digital gadget with a mathematical model. This model is used to recognize, examine, and make predictions approximately the machine. In many cases, the mathematical equations describing the actual or hypothetical system can't be solved analytically. In these instances, numerical approximation strategies can be used to resolve those equations Numerical approximation techniques are a collection of methods used to approximate the answer of mathematical equations or different mathematical troubles. These techniques rely upon dividing a problem into a chain of small steps after which fixing every step using easy analytical techniques (Cohen, H. 2011) [1].

There are many varieties of lenticular approximation techniques, which include Linear approximation strategies: These techniques depend upon dividing the trouble into a sequence of small steps after which solving each step the use of linear equations. Approximation techniques the use of differential equations: These strategies depend on changing the trouble right into a differential equation and then solving this equation the use of numerical approximation methods. The significance of numerical approximation methods is because of them, they permit us to clear up mathematical equations that can't be solved analytically. It enables us to resolve mathematical equations with big or complicated numbers. It allows us to solve mathematical equations that require excessive accuracy (Jazar, R. N. 2020) [2].

Numerical approximation methods are widely used in many scientific and engineering fields, such as physics, chemistry, mechanical engineering, electrical engineering, and civil engineering. Here are some common applications of numerical approximation methods in mathematical modeling: Solving differential equations: numerical approximation methods are used to solve differential equations that describe Many natural phenomena, such as movement, heat, and diffusion. Solving integral equations: numerical approximation methods are used to solve integral equations that describe many natural phenomena, such as thermodynamics and electromagnetism (Abdel Radi, A. R. and. Marwa, E. A. 2023) [3].

Calculus: numerical approximation methods are used to calculate the calculus of complex mathematical function (Simpson, T s, et, al,2884) [4] (Epperson, J. F. 2021). [5]. Solving other mathematical problems: numerical approximation methods are used to solve many other mathematical problems, such as opti numerical approximation methods have several advantages that make them a powerful tool in mathematical modeling. However, it is important to understand the disadvantages of these methods to avoid error. Among the most common mistakes are not choosing the appropriate method for proportional variables, also not evaluating the results, relying on the human element only without resorting to computers, and despite the advantages of approximation methods, which are flexibility and accuracy (Fabio, Z. & Maria, S. & Gianluigi, R. 2024) [6]. Solving complex problems, and the possibility of using them to solve a wide range of mathematical problems; However, it has some disadvantages, including that these methods are inaccurate if not used correctly, somewhat slow, sometimes complicated, and sometimes complex. Mization problems and dynamic systems theory (Jazar, R. N. 2022) [2].

Studying numerical approximation methods and mathematical modeling Some problems: A high level of mathematics is required. Studying lenticular approximation methods and mathematical modeling requires a good understanding of basic mathematics rules, such as calculus and the theory of differential equations. The complex nature of materials numerical approximation and mathematical modeling methods involve a wide range of complex methods, which can be difficult to understand and apply (Meinardus, G. (2012) [7]. The need for powerful computers: Some numerical approximation methods require the use of powerful computers to process complex data. This study aims to introduce the reader to numerical approximation methods in mathematical modeling. In this study, the types of numerical approximation methods and their applications in mathematical modeling will be discussed. This study is important because it shows how numerical approximation methods and mathematical modeling play an important role in many scientific and engineering fields. Therefore, studying these topics is important for students and researchers in these fields.

This paper's goal is to give an in-depth review of numerical approximation techniques while examining their preciseness and applications in a range of engineering and scientific domains. In light of particular application specifications, accuracy standards, and computing limitations, the study seeks to assist practitioners in choosing the best approaches. By use of error analysis and validation by experimentation, this research aims to improve knowledge of numerical approximation methods and their usefulness in resolving challenging issues.

The methodology for studying approximation and mathematical modeling is a mixture of multiple methodologies, where the scientific methodology is used side by side with the descriptive methodology in addition to the analysis and comparison methodology, by studying a mathematical model, choosing the appropriate method for approximation, and then calculating the error rate, the approximate error, and checking the approximation (Bouhamidi, A., & Jbilou, K. 2008) [8].

The paper introduces a new evaluation framework that examines error sources more comprehensively, including discretization and approximation errors. This framework allows for a deeper understanding of each method's limitations and strengths.

2 Related Works

Recent trends in the study of numerical approximation literature and its application in mathematical modeling in the following areas: development of new methods for approximate approximation of numerical, where new methods for approximate approximation of numerical are developed. Application of numerical approximation methods in new fields. Numerical approximation methods are applied in new fields, such as artificial intelligence and smart learning. Study of the effect of numerical convergence on mathematical modeling: the effect of numerical convergence on mathematical modeling is studied to understand how the effect of numerical convergence affects the modeling results.

In a study by (B Adcock, D Huybrechs Siam Review - 2019) [9], offer an overview of the concept of frames and the way they may be used as a framework for numerical approximation tactics. They pointed out the challenges that may be faced even as the usage of numerical approximation techniques in mathematical modeling of its various kinds: linear, non-coverage, differential, and integral, with the aid of reviewing applications of numerical approximation using superior principles of frameworks. The study affords indications of destiny guidelines and the significance of this approach in regions which includes sign processing and applied arithmetic.

In a study by (Seyeon, l. & Hyunju, K., & Bongsoo, J 2024) [10]. Which is a presentation of an approximate approach for solving Fractional Differential Equations, which represent a sort of differential equation that consists of derivatives of an incorrect order. The proposed approach is primarily based on making use of the numerical algorithm for approximation to those equations. The proposed method turned into used to calculate approximations of the answers to a number of discretization differential equations, which lets in for managing issues in numerous fields, consisting of signal processing. The effectiveness of the proposed approach was verified by applying it to numerical examples. And examine it with other methods used in solving partition differential equations. The outcomes show that the proposed method gives a powerful and powerful answer for this magnificence of equations. The take a look at has supplied a powerful numerical tool for fixing partition differential equations, which contributes to statistics technological know-how and sign processing packages, which means that the importance of approximation strategies in solving issues and the opportunity of finding and providing effective solutions (A. Quarteroni, et, al,2021) [11].

In a study by, (Steinbach, O. (2023) [12], she mentioned the use of numerical approximation methods in mathematical imaging of the human cardiovascular machine. The study used numerical approximation strategies to effectively represent the complex physical phenomena associated with the cardiovascular machine. A comprehensive overview is supplied on how to investigate and interpret medical records the usage of mathematical equipment. The results indicate that employing these mathematical models in various medical applications complements the information and medical interaction between mathematical modeling and reasonable medical software and that numerical approximation has contributed to the medical interaction of mathematical modeling.

In another study) Dzyadyk, V. K. 2018) [13]. The author provides a complete review of numerical approximation techniques used in solving boundary cost problems related to differential equations with skewed graphs. The study focused on unique approximation techniques based on finite factors and boundary elements. The study addressed the mathematical concepts of boundary graph problems, then provided details on how to estimate these problems using numerical approximation techniques, with special awareness of finite element-based methods. And boundary elements. It provides a discussion of the advantages and challenging situations of using these techniques, illustrating the applied contexts in which these techniques and finite element strategies can be implemented. It provides deep expertise in the techniques used in fixing boundary graph problems for differential equations and the importance of numerical approximation strategies in facilitating the solution of differential equations. In a look at that centered on mathematical modeling, mathematical evaluation, and numerical approximation of second-order Elliptical troubles that consist of internal elements, the author (Köppl, T., et, al,2018) [14], provided mathematical fashions for troubles with graphical deviation and analyzed them mathematically to recognize the principal houses of those equations. In addition, emphasis is placed on the way to approximate those hard equations numerically, whilst verifying the effectiveness of the strategies used. The research furnished, A comprehensive evaluation of the challenges that rise up whilst coping with skewed differential troubles and a way to reap an accurate numerical approximation to them. This paper is a valuable useful resource for researchers and those interested by applied mathematics and techniques for approximating differential equations.

3 Methodology

The methodology for studying approximation and mathematical modeling is a mixture of multiple methodologies, where the scientific methodology is used side by side with the descriptive methodology in addition to the analysis and comparison methodology, by studying a mathematical model, choosing the appropriate method for approximation, and then calculating the error rate, the approximate error, and checking the approximation. In this section, we will present a practical framework for the study, and the procedures that were taken, starting with data collection, passing through the formulation of the mathematical model, and ending with the results and their scientific analysis.

3.1 The Practical Framework

The applied framework of the study indicates the steps that will be followed during the study, starting from collecting data until presenting and analyzing the results. Care has been taken in choosing the appropriate approximation method to calculate the error rate in the mathematical model. See Fig. 1.

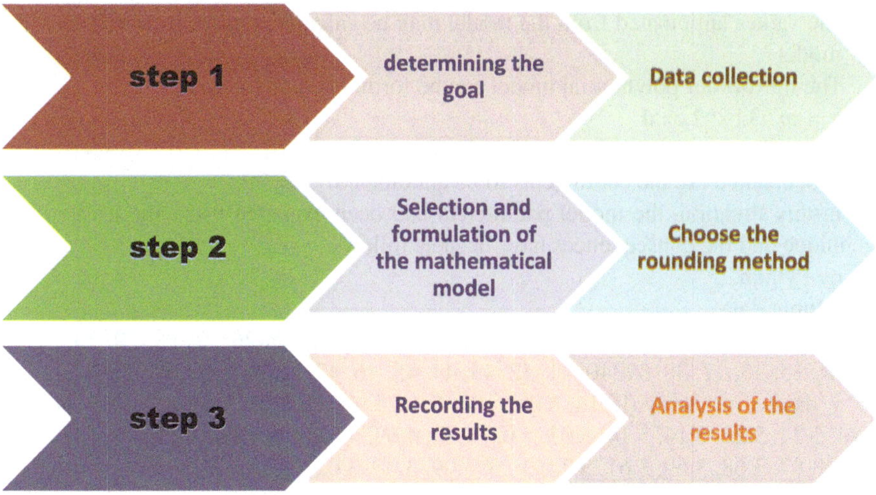

Fig. 1. The practical framework(author)

3.2 The Procedures

The degree of the polynomial determines how accurate the polynomial approximation is in maximum mathematical modelling approximation techniques. The accuracy of the approximation increases with the degree of the polynomial. The following are the tactics:

1. Determine the purpose, that is to remedy a mathematical model problem the usage of numerical approximation techniques, the mathematical model for a laboratory test.
2. Collect information from the supervisory committee, preceding studies, and the Internet
3. Choose the rounding technique

   Most of the strategies are in the main methods that depend upon first-degree linear approximation, so we can model the values mathematically via writing a 2nd-diploma polynomial as follows:

   Where:

   a, b, and c are polynomial coefficients.

   To decide the values of a, b, and c, we will use the subsequent spinoff equations:

   where:

   x0 is a given factor.

   Y0 is the factor cost at point x0.

   Y'zero is the price of the derivative at factor x0.
4. Determining the mathematical model and the usage of numerical approximation methods to calculate the mistake rate

A third-order polynomial version may be formulated as follows:

wherein:

are the coefficients to be anticipated.

Calculate the distinction of values The difference values among the genuine values and the values anticipated from the model may be calculated as follows: diff = y_data - y_model

The third-order polynomial model can be formulated as follows:

y = ax^3 bx^2 cx d

in which:

a, b, c, and d are the coefficients to be anticipated.

In this situation, the model parameters had been expected using the least squares technique and the consequences have been as follows.

by Python.

Input data.

x_data = np. Array ([8, 7, 9, 10, 20, 21, 22, 23, 24, 25, 26, 27, 28, 29, 30, 31, 32, 33, 34, 35, 36, 37, 38, 39, 40, 41, 42, 43, 44, 45, 46, 47, 48, 49, 50, 51, 52]).

y_data = np. Array ([6.14, 6.08, 6.06, 6.04, 5.93, 5.87, 5.77, 5.52, 5.39, 5.24, 5.24, 5.24, 5.24, 5.22, 5.14, 5.14, 5.03, 5.03, 4.82, 4.69, 4.53, 4.46, 4.4, 4.31, 4.3, 4.03, 4.03, 3.99, 3.87, 3.68, 3.69, 3.61, 3.74, 3.26, 3.09, 3.05, 3.02, 2.5]).

Fit a third-degree polynomial using least squares.

coefficients = np. Polyfit (x_data, y_data, 3).

Define the polynomial function.

polynomial = np. Poly1d (coefficients).

```
# Calculate model values at the given x values.
y_model = polynomial(x_data).
# Calculate the difference between the observed and predicted values.
diff = y_data - y_model.
# Plot the data and model.
plt. Plot (x_data, y_data, 'o', label = 'Experimental Data').
plt. Plot (x_data, y_model, '-', label = 'Model').
plt. Plot (x_data, diff, 'r', label = 'Difference').
plt. Legend ().
plt. Xlabel ('x').
plt. Ylabel ('y').
plt. Title ('Third-Degree Polynomial Model').
plt. Show ().
coefficients.
[−0.000002 0.000001 −0.000001 5.819000].
```

4 Result Analysis and Discussion

In this section, and according to the applied framework of the study, we will present the results (the fifth procedure) and we will analyze the results (the sixth procedure).

4.1 Results

The distinction of values between the proper values and the values expected from the version may be calculated as follows:

Table 1. Analysis of model values, expected values, and error

NO	MODEL VALUES	EXP.VALUES	ERROR
8	6.139	6.14	0.001
7	6.079	6.08	0.001
9	6.059	6.06	0.001
10	6.039	6.04	0.001
20	5.929	5.93	0.001
21	5.869	5.87	0.001
22	5.769	5.77	0.001
23	5.519	5.52	0.001
24	5.389	5.39	0.001
25	5.239	5.24	0.001
26	5.239	5.24	0.001

(continued)

Table 1. *(continued)*

NO	MODEL VALUES	EXP.VALUES	ERROR
27	5.239	5.24	0.001
28	5.239	5.24	0.001
29	5.229	5.22	0.009
30	5.139	5.18	0.041
31	5.139	5.14	0.001
32	5.029	5.14	0.111
33	5.029	5.03	0.001
34	4.819	5.03	0.211
35	4.689	4.82	0.131
36	4.529	4.69	0.161
37	4.459	4.53	0.071
38	4.399	4.47	0.071
39	4.309	4.4	0.091
40	4.299	4.33	0.031
41	4.029	4.3	0.271
42	4.029	4.03	0.001
43	4.029	3.99	0.039
44	3.989	3.87	0.119
45	3.869	3.68	0.189
46	3.679	3.64	0.039
47	3.689	3.61	0.079
48	3.609	3.47	0.139
49	3.739	3.26	0.479
50	3.259	3.09	0.169
51	3.08	3.05	0.03
52	2.52	2.5	0.02

The model standard deviation error was 0.001, indicating that the model provides an accurate estimate of the data:

See Fig. 2 which shows relation of error &The squares of the difference in values and the arithmetic mean

4.2 Discussion

One way to reduce discretization error is to decrease the step size, though this comes at the cost of increased computational demands. Alternatively, adaptive methods that adjust

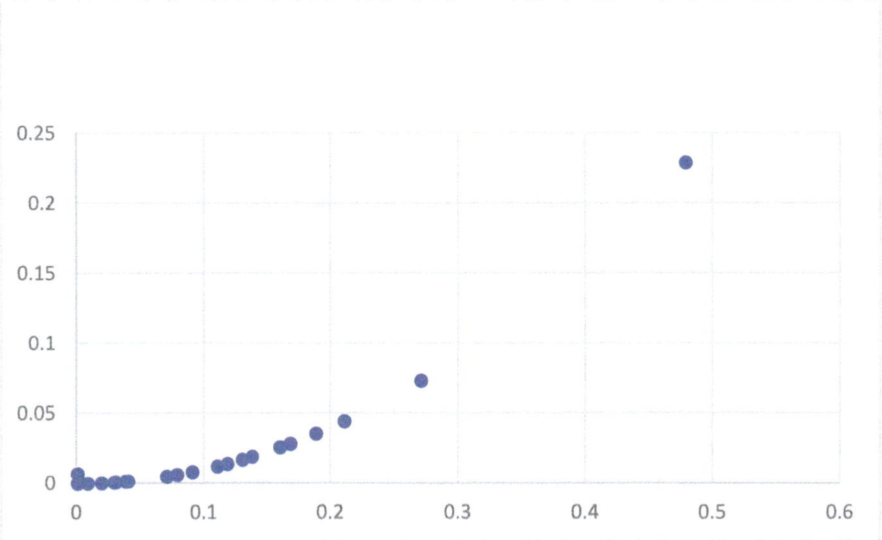

Fig. 2. Relation of error & The squares of the difference in values and the arithmetic mean

Table 2. Comparison of Numerical Approximation Methods Based on Practical Considerations

Method	Accuracy Level	Computational Complexity	Execution Speed	Ideal Applications
Midpoint	Moderate	Low	High	Quick estimations, real-time analysis
Trapezoid	High	Moderate	Moderate	High-precision models
Simpson's Rule	Very High	High	Slow	Detailed scientific calculations

the step size based on the rate of change in the function can provide a balance between accuracy and efficiency. For approximation error, choosing higher-order approximation methods may yield more precise results but could introduce complexity.

According to Table 1, we discover that the virtual approximation solved the trouble of the mathematical version with such accuracy that the highest percent of mistakes became = 0.11 and the lowest percent of mistakes changed into = 0.01.

According to Table 2, we discover that the usual deviation is a degree of the dispersion of the records across the mean. The extra the same old deviation, the extra the dispersion of the facts. In this case, the standard deviation of the values is 0.01. This way that the statistics is slightly scattered across the average.

In this case, the standard deviation is small, because of this that the approximate solution is very close to the exact solution. We can expect that the approximate answer is correct enough for use in mathematical modeling. The empirical results obtained align closely with the theoretical predictions, with only minor deviations observed. These differences could be attributed to factors such as rounding errors in the computational process or minor experimental inaccuracies. Overall, the consistency between theoretical and empirical results highlights the reliability of the numerical approximation methods for practical applications.

The well-known deviation may be used to decide the percentage error in numerical approximation methods in mathematical modeling. For instance, if we've got a mathematical feature that has an actual solution, we can use numerical approximation methods to discover an approximate solution. Then we will use the standard deviation to measure how accurate the approximate answer is. If the usual deviation is small, it method that the approximate answer may be very close to the exact answer. If the same old deviation is large, it method that the approximate solution is less accurate.

5 Conclusion

1. Numerical approximate strategies are very effective in solving mathematical modeling troubles
2. The better the diploma of approximation, the greater the accuracy (Owhadi, H., Scovel, C., & Schäfer, F. 2019). [15].
3. Standard deviation is a measure of the dispersion of facts across the imply.
4. The popular deviation can be used to determine the proportion error in numerical approximation strategies in mathematical modeling.
5. If the standard deviation is small, it manner that the approximate solution could be very close to the precise solution (Wan, X., Wang, W., Liu, J., & Tong, T. 2014). [16].
6. In the case of values for which the same old deviation is calculated, the same old deviation is small, which means that that the approximate answer is very close to the exact solution. We can anticipate that the approximate solution is accurate sufficient for use in mathematical modeling
7. To determine the share mistakes using the usual deviation:

- Make sure the data is uniform.
- Compare the standard deviation to the standard deviation of other data.
- Use linear regression to determine the relationship between the exact solution and the approximate solution.

Declaration of competing interest. I declare that there is no competing interest with any other authors and that there is no interest or benefit that influenced the results of the research.

References

1. Cohen, H.: Numerical approximation methods, p. 485. Springer, New York (2011)
2. Jazar, R.N.: Approximation methods in science and engineering. Springer, New York, NY (2020)

3. Abdel Radi, A.R., Marwa, E.A.: The Numerical methods for solving nonlinear integral equations, Vol. 9. IJRDO - Journal of Mathematics (12–01) (2023)
4. Simpson, T.W., Booker, A.J., Ghosh, D., Giunta, A.A., Koch, P.N., Yang, R.J.: Approximation methods in multidisciplinary analysis and optimization: a panel discussion. Struct. Multidiscip. Optim. **27**, 302–313 (2004)
5. Epperson, J.F.: An introduction to numerical methods and analysis. John Wiley & Sons (2021)
6. Fabio, Z., Maria, S., Gianluigi, R.: A Streamline upwind Petrov-Galerkin Reduced Order Method for Advection-Dominated Partial Differential Equations under Optimal Control, Vol. 2. Springer Science & Business Media (2024)
7. Meinardus, G.: Approximation of functions: Theory and numerical methods, Vol. 13. Springer Science & Business Media (2012)
8. Bouhamidi, A., Jbilou, K.: A note on the numerical approximate solutions for generalized Sylvester matrix equations with applications. Appl. Math. Comput. **206**(2), 687–694 (2008)
9. Frames and numerical approximation B Adcock, D Huybrechs Siam Review (2019). epubs.siam.org
10. Seyeon, l., Hyunju, K., Bongsoo, J.: A Novel Numerical Method for Solving Nonlinear Fractional-Order Differential Equations and Its Applications, Vol. 8. Special Issue Advances in Fractional Integral and Derivative Operators with Applications (2024)
11. Quarteroni, A., Manzoni, A., Vergara, C.: Mathematical modelling of the human cardiovascular system: data, numerical approximation, clinical applications, Vol. 33. Cambridge University Press (2021)
12. Steinbach, O.: An efficient computation of the inverse of the single layer matrix for the resolution of the linear elasticity problem in BEM, Vol. 49. Springer Science & Business Media (2023)
13. Dzyadyk, V.K.: Approximation methods for solutions of differential and integral equations. Walter de Gruyter GmbH & Co KG (2018)
14. Köppl, T., Vidotto, E., Wohlmuth, B., Zunino, P.: Mathematical modeling, analysis and numerical approximation of second-order elliptic problems with inclusions. Math. Models Methods Appl. Sci. **28**(05), 953–978 (2018)
15. Owhadi, H., Scovel, C., Schäfer, F.: Statistical numerical approximation. Notices of the AMS (2019)
16. Wan, X., Wang, W., Liu, J., Tong, T.: Estimating the sample mean and standard deviation from the sample size, median, range and/or interquartile range. BMC Med. Res. Methodol. **14**, 1–13 (2014)

Integrating Connected Devices and Advanced Machine Learning Models for Optimized Waste Classification and Management

S. Loganayagi(✉), D. Usha, and K. Kavitha

Department of Computer Science, Mother Teresa Women's University, Kodaikanal, India
loganayagi.shanmugam@gmail.com

Abstract. Managing the waste serves crucial for ecological balance, community well-being and economic productivity. The research work examines the assimilation of connected devices and advanced machine learning models to optimize the classification and management of solid waste in Indian cities. Utilizing advanced machine learning algorithms including convolutional neural networks and deep learning models are used in analyzing and classifying the waste producing high accuracy. The proposed system overcomes the disadvantages of manual sorting and classifies the images of solid waste into three labels such as organic, recyclable and non-recyclable wastes. The solid wastes after mere classification are processed separately were organic waste can be used as fertilizers, recyclable waste are recycled used in reprocessing industries whereas the non-recyclable waste can be processed by incineration and pyrolysis technologies which converts the non-recyclable waste to energy, thus reducing land fill use and generating electricity, such that the environmental impact can be minimized. The particulars of the waste is sent to through IoT devices to the mail for decisive purpose. The initial objective was to develop a Custom CNN model which produced the accuracy of 77% when compared with the hybrid algorithms, whereas Customized InceptionNet model when incorporated with IoT produced an accuracy of 94%. This paper puts forth the challenges in the implementation, usage of hybrid algorithms, environmental benefits and future research opportunities, highlighting the impact of IoT and advanced machine learning towards waste management.

Keywords: Waste Management · Advanced Machine Learning algorithms · Custom CNN · Customized InceptionNet · Hybrid algorithms

1 Introduction

With the advent of CNN (Convolutional Neural Networks) and DNN (Deep Neural Networks) the identification and classification of objects has become simple and easier. Machine Learning utilizes the image dataset from the internet for classifying the waste into several classes or labels, and it also uses the XceptionNet, ResNet, VGGNet, GoogleNet and various composite models.

The object detection has realized its own importance in various sectors for example in identifying and recognizing humans using facial identification algorithm [2], automated

S. Pal and Á. Rocha (Eds.): ICMMCS 2025, LNNS 1398, pp. 358–368, 2025.
https://doi.org/10.1007/978-3-031-90998-6_33

vehicle control [3], image recognition [4]. Several ML architectures are available for researchers which could be exploited based on their needs. Based on the requirements the machine learning model can be developed, which could be used in the classification of waste from the images, the model once built has to be trained with the images from the dataset, examined and authenticated [6]. The classifier such as ResidualNetwork-50, GoogleNet, VGG-16 and quite a lot of representations can be used for object localization in the image using any of the object identification methods [7].

1.1 Objective of the Paper

The scientists gain enhanced precision in Neural network model, Convolutional model and several other computational learning models utilizing the object localization algorithms such as Probabilistic classifier, Linear model, Ensemble Tree model, Maximum margin classifier. The Existing research work focusses on models such as ResNet-50, VGG-16, InceptionNet model and the newly developed custom CNN combined with the classifier algorithms such as XGBoost algorithm, Random Forest and Fully connected DNN algorithm producing different accuracy rates. The proposed model utilizes customized InceptionNet produces higher accuracy. This model when incorporated into IoT produces the classification results.

2 Literature Review

The research work by [8] engrossed completely on object identification in images, they also put forth their ideas that models that are designed for object identification depends on pattern analysis, correlation methods and cognition. Features such as intensity of images, luminosity, number of objects, hue, scale, pixel attributes are obtained from the input. The features obtained are resized and pre-processed.

The study by Deng and colleagues in 2020 found the identification of several objects in artificial intelligence frameworks using two stage of the visual recognition algorithms. Moreover [10] put forth the variations between the outdated and recent deep learning models and suggested that there exists several stages in object identification that consists of extracting the features, classification and pre-processing and they divided the visual recognition algorithms into one and two stages approach.

Models like SPP-Net by He in 2015, regions with ConvNet by [11], fast RCNN and faster RCNN by [12] and [13] as one step algorithms. Moreover Yolo-v1, Yolo-v2, Yolo-v3 by Redmon in the successive years from 2016 are considered as two step algorithms. The initial phase (single step algorithm) depends on region proposal whereas the latter (two step) depends on regression. They concluded that single-step algorithm is efficient for dependent datasets whereas two-step algorithm even though is more effective, but suffer from multi-category object detection.

The table characterizes the artificial intelligence systems established to forecast and categorize the images (Table 1).

Table 1. Architecture and Functionality of Neural Network Models in Waste Management [19]

S. No.	Reseachers	Network Structure	Performance	Precision
1	Ziouzios and colleagues	PortableNet	The factories waste are recycled into 5 classes	96.57%
2	Gomez	ConvNet	Identification and sorting of wastes through Bag-of-Features and pixel assessment	84%
3	Sunny And colleagues	AlexanderNetwork + ConvNet	ConvNet is used for waste identification and sorting. The system categorizes waste into two classes organic and non-organic	96%
4	Amin et al.,	DenseNet121	The size of dataset is 2800 and it classifies the objects into 2 classes	93%
5	Proposed CIV3	InceptionNet V3	The dataset has 25474 images and it classifies the images of waste into three labels such as Organic, Recyclable and Non-Recyclable waste	94%

3 Existing System

The Existing system focuses on segregating and classifies the waste into organic, recyclable and non-recyclable using the CNN approach. The dataset was obtained from Kaggle having 25474 images 22564 images used for training and 2910 for testing based on Pareto principle using which model is trained and tested. At times it is considered that the substantial number, may not be the representation of the vast diversity of waste types and conditions found in real-world scenarios. This drawback is overcome by implementing few strategies such as Data Augmentation techniques such as scaling, rotation, flipping and cropping in order to introduce variations of dataset images thus enhancing diversity. Similarly Synthetic data generation using GANs, dataset expansion by capturing a wide range of waste type and conditions. Thus the above mentioned strategies enhance the dataset's diversity and tunes the model's ability to generalize various types of waste and thus better performances in real-world scenarios. The study emphases on separating and sorting images of waste into three labels such as organic, recyclable and non-recyclable. This is attained by means of a customized Convolutional Neural Network (CNN) model built on prevailing waste management backgrounds. Primarily, the Custom CNN model is verified for accurateness in contrast to current image augmentation and sorting models.

The results thus obtained are associated with those from the ResNet, VGGNet, and GoogleNet models to govern the most active model for waste identification. Using any of the above algorithm the waste management model classifies the items as Organic wastes-'O' which includes items such as food(like fruits, vegetable peels, meat, pulses, grains and so on), Recyclable waste-'R' items such as newspapers, carpets, aluminum foils, books, paper boxes, glass containers and bottles made from degradable components and Non-recyclable waste 'N' such as plastic wraps, bubble wraps, kitchenware, ceramics and various food-tainted products. The learning used precision rate as the system of measurement to estimate the presentation of several models.

The results exposed that the ResNet-50 model, when balancing with the Random Forest classifier, attained a remarkable precision of 72.44%, outdoing further models in terms of effectiveness and consistency. Although the custom CNN model demonstrated strong capabilities in image identification, the blend of ResNet and Random Forest developed as the higher approach in this study, producing the utmost exact sorting for waste management (Figs. 1, 2, 3 and 4).

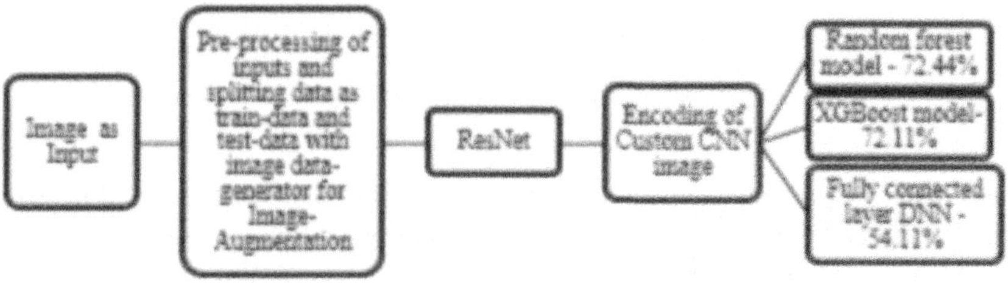

Fig. 1. Flowchart of ResNet with classifiers/Accuracy %

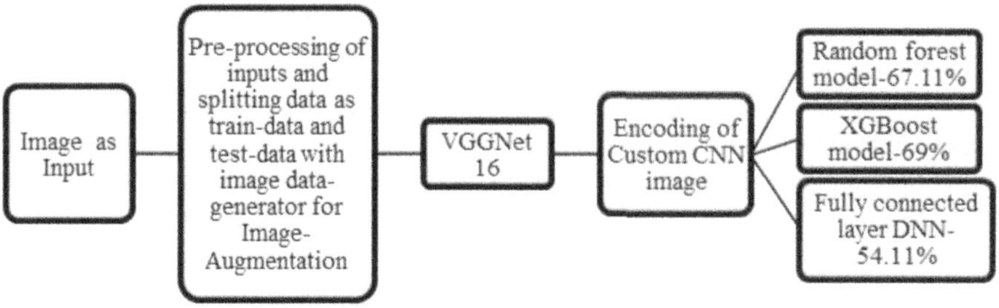

Fig. 2. Flowchart of VGGNet 16 with classifiers/Accuracy %

The precision value was obtained using the confusion matrix and tabulated. The custom CNN model established obtained smaller precision, but, it could be detected that the precision rate of the Inception model (XGB and RF) is lesser than the custom CNN model. Henceforth, there is scope in the Custom CNN model developed in near future. For example, In the traditional methods of waste management, depending on the size of collection trucks which would take $1m annually for labour, fuel and maintenance,

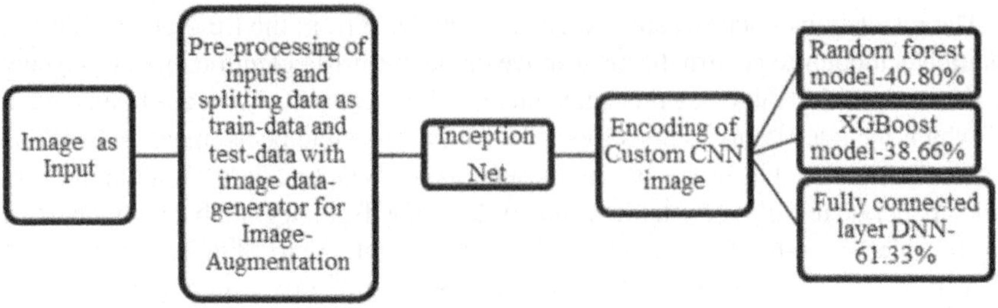

Fig. 3. Flowchart of InceptionNet with classifiers/Accuracy %

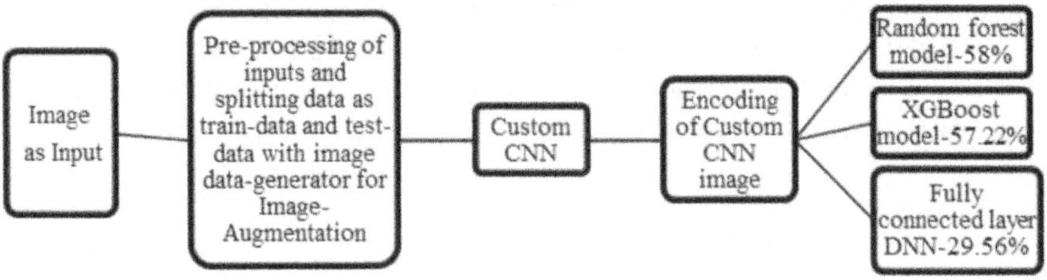

Fig. 4. Flowchart of Custom CNN with classifiers/Accuracy %

an automated waste management system may cost $800,000 for the initial investment of the system and $100,000 annually for additional costs, thus saving one $200,000 in overall labour and operating costs every year.

4 Proposed System

Computerized waste management technologies enhance efficiency through application of sensors and artificial intelligence for identification of waste, collection schedule and precise collection points hence making the management less expensive and prone to human influence. As opposed to conventional approaches that involve manual sorting and set-timed collection and separation, automated systems improve resource utilization, reduce negativity effect, and adjust to constantly fluctuating waste characteristics.

Figure 5 gives the flowchart of the proposed model. The Customized InceptionNet V3 model poses the layers of ConvNet, down sampling, multi -scale modules, spatial compression along with the combination of regression analysis and regularization layers which can avoid excessive fitting and improve precision. The auxillary classifier is used to facilitate the process of training by enhancing gradient flow, it acts as a regularizer and aid in faster convergence. The involvement of the user or the model developer takes place, after classifying the waste by the model and through IOT it is mailed to the user, where one can make proper measures and decisions to process the waste. The idea is shown in Figs. 7 and 8.

The developed proposed system combines InceptionNet layers within the CNN architecture, resulting in the customized InceptionNet V3. Cross entropy is used for loss estimation and Randomized Gradient Descent as an optimizer. The flowchart of the proposed model is depicted and the precision is 94% whereas the basic InceptionNet model produced an accuracy of 77%.The deep leaning model has incorporated into IoT and the results can be obtained through mail and the decisions can be made further.

As for the Convolutional Neural Network (CNN) model Fig. 6 created based on this study's requirements, pre-processing has to be performed on the waste images before they are fed into the network. The procedure involves directly inputting the images into the first convolution layer with 896 filters. Subsequently, the images undergo two more 2D convolutional layers: one of them has 9248 filters.

After that, the input passes through the pooling layer where dimensionality is reduced and dropout layer used to eliminate overfitting by for a certain level of inputs to zero. The images are then convoluted with weights of 18496 the second convolutional layer has weights of 36928, and the last convolutional layer also has a same weight of 36928. There is another dropout layer prior to the flattening layer that brings out the features in a form of a vector instead of a matrix.

After flattening the data is feed to a dense (fully connected) layer and this is done twice more along with the participation of two dropout layers (dropout-2 and dropout-3) and two more dense layers—dense-1 and dense 2. Last but not the least, to help in recognition and labeling the images are passed through the final fully connected convolutional layer.

The CNN model's effectiveness is evaluated using three different methods: To compare the performances of the AI models, Random Forest (RF), XGBoost (XGB), and a Fully Connected Deep Neural Network (FCDNN) were selected. The measure applied to compare these methods is defined as "accuracy" and calculates the ratio of the correctly classified images to the overall amount of the images for assessment. Employing dynamic resource allocation and modular design can enhance the system's ability to handle large-scale urban environments effectively.

5 Experimental Results

The tables gives a clear understanding of the input images and the actual class the image belongs and the predicted class of the images as obtained from Basic InceptionNet model and the proposed Customized InceptionNet (CIV3) (Table 2).

The Confusion matrix obtained as the outcome from Basic InceptionNet and Customized InceptionNet is represented in Figs. 9 and 10. The Basic model produced an accuracy of 77% whereas the Customized proposed model produced an accuracy of 94%. The inception model produces higher accuracy and it can also be trained with hybrid algorithm which may produce efficient results. Conferring to [14] the multi-scale modules in ConvNet generates negligible errors and superior precision levels. Whereas, logarithmic loss can be used to reduce the loss which was scrutinized by researchers [15]. The usage of Stochastic Optimization has provided precise results for the proposed model(CIV3). In order to provide the holistic evaluation of the model performance, we have included precision, recall, f1-score,support, micro average and macro average. Precision refers to the measuring of accuracy of the positive predictions found by the model. Recall (or Sensitivity) refers to the capability of the model to find out all the positive

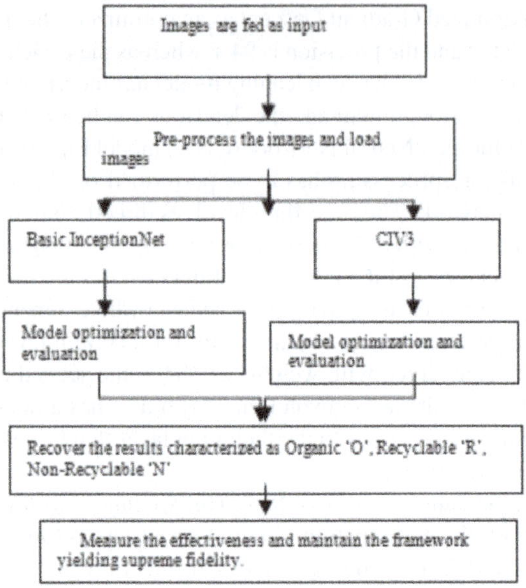

Fig. 5. Flowchart of the Proposed Model

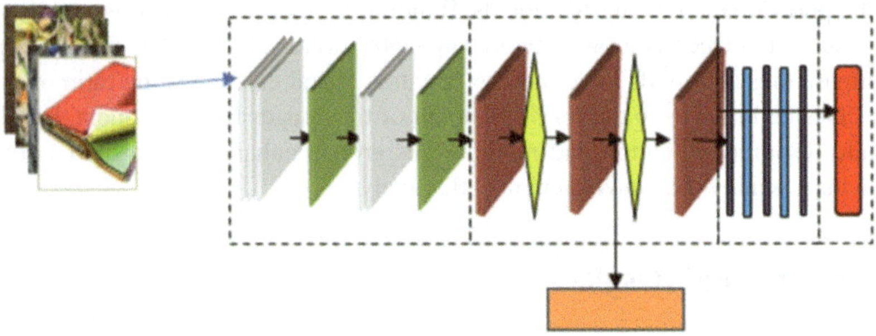

Fig. 6. CIV3- Customized InceptionNet V3(Proposed System)

cases. It is the ratio of correctly predicted positive observations to the all observations in the actual class. F1-Score: The F1-score is the harmonic mean of Precision and Recall. It provides a single metric that balances both precision and recall, especially useful when dealing with imbalanced datasets. Support: It is the number of actual occurrences of the class in the dataset. It provides context on how many instances of each class exist in the dataset, where TP: True Positives, FN: False Negatives,TP: True Positives. FP: False Positives (Tables 3 and 4).

Additionally, the model developed was benchmarked against state-of- the-art models to provide a good understanding of the model effectiveness. The results produced indicated strongly the performance with respect to all metrics, thus demonstrating the model's effectiveness in various real- world scenarios.

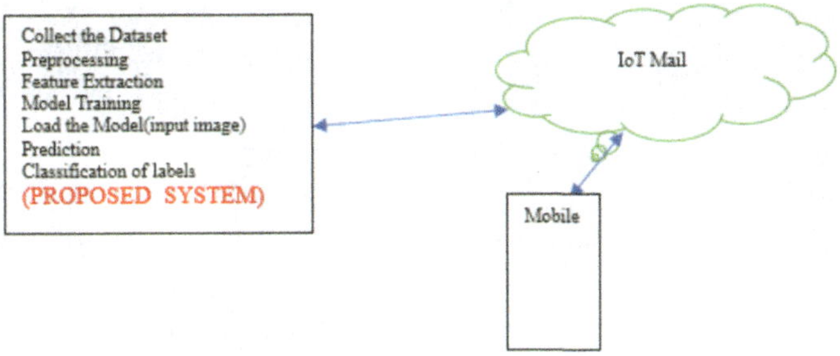

Fig. 7. Incorporation of Deep learning model and IoT

Fig. 8. Images of waste sent to mail

Table 2. Architecture and Functionality of Neural Network Models in Waste Management

Input Images	Actual Class of the Images	Predicted Class of the Images (Basic InceptionNet)	Predicted Class of the Images (CIV3)
Organic Image	Organic	Organic	Organic
Recyclable Image	Recyclable	Non-Recyclable	Recyclable
Non-Recyclable Image	Non-Recyclable	Non-Recyclable	Non-Recyclable
Recyclable Image	Recyclable	Recyclable	Recyclable
Non-Recyclable Image	Non-Recyclable	Recyclable	Non-Recyclable

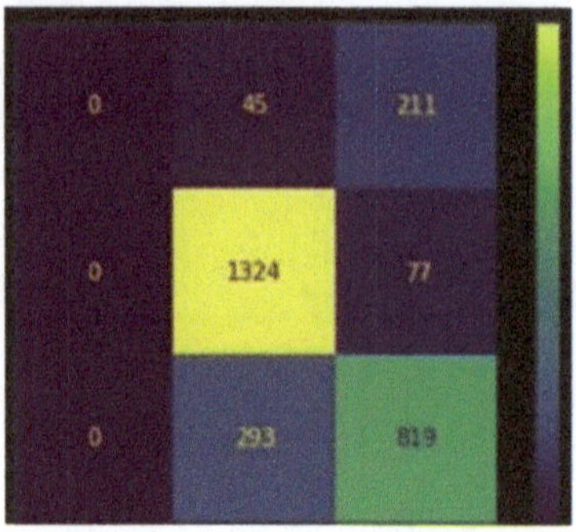

Fig. 9. Basic InceptionNet

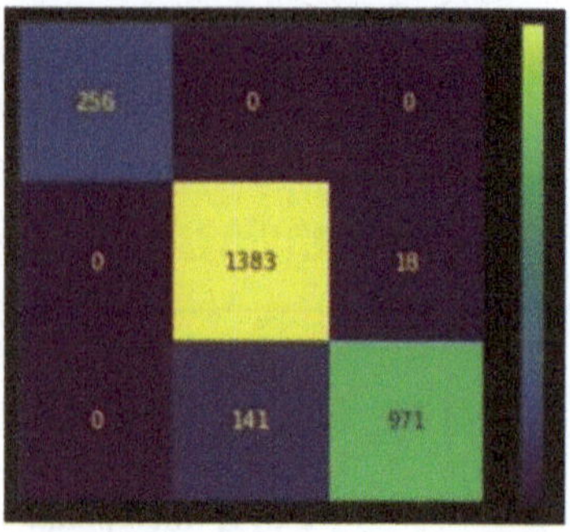

Fig. 10. Customized InceptionNet

Table 3. Evaluation Metrics

Metric	Precision TP/(TP + FP)	Recall TP/TP + FN	F1-score 2 × Precision × Recall/ (Precision + Recall)	Support TP + FN
Predicted 1-'N'	1.00	1.00	1.00	256
Predicted 2-'O'	0.91	0.99	0.95	1401
Predicted 3-'R'	0.98	0.87	0.92	1112
Accuracy			0.94	2769
Macro avg	0.96	0.95	0.96	2769
Weighted avg	0.95	0.95	0.95	2769

Table 4. Comparison between Basic vs Customized InceptionNet and its Benefits

Aspect	Basic InceptionNet V3	Customized InceptionNet V3	Benefits of Customized InceptionNet V3
Model Capacity	Standard capacity	Increased with added regression layers	Enhanced ability to capture complex patterns and relationships in the data
Regularization	Basic dropout	Advanced dropout techniques	Improved prevention of overfitting by dropping more layers
Prediction Type	Categorical outputs	Fine-tuned continuous values with regression	Allows for more precise and accurate predictions

(continued)

Table 4. *(continued)*

Aspect	Basic InceptionNet V3	Customized InceptionNet V3	Benefits of Customized InceptionNet V3
Generalization	May overfit on specific features	Better generalization to unseen data	Helps reduce overfitting and improves model performance
Flexibility	Fixed architecture	Customizable architecture with additional layers	Combines strengths of the original model with new capabilities
Performance	Strong baseline performance	Potentially improved performance	Enhanced results by leveraging additional layers and techniques
Application Scope	General use	Tailored for specific tasks	Customization meets specific needs and requirements of the task

6 Discussion and Conclusion

Waste administration in several nations ought to be measured. It's a key problem deprived of the humanoid interference that contaminates the location and origins wellbeing problems. To address these primary concerns and combat severe health risks, the computational intelligence approach towards waste administration is integrated along-with universal waste disposal organization [16]. The Approach proposed by [17] suggested using GoogleNet together with VGG-16, Densely Connected Network, Alexandra Network. The researchers [18] have suggested that Inception v3 can be used to produce precised result and minimized data loss.

References

1. Zhao, Z,-Q,, Zheng, P., Xu, S.-T., Wu, X.: Object detection with deep learning: a review. IEEE Transactions On Neural Networks And Learning Systems For Publication, 1–21 (2019). https://doi.org/10.1109/TNNLS.2019.2958912
2. Yang, Z., Nevatia, R.: A multi-scale cascade fully convolutional network face detector," in 23rd International Conference on Pattern Recognition (ICPR), pp. 1–6. IEEE (2016). https://doi.org/10.1109/ICPR.2016.7899757
3. Chen, X., Ma, H., Wan, J., Li, B., Xia, T.: Multi-view 3d object detection network for autonomous driving. In: Computer Vision and Pattern Recognition - IEEE, pp. 1907–1915 (2017). https://doi.org/10.1109/CVPR.2017.206
4. Jia, Y., et al.: Caffe: Convolutional architecture for fast feature embedding. In: 22nd MM: Proceedings of the ACM international conference on multimedia, Orlando, Florida- USA, pp. 675–678 (2014). https://doi.org/10.1145/2647868.2654889
5. Krizhevsky, A., Sutskever, I., Hinton, G.E.: Imagenet classification with deep convolutional neural networks. Adv. Neural. Inf. Process. Syst. **25**, 1–9 (2012). https://doi.org/10.1145/3065386

6. Cao, Z., Simon, T., Wei, S.-E., Sheikh, Y.: Realtime multi-person 2d pose estimation using part affinity fields. in Computer Vision and Pattern Recognition, 1–9 (2017). https://doi.org/10.1109/CVPR.2017.143
7. Dino, H.I., Abdulrazzaq, M.B.: A Comparison of Four Classification Algorithms for Facial Expression Recognition. Polytechnic Journal **10**(1), 74–80 (2020). https://doi.org/10.25156/ptj.v10n1y2020.pp74-80
8. Kumar, S., Balyan, A., Chawla, M.: Object detection and recognition in images. Int. J. Eng. Develop. Res. **5**(4), 1029–1034 (2017). https://www.ijedr.org/papers/IJEDR1704161.pdf
9. Deng, J., Xuan, X., Wang, L.Z., Yao, H., Wang, Z.: A review of research on object detection based on deep learning. Journal Physics: Conference Series **1684**(012028), 1–7 (2020). https://doi.org/10.1088/1742-6596/1684/1/012028
10. Jadhav, P., Koli, V., Shinde, P., Pawar, M.M.: Object Detection using Deep Learning. Int. Res. J. Eng. Technol. (IRJET) **7**(9), 275–279 (2020). https://www.irjet.net/archives/V7/i9/IRJET-V7I946.pdf
11. Girshick, R., Donahue, J., Darrel, T., Malik, J.: Rich feature hierarchies for accurate object detection and semantic segmentation. In: Computer Vision and Pattern Recognition, Columbus, pp. 580–587 (2014). https://doi.org/10.1109/CVPR.2014.81
12. Redmon, J., Divvala, S., Grishick, R., Farhadi, A.: You Only Look Once: Unified, Real-Time Object Detection. In: Computer Vision and Pattern Recognition. Las Vegas, pp. 779–788 (2016). https://doi.org/10.1109/CVPR.2016.91
13. Bochkovskiy, A., Wang, C.Y., Liao, H.Y.M.: YOLOv4: Optimal Speed and Accuracy of Object Detection. Computer Vision and Pattern Recognition (2020). https://arxiv.org/abs/2004.10934
14. Ferronato, N., Torretta, V.: Waste mismanagement in developing countries: a review of global issues. Int. J. Environ. Res. Public Health **16**(1060), 1–28 (2019). https://doi.org/10.3390/ijerph16061060
15. Ho, Y., Wookey, S.: The Real-World-Weight Cross-Entropy Loss Function: Modeling the Costs of Mislabeling. IEEE Access **8**, 4806–4813 (2020). https://doi.org/10.1109/ACCESS.2019.2962615
16. Cao, J., Yan, M., Jia, Y., Tian, X., Zhang, Z.: Application of a modified Inception-v3 model in the dynasty-based classification of ancient murals. EURASIP Journal on Advances in Signal Processing **2021**(49), 1–25 (2021). https://doi.org/10.1186/s13634-021-00738-8
17. Feng, J.-W., Tang, X.-Y.: Office Garbage Intelligent Classification Based on Inceptionv3 Transfer Learning Model. In: IOP Conf. Series: Journal of Physics: Conf. Series, **1487**(012008), 1–9 (2020). https://doi.org/10.1088/1742-6596/1487/1/012008
18. Rai, D.M., Maggu, S.: Waste Management and Classification using Deep Learning CNN. Int. J. Adv. Eng. Manage. (IJAEM) **3**(6), 981–984 (2021). https://doi.org/10.35629/5252-0306981984
19. Loganayagi, S., Usha, D.: An Automated Approach to Waste Classification Using Deep learning, ICECCT-2023 (2023). https://doi.org/10.1109/ICECCT56650.2023.10179743

Automating the Future: The Application of Machine Learning in Manufacturing

Amena Mahmoud[1](✉), S. Ruban[2], and Eiko Takaoka[3]

[1] Department of Computer Science, Faculty of Computers and Information, Kafrelsheikh University, Kafr El-Shaikh, Egypt
amena_mahmoud@fci.kfs.edu.eg

[2] Department of Software Technology, St Aloysius Deemed to be University, Mangalore, India

[3] Department of Information and Communication Sciences, Faculty of Science and Technology, Sophia University, Tokyo, Japan
m-g-eiko@sophia.ac.jp

Abstract. The manufacturing industry is undergoing a profound transformation driven by the powerful integration of machine learning (ML) technologies. This paper explores the diverse applications and far-reaching impact of machine learning in revolutionizing modern manufacturing processes. At the core of this revolution is the ability of machine learning algorithms to extract valuable insights from the vast amounts of data generated throughout the manufacturing ecosystem. By leveraging advanced data analytics and predictive modeling, ML-powered systems can identify patterns, optimize production workflows, and anticipate potential issues before they arise.

Keywords: Manufacturing · Machine Learning · Predictive Maintenance · Quality Control

1 Introduction

Manufacturers are increasingly finding themselves creating "smart plants" where more and more decision-making processes can be made by the machines themselves. One of the primary technologies driving this innovation is the development and application of machine learning tools. These techniques have not only broadened the array of manufacturing and warehousing activities that can be efficiently automated but also enabled major productivity enhancements in industries where the physical demands of the tasks have stymied advances. As we demonstrate in our analysis, though, machine learning is not used uniformly. It has been applied more often in certain segments of the manufacturing industry where some tasks are, for physical as well as economic reasons, more suitable datasets for machine learning. This has implications for progress in automation as well as potentially affecting what production structures are economically viable in the future [1].

S. Pal and Á. Rocha (Eds.): ICMMCS 2025, LNNS 1398, pp. 369–380, 2025.
https://doi.org/10.1007/978-3-031-90998-6_34

2 Foundations of Machine Learning

A model has a goodness of fit to the data, and an inferred rule can predict new cases – it can generalize to previously unobserved inputs. Goodness of fit and generalization are opposing goals. A model that exactly fits the training cases can be guaranteed to generalize to new cases, but the learner has no reason to expect a particular relationship between inputs. A model which makes strong assumptions and which generalizes to all possible cases is at risk of overfitting the data. In practice, learning is feasible because the assumption that it is possible to find a model that characterizes the training data and generalizes to new cases is an assumption that the input data contains information useful for inferring the behaviors of interest. The extent to which one can generalize to unseen test cases given a particular observation set and a chosen learning algorithm is not well-studied, rather application in practice has served as a demonstration [2].

Machine learning is a data-driven method based on models that, given inductive bias in the form of sample data, can be used to make non-trivial inferences about new data. At the lowest level of abstraction, a model is realized from a set of parameters. A given model trained on a set of observations has a fitness concerning those observations. Certain properties of the model imply that it may be more or less well-suited as a model for a given problem. Generalization refers to heuristics that we use to select a particular model which best fits the observed data, and which can be used to infer a rule that generalizes beyond its application data set. The role of inductive bias in selecting a model is formalized in the learning-theoretic study of generalization which seeks to find a model consistent with a particular data set, even though the set of models which are consistent given a finite amount of data may be quite large. Different learning theories correspond to different inductive biases. A model is learned from a set of training cases without any prior knowledge about the physics of the system other than training data [3].

2.1 Supervised Learning

To address a simplified version of a real problem, suppose the dependent variable can only have a level value, and it is possible to know the number of possible classes ahead of time, as illustrated in Fig. 1. In this case, we have a classification problem for which our dependent variable contains a finite number of classes. With the set of inputs, we aim to determine which classes these instances belong to. Some of the most popular models that we use in this case are logistic regression and its extensions. In the case of two classes, usually, we favor the probabilities. Given a set of outputs and a set of weights, the logistic regression computes the probabilities using the weights [4].

Sometimes issues that we want to solve have clear patterns that we are aware of. These are regarded as supervised learning problems. We know input values and the corresponding output values, and we can use these as instances for our model. The main aim of linear regression, which is used to make predictions, is to establish a linear combination of inputs in such a way that it produces the output for the given inputs. This method assists in single-output and real-valued outputs, and can also lay the foundations for some classification problems. However, it could be greatly affected by outliers,

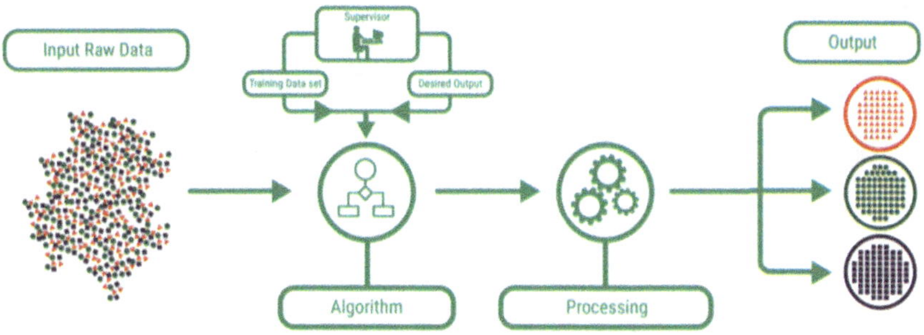

Fig. 1. Supervised learning architecture

which can cause considerable alterations in the sample mean. Thus, other mitigated error objective functions such as a median or a piecewise objective can be more effective [5].

2.2 Unsupervised Learning

Given that the majority of customer problem reports on a new vehicle/software release are binary (e.g. fit for use or not; directly safety-critical or not), terms to refer to this supervisory style are labeled with additional notes about which method can be useful in this type of study. The opposite would be considered as teaching a model to learn through the label and to describe where the best set of possible prediction terms might be discharged, as shown in Fig. 2. A decision maker would be much better assisted by such a system in the context of OEM N/V evaluations than if one is to present a probability. The system, if fitted well, could be used more directly to decide whether the project was ready to be made [6].

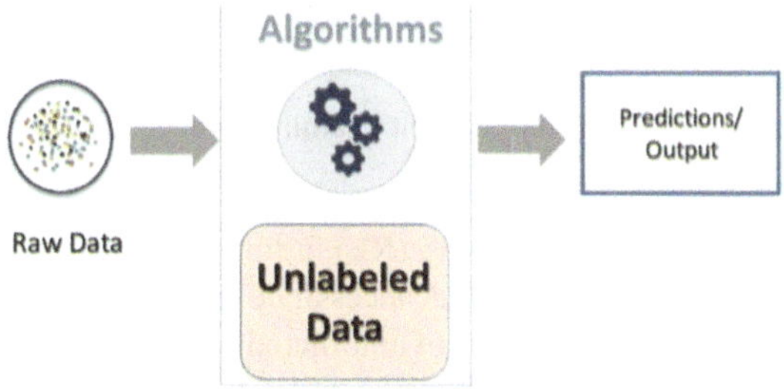

Fig. 2. Unsupervised learning architecture

Unsupervised learning is the process of learning about features without a reference output, such as mill loading, stock usage, and tool life. The algorithms attempt to study the distribution of the data to create different groups thereof. Many of the strengths of unsupervised learning apply to the countless problems that exist when manufacturing data is analyzed and organized. Researchers can discover groups and persuasively understand and visualize the structure of the data. It has the possibility of dimension reduction, feature extraction, and clustering through training on such methods. Using the unsupervised approach, researchers in manufacturing can build and interpret latent groups of different natures fairly easily and efficiently as many are sufficiently knowledgeable to make sense of the information they possess. In contrast to unsupervised learning, it is used to find structure to learn the distributions of a series of training data and to find out how the samples occur without the knowledge of the true outcome.

2.3 Reinforcement Learning

In a simpler perspective, where all the underlying processes, namely the state, the actions, and the agent, are known, the sole questions are: which action should I take to maximize the reward, or which agent optimizes a fixed reward. In this case, one would only attempt to identify a rule that can be implemented as-is as shown in Fig. 3. However, RL is closer to our reality than previous ML models. It allows learning the rules online, and the inventory and tactical decisions can be implemented as an external decision. This can result in a better integration with SAS models and machine learning [7].

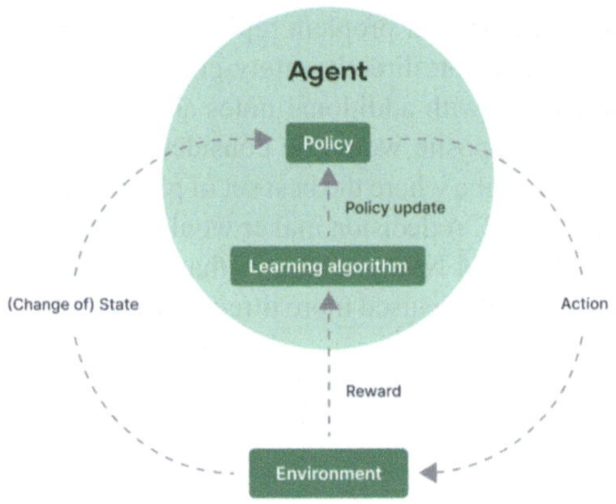

Fig. 3. Reinforcement Learning architecture

The interest in reinforcement learning (RL), a form of machine learning (ML) that is especially useful in environments that necessitate autonomous decision-making, is increasing. In simple terms, RL is a mix of automated control and prediction, or estimation. The goal is to train an agent which has to understand what actions it can take in an environment to maximize some sort of notion of cumulative reward.

3 Machine Learning Techniques in Manufacturing

Machine learning is a field of computer science that enables systems to learn data as opposed to explicitly programmed ones. This crucial area has seen vast success in automating tasks and has the potential to transform manufacturing and related production processes, as machine learning can use data and identify patterns to make accurate decisions, which is necessary for advanced manufacturing. The data usually comes from various sensors monitoring the manufacturing processes and from monitoring the sensors and input data coming from multiple production processes and infrastructure. The entire manufacturing process is also a potential source of data, as it contains information about, e.g. product specifications, geometrical dimensions, and location or energy consumption. The dimension of the data may grow even further when considering heterogeneous data obtained from simulations, historical data, and partners in the value chain [8].

As products become more complex, so does the manufacturing process. This has created a surge in interest in applying machine learning techniques to improve manufacturing processes, especially in bringing automation to the physical fabrication, maintenance, and management domains. Through data analysis, machine learning algorithms can be trained to make accurate predictions for machines, robots, and processes in the entirety of the production systems. Here we give an overview of the different techniques of machine learning that are being implemented or developed to bring revolution in the manufacturing industries.

Techniques for model interpretability and human-in-the-loop approaches address the black-box challenge in machine learning applications in manufacturing by making models more transparent and by involving human experts in the decision-making process.

Human-in-the-loop approaches integrate human expertise and oversight into the machine learning pipeline. This allows for continuous interaction between humans and machines, improving both decision-making and system trustworthiness.

Interactive Model Training:

- Method: In interactive model training, human experts provide real-time feedback during the training process. The model is updated based on human feedback, correcting errors or improving performance in areas where the model might be weak.
- Application in Manufacturing: Anomalies detected by an ML model in a manufacturing process could be flagged for human review. If the anomaly is misclassified, the human expert can correct it, refining the model's future predictions. This is particularly useful in defect detection or quality assurance, where some types of defects may be rare or ambiguous.
- Benefit: Interactive training allows models to continuously learn and adapt based on expert feedback, reducing the risk of erroneous decisions and increasing trust in the model's reliability.

Active Learning.

- Method: Active learning involves the model identifying uncertain or difficult examples and querying human experts for labels. This process improves the model's performance by focusing on the most informative data points.
- Application in Manufacturing: In a predictive maintenance scenario, the model could identify instances where it is uncertain whether a machine component is nearing failure. An engineer could then review the sensor data and label the instance, helping the model to better distinguish between normal and failing conditions.

3.1 Predictive Maintenance

There are several reasons why predictive maintenance is important and useful. Scheduled maintenance is a more cost-effective and less disruptive approach than either a traditional hourly-based schedule or a repair-after-failure schedule. Even in the US alone, avoiding unscheduled downtime can save up to $2–3 trillion. Furthermore, it is now more cost-effective to implement real-time condition monitoring due to advancements in IoT (Internet of Things) concerning creating the sensors required for data collection [9].

Another popular predictive analytics use case in manufacturing is that of predicting when a machine is likely to fail. This intelligent prediction is called predictive maintenance, and it is instrumental in improving the reliability of a developed product. There are two common types of data involved in predictive maintenance—the sensors on the machine (e.g., temperature, vibration, acoustic sensors), and the repair history of the machine. As outlined earlier, the machine learning models typically use the sensor data as predictors and use the repair history as the target variable to train. The output of the machine learning prediction is usually the number of hours/days before a machine repair is likely to be needed.

3.2 Quality Control and Defect Detection

Due to the significance of long-time investment and low product loss and recall, not only deep learning but also current technologies that deal with small-sample-size problems and tighten time constraints, such as benchmark construction and the performance comparison and analysis of feature extraction, pattern recognition, novelty, and outlier detection, transfer learning and data augmentation, are also introduced [10] (Fig. 4).

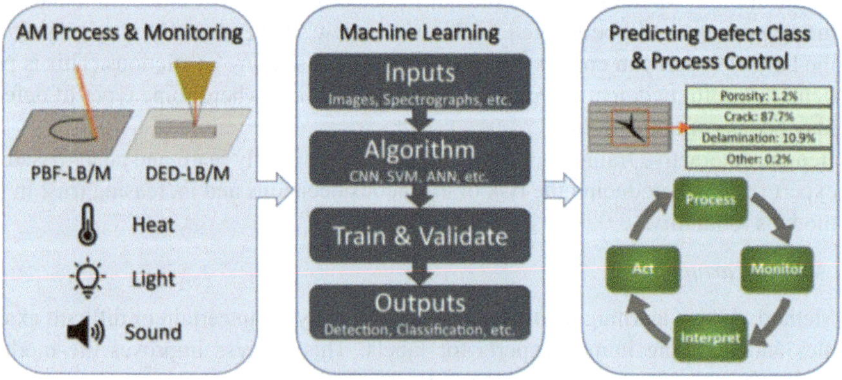

Fig. 4. Process monitoring and machine learning for defect detection [10]

In this section, we classify the important production objectives of manufacturing quality control and defect detection according to their difficulty levels and complexity and then provide the state-of-the-art deep learning-based advancements for each classification and discuss the future challenges and trends, which cover defect detection within

2D or 3D object surface, 2D object point features or point clouds, and multisensory data, particularly the fusion of 2D and 3D aspects with traditional manual detection techniques.

Furthermore, most existing work focuses on two-dimensional (2D) defect detection, while the detection of three-dimensional (3D) objects or point clouds remains a challenge in practice and has not been studied thoroughly.

Deep learning models, especially the convolutional neural network (CNN), have demonstrated significant advantages and successes for quality control and defect detection in recent years. However, when the number of defects per defective sample or the number of defect types is large, deep learning still faces challenges, such as small quantifiable training data, because of being integrated with manual detection methods, potentially high labor costs, and limited noise and data augmentation techniques associated with feature variations and occlusions [11].

Robust quality control is essential in manufacturing, as it affects the durability of the products and ultimately customer satisfaction. Therefore, the industry seeks automatic systems that can promptly detect and identify defects on the production line to provide timely feedback to human operators.

3.3 Optimization of Production Processes

As methods of machine learning become more powerful, we are increasingly able to sidestep the problem of formulating a valuation model by taking data and directly setting control parameters to optimize a response of interest. Such models are of particular value when we have no convincing description of the effects we hope to achieve.

Some of the systems and methods for scheduling or rescheduling production within a manufacturing environment include the implementation of artificial intelligence, statistical process control, the theory of constraints, and the tool of six sigma, among others. Although many automated techniques for production optimization currently exist, it is a difficult and complex process to find the right method for a given individual manufacturing system and to apply the same efficiently and effectively [12].

Research has found that waste, either energy or material, is a concept that is central to this problem because the main objective is energy reduction, which is related to optimal production. The research literature contains several approaches to achieving energy optimization in production contexts, ranging from heuristics in manufacturing to exact algorithms in high-performance computing.

Optimization of production processes is an area of great interest for manufacturers. Firms strive to use their existing production capacities to make as much high-quality output as possible, as quickly and efficiently as possible. The particular problem of allocating energy in the most economically efficient way, to ensure the production of the highest possible amount of the highest quality product, is one we come across often.

4 Challenges and Limitations

The requirement for the sophisticated tuning of additional model hyperparameters will increase the need for specialized software engineers or data analysts. Recent work has proposed algorithms and heuristics that automate or simplify the requirements of hyperparameter tuning, but only feature a limited and conservative improvement or prevent aggravating the problem. Currently, task-specific embedded model optimization during training and robust environmental learning self-adaptation during testing is not common or standardized. Production can be influenced by too small and widely varying learning performance, unreliability, drift, low confidence, and unexpected behavior increase. Technical and other economic limitations currently prevent the application of zero-shot or few-shot learning setups and other new machine-learning-based detection techniques. Machine learning lifetime tends to make industry personnel rely on conservative detections, causing them to miss premature historical trends or the appearance of previously unobserved or unforeseen fault categories [13].

Pre-processing, during the preparatory phase, feature extraction, and feature selection are also time-consuming machine learning steps. However, fault detection in manufacturing does not usually require the identification of highly specific patterns. Furthermore, fault patterns vary with products, process setups, or other conditions. Many feature extraction methods apply only to specific optically or acoustically analyzing sensor devices or transducers. Standardized or low-cost sensor devices may yield limited local quality information. Also, too frequent sensor sampling may compromise the sensor lifetime and increase the amount of data. Other feature extraction methods may require filters, transformations, transformation orders, filter or transformation parameters, which are difficult to choose for a specific application. Attribute reduction for ordering or noisy time series data and feature elimination are already difficult and time-consuming machine learning challenges on their own. Attribute processing and feature selection typically reduce noise, overfitting, and computation time without altering model generalization performance too much. Attribute processing may help to obtain optimal results during clustering, visualization, or regression with fewer regressors too. Any additional effort in this regard and additional insight may help to improve detection performance, trust, and reliability of manufacturing quality control [14].

Currently, the application of machine learning in smart manufacturing is restricted by some challenges. The description of the supervised learning algorithms above already requires the existence of high-quality labeled data. It may take an expert several hours to label a single experimental data set. Furthermore, many fault categories of interest do not correspond to standardization, thus many labels would be required, which tends to be very costly. As mentioned before, the allocation of expert personnel to this manual work may also compromise the preferences of valuable skilled employees. Furthermore, expert evaluation of large industrial datasets remains challenging. Manual expert labeling is also prone to errors, noise, or conflicts, which can be mitigated by heuristics or algorithms. Unfortunately, many imperfections in source data are likely to negatively affect the learning outcomes of algorithms and confidence in task generalization. Recently, solutions affected end-to-end deep learning architectures, which try to map raw perceptual inputs directly to the task to be solved or labeled, to be reliable as an alternative. However, current deep learning is prone to overfitting or generating unreliable confidence estimates in case of noisy or non-representative data [15].

To overcome the challenges of data labeling and the limitations of expert-based approaches in manufacturing, semi-supervised learning and data augmentation techniques offer practical solutions. Semi-supervised learning helps reduce the reliance on fully labeled datasets by utilizing a mix of labeled and unlabeled data, while data augmentation expands the dataset through various transformations, enhancing the model's ability to generalize.

Future research could explore **active learning** methods, where the ML model actively queries experts for labels on only the most informative or uncertain examples, reducing the labeling burden. Additionally, hybrid approaches that combine generative models with expert validation could be developed to create more reliable synthetic data for training. By advancing these methodologies, the application of machine learning in manufacturing can continue to evolve, becoming more scalable, accurate, and adaptable to real-world challenges.

4.1 Security and Privacy Considerations

Manufacturing data in machine learning environments faces numerous security risks, from data breaches and insider threats to adversarial attacks and cloud vulnerabilities. By employing a multi-layered security approach—including encryption, access control, adversarial robustness, and cloud security best practices—companies can significantly mitigate these risks, ensuring that sensitive data remains protected and that ML models continue to function reliably in the manufacturing context.

Secure Machine Learning Frameworks:

- Approach: Use machine learning frameworks that prioritize security, such as differential privacy techniques, which add noise to the data to prevent leakage of individual data points, or homomorphic encryption, which allows computations on encrypted data without exposing the raw data.
- Application: In manufacturing, differential privacy can be used to train models on sensitive data (e.g., proprietary designs or operational secrets) without risking exposure during the training process.

Data Integrity Checks:

- Approach: Implement data validation mechanisms to ensure that the data used for training or inference has not been tampered with. Hashing and checksums can be used to verify the integrity of data before it is processed by the ML models.
- Application: Before feeding data into a model for production monitoring or quality control, ensure that integrity checks are performed to confirm that the data has not been altered maliciously.

5 Future Directions and Opportunities

Second, machine learning enables the manufacturer to attain a greater level of quality or performance with less human introspection. We identify two ways to further increase the automation level in precision manufacturing. Since the manufacturer is often faced with a highly nonlinear, complex, and stochastic system, developing autonomous methods using probability models merits special attention. The methods are capable of minimum human intervention. Model interpretation, i.e. explaining why a model makes certain decisions or predictions, can advance the application of probability models to precision manufacturing. The second pathway is model interaction: how to embed domain physics or human expertise within machine learning models for improved performance. Although any form of interaction is challenging, it is particularly useful for conferring learnable concepts to Bayesian algorithms. These methods not only exhibit strong performance but also enjoy excellent properties that are not present in traditional machine-learning models [16].

Novel Applications and Future Research Directions**

As machine learning continues to evolve, there are several novel applications and research directions that hold promise for the future of manufacturing:

1. Generative Design:
    - Emerging Trend: Generative design is an AI-driven design process where algorithms automatically create optimized design solutions based on specified constraints (e.g., materials, cost, manufacturing techniques).
    - Novel Application: Machine learning could be used to refine and evolve these designs, continuously improving product performance and manufacturability. This approach can lead to the creation of innovative product designs that traditional methods may not conceive.
    - Future Exploration: Research could focus on integrating generative design with ML-driven optimization models to develop autonomous systems capable of both designing and manufacturing complex products in a fully automated cycle.
2. Human-Robot Collaboration:
    - Emerging Trend: Collaborative robots (cobots) that work alongside human workers are becoming increasingly prevalent in manufacturing. Machine learning can enhance this collaboration by allowing robots to learn from human behaviors and adapt accordingly.
    - Novel Application: By using reinforcement learning and natural language processing, cobots could adjust their actions based on verbal or non-verbal cues from human workers, leading to safer and more intuitive human-robot interaction.
    - Future Exploration: Researchers could explore the development of more sophisticated algorithms for understanding human intent and optimizing shared tasks in real-time. This could significantly enhance productivity in complex, mixed-skill environments.
3. Real-Time Process Control:
    - Emerging Trend: Real-time process control systems powered by ML can dynamically adjust manufacturing processes to optimize efficiency and quality.
    - Novel Application: For example, in chemical manufacturing, machine learning models could continuously adjust temperature, pressure, and other variables to optimize reactions in real-time, improving yields and reducing energy consumption.

- Future Exploration: Future research could explore the application of ML to even more complex manufacturing processes, such as additive manufacturing or bio-manufacturing, where real-time adjustments are critical to maintaining product quality and consistency.

4. Sustainable Manufacturing:

- Emerging Trend: Sustainability is becoming a key focus in manufacturing, with efforts to reduce waste, energy consumption, and environmental impact.
- Novel Application: Machine learning can optimize resource use across manufacturing processes, from reducing material waste in production to optimizing energy consumption in factories.
- Future Exploration: Further research could focus on integrating ML with life cycle analysis (LCA) tools to automate sustainability assessments and suggest optimizations for more eco-friendly production practices. This could lead to smarter, greener factories that meet both regulatory requirements and consumer expectations for sustainability.

6 Conclusion

The application of machine learning in the manufacturing industry has undoubtedly revolutionized various aspects of production, quality control, and operational efficiency. As this technology continues to evolve and become more widely adopted, the manufacturing sector will witness a profound transformation in the way it approaches problem-solving, decision-making, and process optimization. The key advantages of implementing machine learning in manufacturing include enhanced predictive maintenance, improved quality control, increased production yield, and optimized resource utilization. By leveraging the power of data-driven insights and predictive analytics, manufacturers can identify potential issues before they arise, proactively address bottlenecks, and make informed decisions to enhance overall operational performance.

References

1. Sharma, A., Zhang, Z., Rai, R.: The interpretive model of manufacturing: a theoretical framework and research agenda for machine learning in manufacturing. Int. J. Prod. Res. **59**(16), 4960–4994 (2021). https://doi.org/10.1080/00207543.2021.1930234
2. Robey, A., Pappas, G.J., et al.: Model-based domain generalization. Adv. Neural Inf. Process. Syst. (2021)
3. Samanta, A., Chowdhuri, S., Williamson, S.S.: Machine learning-based data-driven fault detection/diagnosis of lithium-ion battery: a critical review. Electronics, MDPI.com (2021)
4. Garcia-Lamont, F., Rodríguez-Mazahua, L., Lopez, A.: A comprehensive survey on support vector machine classification: Applications, challenges, and trends. Elsevier, Neurocomputing (2020)
5. Zhu, X., Goldberg, A.B.: Introduction to semi-supervised learning (2022)
6. Bazzi, A., Shaout, A., Ma, D.: A novel variability-rich scheme for software updates of automotive systems. IEEE Access **12**, 79530–79548 (2024). https://doi.org/10.1109/ACCESS.2024.3409629

7. Xia, K., Sacco, C., Kirkpatrick, M., Saidy, C., et al.: A digital twin to train deep reinforcement learning agent for smart manufacturing plants: Environment, interfaces, and intelligence. J. Manuf. Syst. 60, pp. 811–822, Elsevier (2021)
8. Sarker, I.H.: Machine learning: algorithms, real-world applications, and research directions. SN Comput. Sci. (2021)
9. Countdown, L.: The 2022 report of the Lancet countdown on health and climate change: health at the mercy of fossil fuels (2022)
10. Herzog, T., Brandt, M., Trinchi, A., et al.: Process monitoring and machine learning for defect detection in laser-based metal additive manufacturing. J. Intell. Manuf. **35**, 1407–1437 (2024)
11. Hu, C., Wang, Y.: An efficient convolutional neural network model based on object-level attention mechanism for casting defect detection on radiography images. IEEE Trans. Ind. Electron. **67**(12), 10922–10930 (2020). https://doi.org/10.1109/TIE.2019.2962437
12. Parente, M., Figueira, G., Amorim, P., et al.: Production scheduling in the context of Industry 4.0: review and trends. Int. J. Prod. Res. **58**(17), 5401–5431 (2020). https://doi.org/10.1080/00207543.2020.1718794
13. Baltes, S., Ralph, P.: Sampling in software engineering research: a critical review and guidelines. Empir. Softw. Eng. (2022). https://doi.org/10.1007/s10664-021-10072-8
14. Peng, S.-L., Jhanjhi, N.Z., Pal, S., Amsaad, F. (eds.): Proceedings of 3rd International Conference on Mathematical Modeling and Computational Science: ICMMCS 2023. Springer Nature Singapore, Singapore (2023)
15. Peng, S.-L., Lin, C.-K., Pal, S. (eds.): Proceedings of 2nd International Conference on Mathematical Modeling and Computational Science. Advances in Intelligent Systems and Computing. Springer Nature Singapore (2022). https://doi.org/10.1007/978-981-19-0182-9
16. Peng, S.-L., Hao, R.-X., Pal, S. (eds.): Proceedings of First International Conference on Mathematical Modeling and Computational Science. Advances in Intelligent Systems and Computing. Springer Singapore (2021). https://doi.org/10.1007/978-981-33-4389-4

Neutrosophic Semi β^F-Connectedness and Neutrosophic Semi β^F-Feebly Connectedness in Neutrosophic Topological Space

N. Kalaivani(✉) and K. Fayaz Ur Rahman

Department of Mathematics, Vel Tech Rangarajan Dr. Sagunthala R & D Institute of Science and Technology, Chennai, India
{drkalaivani,vtd675}@veltech.edu.in

Abstract. The central ambition based on this article inhabit to nominate the conception based on a neutron S_β^F-conn spc in NTS. Moreover, some properties of the neutron S_β^F-conn spc have been proposed with suitable illustrations. The theorems like "If a NTS $(X_{\mathcal{N}eut}, \tau)$ is neutron S_β^F-conn spc between the NSs A together with B, then it is neutro conn spc between A together with B". Further the inverse of the above theorem is not accurate in general are proved with suitable illustration. Moreover, many essential theorems based on the neutron S_β^F-conn spc have been proved. In addition, various relevant theorems based on the operators like neutron S_β^F-closure have been demonstrated. Further, a new concept based on a neutron S_β^F-FCS in a NTS have been proposed and studied with proper illustration. Moreover, using the theory of neutron S_β^F-CS a new concept based on a neutron S_β^F-FSCS is introduced and studied with proper illustration. Further the theorems connecting neutron S_β^F-FCS and neutro S_β^F-FSCS have been proved and illustrated. In addition, a relationship between a neutron S_β^F-conn spc, neutron S_β^F-FCS and neutron S_β^F-FSCS have been studied with proper illustration as shown below in Fig. 1. Several theorems based on the above concepts have been justified.

Keywords: NTS · Neutro S_β^F-CS · Neutro S_β^F-OS · Neutro S_β^F-conn spc · Neutro S_β^F-FCS · Neutro S_β^F-FSCS

Mathematics Subject Classification: AMS(2000)54A05 · 54A10 · 54A35

Nomenclatures

1 Neutrosophic topological space	NTS
2 Closed set	CS
3 Open set	OS

S. Pal and Á. Rocha (Eds.): ICMMCS 2025, LNNS 1398, pp. 381–393, 2025.
https://doi.org/10.1007/978-3-031-90998-6_35

4 Connected space	Conn spc
5 Feebly connected space	FCS
6 Feebly disconnected space	FDS
7 Feebly strongly connected space	FSCS
8 Topological space	TS
9 Neutrosophic	Neutro
10 Neutrosophic set	NS
11 Neutrosophic sets (Plural)	NSs
12 Topology	Topo

1 Introduction

NTS inhabit an expansion based on intuitionistic TS together with every triplet set in NTS contains membership, indeterminacy together with non-membership values. The NS set inhabit regard with origin, nature together with scope based on neutrality. Research beside the NS inhabit essential therefore it provides contact to a collection based on scientific together with technical applications. The universe inhabit full based on uncertainty, thus a NS can locate a relevant location for inspection. The NS beside the real line inhabit differentiated by single-valued neutro numbers (SVN) together with single-valued bipolar neutro numbers (SVbN). When a human opinion inhabit based on positive together with negative ideas, the SVbN number inhabit essential in opinion-making problems. Every SVbN number membership functions (truth, indeterminacy, together with falsity) are accommodated in the positive together with negative category between -1 together with 0, 0 together with 1, correspondingly. With its probable application in practical settings, neutro topo has now become a important field for inquiry.

Acikgoz and Esenbel [1] proposed the approach of neutro connectedness and some of its characterizations. Further, the concept of neutro super-conn spcs, neutro strongly conn spcs were also introduced and some of its properties were studied. Arokiarani et al. [2] derived the notion based on the neutro semi open function and reviewed some of its inheritance. Also, the condition for a function based on the NTS to be a neutro α-continuous function was given and investigated. Bageerathi and Puvaneswari [3] introduced the notion based on NFCS, NFDS and neutro feebly C_i-disconnected (i = 1,2, 3,4) spaces in NTSs. In addition, some characterizations had been discussed with proper examples. Çoker [4] suggested the basic the approach based on "intuitionistic fuzzy TS". Moreover, the theory based on fuzzy continuity, fuzzy compactness together with fuzzy connectedness were also introduced and inspected. Das et al. [5] originated the notion based on neutro soft set approach with roughness without using full soft set. Furthermore, few inheritance had been inspected beside roughness with neutro soft set. Das and Tripathy [6] examined some properties based on neutro multiset topo. Moreover, the attitude based on the compactness along with connectedness in neutro multiset topo had been investigated. In addition the interpretations based on "neutro multiset TS" had been reviewed. Savithiri et al. [7] prefaced a new class belonging to the sets named neutro vague rw CS, neutro vague rw OS together with neutro vague rw $T_{\frac{1}{2}}$ spc in NTS. Moreover, the properties based on this concept was examined. Ebenanjar et al. [8] prefaced the notion based on neutro b-OS, neutro b-CS along with their properties were investigated. In addition, some operators namely neutro b-interior and neutro b-closure was defined and some relationship between the other operators were also studied.

Gündüz et al. [9] prefaced the notion based on neutro soft set theory. In addition, neutro soft point notion had been prefaced. In addition, the notion based on neutro soft T_i-spaces were prefaced together with the association among them were reviewed in detail. Iswarya and Bageerathi [10] prefaced the notion based on neutro semi-conn spc whichever were solved by adopting neutro semi-OS along with neutro semi-CS. Furthermore, the association among neutro semi-frontier along with neutro semi-irresolute function had been reviewed. Kalaivani et al. [11, 12] prefaced the notion based on $(\gamma, \gamma')^{\alpha}$-regular spaces together with $(\gamma, \gamma')^{\alpha}$-normal spaces with the help of (γ, γ')-OS and (γ, γ')-CS. Furthermore some of its properties had been reviewed with proper illustration. In addition, the notion based on GS_{β}-compact space, GS_{β}-compact set relative to space together with countable GS_{β}-compact space had been prefaced with the idea of GS_{β}-OSs. In addition, the concepts of GS_{β}-compact space together with countable GS_{β}-compact space had been interrelated with suitable illustration. Madhumathi and NirmalaIrudayam [13] prefaced the notion based on a neutro orbit TS whichever was indicated by (X, t_{NO}). Likewise, few of the significant representative based on neutro orbit OSs were reviewed. Nagalakshmi et al. [14] prefaced the notion based on hexagonal fuzzy numbers [HFN] together with neutro triangular fuzzy numbers[NTFN]. Furthermore, prevailing ranking methods are devoted to both the HFNs together with NTFNs for remodeling the fuzzy-based transportation dilemma to crisp-based transportation dilemma. Parimala et al. [15] introduced a new type of generalized CS, generalized OS called nIg-CS and nIg-OS in nano ideal TS and examined few based on their significant inheritance. Parimala, M together with Karthika [16] prefaced the concept of neutro $\alpha\psi$-connectedness in NTSs and examined few properties based on $\alpha\psi$-connectedness between sets along with subsets of two sets. Parvathi et al. [17] prefaced a linear regression analysis in an intuitionistic fuzzy environment provoking intuitionistic fuzzy linear models with symmetric triangular intuitionistic fuzzy number coefficients. Pushpalatha and Nandhini [18] prefaced the notion based on neutro CS in a NTS and reviewed few of the basic properties. Salama and Alblowi [19] proposed a concept based on fuzzy TS along with intuitionistic fuzzy TS together with studied few of their significant properties.

Santhi and Prakash [20] originated the concept based on intuitionistic fuzzy semi-generalized CS together with intuitionistic fuzzy semi generalized OS in intuitionistic fuzzy TS. In addition, the application of intuitionistic fuzzy semi-generalized CS was also examined. Santhi and Udhayarani [21] proposed the theory based on neutro ω-CS in the NTS. Moreover, some characterizations of a neutron ω-CS has been studied. Subasree [22, 23] prefaced the approach based on neutron $\psi\beta$ together with neutron $\beta\psi$-CS in NTS. Further, some of their properties were also studied. Moreover, neutron β^*-continuity and neutro β^*-contra continuity was introduced and some of its characterizations were also discussed. Thomas and Anila [24] prefaced the notion based on neutro semi-pre-OS along with neutro semi-pre-CS in NTS. Further, some of its characterization had been studied. Tran et al. [25] prefaced an operator on NS together with some of its inheritances were studied. In addition, the association between the topo beside NS together with fuzzy soft topologies were reviewed and examined. Zadeh [26] proved a separation theorem for a convex fuzzy set without requiring that the fuzzy set be disjoint.

Objective 1.1. The notion based on neutron S_β^F-connectedness between NSs are generally used together with played a important aspect in sensor networks, robotics, physics, mechanical engineering together with to find the finest fit in data exacting thousands based on variables. The major benefit based on this idea is that one can formulate endless connected networks together with likewise we can find the finest involving thousands based on variable where each variable have acceptance, rejection together with indeterminacy. Going to the literature review Parimala M and Karthika M defined and studied neutron $\alpha\psi$-connectedness in a NTS and also studied some of its properties. This paper motivates me to define a new concept of neutron S_β^F-conn spc in an NTS. In addition, the main disadvantage of the neutron $\alpha\psi$-connectedness paper is that we can't able to define the concept of FCS and FSCS. But using neutron S_β^F-conn spc it is possible to define the concept of FCS and FSCS.

2 Preliminaries

Definition 1. Accredit X prevail a non-empty set along with I prevail the closed interval $[0, 1]$. A NS D inhabit characterized by $D = \{< m, \mu_D(m), \sigma_D(m), \gamma_D(m) >: m \in X\}$, where the mapping $\mu_D : X \to I$, $\sigma_D : X \to I$ and $\gamma_D : X \to I$ denote the degree based on membership, degree based on indeterminacy together with degree based on non-membership function correspondingly for every element $m \in X$ to the set D and $0 \leq \mu_D(m) + \sigma_D(m) + \gamma_D(m) \leq 3$ for each $m \in X$.

Definition 2. Accredit P along with Q are NSs of the form $P = \{< m, \mu_P(m), \sigma_P(m), \gamma_P(m) >: m \in X\}$ and $Q = \{< m, \mu_Q(m), \sigma_Q(m), \gamma_Q(m) : m \in X\}$. Then,

(i) $P \subseteq Q$ in the case that $\mu_P(m) \leq \mu_Q(m)$,$\sigma_P(m) \leq \sigma_Q(m)$ and $\gamma_P(m) \geq \gamma_Q(m)$
(ii) $\overline{P} = \{<\gamma_P(m), \sigma_P(m), \mu_P(m) >: m \in X\}$
(iii) $P \bigcap Q = \{< m, \mu_P(m) \cap \mu_Q(m), \sigma_P(m) \cup \sigma_Q(m), \gamma_P(m) \cup \gamma_Q(m) >: m \in X\}$;
(iv) $P \bigcup Q = \{< m, \mu_P(m) \cup \mu_Q(m), \sigma_P(m) \cap \sigma_Q(m), \gamma_P(m) \cap \gamma_Q(m) >: m \in X\}$.

Definition 3. A neutro topo beside a non-empty set X is a class ρ based on NSs in X, if

(i) $0_N, 1_N \in \rho$
(ii) $A \bigcap B \in \rho$ for all $A, B \in \rho$
(iii) $\bigcup A_i \in \rho$ for all discretionary family $\{A_i : i \in J\} \subseteq \rho$.

Definition 4. Accredit L prevail a NSs in NTS X. Formerly

$N\text{-}int(L) = \bigcup\{G : G$ is a neutro OS in X together with $G \subseteq L\}$ is labelled as a neutro interior based on L.

$N\text{-}cl(L) = \bigcap\{G : G$ is a neutro CS in X together with $G \supseteq L\}$ is labelled as a neutro closure based on L.

Definition 5. $0_N = \{< g, 0, 0, 1 >: g \in G\}$ along with $1_N = \{< g, 1, 0, 0 >: g \in G\}$.

Definition 6. Accredit $(X_{\mathcal{N}eut}, \tau)$ be a NTS and a subset M of $(X_{\mathcal{N}eut}, \tau)$ is termed as

(i) a neutro pre-OS, if $M \subseteq Nint(Ncl(M))$ and a neutro pre-CS if $Ncl(Nint(M)) \subseteq M$.
(ii) a neutro semi-OS, if $M \subseteq Ncl(Nint(M))$ and a neutro semi-CS if $Nint(Ncl(M)) \subseteq M$.
(iii) a neutro β-OS, if $M \subseteq Ncl(Nint(Ncl(M)))$ and a neutro β-CS if $Nint(Ncl(Nint(M))) \subseteq M$.

Definition 7. A subset F of a NTS $(X_{\mathcal{N}eut}, \tau)$ is termed as a

(i) a neutro semi-generalized CS if $scl(F) \subseteq U$ when $F \subseteq U$ together with U inhabit a semi-OS in $(X_{\mathcal{N}eut}, \tau)$
(ii) a neutro S^F_β-CS if $S^F_\beta\text{-}cl(F) \subseteq U$ when $F \subseteq U$ together with U inhabit a neutro S^F_β-OS in $(X_{\mathcal{N}eut}, \tau)$.

Definition 8. Accredit $(X_{\mathcal{N}eut}, \tau)$ be a NTS together with F prevail a NSs in NTS $(X_{\mathcal{N}eut}, \tau)$. Formerly,

(i) $NS^F_\beta\text{-}int(F) = \bigcup\{G : G$ is a NS^F_β-OS in X and $G \subseteq F\}$ is labeled as a neutro S^F_β-interior based on F.
(ii) $NS^F_\beta\text{-}cl(F) = \bigcap\{G : G$ is a NS^F_β-CS in X and $G \supseteq F\}$ is labeled as a neutro S^F_β-closure based on F.

Definition 9. Two NSs A along with B based on X are forenamed to previal q-coincident (AqB for short) in the case particular there occurs an element $y \in X$ so that $A(y)+B(y) > 1$.

Lemma 1. For any two NSs L and M of X, $\neg(LqM)$ if and only if $L \subset M^c$.

Definition 10. A NTS is forenamed to be neutro conn spc if no non-empty NSs is both neutro OS and neutro CS.

Remark 1. Each one of the neutro CS (resp. Neutro open) is a neutro S^F_β-closed (resp. Neutro S^F_β-open) but the inverse may not be accurate.

Illustration 1. Accredit $X = \{a, b\}$, $0_N = \{(0,0,1)(0,0,1)\}$, $1_N = \{(1,0,0)(1,0,0)\}$, $\tau_N = \{0_N, 1_N, G, H, G \bigcup H, G \bigcap H\}$ where $G = \{(0.5, 0.3, 0.2)(0.8, 0.6, 0.4)\}$, $H = \{(0.7, 0.6, 0.3)(0.9, 0.4, 0.2)\}$. Here $F = \{(0.1, 0.2, 0.8)(0.1, 0.5, 0.8)\}$ inhabit a neutro S_{β^F}-CS, but F is not a neutro CS.

Definition 11. A NTS $(X_{\mathcal{N}eut}, \tau)$ is forenamed to be neutro connected space between NSs A and B if there is no neutro closed OS F in X so that $A \subset F$ and $\neg(FqB)$.

Definition 12. A neutro subset F of a NTS $(X_{\mathcal{N}eut}, \tau)$ is forenamed to be a neutro S_{β}^{F}-feebly OS if there is a neutro S_{β}^{F}-OS U in X such that $U \leq F \leq NScl(U)$.

A neutro subset F of a NTS $(X_{\mathcal{N}eut}, \tau)$ is forenamed to be a neutro S_{β}^{F}-feebly CS if there is a neutro S_{β}^{F}-CS U in X such that $NSint(U) \leq F \leq U$.

Definition 13. Accredit $(X_{\mathcal{N}eut}, \tau)$ be a NTS along with $P =< m, \mu_P(m), \sigma_P(m), \gamma_P(m) >$ be a NS in X. Then a neutro S_{β}^{F}-feebly interior of P is characterized by $NS_{\beta}^{F}\text{-}Fint(P) = \bigcup\{G : G$ is a neutro S_{β}^{F}-feebly OS in X along with $G \leq P\}$.

Definition 14. Accredit $(X_{\mathcal{N}eut}, \tau)$ be a NTS along with $P =< m, \mu_P(m), \sigma_P(m), \gamma_P(m) >$ be a NS in X. Formerly a neutro S_{β}^{F}-feebly closure of P is characterized by $NS_{\beta}^{F}\text{-}Fcl(P) = \bigcap\{G : G$ is a neutro S_{β}^{F}-feebly CS in X and $P \leq G\}$.

3 Neutrosophic Semi β^F-Connectedness

Definition 15. A NTS $(X_{\mathcal{N}eut}, \tau)$ is forenamed to be a neutro S_{β}^{F}-conn spc between NSs A and B if there is no neutro S_{β}^{F}-CS S_{β}^{F}-OS F in X such that $A \subset F$ and $\neg(FqB)$.

Illustration 2. Accredit $X = \{a, b\}$ together with $U =< (0.2, 0.5, 0.9)(0.1, 0.3, 0.5) >$,$A =< (0.1, 0.2, 0.7)(0.3, 0, 0.8) >$ and $B =< (0.5, 0.7, 0.8)(0.3, 0.6, 0.8) >$ be a NSs beside X. Accredit $\tau = \{0_N, 1_N, U\}$ be a neutro topology beside X. Formerly $(X_{\mathcal{N}eut}, \tau)$ be a neutro S_{β}^{F}-conn spc between A together with B.

Theorem 1. If a NTS $(X_{\mathcal{N}eut}, \tau)$ is a neutro S_{β}^{F}-conn spc between NS A together with B, formerly it is a neutro conn spc between A together with B.

Proof. In case that $(X_{\mathcal{N}eut}, \tau)$ is not a neutro conn spc between A together with B, then there occurs a neutro CS OS F in X such that $A \subset F$ and $\neg(FqB)$. Then by remark 2.1 (Each one of the neutro CS (resp. Neutro open) is a neutro S_{β}^{F}-closed (resp. Neutro S_{β}^{F}-open) but the inverse may not be accurate),F is a neutro S_{β}^{F}-CS S_{β}^{F}-OS in X so that $A \subset F$ and $\neg(FqB)$. Therefore $(X_{\mathcal{N}eut}, \tau)$ is not a neutro S_{β}^{F}-connected space among A together with B, which conflict our assumption.

Remark 2. The inverse based on the theorem 1 may not be accurate, as exhibited in the subsequent illustration.

Example 3. Let $X = \{a, b\}$ and $U =< (0.3, 0.5, 0.6)(0.2, 0.4, 0.5) >$,$A =< (0.1, 0.2, 0.9)(0, 0.2, 0.9) >$ and $B =< (0.5, 0.6, 0.7)(0.3, 0.5, 0.7) >$ be a NSs beside X. Accredit $\tau = \{0_N, 1_N, U\}$ be a neutro topology beside X. Formerly $(X_{\mathcal{N}eut}, \tau)$ be a neutro conn spc between A together with B but it is not a neutro S_{β}^{F}-conn spc between A together with B.

Theorem 2. A NTS $(X_{\mathcal{N}eut}, \tau)$ is a neutro S_β^F-conn spc between NSs A and B in the case that there is no neutro S_β^F-CS S_β^F-OS F in X so that $A \subset F \subset B^C$.

Proof

(Necessary): Accredit $(X_{\mathcal{N}eut}, \tau)$ is a neutro S_β^F-conn spc between NSs A and B. Presume beside the contradictory, that F is a neutro S_β^F-CS S_β^F-OS in X such that $A \subset F \subset B^C$. Now $F \subset B^C$ which entail that $\neg(FqB)$. Since F is a neutro S_β^F-CS S_β^F-OS in X so that $A \subset F$ together with $\neg(FqB)$. Consequently $(X_{\mathcal{N}eut}, \tau)$ is not a neutro S_β^F-conn spc between NSs A together with B, which disclaim our assmption.

Sufficiency: Presume beside the contradictory, that $(X_{\mathcal{N}eut}, \tau)$ is not a neutro S_β^F-conn spc between NSs A together with B. Formerly there occurs a neutro S_β^F-CS S_β^F-OS F in X such that $A \subset F$ and $\neg(FqB)$. Now,$\neg(FqB)$ which entail that $F \subset B^C$. Since F is a neutro S_β^F-CS S_β^F-OS in X so that $A \subset F \subset B^C$, which disclaim our hypothesis.

Theorem 3. If a NTS $(X_{\mathcal{N}eut}, \tau)$ is a neutro S_β^F-conn spc between NSs A together with B, formerly A together with B are non-empty.

Proof. On the assumption that NSs A abide empty, formerly A abide a neutro S_β^F-CS S_β^F-OS in X together with $A \subset B$. Now we plea that $\neg(AqB)$. If AqB, formerly there occurs an element $y \in X$ so that $\mu_A(y) > \gamma_B(y)$ or $\gamma_A(y) < \mu_B(y)$. But $\mu_A(y) = 0_N$ and $\gamma_A(y) = 1_N$ for entire $y \in X$. So, no point $y \in X$ for which $\mu_A(y) > \gamma_B(y)$ or $\gamma_A(y) < \mu_B(y)$ or $\sigma_A(y) > \sigma_B(y)$, which is a conflict. Therefore $\neg(AqB)$ together with $(X_{\mathcal{N}eut}, \tau)$ is not a neutro S_β^F-conn spc between NSs A together with B.

Theorem 4. If a NTS $(X_{\mathcal{N}eut}, \tau)$ is a neutro S_β^F-conn spc between NSs A and B,$A \subset A_1$ and $B \subset B_1$, formerly $(X_{\mathcal{N}eut}, \tau)$ is a neutro S_β^F-conn spc between A_1 together with B_1.

Proof. Presume $(X_{\mathcal{N}eut}, \tau)$ is not a neutro S_β^F-conn spc between NSs A_1 together with B_1. Then there occurs a neutro S_β^F-CS S_β^F-OS F in X so that $A_1 \subset F$ and $\neg(FqB_1)$. Obviously,$A \subset F$. Now we plea that $\neg(FqB)$. If FqB, formerly there occurs a point $y \in X$ so that $\mu_F(y) > \gamma_B(y)$ or $\gamma_F(y) < \mu_B(y)$. Without loss of generality presume a point $y \in X$ so that $\mu_F(y) > \gamma_B(y)$. Now $B \subset B_1$,$\gamma_B(y) \geq \gamma_{B_1}(y)$. Also,$\mu_F(y) > \gamma_{B_1}(y)$ together with FqB_1, a dispute. Subsequently $(X_{\mathcal{N}eut}, \tau)$ is not a neutro S_β^F-conn spc between NSs A together with B.

Theorem 5. A NTS $(X_{\mathcal{N}eut}, \tau)$ is a neutro S_β^F-conn spc between NSs A and B in the case that it is a neutro S_β^F-conn spc between $NS_\beta^F\text{-}cl(A)$ and $NS_\beta^F\text{-}cl(B)$.

Proof

(Necessary): Follows from the theorem 3.4 (If a NTS $(X_{\mathcal{N}eut}, \tau)$ is a neutro S_β^F-conn spc between NSs A and$B$, $A \subset A_1$ and$B \subset B_1$, formerly $(X_{\mathcal{N}eut}, \tau)$ is a neutro S_β^F-conn spc between A_1 together withB_1), because $A \subset NS_\beta^F\text{-}cl(A)$ and $B \subset NS_\beta^F\text{-}cl(B)$.

Sufficiency: Presume $(X_{\mathcal{N}eut}, \tau)$ is not a neutro S_β^F-conn spc between NSs A and B. Then there occurs a neutro S_β^F-CS S_β^F-OS F based on X such that $A \subset F$ along with $\neg(FqB)$. Therefore F is a neutro S_β^F-closed along with $A \subset F, NS_\beta^F$-$cl(A) \subset F$. Now,$\neg(FqB)$ which implies that $F \subset B^C$. Since $F = NS_\beta^F\text{-}intF \subset NS_\beta^F$-$\left(B^C\right) = (NS_\beta^F cl(B))^C$. Hence $(FqNS_\beta^F\text{-}cl(B))$ along with X is not neutro S_β^F-connected space between $NS_\beta^F\text{-}cl(A)$ and $NS_\beta^F\text{-}cl(B)$.

Theorem 6. Accredit $(X_{\mathcal{N}eut}, \tau)$ be a NTS together with A, B be two NSs in X. If AqB formerly $(X_{\mathcal{N}eut}, \tau)$ is a neutro S_β^F-conn spc between A together with B.

Proof. If F is any neutro S_β^F-CS S_β^F-OS in X so that $A \subset F$, then AqB and therefore FqB.

Remark 3. The converse based on the theorem 6 may not be accurate, as demonstrated in the subsequent illustration.

Illustration 4. Let $X = \{a, b\}$ and $U =< (0.3, 0.5, 0.7)(0.1, 0.3, 0.8) >$,$A =< (0.2, 0.4, 0.7)(0.3, 0.5, 0.8) >$ and $B =< (0.1, 0.3, 0.6)(0.2, 0.5, 0.9) >$ be a NSs on X. Let $\tau_N = \{0_N, 1_N, U\}$ be a neutro topo beside X. Formerly $(X_{\mathcal{N}eut}, \tau)$ is a neutro S_β^F-conn spc between NSs A together with B but not $\neg(AqB)$.

Theorem 7. A NTS $(X_{\mathcal{N}eut}, \tau)$ is a neutro S_β^F-conn spc in the case that it is a neutro S_β^F-conn spc between each pair based on its non-empty NSs.

Proof.
(Necessity): Accredit M, N be any pair of neutro subsets based on X. Presume $(X_{\mathcal{N}eut}, \tau)$ is not a neutro S_β^F-conn spc between NSs M and N, then there occurs a neutro S_β^F-CS S_β^F-OS F of X so that $M \subset F$ and $\neg(FqN)$. Therefore NSs M and N are non-empty, it ensues that F abide a non-empty proper neutro S_β^F-CS S_β^F-OS of X. Therefore $(X_{\mathcal{N}eut}, \tau)$ is not a neutro S_β^F-conn spc.

(Sufficiency): Presume $(X_{\mathcal{N}eut}, \tau)$ is not a neutro S_β^F-conn spc, then there occurs a non-empty proper neutro S_β^F-CS S_β^F-OS F of X. Subsequently X is not a neutro S_β^F-conn spc between F together with F^C, a conflict.

Remark 4. If a NTS $(X_{\mathcal{N}eut}, \tau)$ is a neutro S_β^F-conn spc among a pair based on its neutro subsets it is not significant that $(X_{\mathcal{N}eut}, \tau)$ abide a neutro S_β^F-conn spc among every pair based on its neutro subsets and it is not significant neutro S_β^F-conn spc, as demonstrated in the subsequent illustration.

Example 5. Accredit $X = \{a, b\}$ together with $U =< (0.2, 0.4, 0.7)(0.3, 0.5, 0.8) >$, $A =< (0.1, 0.5, 0.9)(0.1, 0.4, 0.9) >$, $B =< (0.1, 0.4, 0.7)(0.2, 0.6, 0.8) >$, $C =< (0.2, 0.5, 0.8)(0.1, 0.3, 0.9) >$ and $D =< (0.2, 0.4, 0.8)(0.1, 0.5, 0.9) >$ be a NSs on X. Let $\tau = \{0_N, 1_N, U\}$ be a neutro topo based on X. Formerly $(X_{\mathcal{N}eut}, \tau)$ is a neutro conn spc between NSs A and B but it is not a neutro conn spc between NSs C together with D. Likewise $(X_{\mathcal{N}eut}, \tau)$ is not a neutro S_β^F-conn spc.

Theorem 8. Let (Y, τ_Y) previal a subspace based on a NTS $(X_{\mathcal{N}eut}, \tau)$ together with A, B be a neutro subsets based on Y. If (Y, τ_Y) is a neutro S^F_β-conn spc between A together with B then so is $(X_{\mathcal{N}eut}, \tau)$.

Proof. Presume, on the contradictory that $(X_{\mathcal{N}eut}, \tau)$ is not a neutro S^F_β-conn spc between a NSs A together with B. Formerly there occurs a neutro S^F_β-CS S^F_β-OS F of X so that $A \subset F$ and $\neg(FqB)$. Put $F_Y = F \bigcap Y$. Then F_Y is a neutro S^F_β-CS S^F_β-OS in Y so that $A \subset F_Y$ and $\neg(F_Y qB)$. Therefore (Y, τ_Y) is not a neutro S^F_β-conn spc between A and B, a conflict.

Theorem 9. Let (Y, τ_Y) be a neutro closed open subspace based on a NTS $(X_{\mathcal{N}eut}, \tau)$ and A, B be a neutro subsets based on Y. If $(X_{\mathcal{N}eut}, \tau)$ is a neutro S^F_β-conn spc between NSs A and B, formerly therefore is (Y, τ_Y).

Proof. If (Y, τ_Y) is no more a neutro S^F_β-conn spc between NSs A together with B, formerly there occurs a neutro S^F_β-CS S^F_β-OS F of Y so that $A \subset F$ and $\neg(FqB)$. Since Y is a neutro CS OS in X,F is a neutro S^F_β-CS S^F_β-OS in X. Therefore X cannot be neutro S^F_β-conn spc between NSs A and B, a conflict.

4 Neutrosophic S^F_β-Feebly Connected Space

Definition 16. A NTS $(X_{\mathcal{N}eut}, \tau)$ is a neutro S^F_β-FDS if there occurs a neutro S^F_β-feebly OS E, F in X,$E \neq 0_N, F \neq 0_N$ so that $E \cup F = 1_N$ together with $E \cap F = 0_N$.

(i.e.) (i) $< x, \mu_E \cup \mu_F, \sigma_E \cap \sigma_F, \gamma_E \cap \gamma_F >= 1_N$
(ii) $< x, \mu_E \cup \mu_F, \sigma_E \cup \sigma_F, \gamma_E \cap \gamma_F >= 1_N$
(iii) $< x, \mu_E \cap \mu_F, \sigma_E \cap \sigma_F, \gamma_E \cup \gamma_F >= 0_N$
(iv) $< x, \mu_E \cap \mu_F, \sigma_E \cup \sigma_F, \gamma_E \cup \gamma_F >= 0_N$

The subsequent examples display that X inhabit a neutro S^F_β-FDS and neutro S^F_β-FCS.

Example 6. Accredit $X = \{a, b, c\}, \tau = \{0_N, 1_N, A, B, C, D\}$ where

$A =< (0, 0, 0)(1, 1, 0)(0, 0, 1) >$
$B =< (1, 0, 0)(0, 1, 0)(1, 1, 0) >$
$C =< (1, 0, 0)(0, 1, 0)(1, 0, 0) >$
$D =< (0, 0, 0)(0, 1, 0)(0, 1, 1) >$

A and B are neutro S^F_β-feebly OS in $X, A \neq 0_N, B \neq 0_N$ and $A \cup B = C = 1_N, A \cap B = D = 0_N$. Hence X inhabit a neutro S^F_β-FDS.

If X is not a neutro S^F_β-FDS then it is forenamed as a neutro S^F_β-FCS.

Example 7. Accredit $X = \{a, b\}, \tau = \{0_N, 1_N, G_1\}$ where $G_1 =< (0.1, 0.3, 0.7)(0.2, 0.5, 0.9) >$. Then $(X_{\mathcal{N}eut}, \tau)$ is NTS. Let $G_2 =< (0.7, 0.4, 0.1)(0.9, 0.5, 0.3) >$. Then G_1 and G_2 are neutro S^F_β-feebly OS in X, $G_1 \neq 0_N, G_2 \neq 0_N$, and $G_1 \cup G_2 \neq 1_N, G_1 \cap G_2 \neq 0_N$. Hence X is a neutro S^F_β-FCS.

Theorem 10. Let $k : (X_{\mathcal{N}eut}, \tau) \rightarrow (Y_{\mathcal{N}eut}, \sigma)$ be a neutro S_β^F-feebly irresolute surjection together with X be a neutro S_β^F-FCS. Formerly Y inhabit a neutro S_β^F-FCS.

Proof. Presume that Y is not a neutro S_β^F-FCS, then there occurs a non-empty neutro S_β^F-feebly OS U and V in Y so that $U \cup V = 1_N$ and $U \cap V = 0_N$. Therefore k is a neutro S_β^F-feebly irresolute mapping, $A = k^{-1}(U) \neq 0_N$,$B = k^{-1}(V) \neq 0_N$, which is a neutro S_β^F-feebly OS in X and $k^{-1}(U) \cup k^{-1}(V) = k^{-1}(1_N) = 1_N$, which entail $A \cup B = 1_N$. Also $k^{-1}(U) \cap k^{-1}(V) = k^{-1}(0_N) = 0_N$, which entails $A \cap B = 0_N$. Hence X is a neutro S_β^F-FDS, which is a conflict to our assumption. Therefore Y is a neutro S_β^F-FCS.

Theorem 11. A NTS $(X_{\mathcal{N}eut}, \tau)$ is a neutro S_β^F-FCS in the case that there occurs a no non-zero neutro S_β^F-feebly OS U together with V in $(X_{\mathcal{N}eut}, \tau)$ such that $U = V^C$.

Proof.
(Necessity): Accredit U together with V be two neutro S_β^F-feebly OS in $(X_{\mathcal{N}eut}, \tau)$ so that $U \neq 0_N, V \neq 0_N$ and $U = V^C$. Consequently V^C is a neutro S_β^F-feebly CS. Therefore $U \neq 0_N, V \neq 1_N$. This entails V abide a proper S_β^F-neutro subset which abide both neutro S_β^F-feebly OS together with neutro S_β^F-feebly CS in X. Hence X is not a neutro S_β^F-FCS. But the indicated inhabit a conflict to our assumption. Formerly, there exists no non-zero neutro S_β^F-feebly OS U together with V in X, so that $U = V^C$.

(Sufficiency): Accredit U be both neutro S_β^F-feebly OS together with neutro S_β^F-feebly CS in X so that $U \neq 0_N, U \neq 1_N$. Accredit $V = U^C$. Formerly V is a neutro S_β^F-feebly OS together with $V \neq 1_N$. This entails $U^C = V \neq 0_N$, which is a conflict to our assumption. Consequently X is a neutro S_β^F-FCS.

Theorem 12. A NTS $(X_{\mathcal{N}eut}, \tau)$ is a neutro S_β^F-FCS in the case that there occurs no non zero neutro subsets U and V in $(X_{\mathcal{N}eut}, \tau)$, so that $U = V^C, V = (NS_\beta^F Fcl(U))^C$ and $U = (NS_\beta^F Fcl(V))^C$.

Proof.
(Necessity): Accredit U together with V be two neutro subsets in $(X_{\mathcal{N}eut}, \tau)$ so that $U \neq 0_N, V \neq 0_N$ and $U = V^C, V = (NS_\beta^F Fcl(U))^C$ and $U = (NS_\beta^F Fcl(V))^C$. Since $(NS_\beta^F Fcl(V))^C$ and $(NS_\beta^F Fcl(U))^C$ are neutro S_β^F-feebly OS in X,U and V are neutro S_β^F-feebly OS in X. This entails X is not a neutro S_β^F-FCS, which inhabit a conflict. Consequently, there occurs no non zero neutro S_β^F-feebly OS U together with V in X, such that $U = V^C$, such that $V = (NS_\beta^F Fcl(U))^C$ and $U = (NS_\beta^F Fcl(V))^C$.

(Sufficiency): Let U be both neutro S_β^F-feebly OS and neutro S_β^F-feebly CS in X so that $U \neq 0_N, U \neq 1_N$. Now by taking $V = U^C$, we attain a conflict to our assumption. Therefore X is a neutro S_β^F-FCS.

Definition 17. A NTS $(X_{\mathcal{N}eut}, \tau)$ is a neutro S_β^F-FSCS, if there occurs no non empty neutro S_β^F-CS A together with B in X so that $\mu_A+\mu_B \geq 1_N, \sigma_A+\sigma_B \geq 1_N, \gamma_A+\gamma_B \leq 1_N$.

In other words, a NTS X is a neutro S_β^F-FSCS, if there occurs no non empty neutro S_β^F-feebly CS A and B in X such that $A \cap B = 0_N$.

Example 8. Accredit $X = \{a, b\}$ together with $\tau = \{0_N, 1_N, A, B\}$ where $A =< (0.6, 0.5, 0.3)(0.7, 0.5, 0.2) >, B =< (0.8, 0.6, 0.2)(0.9, 0.5, 0.1) >$. Formerly A and B inhabit a neutro S_β^F-feebly OS in X, likewise $\mu_A+\mu_B \geq 1_N, \sigma_A+\sigma_B \geq 1_N, \gamma_A+\gamma_B \leq 1_N$. Therefore X abide a neutro S_β^F-FSCS.

Theorem 13. A NTS $(X_{\mathcal{N}eut}, \tau)$ is a neutro S_β^F-FSCS, if there occurs no non empty neutro S_β^F-feebly OS A and B in X, $A \neq 1_N \neq B$ such that $\mu_A + \mu_B \geq 1_N, \sigma_A + \sigma_B \geq 1_N, \gamma_A + \gamma_B \leq 1_N$.

Proof. Accredit A together with B prevail a neutro S_β^F-feebly OS in X so that $A \neq 1_N \neq B$ along with $\mu_A + \mu_B \geq 1_N, \sigma_A + \sigma_B \geq 1_N, \gamma_A + \gamma_B \leq 1_N$. In the case that we take $C = A^C$ together with $D = B^C$, formerly C together with D become a neutro S_β^F-feebly CS in X together with $C \neq 0_N \neq D$, $\gamma_C + \gamma_D = \mu_A + \mu_B \geq 1_N, \mu_C + \mu_D = \sigma_A + \sigma_B \geq 1_N, \sigma_C + \sigma_D = \gamma_A + \gamma_B \leq 1_N$, a conflict.

Theorem 14. Let $k : (X_{\mathcal{N}eut}, \tau) \to (Y_{\mathcal{N}eut}, \sigma)$ be a neutro S_β^F-feebly irresolute surjection,X be a neutro S_β^F-FSCS. Then Y is also a neutro S_β^F-FSCS.

Proof. Presume that Y is not a neutro S_β^F-FSCS, then there occurs a non-empty neutro S_β^F-feebly CS U and V in Y so that $U \neq 0_N$, $V \neq 0_N$ and $U \cap V = 0_N$. Consequently k is a neutro S_β^F-feebly irresolute mapping, $A = k^{-1}(U) \neq 0_N, B = k^{-1}(V) \neq 0_N$, which is a neutro S_β^F-feebly CS in X. Therefore k inhabit a surjection,$k^{-1}(U) \cap k^{-1}(V) = k^{-1}(0_N) = 0_N$, which entails $A \cap B = 0_N$. Thus X is not a neutro S_β^F-FSCS, which is a conflict to our assumption. Therefore Y is a neutro S_β^F-FSCS.

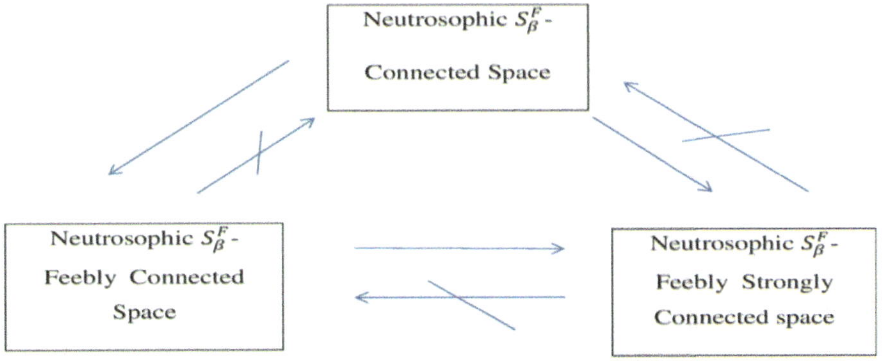

Fig. 1. Diagrammatic representation of various types of connected spaces

5 Conclusion and Future Work

This research paper investigates about the neutro S_β^F-conn spc in a NTS. In addition, many properties of neutro S_β^F-conn spc are proved with the proper illustration. Further the conception based on neutro S_β^F-FCS and neutro S_β^F-FSCS are prefaced and reviewed with proper examples. Furthermore, the interrelationship between neutro S_β^F-conn spc, neutro S_β^F-FCS and neutro S_β^F-FSCS is studied and investigated about these sets and some other existing sets. In addition, various theorems based on the averhead concepts are proved with proper illustrations. Major limitations of this set is that we cannot extend these set to n-NTS. In addition, we cannot use these sets to define neutro S_β^F-compact space and also their properties. In future, the conception based on neutro S_β^F-conn spc, neutro S_β^F-FCS and neutro S_β^F-FSCS can be applied in the field based on neutro orbit TS and neutro star TS. In addition, many properties can be studied in the field based on neutro orbit TS and neutro star TS. Furthermore, using the notion based on neutro S_β^F-compact space we can check whether which is finer and coarser set in neutro orbit TS.

References:

1. Acikgoz, A., Esenbel, F.: A study on connectedness in neutrosophic topological spaces. In: The Fourth International Conference of Mathematical Sciences, March, pp. 020003 (2021). https://doi.org/10.1063/5.0042372
2. Arokiarani, I., Dhavaseelan, R., Jafari, S., Parimala, M.: On some new notions and functions in neutrosophic topological spaces. Neutrosophic Sets Syst. **16**(1), 16–19 (2017)
3. Bageerathi, K., Puvaneswari, P.J.: Neutrosophic feebly connectedness and compactness. IOSR J. Polym. Text. Eng. **6**(3), 7–13 (2019)
4. Çoker, D.: An introduction to intuitionistic fuzzy topological spaces. Fuzzy Sets Syst. **88**(1), 81–89 (1997). https://doi.org/10.1016/S0165-0114(96)00076-0
5. Das, M., Mohanty, D., Parida, K.C.: On the neutrosophic soft set with rough set theory. Soft. Comput. **25**(21), 13365–13376 (2021). https://doi.org/10.1007/s00500-021-06089-2
6. Das, R., Tripathy, B.C.: Neutrosophic multiset topological space. Neutrosophic Sets Syst. **35**(1), 142–152 (2020)
7. Dhavaseelan, R., Savithiri, D., Janaki, C.: Neutrosophic bipolar vague regular weakly closed sets in neutrosophic bipolar vague topological spaces. J. Xi'an Univ. Arch. Technol. **12**(8), 641–649 (2020)
8. Ebenanjar, P.E., Immaculate, H.J., Wilfred, C.B.: On neutrosophic b-open sets in neutrosophic topological space. J. Phys. Conf. Ser. **1139**(1), 1–5 (2018). https://doi.org/10.1088/1742-6596/1139/1/012062
9. Gündüz, C., Oztürk, T.Y., Bayramov, S.: Separation axioms on neutrosophic soft topological spaces. Turk. J. Math. **43**(1), 498–510 (2019)
10. Iswarya, P., Bageerathi, K.: Neutrosophic semi-connected spaces via neutrosophic semi-open sets. Glob. J. Pure Appl. Math. **17**(2), 399–410 (2021)
11. Kalaivani, N., Chandrasekaran, E., Hachimi, H.: A study about (γ, γ) α–regular spaces, normal spaces with (γ, γ)–open sets and α (γ, γ)–open sets. In: Proceedings of 2nd International Conference on Mathematical Modeling and Computational Science: ICMMCS 2021, June, pp. 1–10. Springer Nature Singapore, Singapore (2022). https://doi.org/10.1007/978-981-19-0182-9_1

12. Kalaivani, N., Rahman, K.F.U., Obaid, A.J.: GS β-Compactness of Topological Spaces with Grills. In: Proceedings of 2nd International Conference on Mathematical Modeling and Computational Science: ICMMCS 2021, June, pp. 549–559. Springer Nature Singapore, Singapore (2022). https://doi.org/10.1007/978-981-19-0182-9_56
13. Madhumathi, T., NirmalaIrudayam, F.: Neutrosophic orbit topological spaces. Trends Sci. **18**(24), 1443–1443 (2021)
14. Nagalakshmi, T., Sudharani, R., Ambika, G.: A comparative approach for solving fuzzy transportation problem with hexagonal fuzzy numbers and neutrosophic triangular fuzzy numbers. In: Proceedings of 2nd International Conference on Mathematical Modeling and Computational Science: ICMMCS 2021, June, pp. 315–324. Springer Nature Singapore, Singapore (2022). https://doi.org/10.1007/978-981-19-0182-9_32
15. Parimala, M., Jafari, S., Murali, S.: Nano ideal generalized closed sets in nano ideal topological spaces. Annales Univ. Sci. Budapest **60**, 3–11 (2017)
16. Parimala, M., Karthika, M., Jafari, S., Smarandache, F., El-Atik, A.A.: Neutrosophic αψ-connectedness. J. Intell. Fuzzy Syst. **38**(1), 853–857 (2020)
17. Parvathi, R., Malathi, C., Akram, M., Atanassov, K.T.: Intuitionistic fuzzy linear regression analysis. Fuzzy Opt. Decision Making **12**, 215–229 (2013). https://doi.org/10.1007/s10700-012-9150-9
18. Pushpalatha, A., Nandhini, T.: Generalized closed sets via neutrosophic topological spaces. Malaya J. Matematik **7**(1), 50–54 (2019). https://doi.org/10.26637/MJM0701/0010
19. Salama, A.A., Alblowi, S.A.: Neutrosophic set and neutrosophic topological spaces. IQSR J. Math. **3**(4), 31–35 (2012)
20. Santhi, R., Prakash, K.A.: On intuitionistic fuzzy semi-generalized closed sets and its applications. Int. J. Contemp. Math. Sci. **5**(34), 1677–1688 (2010)
21. Santhi, R., Udhayarani, N.: Nω–closed sets in neutrosophic topological spaces. Neutrosophic Sets Syst. **12**(1), 114–117 (2016)
22. Subasree, R.: A study on Nψβ and Nβψ-closed sets in neutrosophic topological spaces. Baghdad Sci. J. **20**(1(SI)), 0283–0287 (2023)
23. Subasree, R., Basari, K.K.: On Nβ*-closed sets in neutrosophic topological spaces. Neutrosophic Sets Syst. **50**(1), 372–378 (2022)
24. Thomas, R., Anila, S.: On neutrosophic semi-preopen sets and semi-preclosed sets in a neutrosophic topological space. Int. J. Sci. Res. Math. Stat. Sci. **5**(5), 138–143 (2018)
25. Tran, T.B., Ngo, M.P., Bui, Q.T., Snasel, V., Vo, B.: A new approach for operations on neutrosophic soft sets based on the novel norms for constructing topological structures. AIMS Math. **7**(6), 9603–9626 (2022). https://doi.org/10.3934/math.2022534
26. Zadeh, L.A.: Information and control. Fuzzy Sets **8**(3), 338–353 (1965)

On a Certain Class Involving Analytic Univalent Functions Defined by Chebyshiv Polynomials

Adnan Aziz Hussein(✉) and Kassim A. Jassim

Department of Mathematics, College of Science, University of Baghdad, Baghdad, Iraq
adn.altameemi@gmail.com, kassim.jassim@sc.uobaghdad.edu.iq

Abstract. The main purpose of this paper is to define or create certain class $Ch(\alpha, \beta, \gamma, A, B, \tau)$, comprising of function interrelated with Chebyshiv polynomials. We introduce the coefficient bounds also presenting convolution property. Additionally through the utilization of structure of subordinations, two new subclasses $M(\alpha, \beta, \gamma, A, B, \tau)$ *and* $N(\lambda_1, \lambda_2, s)$ of $Ch(\alpha, \beta, \gamma, A, B, \tau)$ are introduced and studied. For these subclasses, we obtain some properties such as coefficients estimate, extreme points, convexity property and geometric exegesis of inclusion outcomes are submitted. Furthermore, we demonstrate that under circumstances some restriction on parameters $Ch(\alpha, \beta, \gamma, A, B, \tau) = N(\lambda_1, \lambda_2, s)$. Chebyshev $Ch(\alpha, \beta, \gamma, A, B, \tau)$ polynomials provide a technique for constructing and investigating new classes of analytic univalent functions that are holomorphic and injective in a given domain. Geometric features of these functions, such as their starlikeness and convexity, are frequently studied since they are crucial in many different applications, such as conformal mapping and dynamic fluid dynamics.

Keywords: Univalent Functions · Coefficient Estimate · Extreme Point · Chebyshev Polynomials

1 Introduction

Let A^o be the class of analytic univalent functions in the open unit disk $D = \{\varsigma \in C : |\varsigma| < 1\}$, with Taylor expansion series of the from.

$$f(\varsigma) = \varsigma + \sum_{j=2}^{\infty} a_j \varsigma^j, \tag{1}$$

also, denoted by G by "*class* of *univlaènt* functions *normalized*" $f(0) = f'(0) - 1 = 0$, [1, 2] moreover assume that N of the subclass Å comprising functions' with negative coefficient of certain type.

$$f(\varsigma) = \varsigma + \sum_{j=2}^{\infty} a_j \varsigma^j, \qquad a_j \geq 0. \tag{2}$$

The importance of Chebyshiv polynomials in numerical analytic is very significant both practically. And theoretically. These polynomials come in four different kinds. Numerical researchers take into consideration and provide many interesting findings.

S. Pal and Á. Rocha (Eds.): ICMMCS 2025, LNNS 1398, pp. 394–403, 2025.
https://doi.org/10.1007/978-3-031-90998-6_36

Instance, both the initial and second forms of Chebyshev polynomials were developed as

$$T_n(t) = cos\theta,\ and\ U_n(t) = \frac{sinsin(n+1)\theta}{sin\theta}, (-1 < t < 1) \tag{3}$$

where t = $cos\theta$, and n is there polynomial of degree. For further details, one many refer to (see [1–6]), Zena. H and Kassm. A, [7], Maryam and Abdulrahman. H [8], Ghanim, Hiba [9] the polynomial in (3) are interconnected through the following relations.

$$\frac{dT_n(t)}{dt} = nU_{n-1}(t) \tag{4}$$

$$T_n(t) = U_n(t) - tU_{n-1}(t)$$

$$2T_n(t) = U_n(t) - U_{n-2}(t), \tag{5}$$

We note that if $T_n(t) = cos\theta$, where $\theta \in \left(\frac{-\pi}{3}, \frac{\pi}{3}\right)$ *then,*

$$H(\varsigma, t) = \frac{1}{1 - 2cos\theta\varsigma + \varsigma^3} = 1 + \sum_{j=2}^{\infty} \frac{sinsin(j+1)\theta}{sin\theta}\varsigma^j, (\varsigma \in D). \tag{6}$$

Also, we can write

$$H(\varsigma, t) = 1 + U_1(t)\varsigma + U_2(t)\varsigma^2 + \ldots (\varsigma \in D. - 1 < t < 1). \tag{7}$$

where

$$U_{n-1}(t) = \frac{sin(narccost)}{\sqrt{1-t^2}}, (n \in N), \tag{8}$$

are the Chebyshiv polynomials of the second kind [10–12], the Hadmered product for functions,

$$f(\varsigma) = \varsigma - \sum_{j=2}^{\infty} a_j\varsigma^j,\ and\ g(\varsigma) = \varsigma - \sum_{j=2}^{\infty} b_j\varsigma^j, \tag{9}$$

is denoted by $f * g$ and defined as follows:

$$(f * g)(\varsigma) = \varsigma - \sum_{j=2}^{\infty} a_j b_j \varsigma^j = (g * f)(\varsigma), (\varsigma \in D), \tag{10}$$

The generating function of the first type of Chebyshev polynomials $T_n(t)$, $(-1 \leq t \leq 1)$, is given by

$$\sum_{n=0}^{\infty} T_n(t)\varsigma^n = \frac{1 - t\varsigma}{1 - 2t\varsigma + \varsigma^2}, (\varsigma \in D), \tag{11}$$

we consider the functions:

$$H_1(\varsigma) = 1 + (2cos\theta + 1)z - H(\varsigma, t),$$

$$H_2(\varsigma) = 1 + (2cos\theta + 1)z - \frac{1 - t\varsigma}{1 - 2t\varsigma + \varsigma^2}, \tag{12}$$

$$\mathrm{V}(\varsigma) = H_1 * H_2) * (H_2 * H_1) * f(\varsigma) \tag{13}$$

where $f(\varsigma) = \varsigma + \sum_{j=2}^{\infty} a_j \varsigma^j \in N$, and "*", denoted the convolution through basic computation, we deduced that $\mathrm{v}(\varsigma) \in N$ and in the following:

$$V(\varsigma) = \varsigma - \sum\nolimits_{j=2}^{\infty} T_j^2(t) \frac{sin^2(j+1)\theta}{sin^2\theta} a_j \varsigma^j, \tag{14}$$

where $\theta \in \left(\frac{-\pi}{3}, \frac{\pi}{3}\right)$, $T_n(t) = cos\theta$.

Let $Ch(\alpha, \beta, \gamma, A, B, \tau)$ denoted the subclass of N consisting of *functions* of from (14) *satisfying* the condition.

$$\left| \frac{\varsigma V''(\varsigma) + V'(\varsigma) - 1}{(A - B)\tau - B\left[2\gamma V'(\varsigma) - \alpha(1 + \beta) + \tau\right]} \right| < 1, \tag{15}$$

where $0 < \alpha \leq 1, 0 < \beta \leq 1, 0 \leq \gamma \leq 1$, $0 < B < A < 1, 0 \leq \tau$ and $\mathrm{V}(\varsigma)$ is given by (14) [11].

2 Main Results

In this subdivision announce sharp coefficient bound for $V(\varsigma) \in Ch(\alpha, \beta, \gamma, A, B, \tau)$. Convolutions is also preserved property under parameters α, β is verified.

Theorem 1. $\mathrm{V}(\varsigma) \in Ch(\alpha, \beta, \gamma, A, B, \tau)$ if and only if

$$\sum\nolimits_{j=2}^{\infty} T_j^2(t) \frac{sin^2(j+1)\theta}{sin^2\theta} j(j + 2B\gamma) a_j \leq (A + B)\tau + B\left[\gamma(2 - \alpha(1 - \beta) + \tau\right]. \tag{16}$$

Proof: Grant inequality (16) hold correct, and assume that, $\varsigma \in \partial \mathrm{D} = \{\varsigma \in D : |\varsigma| = 1\}$, in this case, we attain.

$$\left|\varsigma V''(\varsigma) + V'(\varsigma) - 1\right| - (A - B)\tau - B\left[2\gamma V'(z) - \alpha(1 + \beta) + \tau\right]$$

$$= \left| - \sum\nolimits_{j=2}^{\infty} T_j^2(t) \frac{sin^2(j+1)\theta}{sin^2\theta} j^2 a_j \varsigma^{j-1} \right|$$

$$- \left| (A - B)\tau - B\left[\gamma(2 - \alpha(1 + \beta)) - 2\gamma \sum\nolimits_{j=2}^{\infty} T_j^2(t) \frac{sin^2(j+1)\theta}{sin^2\theta} j a_j \varsigma^{j-1} - \tau\right] \right| \leq 0, \tag{17}$$

Certainly based on the maximum modulus principle we conclude that $V(\varsigma) \in Ch(\alpha, \beta, \gamma, A, B, \tau)$.

Conversely $V(\varsigma)$, defined by (14) be in the class $Ch(\alpha, \beta, \gamma, A, B, \tau)$, *so* condition (15), yields.

$$\left|\frac{\varsigma V''(\varsigma)+V'(\varsigma)-1}{(A-B)\tau-B\left[2\gamma V'(\varsigma)-\alpha(1+\beta)+\tau\right]}\right|$$

$$=\left|\frac{\left|-\sum_{j=2}^{\infty}T_j^2(t)\frac{sin^2(j+1)\theta}{sin^2\theta}j^2a_j\varsigma^{j-1}\right|}{(A-B)\tau-B\gamma(2-\alpha(1+\beta))-2\gamma B\sum_{j=2}^{\infty}T_j^2(t)\frac{sin^2(j+1)\theta}{sin^2\theta}ja_j\varsigma^{j-1}+B\tau}\right|<1. \tag{18}$$

Since for ς |Re ς |<| ς| then.

$$Re\left\{\frac{\sum_{j=2}^{\infty}T_j^2(t)\frac{sin^2(j+1)\theta}{sin^2\theta}j^2a_j\varsigma^{j-1}}{(A-B)\tau-B\gamma(2-\alpha(1+\beta))-2\gamma B\sum_{j=2}^{\infty}T_j^2(t)\frac{sin^2(j+1)\theta}{sin^2\theta}ja_j\varsigma^{j-1}+B\tau}\right\}<1. \tag{19}$$

By letting $\varsigma\to 1^-$ through real values. The proof is complete.

Remark 1. We observed that function,

$$\mathrm{w}(\varsigma)=\varsigma-\frac{(sin^2\theta)\left[(A+B)\tau+B\left[\gamma(2-\alpha(1-\beta)+\tau\right]\right]}{cos^2 2\theta sin^2 3\theta.4(1+\gamma B)}\varsigma^2, \tag{20}$$

Shows that inequality (16) is sharp.

Theorem 2. If $f(\varsigma)=\varsigma-\sum_{j=2}^{\infty}a_j\varsigma^j$, *and* $g(\varsigma)=\varsigma-\sum_{j=2}^{\infty}b_j\varsigma^j$, *are* in the class $Ch(\alpha, \beta, \gamma, A, B, \tau)$, then

i) $(f*g)(\varsigma)$ belong to $Ch(\alpha, \beta, \gamma, A, B, \tau)\left[\alpha^*\right]$, where

$$\alpha^*\le\frac{2+\tau}{1+\beta}-\left(\frac{sin\theta[(A+B)\tau+B[(2-\alpha(1-\beta)+\tau]]}{T_j(t)sinsin(j+1)\theta}\right)^2\times\frac{1}{j(j+2B\gamma)B(1+\beta)} \tag{21}$$

ii) $(f*g)(\varsigma)$ belong to $Ch(\alpha, \beta^*, \gamma)$, where,

$$\beta^*\le\frac{2+\tau}{\alpha}-\left(1+\left(\frac{sin\theta[(A+B)\tau+B[(2-\alpha(1-\beta)+\tau]]}{T_j(t)sinsin(j+1)\theta}\right)^2\right)\times\frac{1}{j(j+2B\gamma)B(1+\beta)}, \tag{22}$$

Proof: **(I)** it is adequate to demonstrate that

$$\sum_{j=2}^{\infty}\frac{T_j^2(t)sin^2(j+1)\theta.j(j+2B\gamma)}{sin^2\theta\left[(A+B)\tau+B(\gamma(2-\alpha(1+\beta)+\tau\right][\alpha^*]}a_jb_j\le 1. \tag{23}$$

Through the application of the Cauchy Schwarz inequality, we derive.

$$\sum_{j=2}^{\infty}\frac{T_j^2(t)sin^2(j+1)\theta.j(j+2B\gamma)}{sin^2\theta\left[(A+B)\tau+B(\gamma(2-\alpha(1+\beta)+\tau\right][\alpha^*]}\sqrt{a_j}\sqrt{b_j}\le 1, \tag{24}$$

hence, we find the largest α^* such that

$$\sum_{j=2}^{\infty} \frac{T_j^2(t) sin^2(j+1)\theta . j(j+2B\gamma)}{sin^2\theta\left[(A+B)\tau + B(\gamma(2-\alpha^*(1+\beta)+\tau\right]} a_j b_j \leq \sum_{j=2}^{\infty} \frac{T_j^2(t) sin^2(j+1)\theta . j(j+2B\gamma)}{sin^2\theta\left[(A+B)\tau + B(\gamma(2-\alpha(1+\beta)+\tau\right]} \sqrt{a_j}\sqrt{b_j} \leq 1, \tag{25}$$

this inequality hold if

$$\left(\frac{sin\theta\left[(A+B)\tau + B(\gamma(2-\alpha(1+\beta)+\tau\right]}{T_j(t) sin(j+1)\theta.}\right)^2 \frac{1}{j(j+2B\gamma)B} \leq 2-\alpha^*(1+\beta)+\tau, \tag{26}$$

$$\alpha^* \leq \frac{2+\tau}{1+\beta} - \left(\frac{sin\theta[(A+B)\tau + B[(2-\alpha(1-\beta)+\tau]]}{T_j(t) sinsin(j+1)\theta}\right)^2 \times \frac{1}{j(j+2B\gamma)B(1+\beta)}, \tag{27}$$

II) we easily illustrate component (II) utilizing identical methods as in part (I) which brings the proof to a close.

3 The Subclass $Ch(\alpha, \beta, \gamma, A, B, \tau)$ Characterized by Geometric Properties

We introduce two new subclass of $Ch(\alpha, \beta, \gamma, A, B, \tau)$ and discuss their geometric features in detail in this study. Once there can be an analytic function w in D such that $w(0) = 0$ and $|w(\varsigma)| < 1$, then for analytic functions $f(\varsigma)$ and $F(w(\varsigma))$ in D, we start f is subordinate to F, if there exists function w analytic in D that $w(0) = 0$, and $|w(\varsigma)| < 1$, such that $f(\varsigma) = F(w(\varsigma))$Ahmmad and Ghaaffer [3], Dziok. J [13], Duren P. [14] Miller. S [15] also Aljuboori and Kassim A. [16], Hiba.F [17] and Wanas and Abdulrahman [18]. is F univalent ? If so, then

$$f \prec F \Leftrightarrow f(0) = \mathrm{F}(0), \quad f(D) \subset F(D) \tag{28}$$

Assume that $\mathrm{M}(\beth_1, \beth_2, s)$ consist of all analytic function $g(\varsigma) \in D$, for which $g(0) = 1$ and

$$g(\varsigma) \prec \frac{1 + (\beth_2 + (\beth_2 - \beth_1)(1-s))}{1 + \beth_2\varsigma}, \tag{29}$$

where $-1 \leq \beth_1 < \beth_2 \leq 1, 0 < \beth_2 \leq 1,$ *and* $0 \leq s < 1$.

Let $\mathrm{N}((\beth_1, \beth_2, s)$ denote the class of all functions $\mathrm{V}(\varsigma) \in Ch(\alpha, \beta, \gamma, A, B, \tau)$ for which

$$\frac{\varsigma V''(\varsigma)}{V'(\varsigma)} + 1 \in N(\beth_1, \beth_2, s), \tag{30}$$

where $V(\varsigma)$ is given by (14)

Theorem 3. V(ς) $\in N((\beth_1, \beth_2, s)$, and only if

$$\sum_{j=2}^{\infty} \frac{j\big[(j-1)(1+\beth_2)+(\beth_2-\beth_1)(1-s)\big]}{(\beth_2-\beth_1)(1-s)} \left(\frac{T_j(t)sinsin(j+1)\theta}{sin\theta}\right)^2 a_j < 1 \tag{31}$$

Proof: Let V(ς) $\in N((\beth_1, \beth_2, s)$, then by (29) and (30) we have

$$\left| \frac{\sum_{j=2}^{\infty} j(j-1)(1+\beth_2)\frac{T_j^2(t)sin^2(j+1)\theta}{sin^2\theta} a_j \varsigma^{j-1}}{(\beth_2-\beth_1)(1-s)\varsigma - (\beth_2-\beth_1)(1-s)\sum_{j=2}^{\infty} j\frac{T_j^2(t)sin^2(j+1)\theta}{sin^2\theta} a_j \varsigma^j} \right| < 1$$

which implies that

$$Re\left\{ \frac{\sum_{j=2}^{\infty} j(j-1)(1+\beth_2)\frac{T_j^2(t)sin^2(j+1)\theta}{sin^2\theta} a_j \varsigma^{j-1}}{(\beth_2-\beth_1)(1-s)\varsigma - (\beth_2-\beth_1)(1-s)\sum_{j=2}^{\infty} j\frac{T_j^2(t)sin^2(j+1)\theta}{sin^2\theta} a_j \varsigma^j} \right\} < 1.$$

We are Now choosing the ς constants on the real axis and allow $\varsigma \to 1^-$, we obtain the desired outcome.

On the other hand, suppose condition that (31) holds true we need to the show that V(ς)$\in$, N $((\beth_1, \beth_2, s)$ or equivalently.

$$|\Omega| = \left| \frac{\varsigma V''(\varsigma) + \beth_2 \varsigma V''(\varsigma)}{(\beth_2-\beth_1)(1-s)\varsigma V'} \right| < 1.$$

However, we have

$$|\Omega| = \left| \frac{\sum_{j=2}^{\infty} j(j-1)(1+\beth_2)\frac{T_j^2(t)sin^2(j+1)\theta}{sin^2\theta} a_j}{(\beth_2-\beth_1)(1-s) - (\beth_2-\beth_1)(1-s)\sum_{j=2}^{\infty} j\frac{T_j^2(t)sin^2(j+1)\theta}{sin^2\theta} a_j} \right|$$

By utilizing (31), we achieve $|\Omega| < 1$, therefore the evidence is finished.

Corollary 1. Let $V(\varsigma)$$\in N((\beth_1, \beth_2, s)$, then.

$$a_j \le \frac{(\beth_2-\beth_1)(1-s)sin^2\theta}{\sum_{j=2}^{\infty} j\big[(j-1)(1+\beth_2)+(\beth_2-\beth_1)(1-s)\big]T_j^2(t)sin^2(j+1)\theta}. \tag{32}$$

Theorem 4. If

$$\frac{1+\beth_2}{(\beth_2-\beth_1)(1-s)} \le \frac{j+2B\gamma-(A+B)\tau - B\big[\gamma(2-\alpha(1-\beta)+\tau\big]}{(A+B)\tau + B\big[\gamma(2-\alpha(1-\beta)+\tau\big]}, \tag{33}$$

Then $Ch(\alpha, \beta, \gamma, A, B, \tau) = N((\beth_1, \beth_2, s)$. $Ch(\alpha, \beta, \gamma, A, B, \tau) = N((\beth_1, \beth_2, s)$.

Proof: By Eq. (30), we have

$$N((\beth_1, \beth_2, s) \subset Ch(\alpha, \beta, \gamma, A, B, \tau), \tag{34}$$

Now assume that V(z)∈ $Ch(\alpha, \beta, \gamma, A, B, \tau)$, then by Theorem 1 we get

$$\sum_{j=2}^{\infty}\left(\frac{T_j(t)sinsin(j+1)\theta}{sin\theta}\right)^2 j(j+2B\gamma)a_j \le (A+B)\tau + B\big[\gamma(2-\alpha(1-\beta)+\tau\big]. \tag{35}$$

According to Theorem 3, it suffices to demonstrate that (31) maintains validity, which is feasible when

$$\frac{j\big[(j-1)(1+\gimel_2)+(\gimel_2-\gimel_1)(1-s)\big]}{(\gimel_2-\gimel_1)(1-s)} \le \frac{j(j+2B\gamma)}{(A+B)\tau + B\big[\gamma(2-\alpha(1-\beta)+\tau\big]}. \tag{36}$$

Or equivalently

$$\frac{(j-1)(1+\gimel_2)}{(\gimel_2-\gimel_1)(1-s)} \le \frac{(j+2B\gamma)-(A+B)\tau - B\big[\gamma(2-\alpha(1-\beta)+\tau\big]}{(A+B)\tau + B\big[\gamma(2-\alpha(1-\beta)+\tau\big]}.$$

Since j starting from 2 then j – 1 ≥1 *and thus from the* last inequality, we derived result.

The following theorem, we establish the notion and the incorporation of attribute of combinations, introducing extreme points.

Theorem 5. Let $0 \le s_2 < s_1 < 1$, then

$$N((\gimel_1, \gimel_2, s)[s_1] \subset N((\gimel_1, \gimel_2, s)[s_2] \tag{37}$$

Proof: Suppose $f \in N((\gimel_1, \gimel_2, s)[s_1]$, then by Theorem 3, we have

$$\sum_{j=2}^{\infty}\frac{j\big[(j-1)(1+\gimel_2)+(\gimel_2-\gimel_1)(1-s)\big]}{(\gimel_2-\gimel_1)(1-s)[s_1]}\left(\frac{T_j(t)sinsin(j+1)\theta}{sin\theta}\right)^2 a_j \le (A+B)\tau + B\big[\gamma(2-\alpha(1-\beta)+\tau\big].$$

We have to prove

$$\sum_{j=2}^{\infty}\left(\frac{j(j-1)(1+\gimel_2)}{(\gimel_2-\gimel_1)(1-s)}+j\right)\left(\frac{T_j(t)sinsin(j+1)\theta}{sin\theta}\right)^2 a_j \le (A+B)\tau + B\big[\gamma(2-\alpha(1-\beta)+\tau\big], \tag{38}$$

However the last inequality holds true if

$$\frac{j(j-1)(1+\gimel_2)}{(\gimel_2-\gimel_1)(1-s)}+j \le \frac{j(j-1)(1+\gimel_2)}{(\gimel_2-\gimel_1)(1-s)}+j, \tag{39}$$

And this inequality follows hypothesis ($s_2 < s_1$), definitely holds true so the proof is complete.

Theorem 6. $N((\gimel_1, \gimel_2, s)$ is convex set.

Proof: We must demonstrate if.

$$V_m(\varsigma) = \varsigma - \sum_{j=2}^{\infty}\left(\frac{T_j(t)sinsin(j+1)\theta}{sin\theta}\right)^2 a_{j,m}\varsigma^j, \quad m = 1, 2\ldots.k \tag{40}$$

is in the class $N((\beth_1, \beth_2, s)$, then the function.

$$L(\varsigma) = \sum_{m=1}^{k} d_m V_m(\varsigma), \tag{41}$$

is in $N((\beth_1, \beth_2, s)$ where $\sum_{m=1}^{k} d_m = 1$. *However* we have

$$L(\varsigma) = \varsigma - \sum_{j=2}^{\infty}\left(\sum_{m=1}^{k} d_m a_{j,m}\right)\varsigma^j, \tag{42}$$

We have prove that if $f_m(\varsigma)$, $(m = 1, 2, \ldots .k)$ is in the class $N((\beth_1, \beth_2, s)$. Then the function $L(\varsigma) = \sum_{m=1}^{k} d_m = 1$, we have

$$L(\varsigma) = \varsigma - \sum_{j=2}^{\infty}\left(\sum_{m=1}^{k} d_m a_{j,m}\right)\varsigma^j.$$

Since by Theorem 3

$$\sum_{j=2}^{\infty}\left(\frac{j(j-1)(1+\beth_2)}{(\beth_2-\beth_1)(1-s)} + j\right)\left(\frac{T_j(t)sinsin(j+1)\theta}{sin\theta}\right)^2\left(\sum_{m=1}^{k} d_m a_{j,m}\right)$$

$$= \sum_{m=1}^{k}\left[\sum_{j=2}^{\infty}\left(\frac{j(j-1)(1+\beth_2)}{(\beth_2-\beth_1)(1-s)} + j\right)\left(\frac{T_j(t)sinsin(j+1)\theta}{sin\theta}\right)^2 a_{j,,m}\right]d_m$$

$$< \sum_{m=1}^{k} d_m = 1.$$

So $L \in N((\beth_1, \beth_2, s)$, the proof is complete.

Theorem 7. The function $V_1(\varsigma) = \varsigma$ and

$$V_j(\varsigma) = \varsigma - \frac{(\beth_2-\beth_1)(1-s)}{j\left[(j-1)(1+\beth_2)+(\beth_2-\beth_1)(1-s)\right]}\left(\frac{sin\theta}{T_j(t)sinsin(j+1)\theta}\right)^2 a_j\varsigma^j\, j \geq 2,$$

are the extreme point of $N((\beth_1, \beth_2, s)$,

Proof: We need to prove that other $L \in N((\beth_1, \beth_2, s)$ if and only if

$$L(\varsigma) = \sum_{j=2}^{\infty} [1]d_j V_j(\varsigma),$$

$$d_j = \frac{(\beth_2-\beth_1)(1-s)}{j\left[(j-1)(1+\beth_2)+(\beth_2-\beth_1)(1-s)\right]}\left(\frac{sin\theta}{T_j(t)sinsin(j+1)\theta}\right)^2 a_j,$$

we get $d_j \geq 0$, and if we put $d_1 = 1 - \sum_{j=2}^{\infty} d_j$, then we obtain.

$$L(\varsigma) = \varsigma - \sum_{j=2}^{\infty}\frac{(\beth_2-\beth_1)(1-s)}{j\left[(j-1)(1+\beth_2)+(\beth_2-\beth_1)(1-s)\right]}\left(\frac{sin\theta}{T_j(t)sinsin(j+1)\theta}\right)^2 a_j\varsigma^j$$

$$= \varsigma - \sum_{j=2}^{\infty} d_j(\varsigma - V_j(\varsigma)) = \sum_{j=2}^{\infty} d_j V_j(\varsigma).$$

Conversely, suppose

$$L(\varsigma) = \sum_{j=2}^{\infty} [1] d_j V_j(\varsigma), L(z) = d_1 V_1(\varsigma) + \sum_{j=2}^{\infty} d_j V_j(\varsigma)$$

$$= d_1 \varsigma + \sum_{j=2}^{\infty} \left[\varsigma - \frac{(\daleth_2 - \daleth_1)(1-s)}{j[(j-1)(1+\daleth_2) + (\daleth_2 - \daleth_1)(1-s)]} \left(\frac{sin\theta}{T_j(t) sinsin(j+1)\theta} \right)^2 a_j \varsigma^j \right]$$

$$= \varsigma - \sum_{j=2}^{\infty} \frac{(\daleth_2 - \daleth_1)(1-s)}{j[(j-1)(1+\daleth_2) + (\daleth_2 - \daleth_1)(1-s)]} \left(\frac{sin\theta}{T_j(t) sinsin(j+1)\theta} \right)^2 a_j \varsigma^j$$

since

$$\sum_{j=2}^{\infty} d_j \left(\frac{j(j-1)(1+\daleth_2)}{j(\daleth_2 - \daleth_1)(1-s)} + j \right) \frac{(\daleth_2 - \daleth_1)(1-s)}{j[(j-1)(1+\daleth_2) + (\daleth_2 - \daleth_1)(1-s)]}$$

$$\times \left(\frac{sin\theta}{T_j(t) sinsin(j+1)\theta} \right)^2 \left(\frac{T_j(t) sinsin(j+1)\theta}{sin\theta} \right)^2$$

$$= \sum_{j=2}^{\infty} d_j = 1 - d_1 < 1,$$

Therefore, by Theorem 3, we conclude the outcome.

4 Conclusion

Univalent functions have consistently been a focal of extensive research in theory of geometric function. Recent studies on Chebyshiv polynomial have centered around classes of analytic normalized univalent functions. The characteristics of geometry include derived for functions in the generally indicated with $Ch(\alpha, \beta, \gamma, A, B, \tau)$, using Chebyshiv polynomials, linked to a convolution framework. Not too many geometric results in traduced for subclass $Ch(\alpha, \beta, \gamma, A, B, \tau)$.

References

1. Khan, M.G., Ahmad, B., Mashwani, W.K., Shaba, T.G., Arif, M.: Third Hankel determinant problem for certain subclasses of analytic functions associated with nephroid domain. Earthline J. Math. Sci. **6**(2), 293–308 (2021)
2. Ahmad, B., Khan, M.G., Darus, M., Arif, M., Mashwani, W.K.: Applications of some generalized Janowski meromorphic multivalent functions. J. Math. **2021**, 13, Article ID 6622748 (2021)
3. Ahmad, B., Ghaffar Khan, M., Aouf, M.K., Mashwani, W.K., Salleh, Z., Tang, H.: Applications of new Q-difference operator in Janowski type meromorphic convex functions. J. Function Spaces **2021,** 10, Article ID 5534357 (2021)
4. Altinkaya, S., Yalçin, S.: On the Chebyshev coefficients for a general subclass of univalent functions. Turkish J. Math. **42**(6), 2885–2890 (2018)
5. Altinkaya, S., Yalcin, S.: Chebyshev polynomial bounds for certain subclass of univalent functions defined by Komatu integral operator. Afr. Math. **30**(3–4), 563–570 (2019)

6. Bulut, S., Magesh, N.M.: On the sharp bounds for a comprehensive class of Analytic and univalent functions by means of Chebyshev polynomials. Khayyam J. Math. **2**(2), 194–200 (2016)
7. Zena, H.M., Kassim, A.J.: On a subclass of meromorphic univalent functions with fixed second coefficients associate with q-differed operators. Iraqi J. Sci. **64**(10), 5182–5189 (2023)
8. Maryam, K.R., Abdulrahman, H.M.: Generalization of Gamma and Beta Functions with Certain Properties and Statistical Application. Iraqi J. Sci. **64**(7), 4487–4495 (2023)
9. Ghanim, F., Al-Janaby, H.F., Bazighifan, O.: Some new extensions on fractional differential and integral properties for Mittag-Leffler confluent hypergeometric function. Fractal Fractional **5**(4), 143 (2021). https://doi.org/10.3390/fractalfract5040143
10. Altinkaya, S., Yalçin, S.: On the Chebyshev polynomial bounds for classes of Univalent functions. Khayyam J. Math. **2**(1), 1–5 (2016)
11. Doha, E.: The first and second kind Chebyshev coefficients of the moments for the general order derivative on an infinitely differentiable function. Int. J. Comput. Math. **51**(1–2), 21–35 (1994)
12. Juma, A.R., Al-khafaji, N.S., Engel, O.: Chebyshev polynomials for certain subclass of Bazilevic functions associated with Ruscheweyh derivative. Kragujevac J. Math. **45**(02), 173–180 (2021). https://doi.org/10.46793/KgJMat2102.173J
13. Dziok, J., Raina, R.K., Sokół, J.: Application of Chebyshev polynomials to classes of analytic functions. Comptes Rendus. Mathématique **353**(5), 433–438 (2015). https://doi.org/10.1016/j.crma.2015.02.001
14. Duren, P.L.: Univalent functions, Grundlehren der math ematischen. Wissenschaften, vol. 259. Springer, Germany (1983)
15. Miller, S.S., Mocanu, P.T.: Differential Subordinations Theory and Applications. CRC Press, Boca Raton (2000)
16. Aljuboori, O.M., Jassim, K.A.: New subclasses of meromorphically uniformly of multivalent functions with positive and fixed second coefficients. Iraqi J. Sci. **64**(1), 365–372 (2023). https://doi.org/10.24996/ijs.2023.64.1.34
17. Al-Janaby, H.F., Ghanim, F., Agarwal, P.: Geomtric studies on inequalities of harmonic functions in a complex field based on ξ-generalized Hurwitz-Lerch Zeta function. J. Iran. J. Math. Sci. Inform. **18**(1), 73–93 (2023)
18. Wanas, A.K., Majeed, A.M.: Second hankel determinant for a certain subclass of λ-pseudo-starlike bi-univalent functions. Iran. J. Math. Sci. Inform. **16**(2), 49–59 (2021). https://doi.org/10.52547/ijmsi.16.2.49

Evaluation of Multiple Phases of the Server in the Simple Queue, Subject to Customer Impatience

C. T. Dora Pravina[1(✉)], P. Kamala[1], S. Sreelakshmi[2], D. Kanagajothi[1], and R. Shanmugapriya[1]

[1] Vel Tech Rangarajan Dr. Sagunthala R&D Institute of Science and Technology, Avadi, Chennai 600062, India
{tdorapravinac,dkanagajothi,shanmugapriya}@veltech.edu.in

[2] Sir M Visvesvaraya Institute of Technology, Krishnadevaraya Nagar, Bangalore 562157, India
Sreelakshmi_maths@sirmvit.edu

Abstract. A Markovian queue with a single server having customer with impatient and server in different phases is examined. The server will be staying at sleep or active mode randomly and alternatively. The active mode comprises of two phases, the full active mode (FA) and slow active mode (SA). When the server is in full active mode, customers arrive and get service in FCFS queue discipline. However, when the server changes from full active to slow active, it will be able to service only in a lower rate lesser than that of full active mode. When the server is in slow active, none of the customer will be accepted to access the system and each of the customers will activate the impatient timer independently and quits the system if they are not served before the timer ends. In active or slow mode, the server may immediately switch to sleep mode (SL) when the very last available customer in the queue is served. The customers are not able to accept to admit the queue when the server is in SL and SA. After the sleep mode, the server will immediately switch to the fully active mode. We have acquired few metrics for the performance as well formulations for the time-independent state probabilities. Finally, extensive results are presented graphically to highlight the vitality of the system characteristics pertaining to the vital performance matrices.

Keywords: Single server queue · Server in Multiple phases · Probability generating function · Customer impatience

1 Introduction

In queueing systems, customers may become impatient when they experience prolonged waiting in the queue, despite the server being available to provide service. With the Queueing models, customer's impatience has been studied by several authors (Altman and Borovkov [11], Takacs [13], Baccelli et al. [4], Choi et al. [12], and Boxma and de Waal [5]). Eitan Altman et al [14] considered a system with absence of server where customers become impatient upon their arrival. Practical problems and their solutions in

S. Pal and Á. Rocha (Eds.): ICMMCS 2025, LNNS 1398, pp. 404–413, 2025.
https://doi.org/10.1007/978-3-031-90998-6_37

queueing theory, including scenarios with impatient customers are discussed in Queueing Systems: Problems and Solutions by Leonard Kleinrock [15]. Analytical and numerical solutions of the queue having impatient customers and simulation modelling are investigated by Wang, Kangzhou [16]. Ch. Swathi et al [17] studied about the reneging and Balking of the customers that are assumed to happen during vacation and breakdown periods of the server. Probability generating functions are used for both single and multiple vacation scenarios to obtain the Steady state probabilities. In many real time situations, these models were used which includes the treatment of vital patients, info packets transmission, manufacturing inventory systems, call centres and so on. A random timer is allotted to each and every customer who enters the system during a critical interval to study the impatience of them. If the customers were not served before the timer ends, the customers cast aside and leave the system. Queueing systems with impatient customers is studied extensively and literatures relevant to this can be found in numerous articles by Doshi [1], Takagi [2] and a monograph by Tian and Zhang [3]. In all these studies, impatience among customers has consistently been attributed to either a lengthy wait they encounter upon arriving at a queue or a prolonged wait they anticipate upon their arrival. Farej, Z.K., Ali, O.M [21] discussed about the analysis of DL-MU-MIMO for the IEEE 802.11ac Wireless LAN and its performance. In cloud computing environments, virtual machines (VMs) often have different states based on their utilization or demand. A VM might be actively serving requests at high performance during peak times (active phase), slow down during periods of lower demand (slow phase), and even go into a low-power state or sleep mode when not needed (sleep phase). Queuing models help in understanding and optimizing resource allocation and performance prediction under such varying server states. Paz and Yechiali [19] studied the dynamics of an M/M/1 queue in a steady-state, operating on a random environment with disaster, where the n-level continuous-time Markov chain is used to describe the underlying environment. Yet, transient solutions for the queueing model were not obtained by them. Udayabaskaran and Dora Pravina [20] analysed and derived the probabilities with time-dependency in regard to the queueing model of Paz and Yechiali [19]. Ammar et al. [18] further elaborated about the queueing model of Paz and Yechiali [19] via including the customer's impatience amid the time of repair, and provided transient solution for the system's state probabilities. From the aforementioned works, at the time of every disaster all active customers in the system are faded away.

Chakravarthy [6], Sudhesh [7] and U. Yechiali [8] analysed the impatient customers and disasters in the queue. Perel along with Yechiali [9] have learned about the slow working servers and impatient customers in the queue, but they did not discuss about sleep phase of server. Udayabaskaran with Dora Pravina [10] analysed and described a single server queue that operates in three modes: active, maintenance and sleep and their transient analysis, but they did not cover the slow serving server due to physical condition and impatience of customers present in it. To fill this gap, we have analysed a single server queueing system having a server that is operating in multiple phases, having impatient customers in it. This article is constituted as: the model is described in 2nd Section. 3rd Section has time-independent balance equations for system's probabilities. Section 4 has the explicit expression for steady state conditions. The performance measures are mentioned in Section 5.

1.1 Objective

To examine a single server which operates in multiple modes such as active, slow and sleep phases, during the slow phase of the server customer develops impatience and quits the system. This model has applications in wireless communication networks where the server is in energy saving phase (slow phase), which will create impatience and loss of customers.

2 Model Description

We have examined a scenario, where a single server operates across three distinct phases: an active phase, a slow phase dedicated solely to service without new arrivals, and a sleep phase. The Customers arrives at a rate of λ according to Poisson's process in active phase. In active phase as well as slow phase, exponential distribution is followed with rates of μ_1 and μ_2 respectively such that $\mu_2 < \mu_1$. In active phase, the server renders its service to the available customers on FCFS discipline and the server moves to sleep phase at the rate of a if no other customers in the server. When in sleep mode is activated, new customers will not be allowed to join the system. The server will spend an indefinite span of time; it has a mean of 1/ß and an exponential distribution. Immediately after the sleep mode (SL), its witches to active mode to satisfy the customers. When the full active mode (FA) server falls sick, it immediately switches over to the slow phase, that is exponentially distribution at a rate of γ, and the service is provided to the customers at a rate of $\mu_{2.}$ During slow phase no new customers are admitted into the system. Once the last customer has been fulfilled, the server transits to sleep mode. While a server is serving in slow mode (SA), impatience may be developed in the customers at buffer and they might quit the system following an exponential distribution with a rate of ξ.

3 Constitutive Equations

The server's phase is denoted by $S(t)$ and let $X(t)$ denotes the number of customers in queueing system at a time t. We define

$$S(t) = \begin{cases} 0, & \text{the server resides in SL} \\ 1, & \text{the server resides in FA} \\ 2, & \text{the server resides in SA} \end{cases}$$

Then the combined process $\{(S(t), X(t))|t \geq 0\}$ describes a CTMC. The picture of transition rates among the states is presented in Fig. 1

We determined the conditional probability function as follows

$$p(i, n, t) = P[S(t) = i, X(t) = n|S(0) = 0, X(0) = 0], i = 0, 1, 2.n = 0, 1, 2, \ldots$$

Using invariant imbedding technique, we obtain the following integral equations:

$$p(0, 0, \mathrm{t}) = e^{-\beta t} + \alpha \int_0^t p(1, 0, u)e^{-\beta(t-u)}du + \mu_2 \int_0^t p(2, 1, u)e^{-\beta(t-u)}du \quad (1)$$

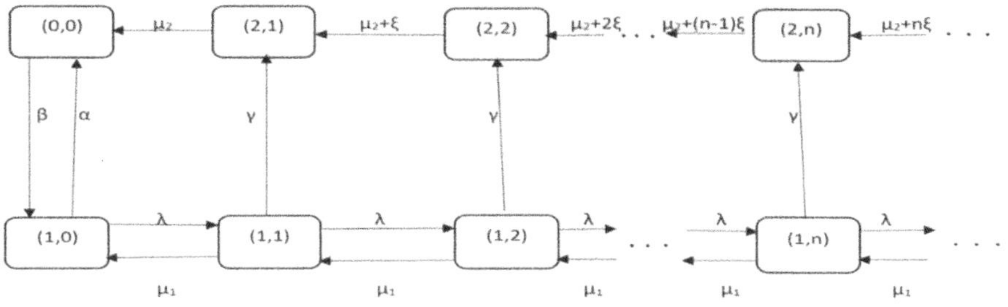

Fig. 1. Transition Rate Diagram

$$p(1,0,\mathrm{t}) = \beta\int_0^t p(0,0,u)e^{-(\lambda+\alpha)(t-u)}du + \mu_1\int_0^t p(1,1,u)e^{-(\lambda+\alpha)(t-u)}du \tag{2}$$

$$p(1,\mathrm{n},\mathrm{t}) = \lambda\int_0^t p(1,n-1,u)e^{-(\lambda+\mu_1+\gamma)(t-u)}du + \mu_1\int_0^t p(1,n+1,u)e^{-(\lambda+\mu_1+\gamma)(t-u)}du, n \geq 1 \tag{3}$$

$$p(2,\mathrm{n},\mathrm{t}) = \gamma\int_0^t p(1,n,u)e^{1(\mu_2+(n-1)\xi(t-u)}du + (\mu_2+n\xi)\int_0^t p(2,n+1,u)e^{1(\mu_2+(n-1)\xi(t-u)}du, n \geq 1 \tag{4}$$

Denoting the Laplace transform a function $f(t)$ by $f^*(t)$, Eq. (1)–(4) give

$$(\mathrm{s} + \beta)\,\mathrm{p}^*(0,0,\mathrm{s}) = 1 + \alpha\,\mathrm{p}^*(1,0,\mathrm{s}) + \mu_2\mathrm{p}^*(2,1,\mathrm{s}) \tag{5}$$

$$(\mathrm{s} + \lambda + \alpha)\,\mathrm{p}^*(1,0,\mathrm{s}) = \beta\,\mathrm{p}^*(0,0,\mathrm{s}) + \mu_1\mathrm{p}^*(1,1,\mathrm{s}) \tag{6}$$

$$(\mathrm{s} + \lambda + \mu_1 + \gamma)\,\mathrm{p}^*(1,\mathrm{n},\mathrm{s}) = \lambda\,\mathrm{p}^*(1,\mathrm{n}-1,\mathrm{s}) + \mu_1\mathrm{p}^*(1,\mathrm{n}+1,\mathrm{s}),\ \mathrm{n} \geq 1 \tag{7}$$

$$(\mathrm{s} + \mu_2 + (\mathrm{n}-1)\xi)\,\mathrm{p}^*(2,\mathrm{n},\mathrm{s}) = \gamma\,\mathrm{p}^*(1,\mathrm{n},\mathrm{s}) + (\mu_2 + n\xi)\mathrm{p}^*(2,\mathrm{n}+1,\mathrm{s}),\ \mathrm{n} \geq 1 \tag{8}$$

The steady – state probabilities are given by

$$\pi_{i,n} = \lim_{t\to\infty} \mathrm{p}^*(1,\mathrm{n},\mathrm{s}),\ \mathrm{i} = 0,1,2;\ \mathrm{n} = 0,1,2,\ldots$$

Consequently, Eqs. (5)–(8) yield

$$\beta\pi_{0,0} = \alpha\pi_{1,0} + \mu_2\pi_{2,1} \tag{9}$$

$$(\lambda + \alpha)\pi_{1,0} = \beta\pi_{0,0} + \mu_1\pi_{1,1} \tag{10}$$

$$(\lambda + \mu_1 + \gamma)\pi_{1,n} = \lambda\pi_{1,n-1} + \mu_1\pi_{1,n+1},\ \mathrm{n} = 1,2,\ldots \tag{11}$$

$$(\mu_2 + (n-1)\xi)\pi_{2,n} = \gamma\pi_{1,n} + (\mu_2 + n\xi)\pi_{2,n+1},\ \mathrm{n} = 1,2,\ldots \tag{12}$$

4 Steady-State Probabilities

On solving Eqs. (9)–(12), we have defined the partially generating functions

$$G_1(\mathrm{u}) = \sum_{n=0}^{\infty} \pi_{1,n}u^n \tag{13}$$

$$G_2(\mathrm{u}) = \sum_{n=0}^{\infty} \pi_{2,n}u^n \tag{14}$$

Multiplying both sides of Eq. (11) by $\mathrm{u^n}$, summing up from 1 to ∞, using Eq. (10) and simplifying, we get

$$G_1(\mathrm{u}) = \frac{\beta\pi_{0,0}u - \left[(\alpha - \mu_1 - \gamma)u + \mu_1)\pi_{1,0}\right]}{(\lambda + \mu_1 + \gamma)\mathrm{u} - \lambda u^2 - \mu_1} \tag{15}$$

For $|u|$, the function $G_1(u)$ converges and hence the RHS numerator of (15) will not appear at all zeros in $|\mathrm{u}| < 1$ of the denominator of Eq. (15). The zeros of the denominator of Eq. (15) are given by

$$r_1 = \frac{(\lambda + \mu_1 + \gamma) - \sqrt{(\lambda + \mu_1 + \gamma)^2 - 4\lambda\mu_1}}{2\lambda} \tag{16}$$

$$r_2 = \frac{(\lambda + \mu_1 + \gamma) + \sqrt{(\lambda + \mu_1 + \gamma)^2 - 4\lambda\mu_1}}{2\lambda} \tag{17}$$

Since, $(\lambda + \mu_1 + \gamma)^2 - 4\lambda\mu_1 = (\mu_1 - \lambda)^2 + \gamma(2\lambda + 2\mu_1 + \gamma) > 0$, the roots r_1 and r_2 are

$$0 < r_1 < 1 < r_2, r_1 + r_2 = \frac{\lambda + \mu_1 + \gamma}{2\lambda}, r_1 r_2 = \frac{\mu_1}{\lambda}$$

$$(\lambda + \mu_1 + \gamma)u - \lambda u^2 - \mu_1 = \lambda(u - r_1)(r_2 - u) \tag{18}$$

There is exactly one zero, namely r_1 lying in $|\mathrm{u}| < 1$. So the RHS of the numerator (14) vanishes at r_1 and it cannot vanish at r_2. Consequently, we get

$$\pi_{1,0} = \frac{\beta r_1 \pi_{0,0}}{(\alpha - \mu_1 - \gamma)r_1 + \mu_1} \tag{19}$$

Equation (15) can be rewritten as

$$G_1(u) = \frac{r_2\pi_{1,0}}{r_2 - u} \tag{20}$$

On expanding the equation's right hand side (20) in region of convergence, we get

$$\sum_{n=0}^{\infty} \pi_{1,n}u^n = \pi_{1,0} \sum_{n=0}^{\infty} \frac{u^n}{r_2^n} \tag{21}$$

Equating the coefficient of u^n on both sides of (21), we get

$$\pi_{1,n} = r_2^{-n}\pi_{1,0}, n = 0, 1, 2, \cdots . \tag{22}$$

Now, multiplying both sides of Eq. (12) by u^n, summing up from 1 to ∞, and simplifying,
we get

$$\frac{\partial}{\partial u}G_2(u) + \frac{k}{u}G_2(u) = \frac{\left[\gamma G_1(u) - \beta\pi_{0,0} + (\gamma - \alpha)\pi_{1,0}\right]}{\xi(u-1)}, \quad \text{where k} = \frac{(\mu_2 - \xi)}{\xi} \tag{23}$$

From Eq. (15), we get

$$\gamma G_1(1) = \beta\pi_{0,0} + (\gamma - \alpha)\pi_{1,0} \tag{24}$$

Substituting Eq. (24) into Eq. (23) and simplifying, we get

$$uG_2^{(1)}(u) + kG_2(u) = \frac{\gamma}{\xi}\left[\sum_{m=1}^{\infty}\left(\sum_{j=m}^{\infty}\pi_{1,j}u^m\right)\right]$$

Differentiating both sides, and putting u = 0, we get

$$G_2^{(n)}(0) = \frac{n!}{n+k}\frac{\gamma}{\xi}\left[\sum_{j=n}^{\infty}\pi_{1,j}\right], \text{ n} = 1, 2, \ldots \tag{25}$$

By Taylor expansion, we get

$$G_2(u) = \sum_{n=1}^{\infty}\pi_{2,n}u^n = \sum_{n=1}^{\infty}\frac{G_2^{(n)}(0)}{n!}u^n = \sum_{n=1}^{\infty}\frac{\gamma}{\xi}\frac{1}{k+n}\left[\sum_{j=n}^{\infty}\pi_{1,j}\right]u^n$$

Equating the coefficients of u^n on both sides, we get

$$\pi_{2,n} = \frac{\gamma}{\xi(k+n)}\left[\sum_{j=n}^{\infty}\pi_{1,j}\right], \text{ n} = 1, 2, \ldots \tag{26}$$

Using Eq. (22) in Eq. (26) and simplifying, we get

$$\pi_{2,n} = \frac{\gamma}{\xi(k+n)r_2^{n-1}(r_2-1)}\pi_{1,0}, \text{ n} = 1, 2, \ldots \tag{27}$$

By total probability law, we get

$$\pi_{0,0} + \sum_{n=0}^{\infty} \pi_{1,n} + \sum_{n=1}^{\infty} \pi_{2,n} = 1 \tag{28}$$

On Substituting Eqs. (22) and (27) in Eq. (28), we obtain

$$\pi_{0,0} + \sum_{n=0}^{\infty} r_2^{-n} \pi_{1,0} + \sum_{n=1}^{\infty} \frac{1}{(k+n) r_2^{n-1} (r_2 - 1)} \pi_{1,0} = 1 \tag{29}$$

Substituting Eq. (19) and simplified we get explicitly,

$$\pi_{0,0} = \left[1 + \frac{r_2}{(r_2 - 1)} \frac{\beta r_1}{\left[\alpha - \mu_1 - \gamma) r_1 + \mu_1 \right]} \left\{ 1 + \frac{\gamma}{\xi} \sum_{n=1}^{\infty} \frac{1}{(k+n) r_2^n} \right\} \right]^{-1} \tag{30}$$

5 Performance Measures

5.1 Stationary Mean Customers Leaves by Impatience

E_M is the number of customers who left due to impatience per unit time of the system. Then, we get

$$E_M = \xi \sum_{n=2}^{\infty} (n-1) \pi_{2,n} = \sum_{n=2}^{\infty} \frac{(n-1)\gamma}{(k+n) r_2^{n-1} (r_2 - 1)} \pi_{1,0} \tag{31}$$

5.2 Stationary Mean Customers in the Active Phase

Assuming E_A, be number of customers expected in the active phase. Then, from Eq. (22) we get,

$$E_A = \sum_{n=0}^{\infty} n \pi_{1,n} = \pi_{1,0} \sum_{n=0}^{\infty} n r_2^{-n} = \pi_{1,0} \left(1 - \frac{1}{r_2} \right)^{-2} \frac{1}{r_2} \tag{32}$$

5.3 Stationary Mean Customers in the Slow Phase

Assuming E_P to be number of customers expected in slow phase. Then, by using Eq. (27), we get

$$E_P = \sum_{n=1}^{\infty} n \pi_{2,n} = \sum_{n=1}^{\infty} \frac{n\gamma}{\xi (k+n) r_2^{n-1} (r_2 - 1)} \pi_{1,0} \tag{33}$$

5.4 Stationary Mean Customers in the System

Taking E_S be mean customers in the system. The

$$E_S = E_A + E_P$$

$$E_S = \pi_{1,0}\left(1 - \frac{1}{r_2}\right)^{-2} + \sum\nolimits_{n=1}^{\infty} \frac{2}{(k+n)r_2^{n-1}(r_2-1)}\pi_{1,0} \tag{34}$$

5.5 Mean Waiting Time in the System

Assuming E_W, be the mean amount of time customers wait in the system. By Little formula, we get

$$E_W = \frac{E_S}{\lambda} = \frac{1}{\lambda}\left[\pi_{1,0}\left(1 - \frac{1}{r_2}\right)^{-2} + \sum\nolimits_{n=1}^{\infty} \frac{2}{(k+n)r_2^{n-1}(r_2-1)}\pi_{1,0}\right] \tag{35}$$

6 Numerical Analysis

Assuming $\lambda = 2$; $\mu_1 = 1.5$; $\mu_2 = 0.5$; $\gamma = 1$; $\beta = 1.2$; $\alpha = 1.1$; and adjusting ξ between 0.1 to 1.0. Calculating the stationary mean count of customers in the slow phase, system, leaves the system by impatience and mean amount of time customers wait in the system, we obtain following Table 1 and Fig. 2:

Table 1. Analysis Values of Stationary mean customers in the active phase, slow phase, impatience, and in the system

ξ	E_P	E_S	E_M	E_W
0.1	0.2484	0.3763	0.1108	0.1881
0.2	0.2335	0.3547	0.1062	0.1773
0.3	0.2187	0.3338	0.1011	0.1669
0.4	0.204	0.3136	0.0908	0.1568
0.5	0.1893	0.2939	0.0845	0.1469
0.6	0.1747	0.2747	0.0746	0.1373
0.7	0.1602	0.256	0.0565	0.128
0.8	0.1458	0.2378	0.0388	0.1189
0.9	0.1315	0.2199	0.0276	0.1099
1.0	0.1174	0.2025	0.0165	0.1012

As the customer's impatience increases number of customers on an average in the system as well as in slow phase decreases. Also, this increase in customer's impatience will decrease waiting time of customers and average number of customers leaving the queue due to impatience.

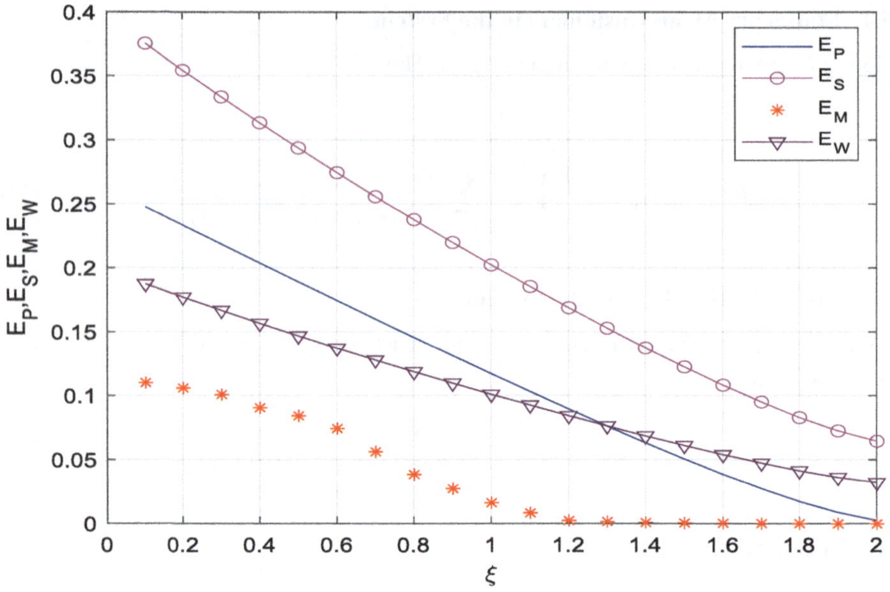

Fig. 2. Graphical analysis of Stationary mean customers in the active phase, slow phase, impatience, and in the system

7 Conclusion

Here, we have analysed the queue with multiple phases of Single server with steady state conditions. In slow phase, where the customers are served, will leave the system immediately due to impatience. Expressions for steady state probability conditions are solved for the system. The measures of the performance such as count of customers expected in both active and slow phase of server, number of times server expected to switch to different phase, count of customers expected to leave the system as an impact of impotency, rate at which the customer joins the system effectively and the waiting time is expected for arrival of the customers in system. This paper has laid a foundation to take up the model by considering the servers to stay in every phase as general distribution by prioritising customers.

References

1. Doshi, B.T.: Queueing systems with vacations - a survey. Queueing Syst. **1**(1), 29–66 (1986)
2. Takagi, H.: Queueing Analysis, A Foundation of Performance Evaluation, Volume 1: Vacation and Priority Systems. Elsevier, New York (1991)
3. Tian, N., Zhang, Z.G.: Vacation Queueing Models: Theory and Applications. Springer, New York (2006)
4. Baccelli, F., Boyer, P., Hebuterne, G.: Single-server queues with impatient customers. Adv. Appl. Probab. **16**, 887–905 (1984)
5. Boxma, O.J., de Waal, P.R.: Multi server queues with impatient customers. ITC **14**, 743–756 (1994)

6. Chakravarthy, S.R.: A disaster queue with Markovian arrivals and impatient customers. Appl. Math. Comput. **214**, 48–59 (2009)
7. Sudhesh, R.: Transient analysis of a queue with system disasters and customer impatience. Queueing Syst. **66**, 95–105 (2010)
8. Yechiali, U.: Queues with system disasters and impatient customers when system is down. Queueing Syst. **56**, 195–202 (2006)
9. Perel, N., Yechiali, U.: Queues with slow servers and impatient customers. Eur. J. Oper. Res. **201**, 247–258 (2010)
10. Udayabaskaran, S., Dora Pravina, C.T.: Transient analysis of an M/M/1 queueing system with server operating in three modes. Far East J. Math. Sci. **101**, 1395–1418 (2017)
11. Altman, E., Borovkov, A.A.: On the stability of retrial queues. Queueing Syst. **26**, 343–363 (1997)
12. Choi, B.D., Kim, B., Chung, J.: M/M/1 queue with impatient customers of higher priority. Queueing Syst. **38**, 49–66 (2001)
13. Takacs, L.: Introduction to the Theory of Queues. Oxford University Press, New York (1962)
14. Altman, E., Yechiali, U.: Analysis of customers' impatience in queues with server vacations. Queueing Syst. **52**, 261–279 (2006)
15. Kleinrock, L., Gail, R.: Queueing Systems: Problems and Solutions (1996). ISBN 978-0-471-55568-1
16. Wang, K., Li, N., Jiang, Z.: Queueing system with impatient customers: a review. In: Proceedings of 2010 IEEE International Conference on Service Operations and Logistics, and Informatics, SOLI 2010 (2010)
17. Swathi, Ch., Vasanta Kumar, V., Hanumantha Rao, S.: M/M/1 queueing system with customer balking and reneging. Int. J. Innov. Technol. Explor. Eng. (IJITEE) **8**(9) (2019). ISSN 2278-3075
18. Ammar, S.I., Jiang, T., Ye, Q.: Transient analysis of impatient customers in an M/M/1 disasters queue in Random environment. Eng. Comput. **37**, 1945–1965 (2020)
19. Paz, N., Yechiali, U.: An M/M/1 queue in random environment with disasters. Asia Pac. J. Oper. Res. **31** (2014). https://doi.org/10.1142/S021759591450016X
20. Udayabaskaran, S., Dora Pravina, C.T.: Transient analysis of an M/M/1 queue in a random environment subject to disasters. Far East J. Math. Sci. (FJMS) **91**, 157–167 (2014)
21. Farej, Z.K., Ali, O.M.: Throughput performance analysis of DL-MU-MIMO for the IEEE 802.11ac wireless LAN. In: Peng, S.L., Jhanjhi, N.Z., Pal, S., Amsaad, F. (eds.) Proceedings of 3rd International Conference on Mathematical Modelling and Computational Science. ICMMCS 2023. Advances in Intelligent Systems and Computing, vol.1450. Springer, Singapore (2023). https://doi.org/10.1007/978-981-99-3611-3_45

Forecasting Financial Markets: A Critical Analysis of Machine Learning and Social Sentiment Analysis

Tom K. Joseph[1(✉)], Vikas Verma[1], Arun Malik[1], Ghadah Naif Alwakid[2], and Manzoor Hussain[3]

[1] Lovely Professional University, Phagwara, Punjab, India
tomkj2252@gmail.com, vermavikas2407@gmail.com, arunmalikhisar@gmail.com

[2] Department of Computer Science, College of Computer and Information Sciences, Jouf University, Sakakah 72341, Al Jouf, Saudi Arabia
gnalwakid@ju.edu.sa

[3] Indus University, Karachi, Pakistan
manzoor.hussain@indus.edu.pk

Abstract. Accurate price prediction is essential for investors, analysts, and policymakers to make informed decisions in the ever-changing world of stock markets. This review paper investigates the relationship between predictive modeling and stock market analysis by utilizing the functions of algorithms. This review assesses the performance of 13 algorithms in stock price forecasting by looking at 15 research articles released between 2018 and 2023. The algorithms that are used in this context include AdaBoost, Artificial Neural Network (ANN), Decision Tree, Gradient Boosting, K-Nearest Neighbors (KNN), Logistic Regression, Naive Bayes, Random Forest, Recurrent Neural Network (RNN), Support Vector Classifier (SVC), the Support Vector Machine (SVM), and XG Boost. This review offers a thorough overview of the advantages, disadvantages, and corresponding effectiveness of multiple algorithms within the context of forecast by combining insights based on these investigations.

Keywords: Stock Market Analysis · Data · Social Media · Machine Learning

1 Introduction

Since the beginning of financial research, investors, analysts, and scholars have been enthralled with the pursuit of precise stock market price prediction. Historically, this endeavor has depended on technical indicators and fundamental analysis, but a paradigm shift has taken place with the introduction of machine learning (ML) approaches. With machine learning (ML), one can possibly find hidden patterns and relationships inside large datasets [4], opening fascinating new potential for accurate market trend prediction. This in-depth investigation delves within the program parameters of differed ML algorithms for predicting the market's prices. By meticulously curating a collection of 19 scientific contents published between 2018 and 2023, the study offers a comprehensive analysis of this evolving field.

S. Pal and Á. Rocha (Eds.): ICMMCS 2025, LNNS 1398, pp. 414–426, 2025.
https://doi.org/10.1007/978-3-031-90998-6_38

Each study within the examined collection provides unique perspectives by leveraging diverse datasets, incorporating a range of sources such as historical stock prices, trading volumes, economic indicators [2], and even sentiment analysis from social media [14, 15]. This multifaceted approach enriches the understanding of market dynamics within the model. Furthermore, scholars employ innovative feature engineering techniques, transforming raw data into structures comprehensible to machine learning algorithms. These techniques extract significant attributes, including volatility metrics, sentiment ratings from social media text analysis [14, 15], and technical indicators. Additionally, a variety of assessment metrics are utilized to evaluate models for projecting stock prices, encompassing measures such as mean square error, root mean square error, and percent error in absolute terms [10]. Beyond these core methodologies, researchers also delve into nuanced analyses, exploring the impact of various algorithmic parameters, the robustness of models across different market conditions, and the scalability of approaches to larger datasets. Through these comprehensive methodologies and additional layers of analysis, researchers aim to enhance the accuracy and applicability of machine learning techniques in forecasting stock market trends.

Through the integration of diverse research endeavours, this review offers an invaluable tool for understanding the advantages, constraints, and relative effectiveness of various machine learning (ML) algorithms in stock market forecasting. By synthesizing and critically evaluating findings from the literature, it not only provides insights into the current state of the art in stock market prediction but also identifies avenues for future research and advancement in this dynamic field. Crucial research questions addressed include identifying machine learning algorithms demonstrating higher prediction accuracy in forecasting stock values, common challenges encountered in applying ML algorithms for stock market prediction, and proposed strategies to address these challenges. Furthermore, the review highlights areas within the body of knowledge on machine learning techniques for stock pricing prediction that require further exploration, suggesting avenues for future study and development.

2 Literature Survey

This review of the literature uses a structured tabular method to examine and assess several research articles that use machine learning algorithms to anticipate stock market movements. Each study's key facts are effectively presented by arranging the data in a table format. These details include references, datasets used, models/algorithms employed, major findings, and weaknesses detected. This methodological framework facilitates a thorough comprehension of the procedures and findings throughout the literature by improving the clarity and effectiveness of synthesizing the multiple research study outcomes. The tabular representation facilitates a rigorous study and interpretation of the corpus of available knowledge by providing researchers in the domain of financial market forecasting with a formal and scientific instrument to identify patterns, trends, and contradictions (Table 1).

This section of the literature review examines sentiment analysis in circumstances other than standard stock price analysis, with a concentration on the importance of sentiment analysis in foreign languages. This research is important even though it is not

Table 1. Comparison between Papers

Paper	Dataset/Input Variables used	Models/Algorithms Used	Key findings	Weakness Identified/Mentioned
[1]	The Bombay Stock Exchange BSE SENSEX the past stock market data	SVM Random Forest Decision Tree Generalized Linear Model Gradient Boost	Gradient Boosted Model provided closest, Prediction as it had the lowest mean absolute error	• Stock market prediction accuracy may be hindered by inherent market volatility and unpredictability • Overfitting and data biases could compromise the reliability of forecasting models • External factors like economic events, political shifts, and global market trends may not be adequately considered in the study, potentially impacting stock price forecasts
[2]	• Historical stock prices • Trading volumes Market • Economic Indicators	Long Short-Term Memory (LSTM), Random Forest, Support Vector Machine (SVM), K-nearest Neighbor (KNN), Logistic Regression, Deep Deterministic Policy Gradient (DDPG), Differential Privacy (DP) - LSTM Random Forest	Differential Privacy (DP) - LSTM gained an exceptional precision of 99.58%	N/A
[3]	• Statistics from Reliance and IBM datasets • The trade volume, peak, low, closing price, and adjusted closing price	Decision Tree Random Forest Naive Bayes SVM Radial Basis Function (RBF) Random Forest LSTM XGBoost Multilayer Perceptron (MLP) Logistic Regression KNN	MLP achieved 98% accuracy on the IBM dataset, while LSTM only reached 67%. Random Forest and Logistic Regression achieved 52%, respectively. Naive Bayes, KNN, SVM, XG Boost, and Decision Tree all obtained 51% or 50% accuracy, respectively. XG Boost has the lowest precision rate (46%). MLP ranked first in recall (97%), followed by LSTM (84%)	• Overfitting: Algorithms that perform extraordinarily well during training may perform inadequately on unseen data due to overfitting • Data Quality: The dataset was extensively reviewed for any gaps, errors, or missing numbers that would skew the conclusions

(*continued*)

Table 1. (*continued*)

Paper	Dataset/Input Variables used	Models/Algorithms Used	Key findings	Weakness Identified/Mentioned
[4]	• The S&P 500 index data, the stock price's seven-day moving mean, the stock high less the low price, and the stock close less the open price • The fourteen-day mean of the stock price • The twenty-one-day mean of the stock price	KNN SVM LSTM Decision Tree (DT) MLP	• Not a single best estimator model for every dataset • In three out of six datasets, the MLP model was deemed to be the best • KNN technique performed well in two datasets, and SVR in one dataset • In four datasets, the LSTM model performed best according to the MAE criterion • Based on both performance criteria, MLP and LSTM were shown to be the most effective way • The features of the dataset determine the best configuration for a machine learning algorithm	• No single machine learning technique is optimal for every dataset • Requirement to choose an acceptable performance metric for evaluating predictive capabilities • The performance metrics that are employed determine how different ML techniques are ranked
[5]	• Energy, mineral stocks, basic metals, and diversified financials are the four different investment groups that make up the Tehran Stock Exchange • A decade of past data for these trading categories	Decision Tree, Random Forest, Adaboost, XGBoost, SVC, Naïve Bayes, Logistic Regression, KNN, ANN, RNN, LSTM	The results show that continuous data models, RNN and LSTM perform better compared to alternative models of prediction, and deep learning techniques greatly enhance binary data performance	• Overfitting is a frequent issue that occurs in Decision Tree models when overly complicated trees are constructed • Both the estimation method and the format of the input data affect the forecasts produced by these models • While every strategy has its own set of drawbacks, using notable traits as input variables might improve prediction models' performance

(*continued*)

Table 1. (*continued*)

Paper	Dataset/Input Variables used	Models/Algorithms Used	Key findings	Weakness Identified/Mentioned
[6]	• Korean Composite Stock Price Index (KOSPI) • Stock indices, Exchange rates, Interest rates, liquidity, economic indices, price indices, and production Indices	LSTM Random Forest SVM	The LSTM model demonstrated superior crisis prediction accuracy compared to RF and SVM models, achieving the highest accuracy among the models tested study demonstrated the link between housing market bubble signals and financial market volatility, showing the effectiveness of the early warning system in detecting risks in both markets	The study suggests potential weaknesses in the stock market's early warning system, including limitations in predictive power and generalizability across different market environments, potentially impacting its applicability in diverse scenarios
[7]	• Fundamental stock data, including volume, peak, valley, and open/close • MACD: Moving Averages Convergence, Divergence, Moving Average • RSI: Relative Strength Index • KDJ: Stochastic Oscillator (KDJ Index) • On-balance volume, or OBV Bollinger Bands (BOLL)	Adaboost, XGBoost, SVC, Naïve Bayes, KNN, logistic regression, ANN, RNN, LSTM, decision trees, random forests	• The effectiveness of the model is evaluated using the following ratios: Sorting Ratio, Annual Return, Compound Return, Maximum Drawdown, Sharpe Ratio, and Calmar Ratio • LSTM model is found to be more accurate in stock price prediction compared to other models	N/A
[8]	• Every day, information covering the Chinese Stock Market's Shanghai and Shenzhen 300 Index component equities during the course of the 18-year time span from 2000 to 2017 • The rate of volume feature, three K-line patterns, and eight inter-day price shift indications • There are 168 stocks in the dataset, and there are 553,028 sections worth of information during that time frame	Logistic Regression, SVM, KNN, Random Forest, Gradient Boosting Decision tree, LSTM	• Decision on Gradient Boosting In most situations, trees and random forests (RF) have demonstrated a strong capacity to anticipate for immediate forecasts • Logistic Regression (LR) required additional features to improve its prediction level • KNN and SVM were found to fit well in some patterns • LSTM did not fully demonstrate its advantage in the scenario studied	• LSTM did not fully demonstrate its advantage in the scenario studied • It was observed that SVM's processing time for large data was lengthy, which limited its applicability for vast-scale stock data forecasting

(*continued*)

Table 1. *(continued)*

Paper	Dataset/Input Variables used	Models/Algorithms Used	Key findings	Weakness Identified/Mentioned
[9]	• The investigation revolves around 55 US financial market companies across 11 distinct business sectors • Lagged Time sequences that are lagging of each stock's past log-returns for 3 months, 6 months, and 1 year	Random Forest, XG Boost, Multilayer Perceptron (MLP), Logistic Regression, and LSTM	• Effective Transfer Entropy (ETE) is used as an additional variable in forecasting the trajectory of the stock market • The study investigates the dynamics of stock prices as they evolve through ETE, as well as the impact on prediction accuracy • It was determined that MLP and LSTM network algorithms were more suited for estimating stock prices	N/A
[10]	• Three market indices from the US (NYSE), India (BSE), and Africa (GSE).For every company, there is information on the Start Price, Top Price, Minimum Price, Closing Price, Modified Closing Price, and Volumes	Linear Regression, MLP, SVM, K-NN, KNN-CCOA (K-Nearest Neighbors with Chicken Swarm Optimization Algorithm)	On a third dataset, where the maximum R2 value was 0.96, the KNN-CCOA model performed better in terms of average absolute error the root mean squared error Mean Absolute Percentage Error, and R2 values	N/A
[11]	• Daily value of shares history from 2015 to 20 for the following five companies: Facebook, Tesla, Amazon, Netflix, and Google • For every company, there is information on the Start Price, Top Price, Minimum Price, Closing Price, Modified Closing Price, and Volumes	SVR, LSTM, ANN, RNN, Random Forest	LSTM outperforms SVR in predicting stock prices, with some studies reporting over 95% accuracy	• The LSTM architecture may not always capture dynamic changes accurately • Efficient Market Hypothesis suggests that properly anticipating price movements in shares is challenging due to market efficiency
[12]	The NASDAQ equity index's historical stock data, which spans the period from March 2018 to March 2023 includes 1258 daily observations and technical indicators like the ROC, S%K, EMA, CCI, and Simple Moving Average	•Gradient Boosting, Random Forest, Gaussian Naïve Bayes, SVM, and KNN	Random Forest (61%) and KNN (59%) models are effective in predicting stock price fluctuations, with the PCA-GS-RF model being recommended for accurate investment decisions	Future work suggests improvements through hyper-parameter tuning and feature selection techniques

(continued)

Table 1. *(continued)*

Paper	Dataset/Input Variables used	Models/Algorithms Used	Key findings	Weakness Identified/Mentioned
[13]	• Historical stock prices • Financial signals	• SVM • Random Forest • LSTM • ANN	Random Forest and SVM outperform LSTM in accuracy and F-score in forecasting, with improved accuracy as time horizon increases	Unexpected events and unforeseen factors can impact stock prices, making it challenging to predict future performance using economic indicators and the Random Forest algorithm
[14]	Tweets sourced from Twitter API, and historical stock market data using Python's yahoo-finance package	• Sentiment Analysis • Naïve Bayes, K-ELM (Extreme Learning Machine), SVM	An effective model, SVM, helps Twitter anticipate stock prices by using tweet sentiments to forecast stock movements. This improves classification speed and solves issues with collecting historical data	Twitter's stock market predictions face challenges in predicting sentiment from tweets due to authentication requirements and filtering out irrelevant data
[15]	Social Media Data and the Taiwanese Stock Market (TSMC, which) Observation Post System	Genetic Algorithm, LSTM, Grey Relational Analysis	The LSTM model, incorporating sentiment from social media, significantly enhanced prediction accuracy by 2.22%, highlighting the significance of integrating traditional quantitative variables with sentiment analysis	The TSMC case study poses an individual obstacle in evaluating forecasting model performance due to the need for a defined benchmark dataset, and the study's focus on specific chip-based indicators and emotion factors may limit generalizability to other markets

related to financial applications as it points out the vital role of sentiment analysis in comprehending public opinion and behaviour.

The sentiment analysis model proposed in this study integrates major techniques including data preprocessing, lemmatization, class label consolidation, and random minority oversampling to achieve class balancing. It utilizes deep learning methods with one-hot encoding for text transformation to improve the analysis of sentiments related to COVID-19 from social media data [16]. Research emphasizes the efficacy of hybrid models in addressing class imbalance, unsupervised learning techniques, and the significance of feature selection during training. In particular, the XGBoost classifier exhibited exceptional performance in tackling class imbalance problems [17]. Also, it was discovered that a comprehensive framework that incorporated resampling methods and extra features with a variety of classifiers, including the Random Forest algorithm, worked well [18]. A remarkable advancement in nonstandard Arabic sentiment analysis is creating a multifactor lexical sentiment analysis (MULDASA) that incorporates elements such as intensifiers, negation, and emojis. This enhances the accuracy of sentiment classification, particularly in colloquial Arabic, and facilitates social media content analysis [19]. Machine learning and sentiment analysis have been widely used in financial forecasting, including banking news extraction [20, 21] and sentiment classification [22, 23].

3 Methodology

The methodology used to collect and analyze research papers on forecast for the stock market using machine learning algorithms included a comprehensive literature survey of key academic databases. Relevant publications published between 2018 and 2023 that focused on machine learning applications in financial analysis were selected using targeted keyword searches and rigorous screening criteria. The evaluation examined the datasets used in these studies, separating those that used broad stock market indices from those that focused on individual stocks or specialized stock groups. Through an examination of the dataset's characteristics, such as past valuations of stocks, the amount of trading, and other market and financial variables, the review gained understanding of the feature engineering and data pre-treatment strategies employed to enhance prediction accuracy.

Furthermore, a thorough comparison of the techniques and outcomes used in the chosen publications was made possible by the evaluation of model performance measures like as accuracy in classification, averaged error in absolute terms, and root mean square error. The use of machine learning algorithms in stock market foreseeing was compared, highlighting trends, advantages, disadvantages, and best practices.

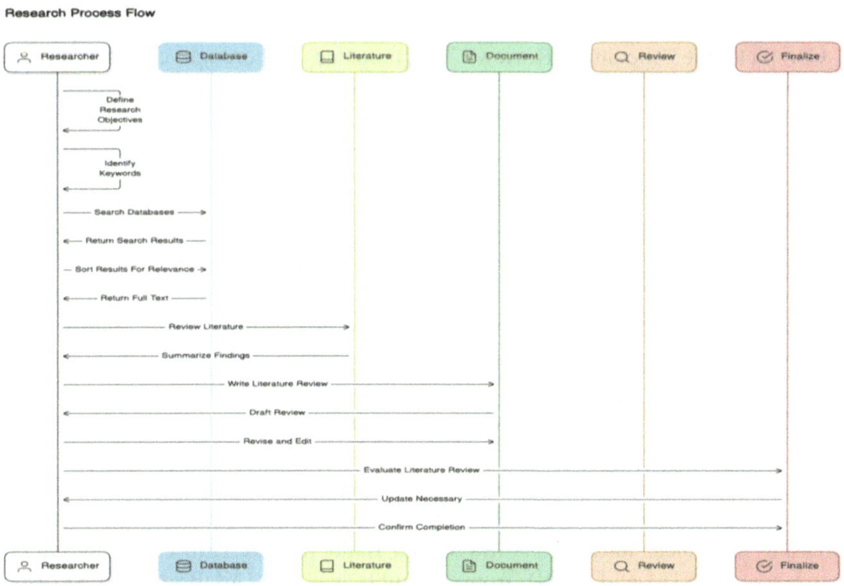

Fig. 1. Sequence diagram of the methodology adapted

The Fig. 1 illustrates a methodical way to performing a literature review on stock market price prediction. This is the explained version:

1. Establish research goals with an emphasis on databases, their impact, knowledge gaps, and widely used techniques for stock market price prediction.
2. Determine your keywords: Choose relevant terms like "stock market," "machine learning," "sentiment analysis," and "price prediction techniques in machine learning."
3. Look up literature databases: Use the terms to do a literature search in academic databases (Used - IEEE Xplore, Google Scholar, Scopus).
4. Filter Query Conclusions: Assess outcomes from searches according to how well they support the goals of the study, giving special attention to papers that address the databases that were utilized, the effect of database size, research gaps, and frequently used algorithms.
5. Obtain the Complete Text: For additional analysis, obtain full-text articles that satisfy the inclusion criteria.
6. Examine the Literature: Examine the articles that were retrieved to determine which databases were utilized, the impact that database size has been discussed, the research gaps that have been found, and the frequently used algorithms for stock market price prediction.
7. Synthesize Findings: Highlight trends, patterns, and author agreement on databases, the impact of database size, research gaps, and widely used algorithms in your summary and synthesizing of the findings.
8. Compose a summary of the literature: By arranging the combined results into logical parts that speak to each study goal. Verify that all relevant sources are properly cited.
9. Edit and Revise: Check for accuracy, coherence, and clarity in the literature review. Verify that citation requirements and academic standards are followed.
10. Complete the Literature Review: To guarantee a thorough covering of every research objective, take advisers' or peers' comments and recommendations into consideration.
11. Update as Required: Update the literature review on a regular basis to reflect new research and advancements in the area.

An excellent basis for the analysis is established by this rigorous research process. This research summarizes the most current developments in machine learning for forecasting the stock market by concentrating on papers from the last five years. Furthermore, the evaluation procedure involves more than just identifying the algorithms that are employed. Through an examination of the used datasets (individual stocks versus wide indexes) and the particular data pre-processing methods (feature engineering), the investigation clarifies the ways in which these variables impact the machine learning models' overall efficacy.

Based on past data and market indicators, a thorough set of 13 machine-learning algorithms was carefully chosen for this study to estimate stock market movements. From addressing non-linear interactions (SVM and Random Forest) to capturing temporal dependencies (LSTM and RNN), from ensemble learning (AdaBoost and XG Boost) to probabilistic classification (Naive Bayes), each method contributes a distinct set of abilities to the predictive modelling process. The purpose of this study is to investigate the advantages and disadvantages of each model in anticipating improvements to the methods for predicting and tweaks to the financial market overall accuracy by utilizing this wide range of algorithms.

Below is a quick overview of each of the algorithms that were discussed (Fig. 2):

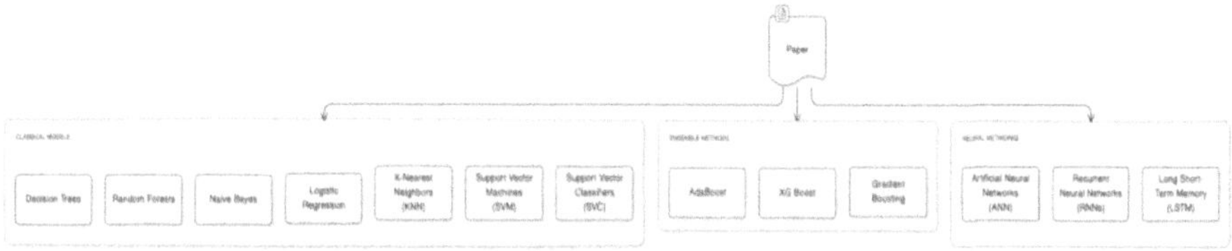

Fig. 2. Techniques Discussed throughout the Survey

- Long Short-Term Memory (LSTM): A specific type of neural network with recurrent architecture that recognizes long-term dependencies in sequential data, making it useful for periodic prediction tasks like stock price prediction.
- Recurrent Neural Network (RNN): It is a family of neural networks that are frequently employed in examination of time-series data and natural language interpretation. RNNs are capable of processing sequential data by remembering prior inputs.
- Decision Tree: This model resembles a tree and divides data into branches based on feature values to provide a predictably easy-to-understand and visually represent model.
- Random Forest: A robust and using a collaborative learning technique, multiple decision chains are constructed during training, with the average forecast from each tree being produced.
- Support Vector Machine (SVM): An efficient supervised learning technique for classification tasks with intricate decision boundaries, SVM determines the best hyperplane to divide data into classes.
- Support Vector Classifier (SVC): A variation of SVM created especially for classification problems, with the goal of improving generalization by maximizing the margin between classes.
- AdaBoost: An ensemble learning technique that iteratively modifies weights to concentrate on examples that are incorrectly classified, combining several weak learners to produce a strong classifier.
- XG Boost: A popular solution for regression, classification, and ranking issues, it is an enhanced version of gradient boosting that improves speed and performance.
- Naive Bayes: A probabilistic classifier that relies heavily on strong independence assumptions across features based on Bayes' theorem. It is frequently used for sentiment analysis and text categorization.
- Logistic regression: It is a homogenous model that estimates based on input features, the probability that an instance will belong to a particular class. It is used for binary classification tasks.
- K-Nearest Neighbors (KNN): Suitable for both regression and classification applications. Using the overall count of their k nearest counterparts, instances are categorized using the irregular KNN approach.

- Artificial Neural Network (ANN): A computational model consisting of interconnected nodes arranged in layers for learning complicated patterns in data, inspired by the neural networks seen in the human brain.
- Gradient Boosting: A method of ensemble learning with high predicted accuracy and adaptability that creates models in a sequential fashion, with one model correcting the mistakes of the next.

4 Conclusion

This study makes use of findings from 19 academic publications, mostly from 2021 and 2023. These papers look at a total of thirteen distinct ML techniques for stock market value prediction. The most widely used algorithms are Random Forest and Long Short-Term Memory (LSTM), but AdaBoost and Support Vector Classifier (SVC) are used less frequently. This wide range of algorithms serves as the foundation for an extensive analysis and comparison of how well they predict stock market trends.

The broad range of assets and marketplaces examined enhances the ability to be generalized of the results. The research keeps itself free from any biases towards certain economic areas or market structures by include both established and growing economies. The study's application to actual investment decisions is improved by its wider scope, which enables investors to make well-informed selections about which strategies would work best for their unique investing objectives and risk tolerance.

The study covers a wide range of stocks and stock markets, including well-known indexes like the NASDAQ index, the Bombay Stock Exchange (BSE) SENSEX, Reliance, IBM, S&P500, the Tehran stock exchange with its diverse sectors, the Korean Composite Stock Price Index (KOSPI), the constituent stocks of the China Stock Market in Shanghai and Shenzhen 300 Index, and individual companies like Amazon, Google, Tesla, Netflix, and Facebook.

Numerous input variables and datasets are used in these analyses. These include historical stock prices, trading volumes, financial indicators, moving averages, lagged time series, inter-day price move indicators, rate of volume feature, price changes, bid and offer prices, indicators (such as the Stock Channel Index, Simple Moving average, Exponential Moving Average, Rate of Change, and Stochastic Oscillator), and social media data from Twitter and other popular platforms. These varied datasets, which cover a range of temporal and geographic areas, enable thorough analysis intended to forecast future stock prices and market patterns. Although forecasting future trends is frequently the main emphasis, this comprehensive examination provides more than a crystal ball. Potential hazards and possibilities can be highlighted by highlighting underlying links found within the data. Investment plans, for example, can be guided by the relationship between sentiment on social media and changes in stock prices. Analysing historical trading volume patterns in conjunction with technical indicators may also assist investors in making well-informed judgments on when to enter and quit the market. This more profound comprehension enables active and planned financial management in addition to predicting.

It is evident that there is still scope for future developments in this stock market price prediction using ML and its derivatives such an example would be the perfect combination of an combined model for market exchange prediction with the essence of social media sentiment analysis that was discussed in [14, 15].

References

1. Rohatgi, S., Kumar Singh, K., Jasuja, D.: Comparative analysis of machine learning algorithm to forecast Indian stock market. In: 2021 International Conference on Advance Computing and Innovative Technologies in Engineering (ICACITE), Greater Noida, India, pp. 278–283 (2021). https://doi.org/10.1109/ICACITE51222.2021.940464
2. Lumoring, N., Chandra, D., Gunawan, A.A.S.: A systematic literature review: forecasting stock price using machine learning approach. In: 2023 International Conference on Data Science and Its Applications (ICoDSA), Bandung, Indonesia, pp. 129–133 (2023). https://doi.org/10.1109/ICoDSA58501.2023.10277318
3. Kotapati, L., Sajjala, R.B., Gudi, S., Hareesh, B.V.N., Tokala, S., Enduri, M.K.: Forecasting stock markets trends using machine learning algorithms. In: 2023 IEEE 15th International Conference on Computational Intelligence and Communication Networks (CICN), Bangkok, Thailand, pp. 278–282 (2023). https://doi.org/10.1109/CICN59264.2023.10402238
4. Medarhri, I., Hosni, M., Nouisser, N., Chakroun, F., Najib, K.: Predicting stock market price movement using machine learning techniques. In: 2022 8th International Conference on Optimization and Applications (ICOA), Genoa, Italy, pp. 1–5 (2022). https://doi.org/10.1109/ICOA55659.2022.9934252
5. Nabipour, M., Nayyeri, P., Jabani, H., Shahab, S., Mosavi, A.: Predicting stock market trends using machine learning and deep learning algorithms via continuous and binary data; a comparative analysis. IEEE Access **8**, 150199–150212 (2020). https://doi.org/10.1109/ACCESS.2020.3015966
6. Park, D., Ryu, D.: A machine learning-based early warning system for the housing and stock markets. IEEE Access **9**, 85566–85572 (2021). https://doi.org/10.1109/ACCESS.2021.3077962
7. Fu, K., Yu, Y., Li, B.: Multi-feature supervised reinforcement learning for stock trading. IEEE Access **11**, 77840–77855 (2023). https://doi.org/10.1109/ACCESS.2023.3298821
8. Lin, Y., Liu, S., Yang, H., Wu, H.: Stock trend prediction using candlestick charting and ensemble machine learning techniques with a novelty feature engineering scheme. IEEE Access **9**, 101433–101446 (2021). https://doi.org/10.1109/ACCESS.2021.3096825
9. Kim, S., Ku, S., Chang, W., Song, J.W.: Predicting the direction of US stock prices using effective transfer entropy and machine learning techniques. IEEE Access **8**, 111660–111682 (2020). https://doi.org/10.1109/ACCESS.2020.3002174
10. Rao, S.S., Kumar Pareek, P., Vishwanatha, C.R., Kumar Bhardwaj, A.: Prediction of technical indicators for stock markets using HPO-CCOA based machine learning algorithm. In: 2023 International Conference on Evolutionary Algorithms and Soft Computing Techniques (EASCT), Bengaluru, India, pp. 1–6 (2023). https://doi.org/10.1109/EASCT59475.2023.10393484
11. Marchai, F.L., Martin, W., Suhartono, D.: Stock prices prediction using machine learning. In: 2021 8th International Conference on Information Technology, Computer and Electrical Engineering (ICITACEE), Semarang, Indonesia, pp. 79–84 (2021). https://doi.org/10.1109/ICITACEE53184.2021.9617222

12. Zouaghia, Z., Kodia Aouina, Z., Ben Said, L.: Stock movement prediction based on technical indicators applying hybrid machine learning models. In: 2023 International Symposium on Networks, Computers and Communications (ISNCC), Doha, Qatar, pp. 1–4 (2023).https://doi.org/10.1109/ISNCC58260.2023.10323971
13. Tiwari, V., Lohani, B.P., Rana, A., Pandey, U.P., Dhariwal, B.: Stock market prediction using different machine learning algorithms. In: 2023 10th IEEE Uttar Pradesh Section International Conference on Electrical, Electronics and Computer Engineering (UPCON), Gautam Buddha Nagar, India, pp. 147–151 (2023). https://doi.org/10.1109/UPCON59197.2023.10434490
14. Mankar, T., Hotchandani, T., Madhwani, M., Chidrawar, A., Lifna, C.S.: Stock market prediction based on social sentiments using machine learning. In: 2018 International Conference on Smart City and Emerging Technology (ICSCET), Mumbai, India (2018)
15. Huang, J.Y., Tung, C.L., Lin, W.Z.: Using social network sentiment analysis and genetic algorithm to improve the stock prediction accuracy of the deep learning-based approach. Int. J. Comput. Intell. Syst. **16**, 93 (2023)
16. Humayun, M., Javed, D., Jhanjhi, N., Almufareh, M.F., Almuayqil, S.N.: Deep learning based sentiment analysis of COVID-19 tweets via resampling and label analysis. Comput. Syst. Sci. Eng. **47**(1), 575–591 (2023)
17. Almuayqil, S.N., Humayun, M., Jhanjhi, N.Z., Almufareh, M.F., Javed, D.: Framework for improved sentiment analysis via random minority oversampling for user tweet review classification. Electronics **11**(19), 3058 (2022)
18. Almuayqil, S.N., Humayun, M., Jhanjhi, N.Z., Almufareh, M.F., Khan, N.A.: Enhancing sentiment analysis via random majority under-sampling with reduced time complexity for classifying tweet reviews. Electronics **11**(21), 3624 (2022)
19. Alwakid, G., Osman, T., Haj, M.E., Alanazi, S., Humayun, M., Sama, N.U.: MULDASA: multifactor lexical sentiment analysis of social-media content in nonstandard Arabic social media. Appl. Sci. **12**(8), 3806 (2022)
20. Dogra, V., Singh, A., Verma, S., Kavita, Jhanjhi, N.Z., Talib, M.N.: Analyzing DistilBERT for sentiment classification of banking financial news. In: Peng, S.L., Hsieh, S.Y., Gopalakrishnan, S., Duraisamy, B. (eds.) Intelligent Computing and Innovation on Data Science. LNNS, vol. 248, pp. 501–510. Springer, Singapore (2021). https://doi.org/10.1007/978-981-16-3153-5_53
21. Dogra, V., Verma, S., Verma, K., Jhanjhi, N.Z., Ghosh, U., Le, D.N.: A comparative analysis of machine learning models for banking news extraction by multiclass classification with imbalanced datasets of financial news: challenges and solutions (2022)
22. Zaman, N., et al.: Stock market prediction based on machine learning and social sentiment analysis. Authorea Preprints (2023)
23. Wassan, S., Chen, X., Shen, T., Waqar, M., Jhanjhi, N.Z.: Amazon product sentiment analysis using machine learning techniques. Revista Argentina de Clínica Psicológica **30**(1), 695 (2021)

The Role of Big Data Analytics in Forensic Computing: An Empirical Analysis of Financial Fraud Detection in Corporate Companies

T. Nathiya[1], Chuah Tong Kuan[2], D. Akila[3](✉), D. Padmapriya[3], K. Madhumathi[4], and Saurabh Adhikari[5]

[1] Department of Computer Science, FSH, SRM Institute of Science and Technology, Kattankulathur, Chennai, India
nathiyat@srmist.edu.in

[2] Department of Digital Health and Health Informatics, School of Business and Technology, IMU University, Kuala Lumpur, Malaysia
TongKuanChuah@imu.edu.my

[3] Department of Computer Applications, Saveetha College of Liberal Arts and Sciences, Saveetha Institute of Medical and Technical Sciences, Chennai, India

[4] Department of Computer Science, Anna Adarsh College for Women, Annanagar, Chennai 40, India

[5] School of Engineering, Swami Vivekananda University, Barrackpore, India
saurabhadhikari@svu.ac.in

Abstract. The research emphasises on the significance of the use of big statistics analytics which will fight towards financial fraud among corporate businesses. Sophisticated analytical strategies can shield groups' monetary well-being, popularity and potential to discover and prevent frauds. For instance, there are numerous impediments to the usage of Big Data analytics in forensic Computing consisting of negative pleasant facts, privacy concerns, complicated technology base and shortage of professional manpower. This paper is a presentation of the Big Data Analytics Framework (BDAF) that turned into advanced especially for detecting economic fraud in Corporate corporations. The framework is an all-inclusive one and it has been designed for handiest use through corporate corporations. In its flip this system employs modern day algorithms used in facts processing, analysis and gadget mastering as a way to understand feasible signs and symptoms of fraud. The BDAF has a couple of programs in forensic Computing which encompass facilitating hazard evaluation, compliance monitoring and proactive identity of fraudulent sports. This method complements their potential to discover fraud instances promptly consequently lowering dangers related to it thru reading specific monetary statistics assets like transaction records, economic reviews or market tendencies. In depth simulation analyses are performed with real economic datasets from Corporate enterprises with a view to examine the effectiveness of BDAF. These simulations investigate whether or not the framework can be able to detect corruption, extortion among different kinds of financial assertion fraud in addition to its accuracy, performance and scalability. Organizations can enhance their structures for detecting deception whilst safeguarding their monetary pastimes within an intricate enterprise environment with the aid of scaling boundaries and employing innovative techniques.

S. Pal and Á. Rocha (Eds.): ICMMCS 2025, LNNS 1398, pp. 427–437, 2025.
https://doi.org/10.1007/978-3-031-90998-6_39

Keywords: Role · Big Data · Forensic Computing · Empirical analysis · Financial · Fraud · Detection · Corporate Companies

1 Introduction

Using large records analytics for forensic Computing poses a complex undertaking, mainly in relation to detecting monetary fraud in Corporate firms [1]. Although there are some of boundaries that need to be triumph over, big statistics analytics do provide a few useful gear for spotting patterns and outliers that might suggest fraudulent diversion [2]. Data accessibility and first-class is a first-level issue. Like many developing international locations, corporate ought to have incomplete or misguided datasets due to a lack of a strong records infrastructure [3]. Analytics technologies aren't as successful in detecting fraudulent behaviour because of this constraint. Building reliable detection algorithms is difficult considering that economic systems are complicated and fraudulent schemes are numerous [4]. Since fraudsters are constantly arising with new techniques, analytical strategies want to be first-level-tuned [5]. Problems are made worse via the shortage of certified individuals who can paintings in forensic Computing and data analytics concurrently [6]. It may be difficult to recover those individuals in Corporate who are well-versed in the utilisation of large-scale statistical analytics with the goal to detect fraudulent activity [7]. Finally, use of big data analytics in forensic Computing is similarly complex using issues concerning records privacy and protection [8]. Improving facts infrastructure, encouraging the improvement of knowledge, and establishing rigorous regulatory frameworks are all essential steps towards overcoming these barriers, and they have to be pursued in tandem with the aid of public and business corporations [9]. The complete ability of large records analytics to fight monetary crime in corporate companies can best be realised thru all-encompassing techniques [10].

Investigating the use of large information analytics in forensic Computing to discover instances of monetary fraud in Corporate establishments brings to light several present day processes and the problems that include them [11]. Data mining, which incorporates analysing huge quantities of financial records for patterns which could imply fraudulent activity, is one common approach [12]. Algorithms are the backbone of this approach, which makes use of them to identify suspicious interest and signals government to analyze suspicious transactions or bills [13]. Another approach is predictive analytics, which entails searching at beyond facts to peer when fraud might happen that can be prepared. Another use of social network evaluation is to discover fraudulent or collusive networks using searching at the links and interconnections between human beings and businesses [14]. These strategies, in the meantime, are not without their difficulties. Because Corporate agencies do not have centralised databases or standardised reporting tactics, the first-rate and accessibility of information might be an issue [15]. This ought to cause facts sets to be inconsistent or lacking crucial facts. It is difficult to construct accurate detection algorithms because of the complexity of monetary systems and the ever-converting nature of fraudulent schemes. The effective adoption of these strategies is further hindered by way of the shortage of able individuals in Corporate who are proficient in forensic Computing and information analytics. To shield sensitive facts and stay according with policies, it's far essential to address issues about information privateness and protection. To triumph over those barriers, people must pour resources into our facts infrastructure, paintings tirelessly to enhance our analytical techniques, and domesticate multidisciplinary competencies in Computing and facts analytics. The promise of massive facts analytics in the combat towards financial crime in Corporate companies can best be realised through combined efforts.

The remaining parts of the paper are structured as follows: Sect. 2 explores the use of big data analytics in forensic Computing, specifically looking at how Corporate companies detect financial fraud. Within Sect. 3, the Big Data Analytics Framework (BDAF) is introduced. The findings and interpretations of the investigations are discussed in Sect. 4, along with parallels to past methodological approaches. The last conclusion is presented in Sect. 5.

2 Literature Survey

When it comes to identifying and avoiding financial crimes, forensic Computing (FA) provides substantial advantages, making it an ever more important weapon in the battle against fraud. There have been a number of studies examining forensic Computing's visibility, difficulties, and regional application.

E. E. Alharasis et al., [16] introduced to gauge the level of familiarity with and thoughts on forensic Computing (FA), a survey will be sent out to Corporate Computing faculty, students, and professionals. Results from the Empirical analysis should show that few people are familiar with FA, that there are strong links between FA and detecting and preventing fraud, and that there are many challenges to implementing FA.

The Empirical analysis by Al-Abbadi, H. M. et al. [17] inquiries into the difficulties of using forensic Computing to fight fraud (FA-FF) in Corporate by means of a questionnaire. Problems include things like the availability of technology, the distribution of fraud crimes, stakeholder cooperation, and the accumulation of expertise. The results suggest that the main impediments are cooperation and experience, followed by the scale of the crime and technology.

A descriptive analytical technique and a questionnaire survey had been used to acquire records from certified public accountants and Corporate legal authorities inside the observe by way of Tapanjeh, A. M. A et al. [18] based on the consequences of the statistical analysis programme (SPSS) and regression tests, forensic Computing is vital in Corporate to lower fraud. This examine exposed discrepancies in perceptions among accountants and felony government.

Researchers Mardjono, E. S. Et al. [19] used a quantitative case take a look at technique (QCSA) to survey 331 auditors in Indonesia and fraud detection is motivated with the aid of forensic Computing awareness, COSO inner control structures, and Big Data Analysis. As a end result of BDA's mediating function, the effects reveal that inner controls are greater effective and fraud detection competencies are progressed. Cloud architectures, security issues, Privacy paradigm [19–22] have been mentioned in the related previous proceedings and previous papers which helps Financial Fraud Detection in our papers.

Based on their examine, a Big Data Analytics Framework (BDAF) is the great possible treatment presently available in coming across fraud and enhancing inner controls' efficiency. This further demonstrates BDAF could be the simplest technique to slash monetary fraud.

3 Proposed Method

Big Data Analytics uses advanced algorithms to comb through large databases for trends, exceptions and threats. As stakeholders seek effective remedies, it is vital to understand how this technology helps detect and prevent financial frauds by learning about its usefulness in detecting such kinds of crimes. The complexity of detecting fraud increases with the pace at which the financial sector changes. The aim of this BDAF is to document and evaluate challenges involved in using customer data for fraud detection. Similarly, it seeks to demonstrate some benefits associated with adopting technology as well as guidance on enhancing banks' and other financial institutions' capabilities of detecting or identifying cases of fraud during digital era..

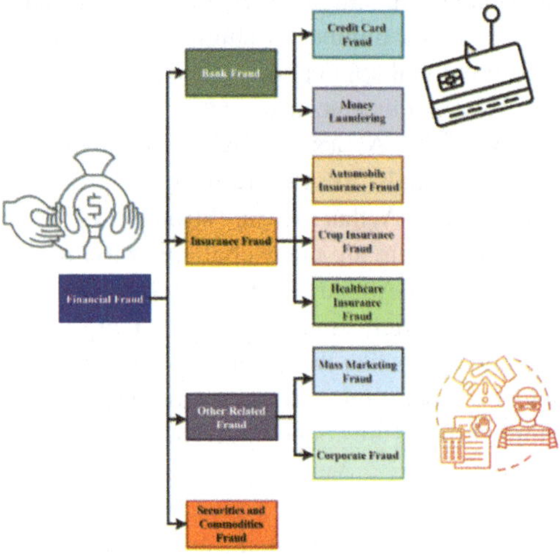

Fig. 1. Financial Fraud Classification

Bank fraud, insurance fraud and corporate fraud are always posing constant danger towards global financial system's integrity and stability. There are many different types of illegal behaviours that fall under the umbrella term "bank fraud," all with the common goal of fooling banks or their clients for personal benefit. Check forgery alongside identity theft results into a lot of loss in terms of money for individuals as well as organizations among others criminal acts. Insurance claims filed falsely or put at exaggerated amounts constitute what is called insurance fraudulent activities. All policyholders see their rates go higher due to this dishonest activity, which damages credibility and effectiveness of insurance systems. Corporate fraud refers to any unethical or illegal acts taken by businesses or their executives that intend deception. Computing fraud, insider trading, and bribery are all examples of wrongdoings that may lead to devastating outcomes, such as financial losses for shareholders, harm to their reputations, and even legal trouble. Big Data-driven real-time analytics allow for the prompt detection of questionable activity, minimizing the amount of time that may be lost and strengthening the institution's capacity for quick action. Challenges with real-time processing include scalability issues and latency. BDAF provide a discreet solution to these problems by protecting consumer information and protecting financial systems and fostering trust and transparency in the global economy from these forms of fraud requires strong regulatory frameworks, cutting-edge analytics tools, and proactive actions from companies and individuals is shown in Fig. 1.

$$\varphi\omega(j) = \int_{-\delta}^{\delta} kv^{u-2} hj^{-sd} \times ty = \frac{mf^{-tk}}{z} \tag{1}$$

The Eq. 1 represents the interplay of a number of parameters, including the transactions (kv), data integrity (hj), and dependent on time variables (ty and mf), which have an effect on the risk of committing fraud (j). Through the process of connecting these aspects, the framework intends to estimate the risk of fraud. This will be accomplished by taking into consideration the variability of the data ($\frac{mf^{-tk}}{z}$) as well as other factors that have an impact, hence improving both the precision and dependability of the procedures that are used to identify fraud.

$$\prod_{j=1}^{-\infty} \left(1 + \frac{k}{f}\right)^{-1} hp^{j/s}, \propto +\forall\exists(1 + \partial) \approx 0.2548 \tag{2}$$

The Eq. 2 takes into account the inverse connection between two important elements (k and f) as well as the rapid expansion of a variable (hp) over an interval (j/s). In the setting of the Big Data Analytic Framework (BDAF), which is used for the identification of financial fraud, this might be interpreted as the repeated method of assessing incremental details $\forall\exists(1 + \partial)$. In this process, each step minimizes the possibility of fraud that has not been discovered by taking into account cumulative risk indicators.

To provide a holistic picture of the financial environment, BDAF takes as inputs a variety of financial data sources, such as records of transactions, financial statements, and market trends. Data ingestion is the process of cleaning and preparing data for analysis by executing Extract, Transform, and Load procedures on data obtained from various sources. Big Data storage options, such as Hadoop Distributed File System or cloud-based storage, store cleaned and processed data, making it accessible and scalable for the system. In Fig. 2, to identify any instances of fraud, the most up-to-date algorithms for data processing, analysis, and machine learning examine the recorded data for any irregularities or suspicious trends. By using the system's risk assessment and compliance monitoring features, firms may bolster their financial security by proactively identifying and mitigating fraud threats. By conducting extensive simulation studies using real-world financial information, we can assess the BDAF's accuracy, efficiency, and scalability in identifying different types of financial fraud. With the use of a straightforward interface, stakeholders, including forensic accountants, can access the results and draw actionable conclusions via the reports and visualizations. By implementing a feedback loop that incorporates user input and continuously monitors system performance, the BDAF may

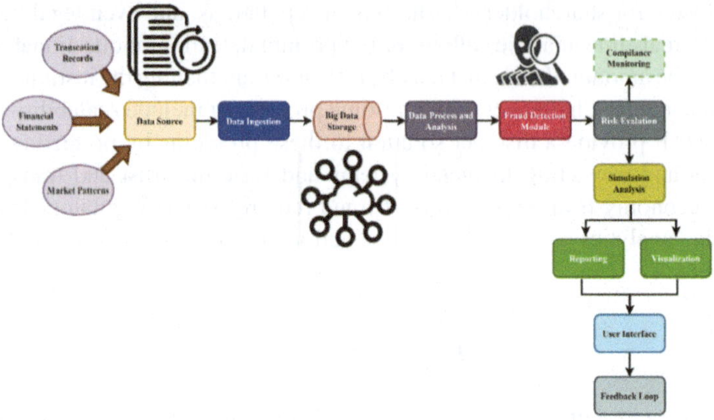

Fig. 2. Block Diagram of Big Data Analytics Framework in Forensic Computing

be improved to react to changing fraud trends and company demands. This helps prevent financial fraud in Corporate businesses.

$$\iiint_{ps} (W \cdot S + W)XL = \oiint_{S} YD \cdot JK(1 + H) \tag{3}$$

The expression in Eq. 3, where the variables W, S, and XL on the far left of the Eq. 3 might possibly represent weighed financial metrics. A number of components of data sizes and the interactions between them are shown by YD, JK, and H. It is likely that this equation illustrates a thorough examination of related financial metrics in the framework's exactitude and reliability.

$$tan\ tan\ f\theta + u\ cot\ cot\ hw\complement = (\beta + i\ tan\ tan\ \rho\sigma\tau)^{h} = g^{yup} \tag{4}$$

The Eq. 4 is about the accuracy analysis of the relationship to the field of financial fraud detection $tantanf\theta + ucotcothw\complement$, , has the potential to symbolize the accuracy analysis. It accomplishes this by merging multiple aspects of data (β) and $itantan\rho\sigma\tau$ to determine the probability of fraudulent occurrences g^{yup}.

4 Results and Discussion

Integrating forensic Computing with the Big Data Analytics Framework (BDAF) has demonstrated potential to improve Corporate organisations' fraud detection capabilities. This Empirical analysis seeks to offer forensic accountants and decision-makers in Corporate a thorough grasp of BDAF's strengths and limits through simulations and real-world data assessments.

The above Fig. 4 shows how forensic Computing applies big data analytics to pinpoint fraudulent activities occurring within Corporate organizations; therefore it relies heavily on accuracy analysis. Given how a great deal there's at stake when it comes to

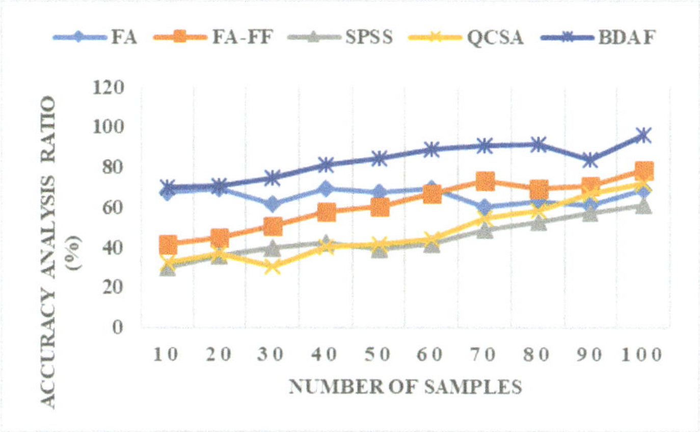

Fig. 4. Accuracy Analysis

uncovering fraudulent schemes, this studies underscores the want for trying out reliability plus precision regarding carried out analytical method. Accuracy evaluation compares BDAF algorithms and models with actual economic information sets from Corporate-based groups to discover how well they perform. For instance, substantial simulation exams on diverse economic fraud categories such as corruption, extortion, and monetary announcement manipulation are accomplished to affirm the framework's accuracy in detecting fraudulent styles and anomalies. This falls under the accuracy analysis that checks whether there is a correlation between the system's detection rate of fraudulent activities, false positive or negative rates and general recall and precision at which it stands at 96.2%. Test its scalability to see if the framework can handle vast amount of data correctly. This Empirical analysis sheds light on the BDAF's dependability and resilience in improving Corporate enterprises' fraud detection skills by performing thorough accuracy studies. In addition, forensic accountants and practitioners can find helpful recommendations on how to choose and apply suitable analytical methods to reduce the likelihood of fraud. As a whole, the accuracy analysis is a crucial yardstick for measuring how well big data analytics work in forensic Computing, which helps improve methods for detecting fraud and boosts financial honesty in Corporate companies.

In the above Fig. 5, Forensic Computing's use of massive statistics analytics to become aware of monetary fraud in Corporate organizations is based heavily on performance evaluation. Considering the complexity and range of financial records worried, this look at highlights the relevance of measuring the efficiency in addition to the correctness of the analytical strategies used. An BDAF performance requires looking at how fast it could scale, how well it uses sources, and what form of data it generates for Corporate companies. Using actual-global financial datasets in enormous simulation studies, people fastidiously take a look at the framework's capability to locate monetary fraud, such as corruption and economic statement manipulation. The devices scalability, processing time, and computational resource requirements are all cautiously examined to look how properly it manages big amounts of information without sacrificing velocity.

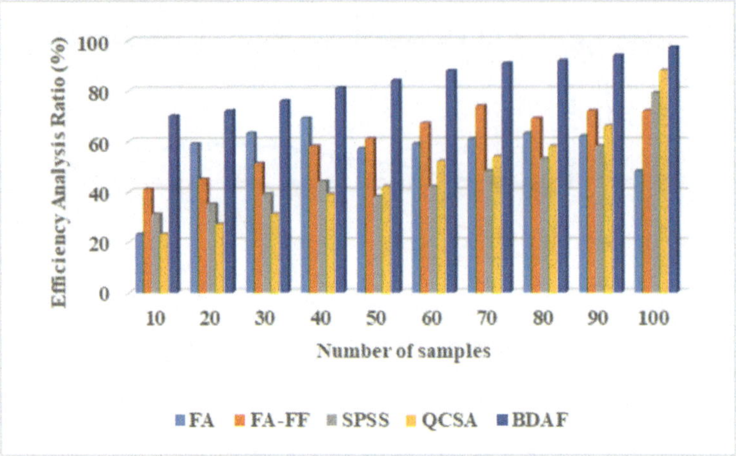

Fig. 5. Efficiency Analysis

In addition, the BDAF's ability to offer forensic accountants speedy insights and actionable intelligence is evaluated, taking into thoughts the critical relevance of responding quickly to fraud worries. This have a look at sheds light at the utility of massive data analytics for forensic Computing in Corporate businesses by performing thorough performance assessments, which reveal how well these equipment paintings and the way to high-quality optimise their sources produces 97.3%. It affords useful recommendation for corporations who want to use analytics to identify fraud, stressing the need to strike a stability between precision and speed for efficient results that might not break the bank. The performance evaluation is an essential step in figuring out how properly the BDAF works and the way it may grow, which will enhance forensic Computing and monetary honesty in Corporate's company world.

In the above Fig. 6, to evaluate the usefulness and efficiency of large facts analytics in forensic Computing for the motive of figuring out monetary crime in Corporate organizations, scalability analysis is essential. This studies highlights the importance of testing the machine's scalability to meet evolving business demands and ever-growing information masses without sacrificing overall performance. To determine the BDAF's scalability, people need to observe how nicely it handles big volumes of financial records. The BDAF became built with Corporate agencies in mind. The scalability of the machine is examined in multiple ways, including information quantity, computational sources, and processing speed, by way of conducting giant simulation checks with real-global datasets.

The machine's ability to hold performance ranges as statistics portions increase, processing time consistent with unit of information, and aid utilisation underneath growing workloads are some of the important thing parameters that are mechanically analysed. Given the ever-changing nature of monetary fraud schemes, people additionally evaluate the BDAF's adaptability and resilience to meet the converting demands of fraud detection and the business panorama produces 94.5%. The examine sheds mild at the system's scalability to fulfil the increasing needs of forensic Computing in Corporate corporations

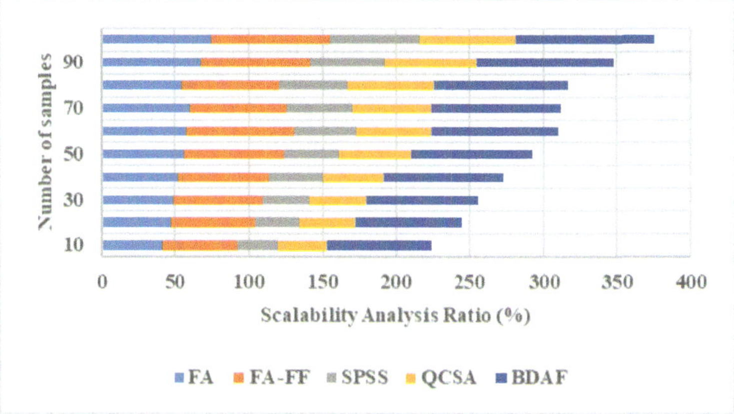

Fig. 6. Scalability Analysis

by means of doing thorough scalability opinions. It additionally highlights the significance of destiny-proofing systems to deal with growing information quantities and new fraud issues, and provides sensible advice for organizations trying to enforce scalable large records analytics solutions for fraud detection. In the end, scalability evaluation is an ought to have for gauging the BDAF's overall performance and sustainability over the long haul, to be able to assist improve economic integrity and forensic Computing in Corporate's company world.

The research evaluates BDAF in forensic Computing by examining how well it scales, how efficient it is, and how accurate it is. Taken as an entire, these exams display how BDAF may additionally significantly gain Corporate establishments' capability to discover fraud and maintain economic integrity.

5 Conclusion

Corporate groups can fortify their resilience towards financial fraud via using huge information analytics in forensic Computing. The paper pressured on the importance of superior analytics to be used to hold monetary health and status of corporations. Corporate's BDAF has been aimed in the direction of neighborhood companies, notwithstanding inherent stressful situations which incorporates records first-rate troubles, privateness concerns, and lack of understanding. Forensic accountants are able to efficiently spot bizarre patterns that can suggest fraud due to a holistic platform prepared with modern algorithms and system getting to know abilities. For example, transactional data similarly to market activities found from numerous financial records assets enable BDAF to assist proactive risk evaluation and early detection of frauds. Through massive simulation tests with real-worldwide financial datasets it indicates that those are capable of effectively find out diverse forms of financial fraud like extortion or corruption. This research explains that this technique is correct, efficient and scalable consequently it holds promise for enhancing fraudulent detection structures inner Corporate groups. Today's complex economic gadget necessitates companies overcoming hurdles by way of adopting modern-day techniques within the direction of limiting fraudulent practices in its realm. In so doing an enterprise should have greater potential in detecting any shape of illegal activity together with fraud for that reason retaining its financial interests and upholding the authenticity among Corporate businesses.

References

1. Alaoubi, A., Almomani, M.A.: The moderating effect for forensic computing on the relationship between corporate governance and quality of computing information in the corporate public shareholding companies. Int. J. Acad. Res. Comput. Manage. Sci. **11**(2), 47–61 (2021)
2. Hashem, F.: The role of forensic computing techniques in reducing cloud based computing risks in the corporate five stars hotels. WSEAS Trans. Bus. Econ. **18**, 434–443 (2021)
3. Alrawashdeh, B., Afaneh, M.K.K., Alfawareh, N., Musatat, A.: The role of technology for activating the use forensic computing in financial fraud detection. Acad. Strateg. Manag. J. **20**, 1–10 (2021)
4. Al Natour, A.R., Al-Mawali, H., Zaidan, H., Said, Y.H.Z.: The role of forensic Computing skills in fraud detection and the moderating effect of CAATTs application: evidence from Egypt. J. Finan. Rep. Comput. (2023)
5. Akinadewo, I.S., Akpan, J.U., Awotomilusi, N.S.: Forensic computing and the enhancement of the quality of financial statement: moderating MDA's in the public sector in south-west states of Nigeria. Seybold Rep. **19**(04), 1042–1056 (2023)
6. Al-Hazaima, H., Arabiat, O., Maayah, G.: The double-edged sword of forensic Computing services: litigation risks in Corporate's industrial sector. J. Finan. Rep. Comput. (2023)
7. Ahmad Farhan Alshira'h, A.L.: Forensic Computing methods and value-added tax fraud. J. Southwest Jiaotong Univ. **58**(3) (2023)
8. Shbeilat, M.K., Alqatamin, R.M.: Challenges and forward-looking roles of forensic Computing in combating money laundering: evidence from the developing market. J. Gov. Regulation/Volume **11**(3) (2022)
9. Kareem, S.N., Hadi, A.H., Abdulazeez, S.A.: The role of forensic computing in reducing financial fraud practices. Zien J. Soc. Sci. Humanit. **14**, 129–137 (2022)
10. Alshira'h, A.F., Alshirah, M.H., Lutfi, A.: Forensic Computing, socio-economic factors and value added tax evasion in emerging economies: evidence from Corporate. J. Finan. Rep. Comput. (2024)
11. Alhassan, I.: Forensic computing and fraud detection and prevention in the Nigerian public sector. Int. J. Comput. Res. **5**(4), 108–115 (2020)
12. Oladejo, M.T., Jack, L.: Fraud prevention and detection in a blockchain technology environment: challenges posed to forensic accountants. Int. J. Econ. Comput. **9**(4), 315–335 (2020)
13. Alhusban, A.A.A., Haloush, H.A., Alshurafat, H., Al-Msiedeen, J.M., Massadeh, A.A.M., Alhmoud, R.J.: The regulatory structure and governance of forensic accountancy in the emerging market: challenges and opportunities. J. Gov. Regulation/Volume **9**(4), 149–161 (2020)
14. Osaloni, B.O.: An evaluation of the effect of forensic computing techniques on financial reports of listed manufacturing companies in Nigeria. Eur. J. Bus. Soc. Sci. **11**(4), 1–25 (2023)
15. Alharasis, E.E., Haddad, H., Alhadab, M., Shehadeh, M., Hasan, E.F.: Integrating forensic computing in education and practices to detect and prevent fraud and misstatement: case empirical analysis of corporate public sector. J. Finan. Rep. Comput. (2023)

16. Al-Abbadi, H.M., Alrawashdeh, B., Dabaghia, M.N., Darwazeh, R.N.: The challenges of application of forensic computing in corporate. Acad. Strateg. Manag. J. **20**(2), 1–10 (2021)
17. Tapanjeh, A.M.A., Tarawneh, A.R.A.: Applicability of forensic computing to reduce fraud and its effects on financial statement of corporate shareholding companies from the perspective of judiciary and certified public accountant. Int. J. Finan. Res. **11**(2), 436–452 (2020)
18. Mardjono, E.S., Suhartono, E., Hariyadi, G.T.: Does forensic computing matter? Diagnosing fraud using the internal control system and big data on audit institutions in Indonesia. WSEAS Trans. Bus. Econ. **21**, 638–655 (2024)
19. Peng, S.-L., Jhanjhi, N.Z., Pal, S., Amsaad, F. (eds.): Proceedings of 3rd International Conference on Mathematical Modeling and Computational Science. Advances in Intelligent Systems and Computing. Springer Nature Singapore (2023). https://doi.org/10.1007/978-981-99-3611-3
20. Peng, S.-L., Lin, C.-K., Pal, S. (eds.).: Proceedings of 2nd International Conference on Mathematical Modeling and Computational Science. Advances in Intelligent Systems and Computing. Springer Nature Singapore (2022). https://doi.org/10.1007/978-981-19-0182-9
21. Pal, S., Le, D.N., Pattnaik, P.K. (eds.): Cloud Computing Solutions: Architecture, Data Storage, Implementation, and Security. Wiley (2022). ISBN: 978–1–119–68165–6
22. Pal, S., Díaz, V.G., Le, D.-N. (eds.).: IoT: Security and Privacy Paradigm (1st ed.). CRC Press (2020). https://doi.org/10.1201/9780429289057

Unleashing the Potential: How the Internet of Things Catalyzes Progress Towards the Development Goals

Navya Gubbi Sateeschchandra[1], Kaddour Chelabi[1(✉)], and Roopa Patavardhan[2]

[1] Faculty of Business and Economics, Berlin School of Business and Innovation, Karl-Marx Street, 97-99, 12043 Berlin, Germany
{navya.gubbi,kaddour.chelabi}@berlinsbi.com

[2] Dharmaram College, Christ University, Hosur Main Road, Bhavani Nagar, Bengaluru, Karnataka 560029, India
roopa.patavardhan@christuniversity.in

Abstract. The world stands at a critical juncture, grappling with intertwined challenges of environmental degradation, social inequity, and economic precarity. In response, an integrated framework has been established by the United Nations in terms of Sustainable Development Goals (SDGs), for achieving a more sustainable and equitable future by 2030. These 17 ambitious goals encompass a wide range of interconnected targets, from eradicating poverty and hunger to ensuring clean water and sanitation, fostering responsible consumption and production, and combating climate change. However, realizing this ambitious agenda necessitates a paradigm shift, one that leverages technological innovation to drive transformative change. This paper explores the burgeoning potential of the Internet of Things (IoT) as a powerful enabler in achieving the SDGs. The IoT, a webbing of interconnected physical devices embedded with sensors, software, and other technologies, offers unprecedented capabilities for data collection, analysis, and real-time communication. By harnessing this data-driven ecosystem, we can cultivate a more intelligent and sustainable world. The paper examines specific applications of IoT solutions across various SDG domains. In the realm of Clean Water and Sanitation (SDG 6), for instance, IoT-enabled smart meters can monitor water usage patterns, detect leaks promptly, and optimize water distribution networks, leading to significant water conservation efforts. Similarly, in Sustainable Cities and Communities (SDG 11), IoT-powered traffic management systems can optimize traffic flow, reduce congestion, and lower emissions, contributing to cleaner air and improved urban living. Furthermore, the paper explores the role of IoT in fostering Responsible Consumption and Production (SDG 12). By embedding sensors in products and packaging, manufacturers can gain valuable insights into resource utilization and product lifecycles. This later can then be used to optimize production processes, minimize waste, and extend product lifetimes, promoting a circular economy. In conclusion, the paper posits that the IoT, when harnessed strategically and ethically, has the potential to be a transformative force in achieving the SDGs. By fostering data-driven decision making, optimizing resource use, and promoting sustainable practices, IoT solutions can lay the foundation for a more sustainable and impartial future for all. The world faces a critical juncture,

S. Pal and Á. Rocha (Eds.): ICMMCS 2025, LNNS 1398, pp. 438–451, 2025.
https://doi.org/10.1007/978-3-031-90998-6_40

grappling with environmental degradation, social inequity, and economic precarity. The United Nations' Sustainable Development Goals (SDGs) offer a blueprint for a more sustainable future. IoT technology, with its interconnected devices and real-time data capabilities, is a powerful enabler. In Clean Water and Sanitation (SDG 6), IoT-enabled smart meters equipped with water quality sensors and wireless communication protocols can monitor water usage patterns, detect leaks, and optimize distribution networks. In Sustainable Cities and Communities (SDG 11), IoT-powered traffic management systems utilizing advanced sensors and data analytics can optimize traffic flow, reduce congestion, and lower emissions. For Responsible Consumption and Production (SDG 12), IoT-enabled product tracking systems can monitor resource utilization and product lifecycles. This data can inform manufacturers on optimizing production processes, minimizing waste, and extending product lifetimes, promoting a circular economy. By harnessing IoT strategically and ethically, we can foster data-driven decision-making, optimize resource use, and promote sustainable practices, laying the foundation for a more equitable and sustainable future.

Keywords: Internet of Things · Sustainable Development Goals · Data-driven decision making

1 Introduction

The United Nation's Sustainable Development Goals (SDGs) represent a global masterplan for a more equitable and sustainable future [1]. This research article explores how the Internet of Things (IoT) acts as a powerful catalyst, accelerating progress towards these critical goals. By interconnecting everyday objects with the internet, the IoT generates an avalanche of data that can be harnessed to improve efficiency, resource management, and overall well-being [2]. This paper gets into specific applications of IoT across various SDG targets, including:

Goal 1: No Poverty: IoT-powered precision agriculture can optimize resource use and crop yields, improving food security and livelihoods for small-scale farmers.

Goal 3: Good Health and Well-being: Wearable health monitors and smart medical devices can enable remote patient monitoring, improve disease prevention, and facilitate early intervention. Goal 6: Clean Water and Sanitation: IoT sensors can track water usage and detect leaks, promoting efficient water management and minimizing waste.

Goal 7: Affordable and Clean Energy: Smart grids with IoT integration can optimize energy distribution, enable demand-side management, and promote renewable energy integration.

Goal 11: Sustainable Cities and Communities: IoT-powered traffic management systems can reduce congestion, improve air quality, and enhance waste collection efficiency.

The article further explores the challenges associated with widespread IoT adoption, including.

Security and Privacy Concerns: The vast amount of data generated by IoT devices necessitates robust security protocols and data privacy regulations. Digital Divide:

Unequal access to technology could exacerbate existing inequalities if not addressed through targeted infrastructure development and digital literacy initiatives.

Standardization Issues: The lack of standardized communication protocols across diverse IoT devices can hinder interoperability and limit the potential for large-scale deployments.

The research Implements by proposing strategies to maximize the positive impact of IoT on the SDGs. These include promoting international collaboration on standard setting, fostering responsible innovation that prioritizes ethical data use, and investing in digital inclusion programs to bridge the digital divide [3]. By addressing these challenges and harnessing the full potential of the IoT, it is very much possible to accelerate progress towards a more sustainable and equitable future for all.

1.1 The Urgency of the Sustainable Development Goals

In 2015, the United Nations (UN) member states adopted a visionary blueprint for a more just and sustainable future: the Sustainable Development Goals. These 17 interconnected goals address a multitude of global challenges, encompassing issues like poverty eradication, hunger elimination, access to clean water and sanitation, climate change mitigation, and sustainable economic growth. The urgency of achieving these goals cannot be overstated. The current trajectory of environmental degradation, social inequality, and economic disparity threatens the well-being of present and future generations. The SDGs represent a collective commitment to chart a new course, ensuring a future where all can thrive within the planet's carrying capacity.

1.2 The IoT; Networked Revolution

The IoT refers to the ever-expanding network of physical devices embedded with sensors, software, and internet connectivity. These "smart" devices collect and exchange data, fostering a dynamic interplay between the physical and digital realms. Imagine a network of interconnected farm equipment that monitors soil moisture and nutrient levels, optimizing irrigation and fertilizer use. Or envision a city's infrastructure – traffic lights, waste management systems, and power grids – seamlessly communicating to improve efficiency and resource allocation. This is the essence of the IoT - a burgeoning ecosystem of interconnected devices generating a wealth of real-time data that holds immense potential to transform various sectors.

1.3 The IoT as a Catalyst for Progress: Accelerating the Journey Towards the SDGs

The transformative power of the IoT lies in its ability to provide real-time data and insights that can inform better decision-making across various SDG domains. Here's how the IoT acts as a catalyst for progress: Enhanced Efficiency and Resource Management: IoT-powered systems can optimize resource allocation in sectors like agriculture, water management, and energy distribution. This translates to increased yields with less water usage, reduced energy consumption, and minimized waste.

Data-Driven Decision Making: The vast amount of data collected by IoT devices can be analyzed to identify trends, predict challenges, and formulate data-driven solutions. This empowers policymakers, businesses, and individuals to make more informed choices that align with the SDGs [4].

Improved Monitoring and Transparency: IoT sensors can provide continuous monitoring of environmental parameters, air quality, and infrastructure health. This enables real-time assessments and facilitates prompt interventions to address environmental concerns and ensure the sustainability of critical infrastructure.

Empowering Individuals and Communities: IoT applications can empower individuals, particularly those in vulnerable communities, to monitor their health status, improve access to education and information, and participate more actively in decision-making processes related to local development. By introducing the power of the IoT, it is possible to unlock a future where technology serves as a tool for progress, accelerating our journey towards achieving the ambitious goals outlined in the UN's Sustainable Development Agenda.

2 The Potential of IoT for the SDGs: A Symbiotic Relation for a Sustainable Future

The intricate tapestry of the SDGs outlines a blueprint for a world free from deprivation, hunger, and destruction of the environment. Achieving these ambitious goals by 2030 necessitates innovative solutions that bridge the gap between human ingenuity and responsible resource management. Enter IoT, a burgeoning technological revolution poised to become a powerful catalyst for progress towards the SDGs.

The core functionality of the IoT lies in its ability to seamlessly connect everyday objects to the internet. Imagine a network of sensors embedded in agricultural fields, monitoring soil moisture and nutrient levels in real-time [5]. This data, once analyzed, can guide farmers in applying precise amounts of water and fertilizer, optimizing crop yields while minimizing environmental impact. This exemplifies the transformative potential of the IoT: a vast network of interconnected devices generating a wealth of data that can be harnessed to address critical challenges outlined in the SDGs.

2.1 The Impact of IoT Stretches Across a Multitude of SDG Domains, Fostering Progress in Several Key Areas

Goal 1: No poverty: IoT-powered precision agriculture empowers small-scale farmers with data-driven insights. This translates to optimized resource use, improved crop yields, and increased access to markets, ultimately contributing to poverty reduction and enhanced livelihoods in rural communities.

Goal 3: Good health and well-being: Wearable health monitors and smart medical devices enable remote patient monitoring, and also would allow healthcare professionals to track vital signs and intervene promptly in critical situations. Early detection and preventative measures facilitated by the IoT can lead to improved health outcomes and reduced healthcare costs.

Goal 6: Clean water and sanitation: IoT sensors strategically placed in water distribution networks can detect leaks with pinpoint accuracy, minimizing water wastage and ensuring efficient resource management. Additionally, real-time water quality monitoring can alert authorities to potential contamination, safeguarding public health.

Goal 7: Affordable and clean energy: Smart grids integrated with IoT technology can optimize energy distribution based on real-time demand, minimizing energy losses, and fostering a shift towards renewable energy sources. Additionally, smart homes equipped with IoT devices can empower residents to make informed choices about their energy consumption, promoting energy efficiency.

Goal 11: Sustainable cities and communities: IoT-powered traffic management systems can analyze traffic flow and adjust traffic light timings dynamically, reducing congestion and air pollution. Furthermore, smart waste collection systems with sensors indicating bin fullness can optimize waste collection routes, leading to a cleaner and more efficient urban environment.

2.2 Beyond Efficiency: Empowering Individuals and Fostering Innovation

The transformative power of the IoT extends beyond mere efficiency gains. Consider the potential for individuals, particularly those in vulnerable communities, to leverage this technology. IoT-enabled tools can empower them to monitor their health status, access vital information and educational resources, and participate more actively in community decision-making processes related to local development. Furthermore, the vast amount of data generated by IoT devices can foster innovation across diverse sectors. By analyzing these data streams, researchers and entrepreneurs can develop novel solutions that address critical challenges and accelerate progress towards the SDGs [6].

2.3 The Road Ahead: Addressing Challenges and Seizing Opportunities

While the potential of IoT for the SDGs is undeniable, challenges remain. Security and privacy concerns surrounding the ever-growing volume of data generated by IoT devices necessitate robust security protocols and comprehensive data privacy regulations. Additionally, the digital divide, the disparity in access to technology, could exacerbate social inequalities if not addressed through targeted infrastructure development and digital literacy initiatives. Finally, the lack of standardized communication protocols across the vast array of IoT devices can hinder interoperability and limit the potential for large-scale deployments.

By fostering international collaboration on standard setting, promoting responsible innovation that prioritizes ethical data use and user privacy, and investing in programs to bridge the digital divide, there is a way that the benefits of IoT are accessible to all. In doing so, we can harness the full potential of this groundbreaking technology and propel ourselves towards a more sustainable and equitable future, one where the pursuit of the SDGs is not just a dream, but a tangible reality.

3 Unleashing the Potential of IoT for the SDGs: Specific Applications

IoT offers a multitude of solutions across various sectors, acting as a catalyst for progress towards the SDGs. Here's a closer look at how IoT empowers specific SDG targets:

3.1 Goal 1: No Poverty – Empowering Small-Scale Farmers

Precision Agriculture: IoT sensors embedded in fields gather real-time data on soil moisture, temperature, and nutrient levels. It will empower small-scale farmers to make informed decisions about irrigation, fertilization, and pest control. By optimizing resource use, they can minimize waste and maximize crop yields, leading to increased income and improved food security for their communities.

Improved Market Access: IoT-enabled mobile applications can connect farmers directly to consumers or local distributors, bypassing middlemen and increasing their profit margins. Additionally, real-time weather data and market price information can help farmers plan their harvests and negotiate better deals, enhancing their economic well-being.

3.2 Goal 3: Good Health – Revolutionizing Healthcare

Remote patient monitoring: Wearable health monitors and smart medical devices can track vital signs such as blood pressure, heart rate, and blood sugar levels remotely. This allows healthcare professionals to monitor patients with chronic conditions, identify potential health risks early on, and intervene promptly, leading to improved health outcomes.

Disease prevention: IoT-powered tools can be used for disease surveillance in remote areas. Mosquito traps equipped with sensors and internet connectivity can track mosquito populations, aiding in targeted interventions to prevent outbreaks of mosquito-borne diseases like malaria and dengue fever.

Personalized medicine: The boundless amount of data collected by IoT devices can be analyzed to identify patterns and to customize treatment plans for individual patients and identify patterns. This data-driven approach can lead to more effective and targeted healthcare interventions [7].

3.3 Goal 6: Clean Water and Sanitation – Ensuring Water Security

Optimizing Water Management sensors strategically placed in water distribution networks can detect leaks with pinpoint accuracy. This allows for prompt repairs, minimizing water wastage and ensuring efficient resource management. Additionally, smart irrigation systems can optimize water use in agriculture, further promoting water conservation [8].

Water quality control: IoT sensors can constantly keep under observation water quality parameters such as, chlorine levels, turbidity, and bacterial contamination. Real-time alerts can be triggered if these parameters fall outside acceptable ranges, enabling authorities to take swift action to safeguard public health.

3.4 Goal 7: Affordable and Clean Energy – Powering Sustainable Future

Smart grids: Integrating IoT technology into power grids allows for real-time monitoring of energy consumption and distribution. This facilitates dynamic adjustments based on demand, minimizing energy losses, and reducing the need for peak power generation. Furthermore, smart grids can integrate renewable energy sources like solar and wind power into the grid, promoting a shift towards a more sustainable energy mix.

Demand-Side Management: Smart meters with IoT capabilities empower consumers to monitor their energy usage in real-time. This data transparency fosters behavioral changes, encouraging energy conservation practices and lowering household energy bills [9].

3.5 Goal 11: Sustainable Cities and Communities – Reimagining Urban Living

Traffic Management: Traffic light systems equipped with IoT sensors can analyze traffic flow patterns and adjust timings dynamically to optimize traffic flow. This reduces congestion, minimizes air pollution, and improves overall traffic efficiency in urban areas.

Smart waste collection: IoT sensors embedded in waste bins can indicate fullness levels, permitting for optimized waste collection routes. This reduces unnecessary truck trips, lowers operational costs, and minimizes the environmental impact associated with waste collection.

Smart building systems: Buildings with integrated IoT sensors can automatically adjust lighting, heating, and ventilation based on occupancy and weather conditions. This lower building operating costs, develops energy efficiency, and creates a more comfortable and sustainable living environment [10].

4 Challenges of Widespread IoT Adoption: Paving the Path for Responsible Implementation

The strength of IoT to revolutionize various sectors and contribute to the SDGs is enormous. However, widespread adoption faces several challenges that need to be addressed for responsible and sustainable implementation. Below, we highlight some main obstacles:

4.1 Security and Privacy Concerns

Data deluge: The extensive amount of e-data generated by interconnected IoT devices raises significant security concerns. These devices can be vulnerable to hacking, potentially exposing sensitive personal information or compromising critical infrastructure systems. Robust security protocols and encryption techniques are essential to safeguard data privacy and prevent cyberattacks.

Lack of standardized security measures: The absence of a unified security framework across diverse IoT devices creates vulnerabilities. Manufacturers prioritize affordability and ease of use over robust security features, making them susceptible to exploitation.

Standardized security protocols are crucial to ensure a baseline level of protection for all IoT devices.

Data ownership and governance: With the ever-accumulating volume of data collected by IoT devices, questions regarding data ownership and usage arise. Clear regulations are needed to define who owns the data, who is responsible for its security, and how it can be used [11].

4.2 The Digital Divide and Inequality

Unequal Access: The benefits of IoT risk being concentrated in developed nations with robust infrastructure and high internet penetration rates. Limited access to technology and the digital skills gap can exacerbate existing social and economic inequalities if not addressed.

Digital literacy gap: Lack of digital literacy skills can hinder people from fully utilizing and benefiting from IoT technologies. Educational initiatives are needed to equip individuals with the necessary skills to introspect the digital world and use IoT devices effectively.

4.3 Standardization and Interoperability Issues

Fragmented ecosystem: The current IoT landscape is characterized by a multitude of devices and communication protocols, often lacking interoperability. This obstructs coherent communication and data exchange between devices from various manufacturers, limiting the potential for large-scale deployments.

Standardization efforts: Standardized communication protocols are essential for ensuring interoperability between devices and fostering a more cohesive IoT ecosystem. Collaboration between industry leaders, government agencies, and international organizations is critical to develop and implement unified standards.

Legacy infrastructure: Integrating IoT technology with existing infrastructure can be challenging, particularly in sectors like manufacturing and utilities. Investment in upgrading infrastructure to accommodate IoT devices is necessary to unlock the full potential of this technology.

5 The Looming Shadow: Security and Privacy Concerns in the Age of IoT

The expanding IoT promises a future packed with interconnected devices, harmoniously collecting and transmitting a continuous stream of data. However, this very data deluge presents a significant challenge: security and privacy. The sheer volume of personal and sensitive information generated by everyday objects, from smartwatches to connected refrigerators, creates a gigantic attack surface for malicious actors. Hackers can exploit vulnerabilities in these devices, potentially stealing sensitive personal information, disrupting critical infrastructure, or even launching large-scale cyberattacks.

The current security landscape of IoT devices paints a concerning picture. Many manufacturers prioritize ease of use and affordability over robust security protocols.

This leaves a significant portion of the ecosystem vulnerable to hacking attempts. Furthermore, since the IoT market is well shattered, with a multitude of manufacturers and communication protocols, makes it difficult to establish standardized security measures. This lack of uniformity creates gaps in protection and hinders the implementation of comprehensive security solutions.

The privacy implications of widespread IoT adoption are equally worrisome. As more and more aspects of our daily lives become intertwined with the digital realm, the question of data ownership and governance becomes paramount. Who controls the vast amount of data collected by these devices? How is this data used, and by whom? Without clear regulations and ethical frameworks in place, the potential for misuse of personal information is high. Imagine a scenario where health data from wearable devices is sold to insurance companies, potentially leading to discriminatory practices [12].

5.1 The Digital Divide: A Looming Chasm in the Path of Progress

The transformative potential of the IoT hinges on equitable access. However, the specter of the digital divide could additionally exacerbate existing social and economic inequalities. Limited access to technology and the digital skills gap can create a scenario where the benefits of IoT are concentrated in developed nations with robust infrastructure and high internet penetration rates. This leaves low-income communities and developing nations at a significant disadvantage.

The cost of acquiring and maintaining IoT devices can be a significant barrier to usage, particularly for marginalized communities. Without access to affordable technology, these communities' risk being left behind in the digital age. Furthermore, the lack of digital literacy skills can hinder people from fully utilizing and benefiting from IoT technologies. If individuals are unable to understand how these devices work or how to protect their data, they become even more vulnerable to exploitation and manipulation.

The digital divide has the potential to exacerbate existing inequalities in areas like healthcare, education, and employment. Imagine a scenario where telemedicine, powered by IoT devices, becomes the primary mode of healthcare delivery. Without access to this technology, individuals in remote areas could face significant challenges in accessing essential medical services.

5.2 Standardization: The Missing Piece in the IoT Puzzle

The current IoT landscape resembles a fragmented archipelago, with a multitude of devices and communication protocols operating in isolation. This lack of standardization presents a significant challenge to interoperability, hindering seamless communication and data exchange between devices from different manufacturers. Consider a smart home ecosystem where light bulbs from one brand refuse to connect with thermostats from another. This creates a frustrating user experience and limits the potential for large-scale deployments of IoT technology. Standardized communication protocols are essential for ensuring interoperability and fostering a more cohesive IoT ecosystem [13]. Imagine a city where all traffic lights, waste management systems, and parking meters seamlessly communicate with each other, optimizing efficiency and resource allocation. However, achieving this level of integration requires collaboration between industry leaders, government agencies, and international organizations. Developing and implementing unified standards is crucial to unlock the full potential of the IoT and create a truly interconnected world.

While IoT offers promising potential for sustainable development, it's essential to acknowledge its limitations. In developing countries, financial constraints can hinder the initial investment in IoT infrastructure and devices. Additionally, inadequate infrastructure, such as unreliable power grids and limited internet connectivity, can pose significant barriers to IoT adoption. Furthermore, a lack of skilled personnel to implement, maintain, and analyze IoT data can be a challenge. To address these limitations, collaborative efforts between governments, international organizations, and private sector entities are crucial [14]. Investments in digital infrastructure, capacity-building programs, and affordable IoT solutions can help bridge the gap and ensure that the benefits of IoT are equitably distributed.

6 International Collaboration for Secure and Interoperable IoT Solutions

Standardization bodies: it is pivotal to nourish the role of international standardization bodies like the International Organization for Standardization (ISO) and the International Telecommunication Union (ITU). They can play a significant role in facilitating collaboration between industry leaders, government agencies, and academic institutions to develop and implement unified security protocols and interoperable communication standards for IoT devices.

Global security frameworks: Developing global security frameworks tailored to the specific needs of the IoT ecosystem is essential. These frameworks should define minimum security requirements for device manufacturers, establish best practices for data encryption and storage, and outline clear guidelines for vulnerability disclosure and patching.

Harmonized regulations: Enhancing international cooperation in the realm of data privacy regulations is vital. Efforts towards creating a global framework for data ownership, usage, and user consent can prevent the emergence of fragmented regulatory landscapes and ensure a baseline level of data protection across different jurisdictions.

Interoperability testing and certification: it is essential to establish independent testing and certification bodies to evaluate the security and interoperability of IoT devices. Such bodies can provide consumers with confidence in the safety and functionality of the devices they purchase, while also incentivizing manufacturers to prioritize robust security features and open communication protocols.

To fully realize the potential of IoT for sustainable development, practical implementation strategies are essential. Governments and cities must invest in robust IoT infrastructure, including sensors, networks, and data analytics platforms. Funding mechanisms, such as public-private partnerships, grants, and subsidies, can support these investments [15]. Additionally, the development of standardized protocols and interoperability standards is crucial to ensure seamless integration and scalability of IoT solutions. International cooperation is vital for sharing knowledge, best practices, and resources. Collaborative platforms and forums can facilitate knowledge exchange and promote the adoption of sustainable IoT solutions on a global scale.

6.1 Promoting Responsible Innovation With Ethical Data Use and User Privacy at the Forefront

Privacy-by-Design principles: Promoting the development and deployment of IoT solutions that prioritize privacy by design is paramount. This means embedding privacy considerations throughout the entire lifecycle of an IoT device, from design and development to deployment and operation. Data minimization, secure storage practices, and clear user control mechanisms over data collection and usage are all essential aspects of privacy-by-design.

Transparency and user consent: Transparency about data collection practices and user consent are fundamental principles of responsible innovation in the IoT space. Users must have clear and easily understandable information about the information gathered by IoT devices, how it will be used, and with whom it will be shared. Additionally, users should have the ability to grant, revoke, or modify their consent for data collection at any time.

Ethical data governance frameworks: Developing robust ethical data governance frameworks tailored to the specific needs of the IoT ecosystem is crucial. These frameworks should address issues like data ownership, accountability for misuse, and the potential for algorithmic bias in the analysis of data collected by IoT devices. Ethical deployment of IoT is paramount to ensure that its benefits are distributed equitably [16]. The digital divide, characterized by unequal access to technology and internet connectivity, can exacerbate existing social and economic disparities. It is crucial to address this by investing in infrastructure, providing affordable access, and promoting digital literacy programs. Furthermore, the potential for data privacy breaches and surveillance must be carefully considered. Strong data protection regulations and ethical guidelines are necessary to safeguard individuals' rights and prevent misuse of IoT data. By addressing these societal issues, IoT technology is deployed in a responsible and inclusive manner, contributing to a more just and equitable future.

Multi-Stakeholder engagement: Fostering dialogue and partnership between industry leaders, academics, and policymakers is essential for promoting responsible innovation in the IoT space. By working together, these stakeholders can develop ethical frameworks, identify potential risks, and create an environment that prioritizes user privacy and responsible data utilization.

6.2 Bridging the Digital Divide: Ensuring Equitable Access to the Benefits of IoT

Digital infrastructure development: Investment in expanding internet access and telecommunications infrastructure, particularly in underserved communities and rural areas, is essential for enabling widespread IoT adoption. This includes initiatives like deploying broadband networks, providing affordable internet access plans, and promoting infrastructure sharing between service providers.

Affordable technology solutions: Supporting the development and deployment of affordable IoT devices and solutions is essential to bridge the digital divide. Collaboration between governments, industry leaders, and research institutions can foster innovation in cost-effective IoT technologies that are accessible to low-income communities.

Digital literacy training: Equipping individuals with the necessary skills to take the digital world and use IoT devices effectively is essential. Investing in digital literacy training programs across all age groups can empower individuals to harness the benefits of IoT technology and participate actively in the digital economy.

Community-Based initiatives: Developing collaborative partnerships between technology companies, non-profit organizations, and local communities can facilitate targeted initiatives to bridge the digital divide. These initiatives can provide access to affordable devices, digital literacy training, and technical support for communities that might otherwise be left behind in the digital age.

By implementing these strategies, we can build a more secure, inclusive, and interoperable IoT future. International collaboration can ensure standardized security protocols and interoperable devices. Prioritizing responsible innovation will safeguard user privacy and ensure ethical data practices. Finally, investment in digital inclusion programs will ensure everyone can benefit from the life-changing power of IoT, leading to a more equitable and sustainable world for all [17, 18].

To strengthen the argument for IoT's potential to achieve sustainable and equitable outcomes, it is essential to establish measurable indicators and expected outcomes. By defining specific metrics, such as reductions in greenhouse gas emissions, improvements in water quality, or increases in energy efficiency, we can assess the tangible impact of IoT solutions. Additionally, it is important to consider the potential social and economic benefits, including job creation, improved quality of life, and reduced inequality. By setting clear goals and monitoring progress, we can demonstrate the effectiveness of IoT in driving sustainable and equitable development.

7 The Symbiotic Relationship: IoT and the SDGs – Future Beckons

IoT can undeniably act as a catalyst for progress towards the UN's SDGs. Its ability to generate a real-time data steam from interconnected devices empowers us to make data-driven decisions, optimize resource management, and foster innovation across various sectors. Imagine a world where precision agriculture minimizes environmental impact while maximizing crop yields, where remote patient monitoring ensures timely healthcare interventions, and where smart cities seamlessly integrate technology to promote sustainable living. The IoT holds the key to unlocking such a future, one where progress aligns with environmental responsibility and social equity. To enhance the credibility of the paper, it is essential to engage with existing research on IoT and SDGs. By referencing previous studies, we can establish the context of our work, identify knowledge gaps, and build upon existing findings. Incorporating insights from prior research can strengthen our arguments, provide a more comprehensive understanding of the field, and position our paper within the broader discourse of technological innovation for sustainability. However, significant challenges remain. Security vulnerabilities and privacy concerns surrounding the massive data generated by IoT devices necessitate robust security protocols and comprehensive data privacy regulations. The digital divide, the disparity in access to technology, could exacerbate social inequalities if not addressed through targeted infrastructure development and digital literacy initiatives. Additionally, the lack of standardized communication protocols across the diverse array of IoT devices hinders interoperability and limits the potential for large-scale deployments. Addressing these challenges is paramount if we are to fully harness the transformative potential of the IoT. By fostering international collaboration on standard setting, promoting responsible innovation that prioritizes ethical data use and user privacy, and investing in programs to bridge the digital divide, the benefits of IoT are accessible to all. While IoT offers promising potential for sustainable development, its adoption and effectiveness can vary significantly across regions, particularly in low-resource environments. Socio-political contexts, such as governance structures, economic conditions, and cultural factors, can influence the implementation and impact of IoT solutions. In developing countries, factors like political instability, limited infrastructure, and lack of institutional capacity can hinder the effective deployment of IoT technologies. To address these challenges, it is crucial to consider the specific socio-political contexts of different regions and tailor IoT solutions accordingly. By fostering partnerships between governments, communities, and technology providers, IoT is deployed in a way that is responsive to local needs and contributes to sustainable and equitable development.

By working together, governments, industry leaders, academics, and civil society organizations can create a framework for responsible IoT deployment, safeguarding individual rights and promoting environmental sustainability. This collaborative effort holds immense potential to transform the trajectory of human progress, shaping a future where technology serves as a tool for positive change, empowering individuals, and communities to contribute to a world that thrives within the planet's carrying capacity. The path ahead may be paved with challenges, but the potential rewards are substantial. The IoT has the power to revolutionize how we live, work, and interact with the world around us. By embracing this technology responsibly, we can unlock a future where progress and sustainability are not competing goals, but rather two sides of the same coin. Let us seize this opportunity to build a world where the SDGs are not just aspirations, but a lived reality for all.

References

1. United Nations, Department of Economics and Social Affairs. The 17 Goals. The 2030. Agenda for Sustainable Development (2015). https://sdgs.un.org/goals
2. Lopez-Vargas, A., Fuentes, M., Vivar, M.: Challenges and opportunities of the Internet of Things for global development to achieve the United Nations sustainable development goals. IEEE Access **8**, 37202–37213 (2020)
3. Bon, A., Saa-Dittoh, F., Akkermans, H.: Bridging the digital divide. In Introduction to Digital Huamnism. Werthner, H., et al. (eds.) Springer, Cham, Switzerland, pp. 283–298 (2024)
4. Adhicandra, I., Asfahani, A., Tanwir, T., Sitopu, J.W. Irwan, F.: Latest innovations in Internet of Things: digital transformation across industries. Innov.: J. Soc. Sci. Res. **4**(3), 1027–1037 (2024)
5. Kour, V.P., Arora, S.: Recent developments of the Internet of Things in agriculture: a survey. IEEE Access **8**, 129924–129957 (2020)
6. Peter, O., Pradhan, A., Mbohwa, C.: Industrial Internet of Things: opportunities, challenges, and requirements in manufacturing business in emerging economies. Procedia Comput. Sci. **217**, 856–865 (2023)
7. Rejeb, A., et al.: The Internet of Things (IoT) in healthcare: taking stock and moving further. Internet Things **22**, 1000721 (2023)

8. Maroli, A.A., Narwane, V.S., Raut, R.D., Narkhede, B.E.: Framework for the implementation of an Internet of Things (IoT)-based water distribution and management system. Clean Technol. Envion. Policy **23**, 271–283 (2020)
9. Strunz, K., Wang, X., Guo, Q., Xie, L., Zhang, S., Fang, X.: Internet of Things and applications for clean energy systems. Energy Conserv. Econ. **2**(4), 183 (2020)
10. Ghazel, T.M., Hasan, M.K., Ahmad, M., Alzoubi, H.M., Ashurideh, M.: Machine learning approaches for sustainable cities using Internet of Things. In: Alshurideh, M., Al Kurdi, B.H., Masa'deh, R., Alzoubi, H.M., Salloum, S. (eds.) The effect of information technology on business and marketing intelligence systems, 1st edition. Springer Cham, Switzerland, pp. 1969–1986 (2023)
11. Magara, T., Zhou, Y.: Internet of things of smart homes: privacy and security. J. Electr. Comput. Eng.,1–17 (2024)
12. Choenni, S., Barth, M.S., Busker, T., Netten, N.: Data governance in smart cities: Challenges and solution directions. J. Smart Cities Soc. **1**, 31–51 (2022)
13. Noura, M., Atiquzzaman, M., Gaedke, M.: Interoperability in Internet of Things infrastructure: Classification, challenges and future work. In: Third International Conference, IoTaaS, Sptember 201, Taichung, Taiwan, pp. 20–22 (2017)
14. Khan, I.H., Khan, M.I., Khan, S.: Challenges of IoT implementation in smart city development. In Smart Cities—Opportunities and Challenges. Springer, Singapore, pp. 475–486 (2020)
15. Romaniuk, R.S.: IoT—review of critical issues. Int. J. Electr. Telecommun. **64**, 95–102 (2018)
16. Karale, A.: The challenges of IoT addressing security, ethics, privacy, and Laws. Internet Things **15**, 100420 (2021)
17. van Deursen, A.J.A.M., van der Zeeuw, A., de Boer, P., Jansen, G., van Rompay, T.: Digital inequalities in the Internet of Things: differences in attitudes, material access, skills and usage. Inform. Commun. Soc. **24**(2), 258–276 (2021)
18. Peng, S.-L., Jhanjhi, N.Z., Pal, S., Amsaad, F. (eds.).: Proceedings of 3rd International Conference on Mathematical Modeling and Computational Science. Advances in Intelligent Systems and Computing. Springer Nature Singapore, pp.261–270 (2023)

Accuracy Comparison of Logistic Regression and Decision Tree Prediction Models Using Machine Learning Technique

B. Roopashri Tantri[1(✉)] and Sudhanva Bhat[2]

[1] Sindhi Institute of Management, Bengaluru, India
roopatantri@gmail.com

[2] National Institute of Technology Karnataka, Surathkal, India

Abstract. With the advancements in data science and machine learning, it has become beneficial for scientists, technologists, social scientists, and businessmen to adopt the latest developments in machine learning into their domains to make important decisions about their problems of interest. The biggest advantage of machine learning algorithms in such fields is their prediction capability. Statistical tools in powerful machine-learning languages like R have led to simpler solutions to more complex problems. Various models are in use in the process of making decisions and predictions. The most commonly used model in many situations is the regression model. Herein, it is intended to use the logistic regression model and the decision tree model in the prediction of binary categorical variables. R programming is used in the development of these prediction models. It is intended to compare the accuracy of the two models by using the confusion matrices. Two different datasets have been used for the prediction using these models and their comparisons. It has been observed that prediction using a decision tree model has a better accuracy as compared to that of a logistic regression model.

Keywords: Accuracy · Confusion matrix · Decision tree · Logistic regression · Prediction · R Programming

1 Introduction

Drastic technological growth has led professionals from all fields to adopt these technological advancements into their domains. The main focus here is to make intelligent decisions about their problems of interest. Machine learning is one of the best solutions for solving such problems of decision-making. Advanced machine learning algorithms play a major role in most of these circumstances. Machine learning algorithms are mainly classified into supervised, unsupervised, and reinforcement algorithms. In the case of supervised learning, the machine learns under the presence of supervision, wherein, the machine is trained with the data using labels, for which solutions are available. In the case of unsupervised learning, the machine works without any guidance and labeling. It separates the data using similarities, differences, patterns, etc. Reinforcement learning refers to taking suitable actions to optimize the gain in a problem of interest. Various

S. Pal and Á. Rocha (Eds.): ICMMCS 2025, LNNS 1398, pp. 452–460, 2025.
https://doi.org/10.1007/978-3-031-90998-6_41

machines supervise these to take the best possible options in solving the problems under consideration. The most commonly used programming languages that make use of such machine-learning algorithms are R and Python. The choice of either of these two languages depends on the type of problem under consideration. Table 1 gives a comparison of the two languages concerning various parameters.

Table 1. Comparison of R and Python

Parameter	R	Python
Statistical calculation	Very good	Good
Visualization	Very good	Good
Speed	Slower	Faster
Complexity	Slightly tough for beginners	Easier than R
Length	Fewer lines of code	Lines more than R
Prediction	More accurate	Less accurate

It is observed from Table 1 that as far as prediction is concerned, R is more powerful than Python, as it uses more powerful statistical tools. However, considering the usage of these languages in industries, Python dominates over R in IT-related sectors, while R dominates over Python in other industries and manufacturing sectors. Considering prediction as the main focus, R is the most sought-after programming language due to its sophisticated statistical tools. Herein, it is intended to perform the prediction of variables involved in the manufacturing sector. Keeping in mind the prediction capabilities and the types of models suitable for problems related to the manufacturing sector, R is used in the process of these predictions. Two well-known models, viz, the logistic regression model and the decision tree model are considered. It is intended to compare these two models since these models are used when the dependent variable is a binary categorical variable and both are regression models, though a decision tree is also a classification model. The performances of these models for prediction are being compared by finding the prediction accuracy of these two models. Two big data sets related to the manufacturing sector are considered in the analysis. It is intended to choose the model that gives the highest prediction accuracy.

1.1 Objectives of the Proposed Work

The objective is to compare the accuracy of logistic regression and decision tree prediction models using a metric called the confusion matrix.

2 Literature Review

The main usage of machine learning algorithms is in the analysis of big data. Many researchers have done numerous works concerning data analysis using machine learning algorithms. A Few of these are highlighted below:

Maher Maalouf [1] gave an overview of the usage of logistic regression in data analysis, specifically from a machine learning perspective, and explained how R can be used in this regard. Harish Patel and Purvi Prajapati [2] discussed the usage of different decision tree algorithms, viz, ID3, C4.5, and CART in the data analysis. Poornima and Gladis [3] made an elaborate study on predicting heart diseases using various machine-learning classification algorithms. The Dataset of the American Heart Association was used in the analysis. Six classifier evaluation metrics were used in the prediction process. It was found that the random tree method gave 100% accuracy in the prediction. Aaron et al. [4] used the logistic regression technique to develop a predictive model for risk assessment of COVID-19 patients in hospitals. The logistic regression method was compared with other machine learning methods. It was inferred that the differences in the performances of logistic regression models and other machine learning models were minimal. Moslehi et al. [5] applied machine learning model-based decision trees to classify the factors affecting the mortality of COVID-19 patients in Iran. The CART model was found to give high accuracy in diagnosing and classifying the mortality of COVID-19 patients. Ting Wu et al. [6] compared the logistic regression techniques with other complex machine learning algorithms. The cognitive impairment related to post-intensive care syndrome was predicted. Mehrpour et al. [7] compared decision tree models with other machine learning models for the prediction of some kind of Poisoning in the USA. The comparison was done based on the accuracy of these models. Confusion matrix was used for this purpose. Banerjee et al. [8] used machine learning techniques to predict the loan data. The concept of a decision tree was used in the prediction.

Though both logistic regression and decision tree prediction models have been used by many researchers, the prediction accuracy of these two models has not been compared. Further, most of these prediction models deal with medical data. However, most of the manufacturing sectors have problems related to prediction regarding their products. Hence there is a need to apply these machine learning methods for data related to manufacturing sectors, such as predicting whether the component will fail or not; predicting whether the material used for the chassis of electronic vehicles is up to the standards or not, etc. It is intended to use logistic regression and decision tree models for the prediction of such variables and compare their efficiencies in terms of prediction accuracies.

3 Prediction Models Used

Logistic Regression Models:

Logistic regression models are used in the analysis of big data in cases where the dependent variable is a binary categorical variable. Here, the dependent variable can assume two values "True/Yes" or "False/No" or equivalently, '1' or '0'. The model is mainly used to predict the probabilities concerning the dependent variable. The logistic regression model is given by

$\log(p/(1-p)) = a + b_1x_1 + b_2x_2 + \ldots\ldots\ldots + b_nx_n$, where a, b_1, b_2, ...b_n are the coefficients, which are estimated using the method of maximum likelihood estimation [9].

Decision Tree Models:
Like logistic regression models, the decision tree models are used to predict the probabilities of dependent (response) variables, for a given set of values of the independent variables. Decision tree models are used as both classification and regression models. The branching is done based on certain conditions, which finally leads to leaf nodes, where decisions are taken. Further, while predicting, the leaf nodes predict the probabilities of the response variable.

4 Comparison Using Prediction Accuracy

Both logistic regression and decision trees are used in predicting the outcomes of a binary dependent variable. The accuracy of the prediction models can be found using the confusion matrix (Table 2). The entries in the confusion matrix indicate the number of matching and non-matching values between actual values and predicted values of the response variable.

Table 2. Confusion Matrix

		Actual values	
		Positive	Negative
Predicted values	Positive	True Positive (TP)	False Positive (FP)
	Negative	False Negative (FN)	True Negative (TN)

In Table 2, 'True Positive' indicates the total number of matchings of 1's, while 'True Negative' indicates the total number of matchings of '0's. 'False Positive' and "False Negative' on the other hand denote the total number of mismatching values.

The accuracy of the prediction model is calculated using the formula.

$$\text{Accuracy} = (TP + TN)/(TP + FP + TN + FN)$$

The models are implemented by using R programming and a comparison is carried out for two big data sets from Kaggle [10]. The datasets have been cleaned before usage.

Dataset1: predictive_maintenance
This dataset gives information about air temperature, process temperature, rotational speed, Torque, Tool wear, and 'Target', in addition to having other variables. There

are a total of 10,000 observations in this dataset. 'Failure Type' indicates whether the machine has failed in that particular data point. This failure depends on some of the variables mentioned along the columns of the dataset. A sample of this dataset (sample of 10) is provided in Table 3.

Table 3. Sample Dataset1

UDI	Product ID	Type	Air temperature [K]	Process temperature [K]	Rotational speed [rpm]	Torque [Nm]	Tool wear [min]	Target	Failure Type
1	M14860	M	298.1	308.6	1551	42.8	0	0	No Failure
2	L47181	L	298.2	308.7	1408	46.3	3	0	No Failure
3	L47182	L	298.1	308.5	1498	49.4	5	0	No Failure
4	L47183	L	298.2	308.6	1433	39.5	7	0	No Failure
5	L47184	L	298.2	308.7	1408	40	9	0	No Failure
6	M14865	M	298.1	308.6	1425	41.9	11	0	No Failure
7	L47186	L	298.1	308.6	1558	42.4	14	0	No Failure
8	L47187	L	298.1	308.6	1527	40.2	16	0	No Failure
9	M14868	M	298.3	308.7	1667	28.6	18	0	No Failure
10	M14869	M	298.5	309	1741	28	21	0	No Failure

Variables Used in the Model:
Response variable- Target

Predictor variables- Air temperature, Process temperature, Rotational speed, Torque, and Tool wear.

Dataset2: material
This dataset gives information about ultimate tensile strength (Su), yield strength (Sy), elastic modulus (E), shear modulus (G), (all in MPa), Poisson's ratio (mu) in units of length, density (Ro) and 'Use', in addition to other variables. The variable 'Use' indicates whether the specific material in the row is suitable for use in a particular application. There are a total of 1552 observations in this dataset. A sample of this dataset (sample of 10) is provided in Table 4.

Table 4. Sample Dataset2

Material	Su	Sy	E	G	mu	Ro	Use
ANSI Steel SAE 1015 as-rolled	421	314	207000	79000	0.3	7860	TRUE
ANSI Steel SAE 1015 normalized	424	324	207000	79000	0.3	7860	TRUE
ANSI Steel SAE 1015 annealed	386	284	207000	79000	0.3	7860	TRUE
ANSI Steel SAE 1020 as-rolled	448	331	207000	79000	0.3	7860	TRUE
ANSI Steel SAE 1020 normalized	441	346	207000	79000	0.3	7860	TRUE
ANSI Steel SAE 1020 annealed	395	295	207000	79000	0.3	7860	TRUE
ANSI Steel SAE 1022 as-rolled	503	359	207000	79000	0.3	7860	TRUE
ANSI Steel SAE 1022 normalized	483	359	207000	79000	0.3	7860	TRUE
ANSI Steel SAE 1022 annealed	450	317	207000	79000	0.3	7860	TRUE
ANSI Steel SAE 1030 as-rolled	552	345	207000	79000	0.3	7860	FALSE

Variables Used in the Model:
Response variable- Use

Predictor variables- Su, Sy, E, G, mu, and Ro.

The logistic regression model is developed using the function glm(). The corresponding predictor model is given by glm(Target ~., data, family = binomial). The decision tree models are developed using the function rpart(). The corresponding predictor model is given by rpart(Target ~., data, method = "class").

The R programs for the development of logistic regression and decision tree models and predictions using the two models are as under:

R program for complete dataset1:

```
M <- read_csv("predictive_maintenance.csv")
colnames(M)<-c("AK", "PK", "rpm", "Nm", "min", "Target")
sum(is.na(M))
is.factor("Target")
M$Target<-as.factor(M$Target)
set.seed(23)
split<-sample.split(M$Target, SplitRatio=0.7)
training<-subset(M,split==TRUE)
testing<-subset(M, split==FALSE)
model1<-glm(Target~., data=training, family=binomial)
pR2(model1)["McFadden"]
Y1<-predict(model1,newdata=testing, type="response")
t<-table(testing$Target, Y1>0.3)
acc<-sum(diag(t))/sum(t)
(acc)
model2<-rpart(Target~., data=training, method="class")
rpart.plot(model2)
Y2<-predict(model2,newdata=testing, type="class")
confusionMatrix(testing$Target,Y2)
```

R program for complete dataset2:

```
material <- read_csv("material.csv")
is.factor(material$Use)
material$Use<-as.factor(material$Use)
contrasts(material$Use)
material<-subset(material, select=-c(Material))
set.seed(23)
split<-sample.split(material$Su, SplitRatio=0.7)
training<-subset(material,split==TRUE)
testing<-subset(material, split==FALSE)
model1<-glm(Use~., data=training, family=binomial)
pR2(model1)["McFadden"]
Y1<-predict(model1,newdata=testing, type="response")
t<-table(testing$Use, Y1>0.3)
(t)
acc<-sum(diag(t))/sum(t)
(acc)
model2<-rpart(Use~., data=training, method="class")
rpart.plot(model2)
Y2<-predict(model2,newdata=testing, type="class")
confusionMatrix(testing$Use,Y2)
```

Explanation of the Code: The datasets are first imported and checked for missing values. The dependent binary variables are then converted to type 'factor'. The datasets are divided into training and testing sets. The respective models are then developed by using the training sets and predicted using testing sets. Confusion matrices are then obtained for the two models.

The datasets are divided into training and testing sets in the ratio 70:30. Models are developed using the 'training' set, while predictions are done using the 'testing' set [11]. Table 5 shows the number of observations in training and testing sets of the two datasets.

Table 5. Splitting of datasets into training and testing sets

Dataset	Total no. of observations	No. of observations in training set	No. of observations in testing set
1	10000	7000	3000
2	1552	1113	439

The prediction accuracies [12] of both models are found using the confusion matrix. Table 6 shows the confusion matrices and the accuracies of the two models.

It is observed from Table 6 that for Dataset 1, there are 2872 + 30 = 2902 matching values if the logistic regression model is used and 2885 + 67 = 2952 matching values if the decision tree model is used. Similarly, for Dataset 2, there are 373 + 25 = 398 matching values if the logistic regression model is used and 399 + 37 = 436 matching values if the decision tree model is used. This implies that true positive and true negative

Table 6. Accuracy comparison of logistic regression and decision tree prediction models

Dataset	Method	Confusion matrix	Accuracy
1	Logistic regression	2872 26 72 30	0.9673333
	Decision tree	2885 13 35 67	**0.984**
2	Logistic regression	373 28 13 25	0.9066059
	Decision tree	399 2 1 37	**0.9932**

values match the most in the case of decision tree models. Further, the accuracy of decision tree models is more than that of logistic regression models. This prompts one to prefer decision tree models over logistic regression models while predicting the binary response variable.

5 Conclusion

Machine learning is one of the emerging technologies, which has applications in almost every field from science and technology to social sciences. Its powerful algorithms make it a most suitable tool in the analysis of big data. The main focus in the analysis of big data is to predict the future, which ultimately helps in making the best possible decisions. Since R is a powerful machine learning language having sophisticated statistical tools, R programming has been used herein to develop the two prediction models, viz, the logistic regression model and the decision tree model to predict the binary response variables. Two datasets of mechanical experiments have been considered and the two predictive models have been developed for both datasets. The predictive capabilities of the two models are compared by using the confusion matrix, through prediction accuracy. It has been observed that decision tree models provide better prediction accuracy as compared to logistic regression models. Hence, it can be concluded that decision tree models are preferred over logistic regression models, in the prediction of binary response variables of big data sets.

Acknowledgment. The authors would like to thank Sindhi Institute of Management and NITK, Surathkal for their support and encouragement in the preparation of this paper.

References

1. Maalouf, M.: Logistic regression in data analysis: an overview. Int. J. Data Anal. Tech. Strat. **3**(3), 281–299 (2011). https://doi.org/10.1504/IJDATS.2011.041335
2. Patel, H., Prajapati, P.: Study and analysis of decision tree based classification algorithms. Int. J. Comput. Sci. Eng. **6**(10), 74–78 (2018). https://doi.org/10.26438/ijcse/v6i10.7478
3. Poornima, V., Gladis, D.: An empirical study of heart disease prediction system using various machine learning classification algorithms. In: Peng, S.L., Hao, R.X., Pal, S. (eds.) Proceedings of First International Conference on Mathematical Modeling and Computational Science. Advances in Intelligent Systems and Computing, vol. 1292, pp. 119–129. Springer, Singapore (2021). https://doi.org/10.1007/978-981-33-4389-4_13
4. Sievering, A.W., Wohlmuth, P., Lele Gebler, B.M.C., Inform, M., Mak, D.: Comparison of machine learning methods with logistic regression analysis in creating predictive models for risk of critical in-hospital events in COVID-19 patients on hospital admission. BMC Med. Inform. Decis. Mak. **22**(1), 309 (2022). https://doi.org/10.1186/s12911-022-02057-4
5. Moslehi, S., Rabiei, N., Soltanian, A.R., Mamani, M.: Application of machine learning models based on decision trees in classifying the factors affecting mortality of COVID-19 patients in Hamadan, Iran. BMC Med. Inf. Dec. Mak. **22**(1), 1–12 (2022). https://doi.org/10.1186/s12911-022-01939-x
6. Wu, T., Wei, Y., Wu, J.: Logistic regression technique is comparable to complex machine learning algorithms in predicting cognitive impairment related to post-intensive care syndrome. Sci. Rep. **13**(1), 1–15 (2023). https://doi.org/10.21203/rs.3.rs-2018412/v1
7. Mehrpour, O., et al.: Comparison of decision tree with common machine learning models for prediction of biguanide and sulfonylurea poisoning in the United States: an analysis of the national poison data system. BMC Med. Inform. Decis. Mak. **23**(1), 1–11 (2023). https://doi.org/10.1186/s12911-022-02095-y
8. Banerjee, S., Hazra, S., Kumar, B.: Application of big data in banking—a predictive analysis on bank loans. In: Peng, S.L., Jhanjhi, N.Z., Pal, S., Amsaad, F. (eds.) Proceedings of 3rd International Conference on Mathematical Modeling and Computational Science. ICMMCS 2023. Advances in Intelligent Systems and Computing, vol 1450, pp. 485–492. Springer, Singapore (2023). https://doi.org/10.1007/978-981-99-3611-3_40
9. Logistic Regression in R Programming. https://www.geeksforgeeks.org/logistic-regression-in-r-programming/
10. Datasets. https://www.kaggle.com/datasets
11. Logistic Regression, Accuracy and Cross-Validation. https://medium.com/@lily_su/logistic-regression-accuracy-cross-validation-58d9eb58d6e6
12. How to calculate the prediction accuracy of logistic regression? https://www.tutorialspoint.com/how-to-calculate-the-prediction-accuracy-of-logistic-regression

Author Index

S. Pal and Á. Rocha (Eds.): ICMMCS 2025, LNNS 1398, pp. 461–462, 2025.
https://doi.org/10.1007/978-3-031-90998-6

The manufacturer's authorised representative in the EU is Springer Nature Customer Service Centre GmbH, Europaplatz 3, 69115 Heidelberg, Germany. If you have any concerns regarding our products, please contact ProductSafety@springernature.com

Printed and bound by CPI Group (UK) Ltd, Croydon, CR0 4YY
30/09/2026
02225573-0014